The Collected Works of Leo Szilard

Scientific Papers

Leo Szilard, 1898–1964
[Photograph 1957]

The Collected Works of Leo Szilard

Scientific Papers

Bernard T. Feld and
Gertrud Weiss Szilard, Editors
with Kathleen R. Winsor

Foreword by Jacques Monod

Introductory Essays
Carl Eckart
Bernard T. Feld
Maurice Goldhaber
Aaron Novick
Julius Tabin

The MIT Press
London, England and Cambridge, Massachusetts

Volume I
All new material except that in the public domain
Copyright © 1972 by
The Massachusetts Institute of Technology

This book was set in "Monophoto" Times New Roman,
Printed on Warren's 1854 Medium Offset
by Halliday Lithograph Corp.,
and bound by Halliday Lithograph Corp.
in the United States of America.

ISBN 0 262 06039 6   (hardcover)

Library of Congress catalog card number: 79-151153

The Collected Works of Leo Szilard

Scientific Papers

Notes on the Bibliography

This bibliography is limited to scientific and technical papers and reports. It includes all known published papers and a selection of the important unpublished work.

Papers preceded by an asterisk are published for the first time. Some of the reports so designated, however, have had limited distribution in duplicated form.

Between the years 1940 and 1945, papers were withheld from publication and issued as classified reports by agencies of the United States government concerned with nuclear energy during the second World War. These reports bear various code numbers. After World War II the reports quoted here were declassified and released as documents of the Atomic Energy Commission.

The bibliography of Leo Szilard's writings on subjects other than science, such as memoranda on operational and political aspects of the Manhattan Project, and other general and political writings, such as his articles in the *Bulletin of the Atomic Scientists*, as well as fiction, will be presented in succeeding volumes.

Key to Abbreviations

Cold Spring Harbor Symp. Quant. Biol. = Cold Spring Harbor Symposia on Quantitative Biology
Phys. Rev. = Physical Review
Proc. Nat. Acad. Sci. = Proceedings of the National Academy of Sciences
Zeits. Physik = Zeitschrift für Physik

Contents and Bibliography

* Asterisk denotes papers published for the first time.

Nuclear Physics

Introduction by Maurice Goldhaber   139

**Part III
Documents relating to the Manhattan Project
(1940–1945)**

Introduction by Bernard T. Feld   171

Correspondence and Memoranda

Declassified Papers and Reports

Physics

Part IV
Published Papers in Biology
(1949–1964)

Appendix to Part IV

Part V
Patents, Patent Applications, and Disclosures
(1923–1959)

Selected Patents and Patent Applications

Lists of Patents and Disclosures

Correspondence Relating to Patents

Foreword

During the Summer of 1947, I was invited to the "Growth Symposium" which was held somewhere in Connecticut. On the way I stopped at Cold Spring Harbor to see my admired friend Max Delbrück, who spent the summer there, preaching the creed of the nascent phage church.

At one of the seminars, I noticed in the crowd a rather unusual person. Unusual in what way? It was hard to say. After all, it is rather usual for scientists to look somewhat unusual. Many of them cultivate extravagance in dress or manners, as a sort of outward expression of their profound originality (whether real or assumed).

There were no such obvious traits about the short fat man whom I noticed in the front row. He seemed to be sound asleep most of the time and, with his round face and potbelly, he certainly expressed little interest and no aggression. Yet, from time to time he would suddenly wake up, his eyes shining with intelligence and wit, to ask sharp, incisive, unexpected questions, which he would impatiently repeat when the answer did not immediately come straight and clear.

We were introduced, but I did not quite catch the name, which meant nothing to me. He spoke English with a heavy accent, probably central European or Balkanic. Nothing unusual there: in these days, more than half the scientific community of America spoke (without any restraint or inhibition) an astonishing brand of pidgin-English.

We had hardly exchanged a handshake when he started shooting question after question at me, concerning my work on the kinetics of bacterial growth and "enzymatic adaptation" as it was known at the time. Many of the questions seemed very unusual, startling, almost incongruous. I was not sure I understood them all, especially since he insisted on redefining the basic problems in his own terms, rather than mine. Yet it was clear that he had carefully read my papers, and even though bewildered, I was delighted by this discussion and by the "new look" in which I myself began to see some of my work. At the time, I was rather tired of it. I could not see how to go beyond the point which I had already reached and I had half a mind to choose another, supposedly more exciting, subject: for instance, joining the phage church. Now however, my problem again appeared fresh, bright, important, challenging. And even though I felt that some of the more disturbing suggestions made by that strange man could not be right, I was forced to think hard to answer and convince him to his satisfaction.

Soon after this conversation, I discovered who this man was I had been talking to: Leo Szilard. Szilard was then an already famous scientist, who had worked in thermodynamics as well as nuclear physics, who with Einstein had launched the Manhattan Project and worked on it through the war, and who had then turned himself into an extraordinarily active and successful lobbyist in Washington, before at last being able to turn to the field which had attracted and fascinated him years before: biology.

I have often asked Leo why, being such an outstandingly successful physicist, he had decided actually to abandon physics, to become exclusively a biologist. He generally brushed the question aside: "Don't you agree that biology is interesting?" he would say, for instance. Or he would express his impatience with huge machines and colossal projects.

I personally believe that there was a more significant and much deeper reason, going back to his early work on the thermodynamics of Maxwell demons. His famous paper on this subject is to be found in Part II; yet there are no Maxwell demons in the physical world, apart from the physicists themselves who, after all, are biological objects. Maxwell demons are, in fact, endowed with properties uniquely characteristic of living beings: choice, intention, and foresight. Yet, as Szilard shows in this paper, Maxwell demons, did they exist, would not, indeed could not, violate the principles of thermodynamics. Thus, in a very deep sense, the old dilemma of mind and matter at last receives its solution. The gap is bridged: the activity of the mind, expressed in an abstract thought, *can* organize matter without violating or superseding any physical principle. There is no contradiction, and the old quarrel between idealists, spiritualists, and materialists is meaningless.

I cannot help feeling that the choice of this question, as one of his first problems, reveals Szilard's early involvement in some of the deepest riddles of biology: he always was a biologist at heart, and it is no accident that his *last* paper should be "on memory and recall." He was fascinated by intelligence, by his own intelligence, and by these mysterious imaginative, cognitive, and creative mechanisms that reside in the brain.

Another reason for Szilard's attraction was his feeling that, because of its very complexity and uncertainties, biology needed ideas, many ideas, to be discussed, tested, rejected, or temporarily accepted; that is, precisely the kind of goods that he knew he could provide in abundance and enjoyed dealing with.

Most scientists of course do not formulate any significant new idea of their own. The few that do are in general inordinately jealous of, and unduly faithful to, their own precious little ideas. Not so with Szilard: he was as generous with his ideas as a Maori chief with his wives. Indeed, he loved ideas, especially his own. But he felt that these lovely objects only revealed all their virtues and charms by being tossed around, circulated, shared, and played with.

His own published papers in biology reflect only a part, a small part, of this constant, intense, although always playful creative activity. But his profound and sometimes decisive influence in the development of many aspects of molecular biology goes way beyond his own published work.

This is surely the case in the field with which I am most intimately familiar, namely, the regulation of cellular metabolism.

Szilard and Novick may not have discovered negative feedback regulation of enzyme activity. They surely invented it, however, and the first I heard of what seemed a rather startling assumption at the time (only later to become obvious and commonplace) was in a discussion with Szilard, back in 1954. I have also recorded, in my

Nobel lecture, how it was Szilard who decisively reconciled me with the idea (repulsive to me, until then) that enzyme induction reflected an antirepressive effect, rather than the reverse, as I tried, unduly, to stick to.

Many other people have had similar experience in their association with Szilard. He had close friends, mostly among younger men, whom he would visit, or summon by phone, at frequent erratic intervals, any time he had an idea to communicate, an experiment to suggest, or he felt he needed to collect more facts to abate temporarily his huge intellectual appetite. His younger friends thus formed a sort of club, whose members, picked by Szilard, enjoyed the envied privilege of being Leo's private correspondents, providing him with new food for his thoughts, which he generously and immediately (mostly by phone or cable) shared with all of them.

Indeed, as I have said, Leo Szilard was too rich in ideas, whether scientific or political, too joyfully familiar with all of them; he derived too much sheer pleasure in playing with them as a child with his toys ever uniquely to pursue only one of them, aggressively reiterating, illustrating, and defending it, as most of us do.

Had he done so, his own specific achievements—written-up, formalized, and stamped—might have *appeared* greater, more definitely significant. Then however he would have been just as good, but no better, than many other highly distinguished scientists.

Szilard was different. He knew that meaningful ideas are more important than any ego, and he lived according to these ethics. To his friends his memory will remain as a unique image of a man to whom science was more than a profession, or even an avocation: a mode of being.

Jacques Monod

Professor, Collège de France
Director, Pasteur Institute
Paris, France

Preface

There are two reasons why Leo Szilard never wrote his memoirs: he was vastly more interested in the future than in the past and much more concerned with the world around him than with himself.

He did have a sense of history, however, and carefully preserved all notebooks, correspondence, and drafts of his writings and even filed documents of special importance in folders marked "History."

He also often said that he wanted to write his memoirs particularly about the Uranium Project in Chicago.* When he discussed this book, I offered to prepare a companion volume, consisting of a collection of the Szilard publications and other interesting writings that he had given to me for safekeeping.

Now we can, with regret, make only this companion offering—the *Collected Works*. In looking at the materials ranging over so many different topics it seems logical to separate them into two categories according to the two dominant and interacting themes of Szilard's life. These two motifs are most vividly expressed by Szilard himself in a letter addressed to Professor Niels Bohr, dated November 7, 1950:

"Dear Professor Bohr: Theoretically I am supposed to divide my time between finding what life is and trying to preserve it by saving the world. At present the world seems to be beyond saving, and that leaves me more time free for biology. Enclosed you will find a manuscript giving our results, some of which might perhaps interest you. Sincerely, Leo Szilard."

Szilard referred to this dichotomy also in a speech which he gave at Brandeis University in November 1953:

"When I was a young boy, I had two great interests in life; one was physics and the other politics. I kept these two interests in two compartments, and it never occurred to me that these two compartments could ever merge into one. To my interest in politics, I probably owe my life; and to my interest in physics, I owe my livelihood. I owe my life to my interest in politics because it enabled me to recognize in 1930 what was going to happen in Germany."

We therefore decided to prepare the collection in separate parts of which this one is the first. *Scientific Papers* is subdivided into different fields and presented mostly in chronological order revealing the development of Szilard's interests. They were so diversified that most people acquainted with his papers in one of the areas of science, are not aware of Szilard's work in other fields. And surely, his ideas expressed

---

* In an unpublished manuscript† he whimsically referred to a conversation he had with Hans Bethe during the time of the project as follows:

Bethe asked Szilard how things were going in the Uranium project in Chicago. "What is going on is so peculiar" Szilard went on, "that I have just about decided to keep a diary. I don't intend to publish it; I am merely going to record the facts for the information of God." "Don't you think God knows the facts?" Bethe asked. "Yes," said Szilard, "He knows the facts, but He does not know *this version of the facts*."

† Leo Szilard: Appendix to an excerpt from *The Voice of the Dolphins*, prepared as a working paper for an international meeting of scientists (Sixth Pugwash Conference) held in Moscow in December 1960.

in patent applications are known only to a very few. In this book the diverse materials are assembled for the first time.

To make the papers more meaningful to the reader, scientist friends of Szilard who had been closely associated with him during various phases of his life have written five introductory essays to this volume, explaining the background of the work covered in the papers.

We have collected and are reproducing all the published papers we are aware of. It is of interest to find that Szilard, during his entire career as a scientist, over some forty years, had only twenty-nine publications in scientific journals. We doubt very much whether in this age of "publish or perish" this number would be considered sufficient for tenure positions in some American universities.

Szilard was very meticulous and highly selective about the quality of the thoughts that he wished to convey in print or in a formal talk. The number of his publications was further limited by the necessity for secrecy during the Uranium Project, first voluntary and self-imposed, then by regulation. Since these restrictions no longer exist, this volume contains in addition to the publications in scientific journals, one major paper (A-55, Divergent Chain Reaction in Systems Composed of Uranium and Carbon, in Part III of this volume) and a number of reports that now have been officially declassified.

In subsequent publications we intend to present Szilard's writings on matters beyond science. Published writings have been gathered from many sources such as the *Bulletin of the Atomic Scientists*, to which he frequently contributed, as well as from other magazines and from his only book of fiction *The Voice of the Dolphins*, a collection of short stories of political satire. (Simon and Schuster, Inc., New York, 1961.) We also plan to add a selection of some unpublished materials, such as memoranda, correspondence, speeches, broadcasts, and interviews. While Szilard never found time to write his autobiography, he enormously enjoyed talking to people, and on such occasions he sometimes reminisced. This happened particularly during a period of serious illness in 1959–1960, when Szilard was confined to a hospital for about a year and had frequent visitors. At that time it seemed, and Szilard believed, that he had only a few more months to live. This awareness did not dampen, but rather accelerated Szilard's intellectual activity and zest for life. There was a tape recorder in his sickroom, and I turned it on when I felt that the conversation was important. Therefore we fortunately have several tapes in which Szilard talked about selected periods of his life. Although the pieces are fragmentary and sometimes overlap—with remarkable consistency—with stories he told on other occasions, they will be included because they reflect his thinking and personality which we want to share with others.

In presenting Leo Szilard in his own words, it is our hope that his voice will continue to be heard. That is the purpose of these "Collected Works."

Gertrud Weiss Szilard, M.D.

School of Medicine,
University of California, San Diego

Acknowledgments

We must begin our acknowledgments by thanking first our many friends who made the physical collection of these papers possible. While Szilard wished to save significant documents and correspondence, he lead a peripatetic life and wanted to be free to travel, unencumbered by possessions. Whenever he went on a trip, he gathered all the papers that he wanted to keep but did not wish to take along, bought several small suitcases, and left them with relatives and friends. It was not until after his death when space was made available in the library at the University of California in La Jolla, that all these suitcases were gathered from various distant locations and finally assembled in one single place. When this became known, several of our friends unexpectedly sent suitcases that they had found in their closets and attics, in addition to those we already knew about. We wish particularly to thank Dr. and Mrs. Efraim Racker who for several years stored the major portion of the collection in their Mt. Vernon home after the files had been transferred from Chicago during Szilard's illness; Mrs. Ruth Vollmer, Mr Bela Silard, Mr. Egon Weiss, Dr. Howard Green, Dr. Maurice Fox, Dr. Robert B. Livingston; also the Council for a Livable World and the Salk Institute who with great care forwarded the personal papers to us.

Through the good offices of Dr. Roger Revelle contact was made with the library of the University of California, San Diego. We are most grateful to Mr. Melvin Voigt, University Librarian, for providing not only suitable space and safe storage but also for giving valuable advice and for putting the facilities of the library at our disposal to process the Szilard papers. We particularly appreciate the help of Mr. George Vdovin, Head, Science and Engineering Library, where the papers are housed, and to the members of his staff for giving cheerful assistance whenever it was needed. Many other members of the faculty of the university helped in several phases of this project.

In conceiving and preparing this book, the contributors to this volume assisted far beyond their written contributions: Professors Jacques Monod, Carl Eckart, Maurice Goldhaber, and Aaron Novick were most helpful advisors throughout this enterprise. We were fortunate in having Dr. Julius Tabin frequently with us in La Jolla. He spent many hours going through the archives to unravel the exceedingly complicated and involved history of the patent applications.

In addition to the scientists who contributed essays to this volume we wish to thank Professors Matthew Meselson, Maurice Fox, and William Doering for their valuable consultation regarding selection of unpublished material.

Mr. W. F. Simpson, Librarian of General Atomics, helped us to obtain the reference list of declassified documents.

Finally we are pleased to acknowledge that permission to reprint copyrighted papers was granted by the following publishers: Springer Verlag (*Zeitschrift für Physik*), Mental Health Research Institute (*Behavioral Science*), Macmillan Ltd. (*Nature*), American Physical Society (*The Physical Review*), American Association for the Advancement of Science (*Science*), Cold Spring Harbor Laboratory of Quantitative Biology (*Symposia Series*), Princeton University Press (*Dynamics of Growth*

Processes) and the Rockefeller University Press (*The Journal of General Physiology*).

Sincere thanks also to Mrs. Laura Fermi for permission to reprint a letter from Enrico Fermi. We are also indebted to Mr. D. K. Miller, photographer of the Salk Institute for providing a number of portraits taken at the Institute; he also was most helpful in processing amateur photographs by others for this volume. We also wish to thank the following for granting permission to use portraits taken by them: Heka Davis, Lotte Meitner-Graf, as well as Esther Bubley and John Loengard of Life Magazine, Time Inc.

The Editors

Part I

Biographical Notes

Curriculum Vitae by Leo Szilard*
(Expanded by Gertrud Weiss Szilard)

I was born in Budapest, Hungary, in 1898. I went through officers' school there during the First World War and studied engineering there.

In 1920 I left Hungary to continue my engineering studies in Berlin. However, the attraction of physics proved to be too great. Einstein, Planck, Von Laue, Schroedinger, Nernst, Haber, and Franck were at that time all assembled in Berlin and attended a journal club in physics which was also open to students. I switched to physics and obtained a Doctor's degree in physics at the University of Berlin under Von Laue in 1922. My thesis (Ref. 1, see attached list of publications) showed that the Second Law of Thermodynamics covers not only the mean values, as was up to then believed, but also determines the general form of the law that governs the fluctuations of the values.

Subsequently, I was a research worker in one of the Kaiser Wilhelm Institutes in Berlin and later joined the teaching staff of the University of Berlin (as *Privatdozent*) where I remained until 1933. Of the papers (1 to 4) published during this period, some are experimental, and some are theoretical. The last one (4) established the connection between entropy and information which forms part of present day information theory.

In 1933 I went to England. I considered at that time becoming a biologist, and A. V. Hill said that he would find a position for me as a demonstrator in physiology. It occurred to me, however, just then that a nuclear chain reaction might be possible if we could find an element that would emit neutrons when bombarded by neutrons. Artificial radioactivity was discovered a few months later by Joliot and seemed to provide an important new research tool in nuclear physics. This decided me to move into nuclear physics.

In the summer of 1934 I started work as a guest in St. Bartholomew's Hospital in London and this work resulted in the establishment of the Szilard–Chalmers Reaction (5) and the discovery that slow neutrons are emitted by beryllium if the beryllium is exposed to gamma rays of radium (6). In 1939, after the discovery of the fission of uranium, the use of these slow neutrons from beryllium made it possible to see that uranium emits neutrons when bombarded by neutrons; the fast neutrons emitted by uranium could be easily distinguished from the bombarding slow neutrons.

In 1935, after a visit to New York, where I spent a few months as research associate at New York University, I accepted a position at the Clarendon Laboratory, Oxford University. During this period I worked in the field of nuclear physics (8 to 11). In 1938 I came to America under arrangement with Oxford University, which permitted me to spend half my time in the United States. I was in the United States during the time the Munich Agreement was negotiated. After Munich I decided to stay in the United States on a full-time basis, and I resigned at Oxford.

* Written by Leo Szilard in a grant application to the General Medical Sciences Division of the National Institutes of Health, Bethesda, Maryland, June 1959.

In January 1939 I learned of the discovery of fission. It seemed important to find out at once if neutrons are emitted in that process, for in that case a chain reaction in uranium had to be regarded as a serious possibility. I, therefore, asked the permission of Columbia University to work there as a guest and perform an experiment in order to settle this question. This experiment (jointly performed with Walter Zinn) led to the discovery of the neutron emission of uranium, upon which the chain reaction is based (12, 13). The same discovery was made independently at about the same time by Fermi and his co-workers and by Joliot and his group.

In July, 1939, I recognized that a chain reaction might be set up in a system composed of graphite and uranium. Because of the serious consequences of this possibility, it seemed that this was a matter in which the government ought to take an interest. I, therefore, went to see Professor Einstein to enlist his help in approaching the government. After several consultations, in which E. P. Wigner and Edward Teller participated, Einstein wrote a letter to President Roosevelt; and in response to this letter, the President appointed a committee under the chairmanship of the Director of the National Bureau of Standards.

In February 1940 I described the chain-reacting uranium–graphite system in a paper I sent to the *Physical Review* (February 1940). For reasons of secrecy, this paper was not published.

In November of 1940 a government contract was given to Columbia University for the development of the graphite–uranium system, and I became a member of Columbia University's National Defense Research Staff. Early in 1942 our group was moved to the University of Chicago; and on December 2, 1942, the chain reaction system was put into action.

Recently a patent was granted to the Atomic Energy Commission on the chain-reacting graphite–uranium system, jointly in the names of Enrico Fermi and myself.

In 1943 I became a naturalized citizen of the United States.

In October, 1946, I joined the staff of the University of Chicago as Professor of Biophysics in the Institute of Radiobiology and Biophysics. This institute never grew as originally intended; it had a succession of directors, and it was recently dissolved. I remained on the staff of the University of Chicago as Professor of Biophysics and was transferred to the Enrico Fermi Institute for Nuclear Studies.

When in 1946 I was faced with the task of converting myself into a biologist, I teamed up with Dr. Aaron Novick, a physical chemist. I had known him from his work in the uranium project. We both got our training in biology through summer courses, such as Dr. Delbrück's course in Cold Spring Harbor in bacterial viruses and Dr. Van Niel's course in bacterial biochemistry at Pacific Grove. Dr. Novick and I worked as a team until the Institute of Radiobiology and Biophysics was dissolved.

A list of publications is attached, containing a short description of each paper. When we started out, we tried to understand a striking phenomenon just then discovered by A. Kelner, who showed that bacteria killed by ultraviolet light can be reactivated by shining visible light at them (17). A detailed analysis of the phenomenon enabled us to interpret it in terms of a "poison" that is produced by ultraviolet light and is decomposed by visible light. This interpretation was at first controversial due to Dulbecco's work on light reactivation of ultraviolet killed bacterial viruses but has in the meantime become widely accepted. My own interest in the subject waned when I could not convince myself that we were dealing with a phenomenon that serves a useful biological purpose in the life of the bacteria.

Next, we turned our attention to the study of bacterial viruses in the assumption that viruses may prove to be much simpler than bacteria. We obtained some very interesting results (18) but decided to shift after a while to the study of bacteria. The two phenomena in which we were particularly interested were (a) mutations and (b) the formation of adaptive enzymes which promised to provide a tool for the study of protein synthesis.

We were dissatisfied, however, with the methods that were available for the study of these phenomena. It seemed to us necessary to study bacterial populations in the growing condition in a stationary state, i.e., we thought we ought to use a continuous flow device. We developed such a device, which we called a "Chemostat." In this particular device the rate of growth of the bacteria can be changed by changing the concentration of one of the growth factors of our choosing which we make the controlling growth factor.

We started out by using the "Chemostat" for the study of mutations and obtained quite unexpected results at the very outset. It turned out, for instance, that the rate at which certain mutations occur does not change when we change the rate at which the bacteria divide; we could vary the rate of growth within a wide range without changing the rate at which these mutations occurred. We found one family of compounds—purines—which may cause an about tenfold increase in the mutation rate of bacteria without any appreciable killing. And we also found antimutagens, which in very small concentrations will fully counteract the effect of purine-type mutagens.

In a bacterial population maintained in the "Chemostat" there occur evolutionary changes (19), and one strain of bacteria is replaced by a mutant strain, which can grow faster in the conditions prevailing in the growth tube of the "Chemostat." We observed successive evolutionary steps of this sort in each experiment of sufficiently long duration and were able to analyze the phenomenon.

After the dissolution of the Institute of Radiobiology and Biophysics I did not maintain a laboratory. In the last few years my interests centered mainly on quantitative studies of general biological phenomena, with strong emphasis on molecular biology. The paper I published most recently (25) attempts to give a quantitative theory of the process of aging which should be applicable to mammals.

Partial Bibliography of Leo Szilard
[with his own annotations]

A. *Physics*

(1) *Zeitschrift für Physik*, **32**, p. 753, 1925. This paper extends the application of thermodynamics to the derivation of the laws of thermodynamical fluctuations. It was accepted as dissertation by the University of Berlin.

(2) *Zeitschrift für Physik*, **33**, p. 688, 1925—jointly with H. Mark. This paper reports experiments which revealed anomalous scattering of x rays.

(3) *Zeitschrift für Physik*, **35**, p. 743, 1926—jointly with H. Mark. This paper reports experiments on polarizing x rays by reflection on crystals.

(4) *Zeitschrift für Physik*, **53**, p. 840, 1929. This paper evaluates the increase of entropy which is connected with operations of an intelligent being on a thermodynamical system if these operations are controlled by measurement of variables which are subject to thermodynamical fluctuations. This paper was accepted as Habilitationsschrift by the University of Berlin.

(5) "Chemical Separation of the Radioactive Element from its Bombarded Isotope in the Fermi Effect"—jointly with Chalmers. *Nature*, **134**, p. 462, 1934. This paper demonstrates a generally applicable process (Szilard–Chalmers reaction) for the concentration of a radioactive element produced by neutrons if the element has to be separated from a mass of a stable element with which it is chemically isotopic.

(6) "Detecting Neutrons Liberated from Beryllium by Gamma Rays," *Nature*, **134**, p. 494, 1934. This paper describes the discovery of radium–beryllium photo neutrons which, being of low energy, represent a useful tool in nuclear research. They were universally used later in the discovery and investigation of neutron emission of uranium on which a chain reaction is based.

(7) "Liberation of Neutrons from Beryllium by X-Rays"—jointly with a group of six others. *Nature*, **134**, p. 880, 1934. Using x rays in place of gamma rays, the threshold for the emission of photo neutrons from beryllium is determined by varying the voltage of an x-ray tube and is found to be somewhat above 1.5, and well below 2 Mev.

(8) "Radioactivity Induced by Neutrons"—jointly with Chalmers. *Nature*, **135**, p. 98, 1935. In this paper a neutron-induced radioactive period of about $3\frac{1}{2}$ hours is reported in indium which does not fit in with the explanations found for other radioactive periods. In a later paper it is shown that it is due to an excited indium nucleus which is isomeric with stable indium nucleus 115.

(9) "Absorption of Residual Neutrons," *Nature*, **136**, p. 950, 1935. This paper reports the discovery of neutron resonances at low energies, gives an estimate of their energies, and states that the energies can be measured by observing the absorption of the residual neutrons in boron or lithium.

(10) "Gamma Rays Excited by Capture of Neutrons"—jointly with Griffiths. *Nature*, **139**, p. 323, 1937. This paper reports on the observation of gamma rays emitted by a

number of odd elements which are strong neutron absorbers. The counts observed per absorbed neutron were found to be within 15% identical for all these elements.

(11) "Radioactivity Induced by Nuclear Excitation"—jointly with Goldhaber and Hill. *Phys. Rev.*, **55**, p. 47, 1939. In this paper the previously reported period in indium is investigated and the conclusion is reached that it is due to nuclear excitation of the stable indium isotope 115.

(12) "Instantaneous Emission of Fast Neutrons in the Interaction of Slow Neutrons with Uranium"—jointly with Zinn. *Phys. Rev.*, **55**, p. 799, 1939. In this paper the discovery of the neutron emission of uranium is reported. It is estimated that two neutrons are emitted per fission. The neutrons from uranium are made visible on an oscillograph screen. As primary neutrons, radium–beryllium photo neutrons were used which, because they are slow, can be easily distinguished from the fast neutrons emitted by uranium. This discovery which was made independently by Fermi in the same year indicated the feasibility of a sustaining nuclear chain reaction.

(13) "Emission of Neutrons by Uranium"—jointly with Zinn. *Phys. Rev.*, **56**, p. 619, 1939. Detailed report of above mentioned experiments, number of neutrons per fission measured as 2.3.

(14) "Neutron Production and Absorption in Uranium"—jointly with Anderson and Fermi. *Phys. Rev.*, **56**, p. 284, 1939. This paper reports an investigation on the chain reaction qualities of a uranium–water system. It is estimated that 1.5 neutrons are emitted for every thermal neutron which is absorbed by uranium.*

B. *Biology*

(17) A. Novick and Leo Szilard, "Experiments on Light-Reactivation of Ultra-Violet Inactivated Bacteria," *Proceedings of the National Academy of Sciences*, **35**, No. 10, pp. 591–600, 1949.

(18) Aaron Novick and Leo Szilard, "Virus Strains of Identical Phenotype but Different Genotype," *Science*, **113**, No. 2924, pp. 34–35, January 12, 1951.

(19) Aaron Novick and Leo Szilard, "Experiments with the Chemostat on Spontaneous Mutations of Bacteria," *Proceedings of the National Academy of Sciences*, **36**, No. 12, pp. 708–719, December 1950.

* [References 15 and 16, left blank in this physics listing of Szilard's, presumably would include Report A-55, see Part III this volume. Dr. Szilard's part in bringing about of the first nuclear chain reaction, in the design of the first nuclear reactor (atomic pile) were first described, insofar as these matters could then be made public, in Henry D. Smythe, *Official Report: Atomic Energy for Military Purposes*, Princeton University Press, 1945, pages 34, 47, etc.—Eds.]

(20) Aaron Novick and Leo Szilard, "Description of the Chemostat," *Science*, **112**, No. 2920, pp. 715–716, December 15, 1950.

(21) Aaron Novick and Leo Szilard, "Experiments on Spontaneous and Chemically Induced Mutations of Bacteria Growing in the Chemostat," Cold Spring Harbor Symposia on Quantitative Biology, Vol. XVI, 1951.

(22) Aaron Novick and Leo Szilard, "Anti-Mutagens," *Nature*, **170**, p. 926, November 29, 1952.

(23) Aaron Novick and Leo Szilard, "Experiments with the Chemostat on the Rates of Amino Acid Synthesis in Bacteria," *Dynamics of Growth Processes*, Princeton University Press, pp. 21–32, 1954.

(24) Maurice S. Fox and Leo Szilard, "A Device for Growing Bacterial Populations under Steady State Conditions," *Journal of General Physiology*, **39**, pp. 261–266, 1955.

(25) Leo Szilard, "On the Nature of the Aging Process," *Proceedings of the National Academy of Sciences*, **45**, pp. 30–45, 1959.

The first of these papers (17) investigates a phenomenon discovered by A. Kelner after the war, who showed that bacteria "killed" by ultraviolet light can be revived by shining visible light on them. Experiments designed to analyze the phenomenon are described in this paper; they lead to the conclusion that the ultraviolet light produces a "poison" which can be inactivated by light and that this "poison," if present when, subsequent to irradiation, the bacteria divide, will cause both death and mutations.

The second paper (18) describes the discovery that, when a bacterium is infected simultaneously with two related viruses which differ from each other both in genotype and phenotype, the virus population emerging from the bacterium contains a class of viruses which have the genotype of one and the phenotype of the other.

The Papers 19 to 23 describe a new way of studying bacteria by maintaining a bacterial population in a stationary (exponentially growing) state indefinitely and controlling the growth rate by controlling the rate of supply of an essential growth factor. An apparatus is described in these papers which will conveniently accomplish this and which is designated as the "Chemostat."

In studying mutations in bacteria or the formation of adaptive enzymes in bacteria, inaccurate and, therefore, misleading results are frequently obtained by studying bacterial cultures in flasks in which the number of bacteria increases exponentially, and today the use of the Chemostat appears to be indispensable.

In the Papers 19 to 22, the Chemostat is used in the study of mutations. It turns out that the rate at which mutations occur in a growing bacterial population under the conditions studied is not proportional to the rate at which cell division occurs; rather the mutation rate is constant per unit time independent of the rate at which the culture is growing. There is found one group of compounds, all purine derivatives, of which caffeine is one, which greatly increases the mutation rate without having an appreciable killing effect on the bacteria.

There is another group of compounds described in these papers, all of them ribosides of purines which in small quantities will completely counteract the action of the above-mentioned purine type mutagens and also reduce the rate of spontaneous mutations.

In Paper 23, the Chemostat is used to study the biosynthesis of amino acids in bacteria and the regulatory mechanisms that are involved in it. The bio-synthetic apparatus of the bacteria responds to amino acid concentrations in the medium, which are exceedingly low. For instance, a bacterium that can make arginine and will do so if there is no arginine in the medium, will stop making arginine if an arginine concentration of 10^{-9} gm/cc is maintained in the medium in the Chemostat. (Novick and Szilard, unpublished paper.)

One way of studying such regulatory mechanisms is based on the use of a mutant which is blocked in the synthesis of an amino acid—in our case tryptophane—and which pours out into the medium a "precursor" of that amino acid. Paper 23 utilizes such a mutant. In the absence of tryptophane in the medium, a precursor of tryptophane is poured out by the mutant into the medium at a rate which is independent of the growth rate of the bacteria. In the presence of tryptophane this "precursor" is *not* poured out by the bacteria. It is conceivable that this indicates a general phenomenon of regulation through a negative feedback of the final product at one of the *early* steps of the metabolic pathway leading to tryptophane.

In Paper 24, there is described a device called a breeder. In this device bacteria may be grown in a continuous flow of nutrient. The flow of the nutrient is controlled by the turbidity of the bacterial culture and the growth is not limited by a growth factor, as is the case in the Chemostat.

This device was developed in order to study mutations in bacteria under conditions of growth at the maximal rate, and such a study was carried out by Maurice S. Fox.

Paper 25 develops a theory of the basic process of aging. According to the theory, the elementary step in the process of aging consists in the random inactivation of whole chromosomes. The differences of longevity of individuals are attributed to the difference of the number of defective "vegetative genes" they have inherited.

The preceding pages were written by Szilard as part of an "Application for a Research Grant" which he filed with the General Medical Sciences Division of the National Institutes of Health in June 1959. His Curriculum Vitae written for this purpose therefore contains only information pertinent to his scientific career and omits other activities of great importance in his life.

In 1933, after Szilard left Germany for England, he established jointly with Sir William Beveridge, Esther Simpson, and others the "Academic Assistance Council," an organization to relocate scholars displaced from Nazi Germany.

While Szilard was instrumental in launching the atomic bomb project under the assumption—later proved to be erroneous—that there was a race against Nazi

Germany, it became clear to him in the Spring of 1945 that Germany had been defeated, and he made desperate efforts to prevent the bomb from being used. Thereafter, a major part of his life was devoted to the prevention of nuclear warfare. In July 1945 he circulated a petition in the Uranium Project, addressed to the President of the United States, asking that the bomb not be dropped. Within the year following the war he fought successfully for civilian control of atomic energy.

In the late 1940s, through the Emergency Committee of Atomic Scientists and numerous publications in the Bulletin of the Atomic Scientists and elsewhere, he launched a perpetual campaign to make people aware of the dangers of nuclear warfare and the arms race. He was a dedicated participant of the international Pugwash Movement, attending all their conferences during his lifetime.

In 1962 he founded the Council for Abolishing War, now called the Council for a Livable World, an organization aimed at influencing the decisions and actions of the U.S. Senate in foreign policy issues.

His writings on these topics will be included in succeeding volumes.

In the Grant Application, mentioned earlier, the title of the project was "Quantitative Studies of General Biological Phenomena" (RG 6876). Because the proposed research plan is unusual indeed, parts of it are excerpted here:

Research Plan and Supporting Data

The purpose of the proposed study is to gain insight into certain *general* biological phenomena rather than to try to understand the functioning of *specialized* biological structures (such as, for instance, of the nerve fiber, of the muscle fiber, or of the specialized sense organs). I am particularly interested in those general biological phenomena where it may now be possible to gain insight into quantitative relationships which can be checked against data obtained from available observations or experiments as yet to be made. The proposed work would take as its starting point preliminary theoretical studies which I carried out in the past three years.

At the University of Chicago, I am holding a Research Professorship. I have neither any teaching duties nor any fixed obligations to be in Chicago at certain fixed periods of time. This freedom has enabled me in the past three years to spend considerable time at various laboratories away from Chicago. It is my understanding that, under the grant here requested, I would have full freedom to move about wherever my research interests may take me. It is anticipated that I may spend nine months of the year away from my home, at various laboratories where work may be pursued in fields in which I am interested.

I anticipate that my work will have a strong theoretical orientation. But in order to be able to function as a theoretical biologist, it is necessary to have intimate knowledge of experiments relating to a variety of biological materials and involving diverse techniques.

Biology has not quite reached the stage which was attained by physics half a century ago when enough relevant facts were established to permit a theoretical physicist to come up with significant insights. Yet in biology we might be very well on the verge of a similar situation, and a few scientists who are so inclined may now perhaps attempt to function as theoretical biologists. Accordingly, these days it might be as well for a few scientists to put less emphasis on their own experiments and spend more time trying to keep in touch with the experiments of others in the hope of being able to recognize new patterns and to try to gain insight into some general biological laws. It may be that the main difference between the theoretical physicists of the past and the would-be theoretical biologist of the present is quantitative rather than qualitative. The would-be theoretical biologist would probably not be able to keep on studying the results of others and thinking about them for a very long stretch of time. Much sooner than a theoretical physicist, he will feel impelled to do further experiments (or to induce someone else to do them) because he will need to cut down *the number* of possible avenues along which his further thinking may be tempted to wander.

How successfully a man may be able to function as a theoretical biologist is likely to depend (apart from his inclinations and abilities) on whether he is put in the position where he can maintain close cooperation with a number of laboratories, initiate some new experiments or observations, and perhaps participate actively in experimental work carried out in laboratories where he holds no position of administrative responsibility.

On a rather limited scale I have tried, in the last few years, to function somewhat as such a theoretical biologist.

The research grant became effective on January 1, 1960, and continued through March 31, 1964. During this time the following papers by Szilard were published. (We continue the numbering system used by Szilard in the preceding pages.)

(26) Leo Szilard, "The Control of the Formation of Specific Proteins in Bacteria and in Animal Cells," *Proceedings of the National Academy of Sciences*, **46**, 277–292, (March) 1960.

(27) Leo Szilard, "The Molecular Basis of Antibody Formation," *Proceedings of the National Academy of Sciences*, **46**, 293–302, (March) 1960.

(28) Leo Szilard, "Dependence of the Sex Ratio at Birth on the Age of the Father," *Nature*, **186**, 649–650, (May 21) 1960. Letter to the Editor.

(29) Leo Szilard, "On Memory and Recall," *Proceedings of the National Academy of Sciences*, **51**, 1092–1099, (June) 1964.

In his progress reports to NIH, Szilard described the first three papers as follows:

A model for the control of the rate of production of repressible enzymes has been developed and this model is described in detail in Paper 26. This model assumes that in bacteria the repressor controls the rate of formation of the enzyme by the enzyme forming site, rather than the rate of formation of the enzyme site itself. Experiments which are at present being conducted in a number of different laboratories, with which the author maintains contact, might elucidate, within a year, whether this "premise" is correct.

The above-quoted paper also assumes that the repressor can attach itself to the enzyme and it is shown that accordingly the cell might have two stable states, a state in which the enzyme level is high and a state in which the enzyme level is low. The validity of this assumption does not depend on the above-mentioned "premise" and the assumption might provide the key to the understanding of a certain type of differentiation, discussed in the paper.

A second paper (27) discusses the possibility that antibody formation—in the primary response—is based on this type of differentiation, triggered by the injection of an antigen into the rabbit. This theory can account for a number of phenomena listed in the paper, including the phenomenon of immune tolerance of the new-born rabbit. The explanation of immune tolerance is, however, again based on the "premise" that the repressor controls the rate at which the protein—in this case the antibody—is formed by the specific protein forming site. If future experiments should show that this "premise" is wrong, then the theory of immune tolerance would have to be modified and it is not as yet clear whether a satisfactory modification of the theory would be possible, in that contingency.

A theory for the dependence of the sex ratio at birth on the age of the father has been presented in Paper 28 which is based on a theory of aging previously presented by the author in Paper 25. The theory accounts for the decrease in the ratio of boys to girls, with increasing age of the father, on the ground that a spermatogonium in which the X-chromosome suffers an "aging hit" may not continue to give rise to sperm, whereas a spermatogonium in which the Y-chromosome suffers an "aging hit" may continue to give rise to the sperm.

Paper 29, Szilard's last effort, was published posthumously. He had prepared a much longer manuscript and sent preprints of the shortened version to his friends accompanied by this memorandum:

May 5, 1964.

Enclosed is a preprint of a paper which will appear in the June issue of the Proceedings of the National Academy of Sciences. Because authors are limited to eight pages in any one issue of the Proceedings, this preprint is but the first of three installments.

Had I merely postulated—as others seem to have done—that if two neurons fire simultaneously, thereafter the synapse bridging these two neurons has a higher efficacy, then I would not be able to account even for Pavlov's experiments on the conditioned salivary reflex of the dog. As it is, it seems conceivable that the two fundamental postulates of my model might be able to account not only for the peculiarities of all of Pavlov's basic experiments but—in conjunction with neuron-networks, as yet to be invented—also for the higher mental functions. This could be true even if the details of the biochemical underpinnings of these two postulates should turn out to be incorrect.

[Signed] Leo Szilard.

In addition to the four published papers, Szilard during this period outlined several theories in the four following reports:

(1) Induction of Mutations in Mammals by Ionizing Radiation (1961).

(2) The Sex Chromatin in Mammalian Cells, "Dosage Compensation" in the Fruit Fly, and the Enzyme Repression in Bacteria (1962).

(3) On the Occasional Dominance of the "Perceptible Phenotype" in Man (1963, rev. 1964).

(4) The aging process and the "Competitive Strength" of Spermatozoa (1963).

He summarized them in his progress reports as follows:

(1) An experimental method has been devised and a theory of the experiment developed which should make it possible to determine the dose of radiation which would raise the mutation rate to twice the value of the spontaneous mutation rate. The method consists in exposing a population of mice to ionizing radiation and subsequently determining, among the first generation off-spring, the proportion of females whose off-spring shows an abnormal sex ratio. The method is described in a paper dated March 10, 1961, *Induction of Mutations in Mammals by Ionizing Radiation* which is being privately circulated to those interested in this type of problem.

(2) In mammals and also in the fruit fly the somatic cells of the female contain two X-chromosomes while the somatic cells of the male contain only one. Accordingly, the cells of the female carry two homologous copies of each sex-linked gene, whereas the cells of the male carry only one copy of each. This difference in "dosage" does *not* usually manifest itself in a phenotypic difference between the male and the female. Recent observations indicate that in the case of mammals, at some point of the embryonal development of the female, one of the two X-chromosomes ceases to be functional in the somatic cells. This, on the face of it, could account for the fact that the double dosage of the sex-linked genes in the female, as compared to the single dosage of the same genes in the male, does not lead to a difference in the phenotype. However, no such difference in phenotype exists in the fruit fly either, and yet I find that the phenomenon of "dosage compensation," which has been studied in the fruit fly by H. J. Muller, cannot be explained on the assumption that only one of the two X-chromosomes is functional in the somatic cells of the female. In those circumstances it is necessary to look for another explanation for "dosage compensation" in the fruit fly. I propose to explain this phenomenon in the fruit fly by assuming that the relevant gene products in the fruit fly are under the control of repressors, in much the same way in which many enzymes are under the control of repressors in bacteria, and by further assuming that in the fruit fly the genes corresponding to the repressors (of those gene products which show "dosage compensation") are located on the X-chromosome. These considerations are described in a paper, *The Sex Chromatin in Mammalian Cells, "Dosage Compensation" in the Fruit Fly and Enzyme Repression in Bacteria*, which is being circulated in preprint among those interested in this kind of problem.

(3) It has been shown that the observed frequent occurrence of a striking resemblance between a child and one of its parents might be explained in one of two ways:

(a) The perceptible phenotype might be determined by a number of different genetic loci, all of which are located on one pair of homologous autosomes. Such an autosome might possess a certain "strength" and a "strong" autosome might suppress the homologous autosome if it is substantially less "strong." (For details see the enclosed paper, *On the Occasional Dominance of the "Perceptible Phenotype" in Man,* dated July 12, 1963.) Revised version dated May 18, 1964.

(b) The genes which determine the perceptible phenotype might all be located within the same operon and different operons might be more strongly or less strongly repressed. This could then account for the observed resemblances, if one assumes that the perceptible phenotype is determined by the ratio of the quantities of the products of these genes in the diploid cell.

(4) An experimental method has been devised for determining whether the competition which exists between spermatozoa for fertilizing an ovum might serve the purpose of protecting the ova against being fertilized by a spermatozoan which might contain genetic material that has deteriorated as a result of the aging process. The method devised consists in inseminating females with a mixture of spermatozoa, derived from two donors, and in determining how the fraction of the offspring which is derived from the older donor, decreases with increasing age differences of the two donors. (For details see the enclosed paper, *The Aging Process and the "Competitive Strength" of Spermatozoa,* dated July 25, 1963.) Arrangements for carrying out experiments of this type are at present under discussion.

After the Institute of Radiobiology and Biophysics at the University of Chicago was dissolved in 1954 Szilard found himself without a laboratory. "I would rather have roots than wings," he said at that time, "but if I cannot have roots I shall use wings." During the last ten years of his life Szilard became a roving sponsor of the newly emerging science of molecular biology. He was visiting professor at the newly created Department of Biophysics at the University of Colorado Medical Center, visiting professor at Brandeis University, and served as consultant to the Basic Research Program of the National Institute of Mental Health in Washington.

In 1963, after reaching the age of 65, Szilard became Professor Emeritus at the University of Chicago but remained in active service until March 31, 1964.

On several visits to Europe between 1957 and 1963 Szilard served as advisor to the research program of the World Health Organization, was consulted by the German government on the organization of postwar scientific research in West Germany, and rounded up the European biologists to explore the possibility of creating a European Laboratory for Molecular Biology. These efforts led to the establishment of EMBO, the European Molecular Biology Organization.*

In 1957, while working on problems of population control, he drafted with his friend Professor William Doering "A Proposal to Create Two Interdependent Research Institutes Operating in the General Area of Public Health, Designated as: 'Research Institute for Fundamental Biology and Public Health' and 'Institute for Problem Studies.'" This proposal contains many far reaching thoughts and is therefore partially reproduced in this volume as an Appendix to Part IV, "Published Papers in Biology."

At that time Szilard contacted a number of scientists to explore whether they might be interested in joining the kind of institute he and Doering had visualized. He was fortunate in finding a sympathetic response among scientists with similar inclinations

* J. C. Kendrew, "EMBO and the Idea of a European Laboratory," *Nature,* **218**, 840–842, (June 1) 1968; Marie Luise Zarnitz, "Molekulare und Physikalische Biologie," Bericht zur Situation eines interdisziplinaren Forschungsgebietes in der Bundesrepublik Deutschland, Erstattet im Auftrag der Stiftung Volkswagenwerk (Vandenhoeck & Ruprecht in Gottingen, 1968).

and "was one of the moving spirits who helped to conceive the idea of the Salk Institute and to bring it into being."*

He joined the Salk Institute as Non-Resident Fellow in July 1963 and became a Resident Fellow on April 1, 1964. After a few short but happy and productive months in La Jolla, he died there suddenly on May 30, 1964.

Awards

During his lifetime, Leo Szilard received the following recognitions:

Fellow, American Academy of Arts and Sciences, 1954.

Humanist of the Year, American Humanist Association, 1960.

Einstein Gold Medal of the Lewis and Rosa Strauss Memorial Fund, 1960.

Atoms for Peace Award, 1960.

Elected Membership, National Academy of Sciences, 1961.

Honorary Doctor of Humane Letters, Brandeis University, 1961.

In 1970, a crater on the far side of the moon (34° N; 106° E) was named "Szilard" by the International Astronomical Union.

* *Topics in the Biology of Aging.* A Symposium held at The Salk Institute for Biological Studies, San Diego, California, 1965. Peter L. Krohn, Ed., Interscience Publishers, John Wiley & Sons, New York, 1966.

Student days in Budapest, 1915.

Early portrait, age 18 years, taken at time of enroll-
ment at the Institute of Technology (Muegyetem)
in Budapest, Hungary, 1916.

England, 1936. (Photograph by Trude Weiss.)

Countryside near Oxford, England, Spring 1936. (Photograph by Trude Weiss.)

Emergency Committee of Atomic Scientists, late
1940s. Back row, left to right, Victor F. Weisskopf,
Leo Szilard, Hans A. Bethe, Thorfin R. Hogness,
Philip M. Morse. Seated, left to right, Harold C.
Urey, Albert Einstein, and Selig Hecht.

First pile scientists, University of Chicago, December 2, 1947, the fifth anniversary of their success. Back row, left to right, Norman Hilberry, Samuel Allison, Thomas Brill, Robert G. Nobles, Warren Nyer, and Marvin Wilkening. Middle row, Harold Agnew, William Sturm, Harold Lichtenberger, Leona W. Libby, and Leo Szilard. Front row, Enrico Fermi, Walter H. Zinn, Albert Wattenberg, and Herbert L. Anderson.

At Cold Spring Harbor phage course, learning
biology, Summer, 1947. (Photograph by Trude
Weiss.)

With Albert Einstein writing letter to President
Franklin D. Roosevelt of August 2, 1939. (This
re-enactment taken in late 1940s.)

With Norman Hilberry at Stagg Field, site of first
nuclear chain reaction, Chicago, 1957.

At the 3rd Pugwash Conference, Kitzbühel, Austria,
1958. Left to right, Bernard T. Feld, Morton
Grodzins, Leo Szilard, Harrison Brown, and Sir
Robert Watson-Watt.

Receiving Atoms for Peace award, Washington,
D.C., May 1960. With James R. Killian and, in the
background, Eugene P. Wigner.

Cold Spring Harbor with Max Delbrück, early
1950s.

Working on tapes with Mrs. Szilard, Memorial
Hospital, New York, June 1960. (Photograph by
John Loengard, LIFE Magazine © Time Inc.)

Lecturing at Cold Spring Harbor symposium,
June 1961. (Photograph by Ester Bubley, LIFE
Magazine © Time Inc.)

At Cold Spring Harbor Symposium, June 1963.
(Photograph by Trude Weiss.)

With William Doering, Washington, D.C., 1963.
(Photograph by Trude Weiss.)

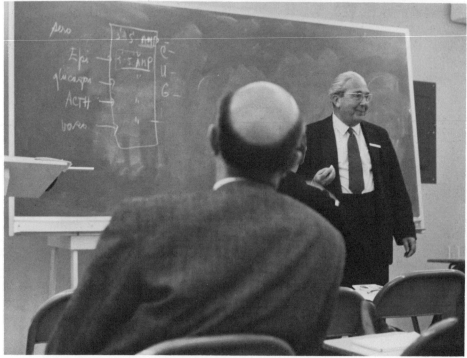

With Jonas Salk, La Jolla, California, February
1964. (Photograph by Trude Weiss.)

Seminar at Salk Institute, February 1964. (Photograph by D. K. Miller, Salk Institute.)

With Francis Crick, Fellows Meeting, Salk Institute, 1964. (Photograph by D. K. Miller, Salk
Institute.)

Fellows' Meeting, Salk Institute, February 1964.
Front row, left to right, Augustus B. Kinzel
(standing), Leo Szilard, Francis Crick, and Renato
Dulbecco. Second row, left to right, Charles M.
Pomerat, Rita Levi-Montalcini, and Melvin Cohn.
(Photograph by D. K. Miller, Salk Institute.)

Part II

Published Papers in Physics
(1925–1939)

Thermodynamics

Introduction by Carl Eckart*

The two papers:

1. Über die Ausdehnung der phänomenologischen Thermodynamik auf die Schwankungserscheinungen,
2. Über die Entropieverminderung in einem thermodynamischen System bei Eingriffen intelligenter Wesen,

are closely interrelated, though each makes a distinct and highly original contribution to thermodynamic theory. These notes will try to place these papers against the background of both earlier and later contributions by others.

There are really two thermodynamic theories: the phenomenological and the statistical or atomistic. They were constructed simultaneously and largely by the same men. The phenomenological theory starts from the principle that heat is energy and always flows from hot to cold regions. By ingenious and abstract arguments, many diverse phenomena are brought under this principle, and remarkable conclusions are reached. An abstract concept of entropy is developed, and the basic principle is reformulated by Clausius as, *the energy of an isolated system remains constant, and its entropy always increases*—or, in ideal cases, remains constant.

The abstract elegance of this theory, together with its great predictive power, has impressed many scientists including Max Planck. As he repeatedly stressed, one will not readily accept any theory unless it is demonstrably in accord with Clausius's principle.

On the other hand, its abstractness has led to the search for alternatives, based on hypotheses concerning the interactions of the atoms that constitute all matter. Heat is then not an abstractly defined form of energy but the random motion of the atoms. Entropy is related, according to Boltzmann, to the probability of a given state of motion of the atoms, and Clausius's principle is seen as a consequence of the tendency of a complex system to reach its most probable state.

Attractive as this seems at first, by the beginning of the twentieth century many had recognized that it merely replaces one abstraction, entropy, by an equally difficult one, probability. Moreover, generality is lost, because the Boltzmann principle is founded upon an atomic model. Hence Planck and others who were actively questioning the applicability of Newtonian mechanics to structures of atomic size, were always careful to integrate their results with the phenomenological thermodynamics.

Probabilistic thermodynamics differs from the phenomenological theory in one important respect. The latter considers thermodynamic equilibrium as static, a state of no change. In the atomistic theory, equilibrium is a state of motion. The modern art form of the *mobile* illustrates this concept: after a mobile has been set in motion and left to itself for a time, its constantly fluctuating configurations illustrate thermodynamic equilibrium.

These fluctuations can also be observed in more interesting circumstances. The botanist R. Brown discovered that the small particles of a colloidal suspension are in constant irregular motion. Einstein[1] developed a theory of this as an example of thermal agitation, and Perrin[2] used his equations to obtain good experimental values of Avogadro's number and of Boltzmann's constant. Smoluchowski[3] studied the thermal fluctuations in density, and W. Schottky,[4] the thermal agitation of electrons in conductors. Today, numerous other examples are known.

While Szilard was a graduate student at the University of Berlin, these phenomena were under active discussion. Perrin and others cited them as "proof" that atoms

* Professor of Geophysics, University of California, San Diego.

"really" exist. It is therefore not surprising that they engaged Szilard's attention. In the above mentioned Paper 1, he showed that these fluctuations can be included within the framework of the phenomenological theory *without making any reference to atomic models*. This surprised many, including Einstein and von Laue; it is not surprising that the latter readily accepted the paper as satisfactory for a Ph.D. thesis.

In an interview recorded in 1963, Szilard mentions that this paper was written during the Christmas recess of 1921, though it was not published until 1925. He also records that Paper 2 was written about six months later; it was not published until 1929. It was accepted as his Habilitationschrift in 1925, when he was appointed Privatdozent at the University of Berlin.

Paper 2 concerns the paradox invented by Maxwell. Maxwell[5] imagined "a being whose faculties are so sharpened that he can follow every molecule in his course, and would be able to do what is at present impossible to us .... Let us suppose that a vessel is divided into two portions A and B by a division in which there is a small hole, and that a being *who can see the individual molecules* opens and closes this hole, so as to allow only the swifter molecules to pass from A to B, and only the slower ones to pass from B to A. He will, thus, without expenditure of work raise the temperature of B and lower that of A, in contradiction to the second law of thermodynamics."

Smoluchowski was, perhaps, the first to recognize the fallacy in this conclusion. The being (or demon) must be considered as a part of the system—otherwise it is not isolated. Moreover, the demon's metabolism must be included among the processes that occur in it. Obvious as this may seem today, forty years elapsed before it was explicitly noted. In the 1920s, this resolution of the paradox still did not immediately convince many physicists.

Szilard seeks for the essential mode of interaction that enables the demon to decrease the entropy of the remainder of the system. He finds it in "a kind of memory," which is inherent in "measurement." Then he concludes that ignorance of the metabolism of the memory process need be no hindrance to his project. One can imagine inanimate mechanisms that have this ability to remember. Their "metabolism," being simpler, can be subjected to detailed analysis—and this is the subject of Paper 2. He postulates that any diminution of entropy in other parts of the system will be compensated by an increase in that of the mechanism. Then this postulate is verified in an example.

During the same decade in which Szilard published these ideas, an apparently unrelated theoretical development began. Communication engineers needed a precise definition of "information." This started with the general notion that information is the commodity that telegraph companies transport from one place to another. (The more refined definitions were developed by H. Nyquist,[6] R. V. L. Hartley,[7] and others.) This led to the recognition that messages with a high probability of occurrence (such as conventional greetings) convey little information.

Perhaps John von Neumann (whose friendship with Szilard dated from their days at the University of Berlin) was the first to recognize this intrusion of probabilistic ideas into information theory, thus connecting that theory with thermodynamics. Moreover, there is a strong analogy between Szilard's formulas for the entropy generated by the demon and those of Nyquist and Hartley for information. N. Wiener[8] had also proposed similar formulas.

This was soon seen to be of fundamental significance. C. E. Shannon[9] then established the complete and detailed relation between information and entropy. Shannon's work was followed by a whole *genre* of papers[10] similar to Szilard's. It is also significant that the inanimate memory imagined in Paper 2 has been realized in the modern computer—the "thinking machine."

By this time, Szilard was occupied with other matters. However, in his last paper "On Memory and Recall," he returned to the metabolism of memory and showed that he intended to take an active part in the study of these processes in living organisms.

References

1. A. Einstein, "Untersuchungen über die Theorie der Brownschen Bewegung," *Annalen der Physik*, **17**, 549 (1905); **19**, 289 (1906); **19**, 371 (1906).

2. J. Perrin, "Brownian Movement and Molecular Reality," F. Soddy (tr.), *Annales de Chimie et de Physique*, London: Taylor and Francis, (1910).

3. M. von Smoluchowski, "Vorträge über die kinetische Theorie der Materie und Elektrizität," Leipzig, 1914, p. 89.

4. W. Schottky, "Über spontane Stromschwankungen in verschiedenen Elektrizitätsleitern," *Annalen der Physik*, **68**, 157 (1922).

5. J. C. Maxwell, *Theory of Heat*, London: Longmans, Green and Co. (1871), p. 328.

6. H. Nyquist, "Certain Factors Affecting Telegraph Speed," *Bell System Technical Journal*, **3**, 324 (1924).

7. R. V. L. Hartley, "Transmission of Information," *Bell System Technical Journal*, **7**, 535 (1928).

8. N. Wiener, *Cybernetics, or Control and Communication in the Animal and Machine*, New York: John Wiley and Sons (1948).

9. C. E. Shannon, "A Mathematical Theory of Communication," *Bell System Technical Journal*, **27**, 379, 623 (1948).

10. See, for example, L. Brillouin, *Science and Information Theory* (especially Chap. 13), New York: Academic Press (1962).

753

Über die Ausdehnung der phänomenologischen Thermodynamik auf die Schwankungserscheinungen.

Von **Leo Szilard** in Berlin-Dahlem.

Mit drei Abbildungen. (Eingegangen am 11. September 1924.)

Es wird gezeigt, daß man durch rein phänomenologisch thermodynamische Betrachtungen zur Beherrschung von Gesetzmäßigkeiten der Schwankungserscheinungen gelangen kann, zu deren Ableitung man sonst andere Prinzipien heranzuziehen pflegt.

In einem System, welches sich im thermodynamischen Gleichgewicht befindet, sind die Parameter, wie etwa der Energieinhalt eines Teiles, zeitlichen Änderungen — Schwankungen — unterworfen. Wir möchten nun hier zeigen, daß der zweite Hauptsatz nicht nur über die mittleren Werte dieser schwankenden Parameter etwas aussagt, sondern auch noch über die Gesetzmäßigkeiten der Abweichungen von diesen mittleren Werten Auskunft gibt. Bei flüchtiger Betrachtung könnte es so erscheinen, als ob dies unmöglich wäre, weil der zweite Hauptsatz gerade dort zu versagen scheint, wo die Schwankungen anfangen, sich bemerkbar zu machen. Dieses Bedenken trifft aber mehr die Form als den Grundgedanken der Thermodynamik, und man kann in der Tat durch rein phänomenologisch thermodynamische Betrachtungen zur Beherrschung jener Gesetzmäßigkeiten gelangen, zu deren Ableitung man sonst das Boltzmannsche Prinzip heranzuziehen pflegt.

Wir müssen zunächst eine durchaus naheliegende Präzisierung des zweiten Hauptsatzes mit Rücksicht auf die Schwankungen vornehmen und knüpfen zu diesem Zwecke an eine bekannte Formulierung des klassischen Postulats der Thermodynamik an, indem wir ein abgeschlossenes System betrachten, welches besteht:

a) aus einem System verschiedener Körper, welches man Kreisprozesse durchlaufen läßt;

b) aus einem Gewicht, das man heben oder senken kann, und das zur Aufspeicherung von Arbeit dient;

c) aus einer Anzahl von unendlich großen Wärmereservoiren, den Hilfsreservoiren, welche zur Aufnahme der bei den Kreisprozessen frei werdenden Wärmemengen dienen.

Das klassische Postulat besagt dann, daß bei einem Kreisprozeß, bei welchem der Wärmeinhalt der Hilfsreservoire mit Ausnahme eines Reservoirs unverändert bleibt, der Wärmeinhalt dieses hervorgehobenen Reservoirs keinesfalls abnehmen kann. Bezeichnen wir die Wärmemenge, welche das i-te Reservoir anläßlich des Kreisprozesses aufnimmt, mit Q_i, so muß also für dieses hervorgehobene Reservoir

$$Q \gtreqless 0$$

gelten, wenn dabei für die anderen Reservoire

$$Q_i = 0$$

gilt.

Abstrahieren wir nun von den Schwankungserscheinungen nicht, so müssen wir folgendes sagen: Durchläuft das betrachtete Körpersystem wiederholt einen ganz bestimmten Kreisprozeß und richten wir unsere Aufmerksamkeit auf die Wärmemenge, die eines der Reservoire, etwa das i-te Reservoir, dabei aufnimmt, dann finden wir infolge der Schwankungen jedesmal einen anderen Wert für Q_i. So gelangen wir, indem wir denselben Prozeß oft wiederholen, zu einer ganzen Statistik für den Wert Q_i. Wie groß im Einzelfall dieser Wert ist, hängt vom Zufall ab, und die Wahrscheinlichkeit dafür, daß bei einmaliger Durchführung unseres Kreisprozesses Q_i bei Q im Intervall dQ liegt, wird durch irgend ein Wahrscheinlichkeitsgesetz

$$W_i(Q)\,dQ$$

gegeben *).

Wir postulieren nun: es muß bei allen Kreisprozessen für die durch ein Reservoir im Mittel aufgenommene Wärmemenge

$$\overline{Q} \gtreqless \int\limits_{-\infty}^{\infty} Q\,W(Q)\,dQ \gtreqless 0$$

gelten, wenn dabei die anderen Reservoire im Mittel keine Wärme aufnehmen, so daß für sie

$$\overline{Q_i} = \int\limits_{-\infty}^{\infty} Q\,W_i(Q)\,dQ = 0$$

gilt.

Wäre dieses Postulat bei irgend einem Kreisprozeß nicht erfüllt, so brauchte man denselben nur hinreichend oft zu wiederholen, um beliebig große Wärmeabgaben des hervorgehobenen Reservoirs fast sicher erwarten zu dürfen. Das Postulat ist also nur ein Ausdruck für die Überzeugung, daß man ein Perpetuum mobile zweiter Art auch bei Heranziehung der Schwankungserscheinungen nicht konstruieren kann.

---

*) Die Behauptung, daß irgend ein Wahrscheinlichkeitsgesetz maßgebend sei, ist eine Annahme, die in der vorliegenden Arbeit zugrunde gelegt wird.

Über die Ausdehnung der phänomenologischen Thermodynamik usw.   755

Exakte Kreisprozesse kann man zwar infolge der Schwankungen im allgemeinen gar nicht durchführen, doch brauchen die Kreisprozesse, die wir hier ins Auge fassen, durchaus nicht exakt im Sinne der klassischen Theorie zu sein.   Man erkennt dies am besten an einem Beispiel:

Wenn ein Körper mit dem unendlichen Wärmereservoir im thermischen Gleichgewicht steht, so schwankt sein Energieinhalt noch um irgend einen mittleren Wert.   Trennt man ihn nun vom Reservoir in irgend einem Augenblick, so wird dabei eine gewisse Energie im Körper abgefangen, deren Betrag vom Zufall abhängt.   Obwohl der Energieinhalt dieses abgetrennten Körpers also gar nicht exakt bestimmt ist, wollen wir seinen Zustand doch als Anfangszustand eines Kreisprozesses ansehen.   Nun verändern wir einen Parameter des Körpers, z. B. sein Volumen, lassen ihn in bezug auf diesen Parameter etwa einen exakten Kreisprozeß durchlaufen und bringen ihn nochmals mit dem zuerst gebrauchten Wärmereservoir in thermische Verbindung, von welchem wir ihn erst nach hinreichend langer Berührung wieder trennen. Nun ist der Endzustand des Prozesses erreicht, und wir sagen, daß der Körper einen Kreisprozeß durchlaufen hat, obwohl der Energieinhalt nach dem Prozeß im allgemeinen ein anderer sein wird als im Anfangszustand, da doch beide Werte vom Zufall abhängig sind.   Für die Anwendbarkeit des zweiten Hauptsatzes genügt es eben, daß der Prozeß unendlich oft immer wieder durchgeführt werden kann, wenn nur dabei für die aufgenommene Wärmemenge jedesmal dieselben Wahrscheinlichkeitsgesetze gelten.

Noch besser erkennt man, worauf es ankommt, wenn man, anstatt an einem Körper wiederholt den Kreisprozeß durchzuführen, ihn an einer Reihe von unendlich vielen Exemplaren desselben Körpers simultan durchgeführt denkt.   Wir reden wieder vom „Kreisprozeß", auch wenn ein Exemplar für sich nicht in den Anfangszustand nach dem Prozeß zurückkehrt, falls nur die Statistik der Reihe vor und nach dem Prozeß dieselbe ist. Für die bei dem Kreisprozeß durch die Reservoire aufgenommenen Wärmemengen, gemittelt über die unendlich vielen Exemplare der Reihe, gilt das Postulat der Thermodynamik ebenso wie für ein Exemplar, gemittelt über viele Versuche.   Da sich die Ausdrucksweise einfacher gestaltet, wenn wir den zweiten Hauptsatz so für viele Exemplare desselben Körpers formulieren, werden wir im folgenden diese Formulierung bevorzugen.

52*

Bei den endlichen Körpern, welche die Kreisprozesse durchlaufen, wollen wir von einer Temperatur hier nicht reden; wir ordnen aber jedem unendlichen Wärmereservoir i eine bestimmte Temperatur t_i zu. Diese Zuordnung ist noch fast ganz willkürlich, es wird nur darauf geachtet, daß das wärmere Reservoir i das größere t_i erhält*).

Nun betrachte man jene reversiblen Kreisprozesse, welche insofern speziell sind, als bei ihnen nur zwei Reservoire, etwa die Reservoire t_1 und t_2 Wärme aufnehmen oder abgeben. Aus dem klassischen Postulat folgert man dann für die Wärmemengen, die bei einem solchen Prozeß von den beiden Reservoiren aufgenommen werden,

$$\frac{Q_1}{Q_2} = -\frac{f(t_1)}{f(t_2)};$$

dabei ist $f(t)$ unabhängig davon, welches der vielen möglichen reversiblen Prozesse man sich bedient, und ist somit eine universelle Funktion der willkürlich definierten Temperatur t.

Ganz analog folgt aus unserem Postulat für solche Prozesse

$$\frac{\overline{Q_1}}{\overline{Q_2}} = -\frac{f(t_1)}{f(t_2)}.$$

Aus dem klassischen Postulat folgt dann für die Wärmemengen, die anläßlich eines nunmehr beliebigen Kreisprozesses durch die Reservoire aufgenommen werden

$$\sum \frac{Q_i}{T_i} \gneqq 0,$$

dabei ist T_i die thermodynamische Temperatur des Reservoirs, welche mit Hilfe der universellen Funktion durch die Festsetzung

$$T_i = f(t_i)$$

definiert wird.

In ganz ähnlicher Weise folgt aus unserem Postulat

$$\sum \frac{\overline{Q_i}}{T_i} \gneqq 0. \tag{1}$$

Es genügt auch, den Begriff der Entropieänderung zunächst nur für die unendlichen Reservoire festzulegen. Nimmt ein unendliches Reservoir

---

*) Werden zwei unendliche Reservoire miteinander wärmeleitend verbunden, und fließt dann im Mittel Energie aus dem einen in den anderen hinüber, so ist der erstere der wärmere.

Über die Ausdehnung der phänomenologischen Thermodynamik usw.   757

von der Temperatur T in einem Einzelfall die Wärmemenge Q auf, so sei der Quotient

$$S = \frac{Q}{T}$$

als seine Entropieänderung bezeichnet.   Bei mehreren Reservoiren sei entsprechend

$$S = \sum \frac{Q_i}{T_i}$$

festgesetzt.

Während nun das klassische Postulat für alle Kreisprozesse

$$S \geqq 0$$

forderte, müssen wir folgendes sagen: Bei wiederholter Durchführung ein und desselben Kreisprozesses werden wir verschiedene Wertsysteme (Q_1, Q_2, ...) und daher auch verschiedene Werte S erhalten. Bei einem bestimmten Kreisprozeß gibt aber eine bestimmte Wahrscheinlichkeitsfunktion

$$W(S) \, dS$$

die Wahrscheinlichkeit dafür an, daß wir bei einmaliger Durchführung des Kreisprozesses eine Entropiezunahme der Reservoire zwischen S und $S + dS$ erhalten.   Es berechnet sich $W(S)$ in einfacher Weise aus den Funktionen $W_i(Q)$, welche für die aufgenommenen Wärmemengen bei diesem Kreisprozeß gelten.   Für den Mittelwert von S

$$\overline{S} = \int_0^\infty S \, W(S) \, dS$$

findet man

$$\overline{S} = \sum \frac{\overline{Q_i}}{T_i}.$$

Unser Postulat fordert also für alle Kreisprozesse

$$\overline{S} \geqq 0.$$

Diese Ungleichung läßt sich auch folgendermaßen interpretieren: Spielt jemand mit Hilfe von Kreisprozessen ein thermodynamisches Glücksspiel mit der Absicht, die Entropie der Wärmereservoire zu verkleinern, so verhält sich die Natur ihm gegenüber wie eine wohleingerichtete Spielbank, bei welcher man zwar ab und zu gewinnen kann, bei der es aber keinen Spielplan gibt, welcher bei Dauerspiel zur Bereicherung führt*).

---

*) Es ist naheliegend, gegen die vorliegende Fassung des zweiten Hauptsatzes, welche auch angesichts der Schwankungserscheinungen strenge Gültigkeit beansprucht, ein Bedenken geltend zu machen, welches schon auf Maxwell zurückgeht:

Wenn ein Dämon zu unseren Diensten stände, der imstande ist, den jeweiligen Wert der schwankenden Parameter zu erraten und dann jeweils passend

758   Leo Szilard,

Recht ähnlich hat auch S m o l u c h o w s k i die Überzeugung ausgedrückt, daß man ein Perpetuum mobile zweiter Art auch bei Heranziehung der Schwankungserscheinungen nicht konstruieren kann.

Es soll nun an dem Beispiel der Energieschwankung gezeigt werden, wie man aus dem Postulat der Thermodynamik, in welches explizite nur gewisse mittlere Werte $\overline{Q}$ eingehen, nicht nur über die mittleren Werte der schwankenden Parameter, sondern darüber hinaus über die Gesetzmäßigkeiten der Abweichungen von diesen mittleren Werten Aufschluß bekommt. Es ergibt sich so die Möglichkeit einer einheitlich thermodynamischen Behandlung der Schwankungserscheinungen.

Die Energieschwankungen.

Wenn ein Körper mit einem unendlichen Wärmereservoir in thermischer Verbindung steht und wir ihn sodann in irgend einem Augenblick von dem Reservoir trennen, so wird er einen bestimmten Energieinhalt zurückbehalten, deren Betrag durch ein Wahrscheinlichkeitsgesetz geregelt wird, welches noch von der Temperatur des Reservoirs abhängt. Die Wahrscheinlichkeit dafür, daß dieser Energieinhalt zwischen u und $u + du$ liegt, sei durch

$$W^*(u;\, T)\, du$$

angegeben. Dieselbe Verteilungsfunktion $W^*(u; T)$ beschreibt auch die Energiestatistik einer Reihe, die aus unendlich vielen Exemplaren desselben Körpers besteht, wenn jedes Exemplar zuvor hinreichend lange mit einem Reservoir von derselben Temperatur T in thermischer Berührung gestanden hat.

Wir wollen diese Energiestatistik als d i e normale Verteilung der Reihe bei der Temperatur T bezeichnen, weil es sich zeigt, daß sie für den betreffenden Körper charakteristisch ist und von Nebenumständen nicht mitbestimmt wird. So ist es z. B. gleichgültig, ob die Berührung durch Wärmestrahlung oder durch Wärmeleitung vermittelt wird. Dies

---

einzugreifen, so könnte man, indem man sich seiner bedient, sicherlich ein Perpetuum mobile zweiter Art konstruieren. Wir Menschen können zwar den Wert der Parameter nicht erraten, aber wir können ihn messen und könnten so, je nach dem Ergebnis dieser Messung, unseren Eingriff passend einrichten. Es erhebt sich so die Frage, ob man nicht auf diese Weise doch zu einem Widerspruch zur dogmatisch strengen Auffassung des zweiten Hauptsatzes kommt.

Wir hoffen, diese Frage in einer bald erscheinenden Arbeit befriedigend beantworten zu können und umgehen in der vorliegenden Arbeit die Schwierigkeit, indem wir zunächst bei den Gedankenexperimenten die Eingriffe nicht mit den Schwankungen koppeln, sondern uns auf Eingriffe beschränken, die ebensogut durch periodisch funktionierende Maschinen getätigt werden könnten.

Über die Ausdehnung der phänomenologischen Thermodynamik usw. 759

hängt mit der Stabilität der normalen Verteilung zusammen, einer Eigenschaft, welche sie vor anderen Verteilungen auszeichnet. Liegt eine Reihe mit irgend einer anderen Energiestatistik $W(u)$ vor, so kann man, indem man jedes ihrer Exemplare einen Prozeß durchlaufen läßt, ihre Verteilung abändern, ohne daß dabei in anderen Körpern eine Veränderung hinterbleiben müßte. Anders ist es, wenn die Reihe mit ihrer normalen Verteilung vorliegt; wie es sich zeigen wird, ist es dann ohne Kompensation nicht möglich, ihre Energiestatistik bei festgehaltenem Energiemittelwert abzuändern. Diese Stabilität hängt damit zusammen, daß eine Reihe mit nicht normaler Verteilung unter Erzwingung einer ganz bestimmten Entropieverminderung der Hilfsreservoire bei festgehaltenem Energiemittelwert in die normale übergeführt werden kann.

Die Bestimmung der Gleichgewichtsstatistik.

Wir wollen weiter unten erörtern, inwiefern man aus dem zweiten Hauptsatz darauf schließen kann, daß bei jedem Energiemittelwert eine Verteilung für die Reihe existiert, welche ohne Kompensation nicht abgeändert werden kann und die deshalb als stabile Verteilung bezeichnet werden soll. Im vorliegenden Abschnitt wollen wir jedoch die Existenz voraussetzen und die Gesetzmäßigkeit ermitteln, welcher die stabilen Verteilungen notwendig gehorchen müssen.

Es ist leicht einzusehen, daß die Verteilung, welche sich einstellt, wenn man eine Reihe gliedweise mit einem unendlichen Wärmereservoir ins thermische Gleichgewicht setzt und die wir als normale Verteilung bezeichnet haben, eine stabile Verteilung sein muß. Um dies zu erkennen, braucht man bloß zu bedenken, daß jene stabile Verteilung, deren Energiemittelwert der Temperatur des Reservoirs entspricht, durch Berührung mit dem Reservoir in die normale Verteilung übergeführt wird, ohne daß dabei das Reservoir im Mittel Wärme aufnimmt. Es würde also eine Abänderung der stabilen Verteilung ohne Kompensation vorliegen, falls die stabile Verteilung mit der normalen Verteilung nicht identisch wäre. Um über die stabilen Verteilungen Näheres zu erfahren, stellen wir nun die folgenden Betrachtungen an.

Es soll eine Reihe des Körpers 1) mit einer beliebigen Energiestatistik $W_1(u)\,du$ und eine Reihe des Körpers 2) mit der Verteilung $W_2(u)\,du$ vorliegen. Wir ordnen nun jedem Exemplar der Reihe 1) ein Exemplar der Reihe 2) zu und bringen sie in thermische Berührung miteinander. Die Wahrscheinlichkeit dafür, daß in einem solchen

760   Leo Szilard,

zusammengesetzten Exemplar dann eine Energie zwischen u und $u + d\,u$ vorhanden ist, wird durch

$$d\,u\,W(u) = d\,u \int\limits_0^u W_1(\xi)\,W_2(u-\xi)\,d\xi \qquad (2)$$

angegeben*). Wenn wir nun nach hinreichend langer Berührung die beiden einander zugeordneten Exemplare wieder voneinander trennen, so ist im allgemeinen die Statistik der beiden Reihen 1) und 2) eine andere geworden. Liegen jedoch die beiden Reihen zu Anfang mit ihrer normalen Statistik vor (beide von derselben Temperatur) und führen wir eine Berührung in der geschilderten Weise durch, so fordert die Stabilität, daß nach der Berührung wieder die ursprünglichen Verteilungen vorhanden sind. Man könnte an diese Forderung allein anknüpfen, um über die Gleichgewichtsverteilung Näheres zu erfahren. Wir wollen aber noch ein Zweites fordern:

Wenn wir jedem Exemplar der Reihe 1) ein Exemplar der Reihe 2) zuordnen, so kann man aus der Kenntnis des Energieinhaltes u_1 eines Körpers 1) selbstverständlich in keiner Weise Schlüsse auf den Energieinhalt des zugehörigen Körpers 2) ziehen, falls sich die Gleichgewichtsverteilung in den beiden Reihen voneinander unabhängig auf irgend eine Weise eingestellt hat; die beiden Werte u_1 und u_2 sind voneinander statistisch unabhängig. Wir fordern nun, daß diese statistische Unabhängigkeit auch nach vorgenommener Berührung erhalten bleibt.

Bringen wir je einen Körper der Reihe 1) mit je einem Körper der Reihe 2) in Berührung, so schwankt die Energie jeweils zwischen den beiden Körpern hin und her. Trennt man ein solches Körperpaar nach hinreichend langer Berührung, in irgend einem Augenblick, so wird die Wahrscheinlichkeit dafür, daß in dem Körper 1) eine Energie bei u_1 im Intervall $d\,u_1$ abgefangen wird, durch irgend ein Wahrscheinlichkeitsgesetz

$$V_{12}(u_1;\,u)\,d\,u_1$$

angegeben. Nach hinreichend langer Berührung kann dieses Gesetz nur noch von der Energie u abhängen, welche in den beiden Körpern insgesamt vorhanden ist, nicht aber davon, in welcher Weise sich die Gesamtenergie ursprünglich aus den Beiträgen der beiden Körper zusammengesetzt hat. Liegt also eine ganze Reihe von Körperpaaren in

———————————

*) Vgl. Anmerkung 1. Die Anmerkungen sind am Schluß der Arbeit zusammengestellt.

Über die Ausdehnung der phänomenologischen Thermodynamik usw. 761

Berührung vor und ist der Energieinhalt bei u, so ist nach dem Spalten die relative Anzahl jener Paare, bei welchen die Energie des Körpers 1) bei u_1 im Intervall du_1 liegt, durch $V_{12}(u_1; u) du_1$ gegeben*).

Wir wollen nun zusehen, wie sich die temperaturunabhängige Funktion V_{12} aus den temperaturabhängigen Funktionen W_1^* und W_2^* berechnen läßt, welche die normalen Verteilungen für die beiden Reihen angeben. Zu diesem Zwecke betrachten wir zwei Reihen, welche mit der Gleichgewichtsverteilung vorliegen. Nehmen wir dann eine Berührung vor, so erhalten wir nach der Berührung wieder die ursprünglichen Verteilungen W_1^* und W_2^* und es ist die Wahrscheinlichkeit dafür, daß nach der Berührung die Energie des Körpers 1) bei u_1 im Intervall du_1 und zugleich die Energie des zugehörigen Körpers 2) bei u_2 im Intervall du_2 liegt, entsprechend der statistischen Unabhängigkeit durch

$$W_1^*(u_1)\ W_2^*(u_2)\, du_1\, du_2$$

gegeben. Wir können den Zustand des Paares ebensogut durch u_1 und die Gesamtenergie $u = u_1 + u_2$ charakterisieren und erhalten dann, da die in Frage kommende Funktionaldeterminante identisch Eins ist, für die Wahrscheinlichkeit q, daß die Energie des Körpers 1) bei u_1 im Intervall du_1 und zugleich die Gesamtenergie bei u im Intervall du liegt:

$$q = W_1^*(u_1)\ W_2^*(u - u_1)\, du_1\, du.$$

Dieselbe Wahrscheinlichkeit q können wir aber auch so berechnen, daß wir zuerst alle Paare auswählen, deren Gesamtenergie bei u im Intervall du liegt. Durch diese Auswahl erhalten wir eine Teilreihe; die Wahrscheinlichkeit dafür, daß bei einem aus dieser Teilreihe herausgegriffenen Paar die Energie des Körpers 1) bei u_1 im Intervall du_1 liegt, ist definitionsgemäß durch $V_{12}(u_1, u) du_1$ gegeben. Andererseits ist die Wahrscheinlichkeit dafür, daß ein aus der ursprünglichen, vollständigen Reihe herausgegriffenes Paar der Teilreihe angehört, durch $W^*(u) du$ gegeben (vgl. Nr. 2). Es ist also die Wahrscheinlichkeit dafür, daß bei einem aus der vollständigen Reihe herausgegriffenen Paar beide Bedingungen erfüllt sind, durch

$$q = V_{12}(u_1, u)\ W^*(u)\, du_1\, du$$

gegeben.

---

*) Für den Körper 2) gilt analog
$$V_{21}(u_2; u)\, du_2,$$
und da $u_1 + u_2 = u$ gilt, so ist
$$V_{21}(u - u_1; u) = V_{12}(u_1; u).$$

Der Vergleich mit dem oben gefundenen Werte liefert

$$V_{12}(u_1,\, u) = \frac{W_1^*(u_1)\, W_2^*(u - u_1)}{W^*(u)};\qquad(4)$$

dabei ist

$$W^*(u) = \int_0^u W_1^*(\xi)\, W_2^*(u - \xi)\, d\xi.$$

Die Funktionen W_1^*, W_2^* und W^* hängen auch noch von T ab, dagegen darf selbstverständlich der Quotient $V_{12}(u_1,\, u)$ von T nicht abhängen. Hieraus folgt nun bereits, daß die Verteilung W_2^* von der Form ist

$$W_2^*(u,\, T) = C(T)\, g(u)\, e^{\varphi(T)\, u}.$$

Um dies einzusehen, geht man am besten zu den Logarithmen über. Bezeichnet man die Logarithmen der in Nr. (4) vorkommenden Größen mit den entsprechenden kleinen Buchstaben, so erhält man

$$v_{12}(u_1,\, u) + w^*(u,\, T) = w_1^*(u_1,\, T) + w_2^*(u - u_1,\, T).$$

Differenziert man nun partiell bei festem T nach u_1 und setzt dann etwa $u_1 = 0$, so sieht man, daß $\dfrac{\partial w_2^*}{\partial u}$ von der Form

$$\frac{\partial}{\partial u}\, w_2^*(u,\, T) = f_1(u) + f_2(\tau).$$

Durch Integration nach u findet man, daß w_2^* von der Form ist

$$w_2^*(u,\, T) = f_3(u) + f_2(T)\, u + f_4(T)$$

und entsprechend muß also die normale Verteilung W_2^* von der Form

$$W_2^*(u,\, T) = C(T)\, g(u)\, e^{\varphi(T)\, u}.\qquad(5)$$

sein. Dasselbe findet man natürlich auch für die normale Verteilung W_1^*, indem man den Quotienten V_{21} betrachtet.

Haben nun umgekehrt W_1^* und W_2^* die angegebene Form und führt man sie in den Quotienten V_{12} ein (vgl. Nr. 4), so sieht man, daß die Unabhängigkeit von T dann und nur dann gewahrt ist, wenn φ für beide Körper dieselbe Funktion von T darstellt (bis auf eine von T unabhängige Konstante, welche man ja immer in die Funktion g mit einbeziehen kann). Da dies für zwei beliebige Körper gilt, so ist φ eine von der besonderen Wahl des Körpers unabhängige, d. h. universelle Funktion der Temperatur. Die Funktion g bleibt unbestimmt, $C(T)$ wird aus g durch die Bedingung

$$\int_0^\infty W^*(u)\, du = 1$$

bestimmt. Es ist

$$C(T) = \frac{1}{\displaystyle\int_0^\infty g(u)\, e^{\varphi(T)\, u}\, du}.$$

Über die Ausdehnung der phänomenologischen Thermodynamik usw.   763

Die normale Verteilung für einen Körper wird somit durch die Angabe der „Gewichtsfunktion" g festgelegt*).

Wir müßten hier auch die Möglichkeit in Betracht ziehen, daß nicht nur ein Energieintervall, sondern auch gewisse Energiewerte, im besonderen die Energie Null, eine endliche Wahrscheinlichkeit haben könnten (quantenhafte Energieverteilung).   Dies müßte um so mehr geschehen, als man sonst bei keiner Wahl von $g(u)$ erreichen kann, daß die Wärmekapazität etwa mit T^3 abfällt, wie es z. B. bei einem Hohlraum der Fall sein müßte, in welchem das Stefan-Boltzmannsche Gesetz herrscht**). Dennoch wollen wir uns auf stetige Verteilungen beschränken, um die Darstellung nicht zu komplizieren.

Näherungsrechnung.   Die Streuung der normalen Verteilung.

Indem man eine Näherungsrechnung zugrunde legt, lassen sich die Betrachtungen leicht auf Grund des zweiten Hauptsatzes allein durchführen, ohne daß man die Existenz einer stabilen Verteilung und dergleichen mehr vorauszusetzen brauchte, was wir in dem vorigen Abschnitt tun mußten.   Im Interesse der Reinheit der Methode könnte man daher im folgenden durchweg von dieser Näherung Gebrauch machen. Wir begnügen uns aber damit, in dem vorliegenden Abschnitt zu zeigen, daß man bei Verwendung der Näherungsrechnung auf Grund des zweiten Hauptsatzes allein die statistische Thermodynamik aufbauen könnte und werden dann in dem restlichen Teile der Arbeit wieder mathematisch streng rechnen.

Die erwähnte Näherung beruht darauf, daß wir die Schwankungen als sehr klein ansehen und dementsprechend annehmen, daß die Verteilungsfunktionen $W(u)$ rechts und links vom Höchstwert sehr rasch abfallen.   Man kann sie dann nämlich durch ihren Mittelwert

$$\bar{u} = \int_0^\infty u\, W(u)\, d\,u$$

und ihre Streuung

$$\eta = \int_0^\infty (u - \bar{u})^2\, W(u)\, d\,u$$

---

*) Ist umgekehrt die normale Verteilung gegeben, so ist dadurch g nicht eindeutig definiert, sondern nur bis auf einen Faktor, der von u nicht abhängt.
**) Vgl. Anm. 2.

764   Leo Szilard,

charakterisieren. Im besonderen gilt, wenn $f(u)$ eine sanft verlaufende Funktion ist:

$$\int_0^\infty f(u)\,W(u)\,du \cong f(\bar{u}) + \left\{\frac{d^2 f}{d\,u^2}\right\}_{u\,=\,\bar{u}} \frac{\eta}{2}; \qquad (6)$$

letzteres erhält man, indem man $f(u)$ an der Stelle $u = \bar{u}$ nach der Taylorschen Formel entwickelt und das Integral des Restgliedes

$$J = \int_0^\infty \frac{d^3 f}{d\,u^3}\,\frac{(u-\bar{u})^3}{3\,!}\,W(u)\,du,$$

wobei $\dfrac{d^3 f}{d\,u^3}$ an Stellen zwischen u und $\bar{u}$ zu nehmen ist, vernachlässigt.

Dieser Näherung entsprechend wollen wir nun die Stabilität der normalen Statistik aus dem zweiten Hauptsatz ableiten, indem wir zeigen, daß es ohne Kompensation nicht möglich ist, die Streuung einer Reihe mit normaler Statistik abzuändern. Aus der Stabilität der normalen Streuung werden wir dann folgern können, daß ihr Wert durch

$$\eta^* = \psi(T)\frac{d\,\bar{u}^*}{d\,T} \qquad (7)$$

angegeben ist. Dabei bedeutet ψ für alle Körper dieselbe Zahl, d. h. eine universelle Funktion der Temperatur. Wir überlegen wie folgt:

Wenn zwei Reihen vorliegen, welche zuvor hinreichend lange mit einem Wärmereservoir in Berührung gestanden haben, so fordert der zweite Hauptsatz unmittelbar, daß man ohne Kompensation den Energiemittelwert der einen Reihe nicht auf Kosten des Energiemittelwertes der anderen Reihe erhöhen kann. Wäre dies nämlich möglich, so brauchte man bloß diese Reihe in reversibler Weise abzukühlen und die andere Reihe entsprechend zu erwärmen, bis die ursprünglichen Energiemittelwerte wieder erreicht sind, um eine Entropieabnahme der dabei benutzten Reservoire zu erzwingen. (Es wird nämlich beim Erwärmen der einen Reihe dieselbe Wärmemenge von den Reservoiren abgegeben, welche bei der Abkühlung der anderen Reihe von ihnen bei höherer Temperatur aufgenommen wird.)

Bringen wir zwei Reihen, deren Energiestatistik durch W_1 bzw. W_2 angegeben ist, miteinander gliedweise in thermische Berührung und trennen wir sie dann nach hinreichend langer Berührung wieder voneinander, so werden nach dem Trennen die beiden Reihen mit neuen Energieverteilungen vorliegen, welche mit W_1' bzw. W_2' bezeichnet seien. Für den mittleren Energieinhalt der Reihe 1) nach dem Trennen

$$\bar{u}_1' = \int_0^\infty u_1\,W_1'(u_1)\,du_1$$

gilt in unserer Näherung*)

$$\bar{u}_1' \cong \bar{u}_{12}(\bar{u}) + \left\{\frac{d^2\,\bar{u}_{12}}{d\,u^2}\right\}_{u\,=\,\bar{u}}\frac{\eta_1+\eta_2}{2}. \qquad (8)$$

---

1) Vgl. Anm. 3.

Über die Ausdehnung der phänomenologischen Thermodynamik usw. 765

Dabei bedeutet

$$\bar{u}_{12}(u) = \int\limits_0^\infty u_1 \, V_{12}(u_1; u) \, du_1.$$

Bringen wir nun zwei normale Reihen in der geschilderten Weise zur Berührung, so muß der Energiemittelwert unverändert bleiben, und es gilt also

$$\bar{u}_1^* = \bar{u}_{12}(\bar{u}^*) + \left\{\frac{d^2\bar{u}_{12}}{du^2}\right\}_{u=\bar{u}^*} \frac{\eta_1^* + \eta_2^*}{2}. \tag{9}$$

Wie man in Nr. (8) sieht, wird der Energiemittelwert nach der Berührung $\bar{u}_1'$ durch die Energiemittelwerte vor der Berührung allein nicht bestimmt, sondern wird in unserer Näherung durch die Streuungen mitbestimmt. Könnte man daher, von zwei normalen Reihen ausgehend, die Streuung der einen Reihe, etwa der Reihe 1), ohne Kompensation abändern, so wäre der zweite Hauptsatz sicherlich verletzt, denn bei der Berührung der beiden Reihen würden die Energiemittelwerte abgeändert werden. Man erhält nämlich für den Wert dieser Abänderung des Energiemittelwertes

$$\bar{u}_1' - \bar{u}_1^* = \left\{\frac{d^2\bar{u}_{12}}{du^2}\right\}_{u=\bar{u}} \left(\frac{\eta_2 - \eta_2^*}{2} + \frac{\eta_1 - \eta_1^*}{2}\right),$$

indem man Gleichung (9) aus der Gleichung (8) subtrahiert. So folgt, wenn etwa $\eta_1 = \eta_1^*$ und $\eta \neq \eta_2^*$ ist, tatsächlich eine Abänderung des Energiemittelwertes. Damit ist die Stabilität der normalen Streuung in unserer Näherung thermodynamisch erwiesen.

Aus der Stabilität folgt nun, daß, wenn man von zwei normalen Reihen ausgeht und eine Berührung in der geschilderten Weise vornimmt, nach der Berührung wieder die ursprünglichen normalen Streuungen vorliegen müssen. Für die Streuungen nach der Berührung

$$\eta_1' = \int\limits_0^\infty (u_1 - \bar{u}_1')^2 \, W_1'(u_1) \, du_1,$$

$$\eta_2' = \int\limits_0^\infty (u_2 - \bar{u}_2')^2 \, W_2'(u_2) \, du_2$$

gilt in unserer Näherung*)

$$\eta_1' = \eta_{12}(\bar{u}) + \left\{\frac{d\bar{u}_{12}}{du}\right\}_{u=\bar{u}}^2 (\eta_1 + \eta_2),$$

$$\eta_2' = \eta_{21}(\bar{u}) + \left\{\frac{d\bar{u}_{21}}{du}\right\}_{u=\bar{u}}^2 (\eta_1 + \eta_2),$$

dabei bedeutet

$$\eta_{12}(u) = \int\limits_0^\infty [u_1 - \bar{u}_{12}(u)]^2 \, V_{12}(u_1, u) \, du_1,$$

und wenn man von zwei normalen Reihen ausgeht, so gilt also

$$\eta_1^* = \eta_{12}(\bar{u}) + \left\{\frac{d\bar{u}_{12}}{du}\right\}_{u=\bar{u}}^2 (\eta_1^* + \eta_2^*),$$

$$\eta_2^* = \eta_{21}(\bar{u}) + \left\{\frac{d\bar{u}_{21}}{du}\right\}_{u=\bar{u}}^2 (\eta_1^* + \eta_2^*). \tag{10}$$

---

*) Vgl. Anm. 4.

766   Leo Szilard,

Berücksichtigt man, daß entsprechend Nr. (3):

$$\eta_{12} = \eta_{21} \tag{11}$$

gilt, so erhält man*) aus Nr. (10):

$$\frac{\eta_1^*}{\dfrac{d\bar{u}_1^*}{dT}} = \frac{\eta_2^*}{\dfrac{d\bar{u}_2^*}{dT}};$$

da dies aber für zwei beliebige Körper gilt, so muß der Quotient

$$\frac{\eta^*}{\dfrac{d\bar{u}^*}{dT}} = \psi(T) \tag{12}$$

für alle Körper dieselbe, d. h. eine **universelle Funktion** der Temperatur sein.

Es ist bemerkenswert, daß schon aus dem phänomenologischen zweiten Hauptsatz die Existenz einer solchen **universellen Funktion** folgt.

Vergleichen wir nun das Resultat der Näherungsrechnung mit dem Ergebnis der strengen Rechnung. Letztere ergab für die normale Verteilung W^*:

$$W^* = C(T)\,g(u)\,e^{\varphi(T)u};$$

bilden wir die Streuung

$$\eta^* = \int_0^\infty (u - \bar{u})^2\, W^*\,du,$$

so finden wir die Identität

$$\frac{d\bar{u}^*}{dT} = \frac{d\varphi}{dT}\eta^*,$$

setzen wir

$$\frac{d\varphi}{dT} = \frac{1}{\psi(T)}, \tag{13}$$

so erhalten wir in Übereinstimmung mit Nr. (7):

$$\eta^* = \psi(T)\frac{d\bar{u}^*}{dT}.$$

„Freie" und „normale" adiabatische Prozesse. „Normale" Kreisprozesse.

Wir betrachten nun einen homogenen Körper, dessen Zustand durch seinen Energieinhalt und sein Volumen bestimmt ist. Man denke etwa an ein Gas**) in einem zylindrischen Gefäß, das durch einen verschiebbaren Kolben abgesperrt ist. Indem man den Kolben hineindrückt oder herauszieht, wird dann das Gas komprimiert bzw. expandiert und es liegt so ein Körper vor, dessen Zustand man durch die Angabe des Energieinhaltes und des Volumens als gegeben ansehen kann.

---

*) Vgl. Anm. 5.
**) Man denke dabei nicht gerade an ein ideales Gas, sondern ganz allgemein an einen Körper, dessen Zustand durch Energie und Volumen bereits gegeben ist.

Über die Ausdehnung der phänomenologischen Thermodynamik usw.   767

Expandiert man nun einen solchen Körper vom exakten*) Volumen v_1 adiabatisch bis zu dem exakten Volumen v, so wird sein Energieinhalt im expandierten Zustande durch seinen Energieinhalt im Anfangszustande exakt bestimmt. Dabei muß man freilich die Expansion hinreichend langsam leiten, sonst macht sich der Einfluß der Druckschwankungen bemerkbar.

Wir betrachten nun eine Reihe, welche aus unendlich vielen Exemplaren eines solchen Körpers besteht. Ist das Wahrscheinlichkeitsgesetz gegeben, welches die Energiestatistik der Reihe im Anfangszustand $v = v_1$ beschreibt, dann kann man leicht berechnen, welches Wahrscheinlichkeitsgesetz nach der Expansion die Statistik angibt, wenn wir nur die Energie eines Exemplars entlang der adiabatischen Expansion als Funktion der Anfangsenergie u_1 und des Volumens v kennen. Bezeichnen wir das Wahrscheinlichkeitsgesetz, welches nach der Expansion bis zu dem Volumen v herrscht, mit $W(u, v)$, so gilt nämlich die lineare partielle Differentialgleichung**)

$$\frac{\partial}{\partial u}\left[p(u, v)\, W(u, v)\right] - \frac{\partial}{\partial v}\, W(u, v) = 0. \qquad (14)$$

Dabei ist $p(u, v)$ mit Hilfe der Adiabatenschar $u = u(u_1, v)$ so definiert, daß längs der Adiabate

$$du = -p\, dv$$

gilt. Es ist also

$$p(u, v) = -\frac{\partial u(u_1, v)}{\partial v}.$$

Es ist dann $p(u)$ nichts anderes als der zeitliche Mittelwert des Druckes, wenn der Energieinhalt des Körpers nicht schwankt, sondern den exakten Wert u dauernd beibehält.

Man findet***) aus Nr. (14)

$$\frac{d\bar{u}}{dv} = -\int_0^\infty p(u, v)\, W(u, v)\, du \qquad (15)$$

oder kürzer

$$\frac{d\bar{u}}{dv} = -\bar{p}.$$

Für die Streuung findet man

$$\frac{d\eta}{dv} = -2\int_0^\infty (u - \bar{u})\, p(u, v)\, W(u, v)\, du. \qquad (16)$$

_____

*) Vgl. Anm. 6.
**) Vgl. Anm. 7.
***) Vgl. Anm. 8.

Wie man aus Nr. (15) ersieht, ist der Abfall des Energiemittelwertes entlang der adiabatischen Expansion durch den jeweiligen Energiemittelwert $\bar{u}$ allein nicht bestimmt, sondern wird im allgemeinen durch die ganze Verteilung $W(u)$ mitbestimmt.

Geht man nun von einer normalen Reihe aus, so wird entlang der adiabatischen Expansion die Statistik keineswegs normal bleiben, sondern wird sich bei dieser „freien" adiabatischen Expansion mehr oder weniger von der normalen Statistik entfernen. Man kann jedoch auch so expandieren lassen, daß die Statistik dauernd normal bleibt und daß trotzdem ein Vorgang vorliegt, den man mit einem gewissen Recht auch als „adiabatisch" bezeichnen kann. Ein solcher „normaler" adiabatischer Prozeß kommt so zustande, daß man die Reihe, von einem normalen Zustand ausgehend, nur um unendlich wenig „frei adiabatisch" expandieren läßt, dann aber mit einem passend temperierten Reservoir in thermische Berührung bringt. Die Temperatur des Reservoirs wird so gewählt, daß es bei der Berührung im Mittel keine Wärme aufnimmt. Die Berührung bewirkt dann lediglich, daß in der Reihe sich wieder genau die normale Verteilung einstellt, von der die Verteilung nach der unendlich kleinen freien adiabatischen Expansion bereits um unendlich wenig abweicht. Nun wird wieder vom Reservoir getrennt und expandiert usw. Indem man so entlang der ganzen Expansion mit entsprechend temperierten Reservoiren Gleichgewicht aufrecht erhält, erzielt man eine Expansion, entlang welcher die Statistik normal bleibt. Wir wollen sie, weil die Reservoire dabei im Mittel keine Wärme aufnehmen, auch als adiabatische, und zwar als „normale" adiabatische Expansion bezeichnen zum Unterschied von der früher besprochenen „freien" adiabatischen Expansion. Geht man von einer Reihe mit normaler Verteilung aus, so läuft im allgemeinen der Energiemittelwert bei der Expansion entlang verschiedener Kurven im (u, v)-Diagramm je nachdem, ob man „frei" oder „normal" adiabatisch expandieren läßt.

Man kann nun eine Körperreihe in reversibler Weise auch erwärmen, indem man sie allmählich mit immer wärmeren Reservoiren in Berührung bringt. Der Energiemittelwert läuft dann im (u, v)-Diagramm entlang einer Ordinate. Indem wir Erwärmung bei konstantem Volumen mit „normaler" adiabatischer Expansion geeignet kombinieren, haben wir es also in der Hand, den Energiemittelwert in reversibler Weise entlang beliebiger Kurven zu führen und haben dabei stets normale Statistik. Bei diesen normalen Prozessen läßt sich die Schlußweise der

Über die Ausdehnung der phänomenologischen Thermodynamik usw.   769

klassischen Thermodynamik ohne weiteres in Anwendung
bringen. Durchläuft der Energiemittelwert in dieser Weise eine geschlossene Kurve, so muß für die dabei beteiligten Reservoire

$$\overline{S} = 0$$

gelten*). Es muß daher bei einem Körper, dessen Zustand durch Energieinhalt und Volumen bestimmt ist,

$$d\,\overline{S} = -\frac{\overline{p}^*\,dv + d\,\overline{u}^*}{T}$$

ein vollständiges Differential sein, was die Beziehung

$$\overline{p}^* + \frac{\partial\,\overline{u}^*}{\partial v} = T\frac{\partial\,\overline{p}^*}{\partial T} \qquad (17)$$

liefert. Dabei ist

$$\overline{p}^*(T) = \int_0^\infty p(u)\,W^*(u,\,T)\,d\,u,$$

so daß entlang der „normalen" adiabatischen Expansion gemäß Nr. (15)

$$d\,\bar{u} = -\,\overline{p}^*\,dv$$

gilt.

Temperaturabhängigkeit der normalen Verteilung. Bestimmung aus der Thermodynamik**)

Da der Quotient

$$\frac{\eta^*}{\dfrac{d\,\overline{u}^*}{d\,T}} = \psi(T)$$

für alle Körper dieselbe Zahl ist, so ist er bestimmt, sobald man für
einen einzigen Körper η^* als Funktion der Temperatur ermittelt hat.

---

*) Es sei an die analoge Betrachtung in der klassischen Thermodynamik er-
innert. Dort ist für alle reversiblen Kreisprozesse

$$S = 0.$$

Dies läuft darauf hinaus, daß

$$d\,S = -\frac{p\,dv + du}{T}$$

ein vollständiges Differential ist. Hieraus ergibt sich die Beziehung

$$p + \frac{\partial u}{\partial v} = T\frac{\partial p}{\partial T}.$$

Dabei ist p der Druck und längs der Adiabate gilt für die zugeführte Arbeit

$$du = -\,p\,dv.$$

**) Vgl. Anm. 12.

770   Leo Szilard,

Liegt nun eine Reihe mit normaler Energiestatistik vor, so weiß man zwar, wie sich die Statistik entlang der „freien" adiabatischen Expansion verändert (vgl. Nr. 14), doch ist damit nicht viel gewonnen, weil entlang dieser Expansion im allgemeinen keineswegs die zu dem jeweiligen Energiemittelwert $\bar{u}$ gehörende normale Statistik vorliegt. Man wird aber sehen, daß bei solchen Körpern, für welche p von der Form ist

$$p(u, v) = u f(v),$$

die Statistik entlang der „freien" adiabatischen Expansion, stets normal bleiben muß, wenn der zweite Hauptsatz zu Recht besteht*). Man sieht zunächst, daß bei diesen Körpern der Abfall des Energiemittelwertes (vgl. Nr. 15)

$$\frac{d\bar{u}}{dv} = f(v) \int_0^\infty u\, W(u)\, du = f(v) \cdot \bar{u}$$

nur von dem jeweiligen Energiemittelwert $\bar{u}$ allein abhängt und nicht durch die ganze Verteilung $W(u)$ mitbestimmt wird. Daraus folgt unmittelbar, daß wenn man von einer Reihe mit normaler Energieverteilung ausgeht, der Energiemittelwert bei der freien adiabatischen Expansion entlang derselben Kurve im (u, v)-Diagramm läuft wie bei der normalen Expansion. Dies sieht man, indem man sich vergegenwärtigt, daß die normal adiabatische Expansion aus unendlich kleinen freien Expansionen zusammengesetzt ist und die Berührung mit dem Reservoir, welche zur Normalhaltung der Verteilung dient, den Energiemittelwert u jeweils unverändert läßt. Im vorliegenden Falle wird dann aber $\dfrac{d\bar{u}}{dv}$ durch diese Berührung auch nicht beeinflußt, so daß wir dieselbe Funktion $\bar{u}(v)$ erhalten wie bei der freien Expansion.

Diese Eigenschaft der hervorgehobenen Körperklasse gestattet es, einen Widerspruch zu dem zweiten Hauptsatz zu konstruieren, wenn man bei den ihr angehörenden Körpern annimmt, daß die ursprünglich normale Verteilung bei der freien adiabatischen Expansion nicht normal bleibt. Um dies einzusehen, kann man von einer normalen Reihe ausgehen und diese bis zu einem bestimmten Volumen frei adiabatisch expandieren lassen. Man nehme nun versuchsweise an, die Verteilung wäre nach der Expansion nicht mehr normal. Dann kann man eine fremde normale Reihe dadurch infizieren, daß man je eines ihrer Exemplare paarweise mit den Exemplaren unserer nicht normalen Reihe in Berührung

---

*) Das ideale Gas und der mit Strahlung erfüllte Hohlraum gehören zum Beispiel in diese Kategorie.

Über die Ausdehnung der phänomenologischen Thermodynamik usw. 771

bringt und nach hinreichend langer Zeit wieder trennt. Hierdurch hat man erreicht, daß nunmehr die Hilfsreihe auch „nicht normal" geworden ist*). Dabei sind die Energiemittelwerte der Reihen bei dieser Berührung unverändert geblieben, falls man die Hilfsreihe von passender Temperatur gewählt hat. Allerdings wurde dabei auch die Statistik unserer expandierten Reihe abgeändert, und dies hat nun zur Folge, daß man im allgemeinen nicht den ursprünglichen Energiemittelwert zurückerhält, wenn man diese nun wieder adiabatisch bis zum Ausgangsvolumen komprimiert. Bei den erwähnten Körpern, für welche

$$p(u, v) = u f(v)$$

gilt, läuft jedoch der Energiemittelwert bei dieser Kompression, trotz abgeänderter Statistik auf derselben Kurve, wie bei der vorangehenden Expansion (und man kann dann selbstverständlich auch noch ihre Statistik zur normalen machen, dadurch, daß man sie mit einem passend temperierten Reservoir in Berührung bringt). So wäre also die einzige Veränderung, welche in diesem Falle hinterbleibt, die, daß die normale Statistik der Hilfsreihe abgeändert wurde. Da dies wegen der Stabilität unmöglich ist, muß also bei diesen Körpern die Statistik entlang der „freien" adiabatischen Expansion stets normal bleiben.

Daher brauchen wir, um $\psi(T)$ zu erhalten, bloß zu berechnen, wie sich bei diesen Körpern die Streuung η entlang der „freien" Expansion verändert, wenn wir von einer normalen Statistik ausgehen; die normale Streuung muß sich dann entlang der „normalen" Expansion ebenso ändern.

Nun ändert sich aber längs der „freien" adiabatischen Expansion die Streuung gemäß Nr. (16)

$$\frac{d\eta}{dv} = -2 \int_0^\infty (u - \bar{u}) \, p(u, v) \, W(u, v) \, du.$$

Für die hervorgehobene Körperklasse, für welche

$$p(u, v) = u f(v)$$

gilt, ist dementsprechend im normalen Anfangszustand der Abfall der Streuung **)

$$\frac{d\eta}{dv} = -2 \psi(T) \frac{\partial \bar{u}}{\partial T} f(v). \tag{18}$$

Bei der „normalen" adiabatischen Expansion ist andererseits der Abfall

$$\frac{d\eta^*}{dv} = \frac{\partial}{\partial v} \eta^* + \frac{dT}{dv} \frac{\partial}{\partial T} \eta^*, \tag{19}$$

---

*) Vgl. Anm. 9.
**) Vgl. Anm. 10.

53*

772　　　　　　　　　　　Leo Szilard,

und dies geht für Körper der hervorgehobenen Körperklasse über in *)

$$\frac{d\eta^*}{dv} = -T\frac{d\psi}{dT}\frac{\partial \bar{u}}{\partial T}f(v).\tag{20}$$

Die Forderung, daß die Streuung bei der „freien" und bei der „normalen" Expansion sich in gleicher Weise ändert, ergibt so gemäß Nr. (18) und Nr. (20) für ψ die Differentialgleichung

$$T\frac{d\psi}{dT} - 2\psi = 0.$$

Es ist also

$$\psi = k.T^2.\tag{21}$$

Und die Formel für die normale Streuung lautet:

$$\eta^* = kT^2\frac{\partial \bar{u}^*}{\partial T}.$$

Dabei ist k eine Integrationskonstante, deren Wert aus der Erfahrung genommen werden muß, da darüber die thermodynamische Theorie nichts aussagen kann. Wohl aber fordert die Thermodynamik, daß der Quotient

$$\frac{\eta^*}{T^2\frac{\partial \bar{u}^*}{\partial T}} = k$$

für alle Körper dieselbe Zahl, also eine universelle Konstante sei. Beim Vergleich mit der molekularen Theorie erkennt man ihre Identität mit der sogenannten Boltzmannschen Konstante.

Für die Funktion φ erhält man gemäß Nr. (13)

$$\varphi = -\frac{1}{kT}.$$

Adiabatische Invarianz des statistischen Gewichts.

Nachdem die Funktion φ mit Hilfe des zweiten Hauptsatzes so ermittelt wurde, gewinnt die normale Energiestatistik der Körper die Form

$$W^* = C(T)g(u)e^{-\frac{u}{kT}},\tag{22}$$

dabei ist

$$C(T) = \frac{1}{\int_0^\infty g(u)e^{-\frac{u}{kT}}du}.\tag{23}$$

*) Vgl. Anm. 11.

Über die Ausdehnung der phänomenologischen Thermodynamik usw.   773

Die Gleichung für den „thermodynamischen" Körper Nr. (17):

$$\bar{p}^* + \frac{\partial \bar{u}^*}{\partial v} = T \frac{\partial \bar{p}^*}{\partial T}$$

lautet also ausführlich geschrieben

$$\frac{\int_0^\infty p\,g\,e^{-\frac{u}{kT}}\,du}{\int_0^\infty g\,e^{-\frac{u}{kT}}\,du} + \frac{\partial}{\partial v}\frac{\int_0^\infty u\,g\,e^{-\frac{u}{kT}}\,du}{\int_0^\infty g\,e^{-\frac{u}{kT}}\,du} = T\frac{\partial}{\partial T}\frac{\int_0^\infty p\,g\,e^{-\frac{u}{kT}}\,du}{\int_0^\infty g\,e^{-\frac{u}{kT}}\,du} \qquad (24)$$

Dabei sind sowohl p wie g Funktionen von u und v. Durch eine einfache Umformung erhält man daraus*)

$$\frac{\partial}{\partial T}\frac{\int_0^\infty \left\{\frac{\partial}{\partial u}[p\,g] - \frac{\partial}{\partial v}g\right\}e^{-\frac{u}{kT}}\,du}{\int_0^\infty g\,e^{-\frac{u}{kT}}\,du} = 0$$

und durch Integration nach T

$$\int_0^\infty \left\{\frac{\partial}{\partial u}[p\,g] - \frac{\partial}{\partial v}g + K(v)g\right\}e^{-\frac{u}{kT}}\,du = 0.$$

Dabei ist $K(v)$ die Integrationskonstante, welche nur noch eine Funktion von v ist. Weil die Gleichung für alle Temperaturen T gilt, so gilt auch

$$\frac{\partial}{\partial u}[p\,g] - \frac{\partial}{\partial v}g + K(v)g = 0. \qquad (25)$$

Bei allen „thermodynamischen" Körpern, für welche also Nr. (17) gilt, gehorchen die Funktionen p und g der Beziehung Nr. (25).

Da die Funktionen $g(u)$ nur bis auf einen vom Volumen noch abhängigen Faktor definiert sind**), so hat man eine gewisse Freiheit, um g geeignet zu normieren. Setzt man

$$G(u) = \gamma(v)\,g(u)$$

und führt man dann G in die Beziehung Nr. (25) ein, so sieht man, daß es sich durch passende Wahl von $\gamma(v)$ stets erreichen läßt, daß die so normierte Gewichtsfunktion G die Gleichung

$$\frac{\partial}{\partial u}[p\,G] - \frac{\partial}{\partial v}G = 0 \qquad (26)$$

---

*) Vgl. Anm. 13.
**) Vgl. Anmerkung auf S. 763.

774   Leo Szilard,

erfüllt*). Es gehorcht also $G(u)$ derselben partiellen Differentialgleichung, welche bei der „freien" adiabatischen Expansion für irgend eine Verteilung $W(u)$ gilt [vgl. Nr. (14)]. Sie wurde oben abgeleitet als hinreichende Bedingung dafür, daß bei der freien adiabatischen Expansion von v_1 bis v_2, durch welche u_1 in u_2 übergeführt wird, die Beziehung gelten soll

$$W(u_1, v_1)\, du_1 \; = \; W(u_2, v_2)\, du_2.$$

Demgemäß gilt auch

$$G(u_1, v_1)\, du_1 \; = \; G(u_2, v_2)\, du_2$$

und es ist also

$$G\, du$$

entlang einer „freien" adiabatischen Expansion in variant.

Die Entropie einer Verteilung.

In manchen Fällen ist es mit Hilfe sehr einfacher reversibler Prozesse möglich, eine nicht normale Verteilung in die normale Verteilung von demselben Energiemittelwert überzuführen. Dabei läßt sich eine ganz bestimmte Entropieverminderung der an diesen Prozessen beteiligten Hilfsreservoire erzwingen, so daß es aus thermodynamischen Gründen angebracht erscheint, einer bestimmten Verteilung eine bestimmte Entropie zuzuordnen, über welche man leicht auch quantitative Aussagen machen kann.

Wir betrachten hierzu wieder einen Körper, dessen Zustand durch Energieinhalt und Volumen bestimmt ist. Wenn eine Reihe solcher Körper bei $v = v_a$ im normalen Zustand vorliegt und wir dieselbe bis zu dem Volumen v_b einmal „normal" adiabatisch, ein andermal „frei" adiabatisch expandieren lassen, so läuft der Energiemittelwert von demselben Punkte A bei v_a ausgehend entlang zweier verschiedener Kurven (Fig. 1). Während wir bei der „normalen" Expansion entlang der unteren Kurve laufend im Punkte B^* ankommen, kommen wir bei der „freien" Expansion entlang der oberen Kurve laufend bei B an. Dabei werden wir bei der „freien" Expansion im Punkte B mit einer Energiestatistik

---

*) Wenn man G in Nr. (25) einführt, so erhält man

$$\frac{1}{\gamma}\left\{\frac{\partial}{\partial u}[p\,G] - \frac{\partial}{\partial v}\,G\right\} + G\left\{K(v)\frac{1}{\gamma} - \frac{\partial}{\partial v}\frac{1}{\gamma}\right\} = 0,$$

wird also $\gamma(v)$ so gewählt, daß

$$\left\{K(v)\frac{1}{\gamma} - \frac{\partial}{\partial v}\frac{1}{\gamma}\right\} = 0$$

gilt, so erfüllt G die Gleichung Nr. (26).

Über die Ausdehnung der phänomenologischen Thermodynamik usw.   775

ankommen, welche von der dem Energiemittelwert $\bar{u}_B$ zugehörigen normalen Statistik abweicht.

Liegt nun umgekehrt von vornherein diese „nicht normale"
Reihe bei B vor, so können wir die Statistik in reversibler Weise in die
dem Energiemittelwert $\bar{u}_B$ zugehörige normale Statistik überführen, indem
wir die Reihe „frei" adiabatisch bis $v = v_a$ komprimieren, sodann die
nunmehr bereits normale Reihe „normal" adiabatisch bis $v = v_b$ expandieren lassen und dann noch eine Erwärmung bei festem Volumen vornehmen. Bei der Kompression läuft dann der Energiemittelwert auf der
oberen Kurve bis A und bei der darauf folgenden Expansion sodann
entlang der unteren Kurve bis B^*; wir brauchen also nur noch bei

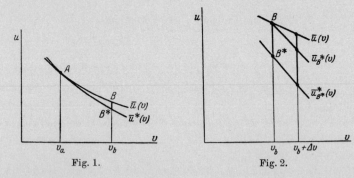

Fig. 1.   Fig. 2.

festem Volumen zu erwärmen, bis der Energiemittelwert von B^* bis B
gelangt. Dann haben wir bei unverändertem Energiemittelwert u_B die
Verteilung in die normale übergeführt, wobei eine Entropieabnahme der
Reservoire lediglich anläßlich der Erwärmung bei festem Volumen erfolgt,
so daß ihr Betrag $\mathfrak{S}$ durch das Integral

$$\mathfrak{S} = \int_{B^\bullet}^{B} \frac{\dfrac{\partial \bar{u}^*}{\partial T}}{T}\, dT$$

angegeben ist. Wir wollen zunächst den Differentialquotienten $\dfrac{d\mathfrak{S}}{dv}$ berechnen. Zu diesem Zwecke ziehen wir die von B ausgehende Kure $\bar{u}_B^*$,
entlang welcher der Energiemittelwert bei der normalen adiabatischen
Expansion abfällt, wenn wir von einer normalen Reihe bei B ausgehen.
Da die Entropie bei der adiabatischen Expansion konstant bleiben muß
und die Entropiedifferenz zwischen zwei normalen Reihen bei demselben

Volumen durch das Integral über $\dfrac{\partial \bar{u}^*}{\partial T}\Big/ T$ gegeben ist, so ist also bei der Expansion

$$\Delta \int\limits_{\bar{u}^*=\bar{u}^*_{B*}}^{\bar{u}^*=\bar{u}^*_{B}} \frac{\dfrac{\partial \bar{u}^*}{\partial T}}{T}\,dT = 0$$

und es ist daher im Punkte B bis auf Glieder höherer Ordnung

$$\Delta \mathfrak{S} = \frac{\Delta \bar{u} - \Delta \bar{u}^*}{T}.$$

Und also

$$\frac{d \mathfrak{S}}{d v} = \frac{1}{T}\left(\frac{d\bar{u}}{dv} - \frac{d\bar{u}^*}{dv}\right)$$

oder kürzer

$$\frac{d \mathfrak{S}}{d v} = \frac{\bar{p}^* - \bar{p}}{T}, \tag{27}$$

wenn man gemäß Nr. (15) die Bezeichnungen

$$\bar{p} = -\frac{d\bar{u}}{dv} = \int\limits_0^\infty p(u)\,W(u)\,d(u),$$

$$\bar{p}^* = -\frac{d\bar{u}}{dv} = \int\limits_0^\infty p(u)\,W^*(u;T)\,du$$

einführt. Es gilt nun, wie man sich leicht überzeugt*), die Identität

$$\frac{\bar{p}^* - \bar{p}}{T} = \frac{d}{dv}\left(k\int\limits_0^\infty W^* \log\frac{g}{W^*}\,du - k\int\limits_0^\infty W \log\frac{g}{W}\,du \right).$$

Es ist also $\mathfrak{S}$ gemäß Nr. (27) und entsprechend der Anfangsbedingung $\mathfrak{S} = 0$ für $v = v_a$ gegeben durch

$$\mathfrak{S} = k\int\limits_0^\infty W^* \log\frac{g}{W^*}\,du - k\int\limits_0^\infty W \log\frac{g}{W}\,du. \tag{28}$$

Wollen wir einer Energiestatistik eine bestimmte Entropie zuordnen, und zwar so, daß die Entropiedifferenz $S_1 - S_2$ zweier Statistiken gleich der Entropieabnahme der Reservoire sei, welche man beim Überführen der Verteilung 2 in die Verteilung 1 pro Exemplar im Mittel erzwingt, so müssen wir schreiben

$$S - S^* = -\mathfrak{S}$$

---

*) Vgl. Anm. 14.

Über die Ausdehnung der phänomenologischen Thermodynamik usw.   777

und also

$$S - S^* = k \int_0^\infty W \log \frac{g}{W} du - k \int_0^\infty W^* \log \frac{g}{W^*} du.$$

Man kann hier leicht zeigen, daß im allgemeinen

$$S_1 - S_2 = k \int_0^\infty W_1 \log \frac{g}{W_1} du - k \int_0^\infty W_2 \log \frac{g}{W_2} du \qquad (29)$$

gilt, wenn die beiden Verteilungen durch Kombination von „freier" adiabatischer Expansion, „normaler" adiabatischer Expansion und Erwärmen bei festem Volumen ineinander übergeführt werden können*) und die Volumina 1 und 2 identisch sind. Die letztere Einschränkung fällt fort, falls man statt g die gemäß Gleichung Nr. (26) normierte Gewichtsfunktion G verwendet, so daß es aus rein thermodynamischen Gründen angebracht erscheint, einer Verteilung die Entropie

$$S = k \int_0^\infty W \log \frac{G}{W} du \qquad (30)$$

zuzuordnen.

———————

Die bisher angeführten Beispiele sollten zeigen, in welcher Weise man den zweiten Hauptsatz handhaben muß, um die Gesetzmäßigkeiten der Schwankungserscheinungen zu erhalten. Es ist offenbar möglich, auf Grund des zweiten Hauptsatzes durch phänomenologische Betrachtungen die statistische Thermodynamik aufzubauen. Immerhin ist auf diesem Wege nicht leicht vorwärts zu kommen und es scheint uns daher vernünftiger, einen von Einstein eingeschlagenen Weg**) zu verfolgen, wenn man die statistischen Gesetze der Thermodynamik gewinnen will. Hier hatten wir allerdings noch ein anderes Ziel im Auge. Wir wollten nämlich zeigen, daß der zweite Hauptsatz auch angesichts der Schwankungserscheinungen nichts von seinem dogmatisch strengen Charakter verliert und keineswegs zu einem nur angenähert gültigen Gesetz herabsinkt, sondern sich nur in einer höheren Harmonie auflöst und so auch noch die Gesetze der Schwankungen festlegt.

———————

*) Für zwei Verteilungen, die sich nicht in solcher Weise ineinander überführen lassen, dürfte es schwerer sein, die Richtigkeit der Formel Nr. (29) mit unseren Hilfsmitteln zu beweisen.

) Da wir a. a. O. Gelegenheit haben, hierauf näher einzugehen, möchten wir hier nur kurz auf die anscheinend nicht hinreichend bekannte Arbeit von Einstein hinweisen. (Verh. d. Deutsch. Phys. Ges. **12, 820, 1914.)

778   Leo Szilard,

Die vorliegende Arbeit ist eng mit meiner Dissertation verknüpft, die im Druck nicht erscheint*). Ich darf wohl die Gelegenheit benutzen, Herrn Professor v. Laue für seine freundliche Anteilnahme und sein wohlwollendes Fördern auch an dieser Stelle wärmsten Dank zu sagen.

Anmerkungen:

[1]) Ist der Energieinhalt des Exemplars aus der Reihe 1) u_1 und des aus der Reihe 2) u_2, so ist der Energieinhalt des aus beiden zusammengesetzten Körpers

$$u = u_1 + u_2.$$

Gilt nun für eine Größe u_1 das Wahrscheinlichkeitsgesetz W_1 und für eine Größe u_2 das Wahrscheinlichkeitsgesetz W_2, so gilt, wenn die beiden Größen voneinander statistisch unabhängig sind, für die Summe u das Wahrscheinlichkeitsgesetz

$$W(u) = \int_0^u W_1(\xi) \, W_2(u-\xi) \, d\xi.$$

[2]) Wir schreiben (vgl. Nr. (19)):

$$\frac{1}{\int_0^\infty g(u) \, e^{-\frac{u}{kT}} \, du} = C(T), \tag{31}$$

es gilt dann die Identität

$$\frac{d}{dT} \log C(T) = -\frac{\overline{u}^*}{k \, T^2},$$

durch Integration erhält man also

$$C(T) = e^{\int^A_T \frac{\overline{u}^{\bullet}(T)}{k|T^2} \, dT},$$

wobei A die Integrationskonstante bedeutet. Nun geht aber $C(T)$, wenn T gegen 0 geht, bei jeder Wahl von $g(u)$ gegen Unendlich [vgl. Nr. (31)], und es geht also auch das Integral im Exponenten gegen Unendlich. Man sieht daraus, daß man bei keiner Wahl von g erreichen kann, daß $\overline{u}^*$ mit T^4 abfällt (wie es bei dem mit schwarzer Strahlung erfüllten Hohlraum sein müßte), denn dann würde ja jenes Integral für $T = 0$ nicht unendlich werden.

[3]) Wenn wir zwei Reihen mit den Verteilungen W_1 und W_2 miteinander in der geschilderten Weise in Berührung bringen, dann gilt für die so entstehende Reihe, deren Exemplare also aus je zwei Körpern zusammengesetzt sind,

$$W(u) = \int_0^u W_1(\xi) \, W_2(u-\xi) \, d\xi.$$

Wir fragen nun nach den neuen Verteilungen in den beiden Reihen, wenn wir sie nach hinreichend langer Berührung wieder trennen. Hätten die zusammengesetzten Exemplare alle denselben Energieinhalt u, so wären diese beim Trennen entstehenden neuen Verteilungen durch $V_{12}(u_1, u) \, du_1$ bzw. $V_{21}(u_2, u) \, du_2$ angegeben. Da jedoch die zusammengesetzten Exemplare nicht alle dieselbe Energie enthalten, vielmehr die Wahrscheinlichkeit dafür, daß die Energie eines solchen Exemplars zwischen u und $u + du$ liegt, durch ein Wahrscheinlichkeits-

---

*) Dissertation Berlin 1922.

Über die Ausdehnung der phänomenologischen Thermodynamik usw.   779

gesetz $W(u)\,du$ geregelt wird, so wird die neue Energieverteilung etwa der Reihe 1) nach dem Trennen angegeben durch das Integral

$$W_1'(u_1) = \int_0^\infty V_{12}(u_1, u)\, W(u)\, du. \tag{32}$$

Für den neuen Energiemittelwert

$$\bar{u}_1' = \int_0^\infty u_1\, W_1'(u_1)\, du_1$$

finden wir, wenn wir W_1' aus Nr. (32) einsetzen und die Bezeichnung

$$\bar{u}_{12}(u) = \int_0^u u_1\, V_{12}(u_1, u)\, du_1$$

einführen,

$$\bar{u}_1' = \int_0^u \bar{u}_{12}(u)\, W(u)\, du$$

und in unserer Näherung [vgl. Nr. (6)]

$$\bar{u}_1' = \bar{u}_{12}(\bar{u}) + \left\{ \frac{d^2\, \bar{u}_{12}}{d\, u^2} \right\}_{u\,=\,\bar{u}} \cdot \eta;$$

dabei bedeutet η:

$$\eta = \int_0^\infty (u - \bar{u})^2\, W(u)\, du$$

und es ist Nr. (3) entsprechend

$$\eta = \eta_1 + \eta_2.$$

Es ergibt sich so Nr. (8).

4) Für die Streuung der neuen Verteilung

$$\eta_1' = \int_0^\infty (u_1 - \bar{u}_1')^2\, W_1'(u_1)\, du_1$$

ergibt sich, wenn man W_1' aus Nr. (32) einsetzt und die Bezeichnung

$$\eta_{12} = \int_0^\infty [u_1 - \bar{u}_{12}(u)]^2\, V_{12}(u_1, u)\, du_1$$

eingeführt, in unserer Näherung [vgl. Nr. (6)]

$$\eta_1' = \eta_{12}(\bar{u}) + \left\{ \frac{d\, \bar{u}_{12}}{d\, u} \right\}_{u\,=\,\bar{u}}^2 (\eta_1 + \eta_2)$$

es sind dabei die Glieder, welche Produkte von Streuungen enthalten, vernachlässigt worden).

5) In Nr. (10) substituieren wir für $\left\{ \dfrac{d\, \bar{u}_{12}}{du} \right\}_{u\,=\,\bar{u}^*}$

$$\left\{ \frac{d\, \bar{u}_{12}}{d\, u} \right\}_{u\,=\,\bar{u}^*} = \frac{\dfrac{d\, \bar{u}_1^*}{d\, T}}{\dfrac{d\, \bar{u}_1^*}{d\, T} + \dfrac{d\, \bar{u}_2^*}{d\, T}},$$

dies ist in unserer Näherung gestattet, denn es ist gemäß Nr. (9)

$$\bar{u}_{12}\,[u^*(T)] \simeq \bar{u}_1^*(T),$$

780 Leo Szilard,

es ist also

$$\left\{\frac{d\,\bar{u}_{12}}{d\,u}\right\}_{u\,=\,\bar{u}^{\bullet}\,(T)} = \frac{d\,\bar{u}_1^*}{d\,T}\frac{d\,T}{d\,\bar{u}^*} = \frac{\dfrac{d\,\bar{u}_1^*}{d\,T}}{\dfrac{d\,\bar{u}^*}{d\,T}}$$

und da

$$\bar{u}^*(T) = \bar{u}_1^*(T) + \bar{u}_2^*(T)$$

gilt, so ist

$$\left\{\frac{d\,\bar{u}_{12}}{d\,u}\right\}_{u\,=\,\bar{u}^{\bullet}} = \frac{\dfrac{d\,\bar{u}_1^*}{d\,T}}{\dfrac{d\,\bar{u}_1^*}{d\,T} + \dfrac{d\,\bar{u}_2^*}{d\,T}}.$$

Ganz analog findet man

$$\left\{\frac{d\,\bar{u}_{21}}{d\,u}\right\}_{u\,=\,\bar{u}^{\bullet}} = \frac{\dfrac{d\,\bar{u}_2^*}{d\,T}}{\dfrac{d\,\bar{u}_1^*}{d\,T} + \dfrac{d\,\bar{u}_2^*}{d\,T}}$$

Es stellt nun Nr. (10) ein homogenes Gleichungssystem dar für η_1^*, η_2^* und $\eta_{12} = \eta_{21}$, und man findet daraus

$$\frac{\eta_1^*}{\eta_2^*} = \frac{\dfrac{d\,\bar{u}_1^*}{d\,T}}{\dfrac{d\,\bar{u}_2^*}{d\,T}}.$$

[6]) Genau genommen, dürfen wir nicht von einem exakten Volumen v ausgehen, sondern von einem Zustand, bei welchem der Kolben noch innerhalb eines endlichen, wenn auch sehr kleinen Intervalls frei beweglich ist. Lassen wir auf den Kolben nämlich eine konstante Kraft wirken, welche dem Druck des Gases im Mittel das Gleichgewicht hält, so wird der an sich freie Kolben nicht in Ruhe verharren, sondern eine Art Brownsche Bewegung ausführen und von der Gleichgewichtslage bald mehr, bald weniger entfernt zu liegen kommen. Wollen wir seine Lage auf ein kleines Intervall einschränken, indem wir etwa von beiden Seiten je einen Stellring langsam heranschieben, so müssen wir Arbeit leisten. Dieser Vorgang ist mit der Kompression eines aus einem einzigen Molekül bestehenden Gases vergleichbar. Die aufzuwendende Arbeit geht gegen Unendlich, wenn das Intervall gegen 0 konvergiert. Wir können aber ein für allemal irgend ein kleines Intervall, im Volumen ausgedrückt etwa $\varDelta$, festsetzen, innerhalb dessen sich der Kolben frei bewegen kann; die Arbeit, die geleistet werden muß, um von einem Volumen zwischen v_1 und $v_1 + \varDelta$ auf ein anderes Volumen zwischen v_2 und $v_2 + \varDelta$ zu kommen, ist dann lediglich die Kompressionsarbeit für das Gas im zylindrischen Gefäß, vorausgesetzt, daß nur $\varDelta$ genügend klein ist.

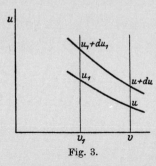

Fig. 3.

[7]) Die Ordinate bei v_1 (Fig. 3) wird durch die Adiabatenschar

$$u = u(u_1, v)$$

auf die Ordinate bei v abgebildet. Ist die Wahrscheinlichkeit dafür, daß ein Punkt auf der linken Ordinate zwischen u_1 und $u_1 + d\,u_1$ zu liegen kommt

$$W(u_1, v_1)\,d\,u_1,$$

so ist die Wahrscheinlichkeit dafür, daß ihr Bild auf der rechten Ordinate zwischen u und $u + d\,u$ liegt:

$$W\,(u,\ v)\ d\,u = W\,(u_1,\ v_1)\,d\,u_1.$$

Es gilt also

$$W\,[u\,(u_1,\ v_1),\,v]\,\frac{\partial\,u\,(u_1,\,v)}{\partial\,u_1} = W\,(u_1,\,v_1).$$

Differenziert man diese Gleichung partiell nach v bei festem u_1 und v_1, und setzt man dann $v = v_1$, so erhält man, wenn man die Bezeichnung

$$\frac{\partial\,u\,(u_1,\,v)}{\partial\,v} = -\,p\,(u,\,v)$$

einführt,

$$\frac{\partial}{\partial\,u}\,(p\,W) - \frac{\partial}{\partial\,v}\,W = 0.$$

[8]) Es ist

$$\bar{u}\,(v) = \int_0^\infty u\,W(u,\,v)\,d\,u$$

und unter den nötigen Stetigkeitsannahmen

$$\frac{d\,\bar{u}}{d\,v} = \int_0^\infty u\,\frac{\partial}{\partial\,v}\,W\,d\,u.$$

Indem man aus der partiellen Differentialgleichung Nr. (14) $\dfrac{\partial}{\partial\,v}\,W$ substituiert,

$$\frac{d\,\bar{u}}{d\,v} = \int_0^\infty u\,\frac{\partial}{\partial\,u}\,[p\,W]\,d\,u,$$

hieraus erhält man durch partielle Integration, wenn man $(p\,W)_0^\infty = 0$ setzt,

$$\frac{d\,\bar{u}}{d\,v} = -\int_0^\infty p\,W\,d\,u.$$

Ähnlich erhält man auch aus der partiellen Differentialgleichung Nr. (14) für

$$\eta = \int_0^\infty (u - \bar{u})^2\ W\,(u)\,d\,u$$

$$\frac{d\,\eta}{d\,v} = -\,2\int_0^\infty (u - \bar{u})\,p\,W\,d\,u.$$

[9]) Wir zeigen zunächst, daß man zu einer nichtnormalen zusammengesetzten Reihe kommt, wenn man je ein Exemplar einer Reihe mit normaler Statistik $W_1^*(u)$ mit je einem Exemplar einer zweiten Reihe mit nichtnormaler Energiestatistik $W_2\,(u)$ thermisch in Berührung bringt. Die Statistik der zusammengesetzten Reihe ist angegeben durch

$$W\,(u) = \int_0^u W_1^*\,(u - \xi)\,W_2\,(\xi)\,d\,\xi.$$

782   Leo Szilard,

Andererseits erhält man, wenn man zwei normale Reihen zusammensetzt, die zusammengesetzte Reihe mit normaler Statistik; es ist

$$W^*(u) = \int_0^u W_1^*(u - \xi) W_2^*(\xi) \, d\xi.$$

Ziehen wir diese beiden Gleichungen voneinander ab, so erhalten wir

$$W^*(u) - W(u) = \int_0^u W_1^*(u - \xi) \{ W_2^*(\xi) - W_2(\xi) \} \, d\xi.$$

Wäre nun $W = W^*$, so würde für alle u

$$\int_0^u W_1^*(u - \xi) \{ W_2^*(\xi) - W_2(\xi) \} \, d u \equiv 0$$

gelten, was nicht möglich ist, wenn

$$W_2 \neq W^*.$$

Um zu beweisen, daß durch die Berührung mit einer nichtnormalen Reihe die ursprünglich normale Reihe in der Tat infiziert wird, müssen wir noch zeigen, daß man beim Spalten einer solchen nichtnormalen zusammengesetzten Reihe wieder nichtnormale Reihen bekommt. Dies ist auch der Fall, denn beim Spalten erhält man die Statistik (vgl. Nr. (32)

$$W_1'(u_1) = \int_0^\infty \frac{g_1(u_1) \, g_2(u - u_1)}{g_{12}(u)} W(u) \, d u.$$

Und da andererseits

$$W_1^*(u_1) = \int_0^\infty \frac{g_1(u_1) \, g_2(u - u_1)}{g_{12}(u)} W^*(u) \, d u$$

gilt, so müßte im Falle $W_1' = W_1^*$

$$\int_0^\infty g_2(u - u_1) \frac{W^*(u) - W(u)}{g_{12}(u)} \, d u \equiv 0$$

für alle u_1 gelten, was nicht möglich ist, wenn

$$W \neq W^*$$

und die in Betracht kommenden Funktionen hinreichend stark für $u = \infty$ verschwinden.

[10]) Die Gleichung Nr. (16) ergibt für die hervorgehobene Körperklasse $[p = u f(v)]$

$$\frac{d\eta}{dv} = - 2 f(v) \int_0^\infty u(u - \bar{u}) W(u) \, d u.$$

Da die Identität besteht

$$\int_0^\infty u(u - \bar{u}) W(u) \, d u = \int_0^\infty (u - \bar{u})^2 W(u) \, d u = \eta,$$

so können wir auch schreiben

$$\frac{d\eta}{dv} = - 2 \eta f(v)$$

Über die Ausdehnung der phänomenologischen Thermodynamik usw.   783

und für den normalen Anfangszustand müssen wir

$$\eta = \eta^* = \psi\,(T)\,\frac{\partial\,\bar{u}^*}{\partial\,T}$$

setzen. Wir erhalten also für den Abfall der Streuung bei der freien Expansion

$$\frac{d\,\eta}{d\,v} = -\,2\,\psi\,(T)\,\frac{\partial\,\bar{u}^*}{\partial\,T}\,f(v).$$

[11]) Setzen wir

$$\eta^* = \psi\,(T)\,\frac{\partial\,\bar{u}^*}{\partial\,T},$$

so erhalten wir

$$\frac{\partial}{\partial\,v}\,\eta^* = \psi\,\frac{\partial^2\,\bar{u}}{\partial\,v\,\partial\,T},$$

$$\frac{\partial}{\partial\,T}\eta^* = \frac{\partial\,\psi}{\partial\,T}\,\frac{\partial\,\bar{u}^*}{\partial\,T} + \psi\frac{\partial^2\,\bar{u}^*}{\partial\,T^2}.$$

Es gilt nun für alle Körper für den Temperaturabfall bei der adiabatischen Expansion

$$\frac{d\,T}{d\,v} = -\,\frac{T\dfrac{\partial\,\bar{p}^*}{\partial\,T}}{\dfrac{\partial\,\bar{u}^*}{\partial\,T}},$$

so daß gemäß Nr. (19)

$$\frac{d\,\eta^*}{d\,v} = \psi\,\frac{\partial^2\,\bar{u}^*}{\partial\,v\,\partial\,T} - \frac{T\dfrac{\partial\,\bar{p}^*}{\partial\,T}}{\dfrac{\partial\,\bar{u}^*}{\partial\,T}}\Big(\frac{d\,\psi}{d\,T}\,\frac{\partial\,\bar{u}^*}{\partial\,T} + \psi\frac{\partial^2\,\bar{u}^*}{\partial\,T^2}\Big),$$

für unsere hervorgehobene Körperklasse, für welche

$$p\,(u, v) = u\,f(v)$$

gilt, ist nun

$$\frac{\partial\,\bar{p}^*}{\partial\,T} = \frac{\partial\,\bar{u}^*}{\partial\,T}\,f(v)$$

und da allgemein für die Körper entsprechend Nr. (17)

$$\frac{\partial^2\,\bar{u}^*}{\partial\,v\,\partial\,T} = T\frac{\partial^2\,\bar{p}^*}{\partial\,T^2}$$

gilt [man erhält dies durch Differentiation aus Nr. (17)], so gilt in unserem Falle

$$\frac{\partial^2\,\bar{u}^*}{\partial\,v\,\partial\,T} = T\frac{\partial^2\,\bar{u}^*}{\partial\,T^2}\,f(v).$$

Es ergibt sich daher

$$\frac{d\,\eta^*}{d\,v} = -\,T\,\frac{d\,\psi}{d\,T}\,\frac{\partial\,\bar{u}^*}{\partial\,T}\,f(v).$$

[12]) Bei der Betrachtung im Text spielten die Körper, für welche

$$p\,(u, v) = u\,f(v)$$

gilt, eine ausgezeichnete Rolle. Man kann aber die Funktion $\varphi\,(T)$ auch ermitteln, ohne eine Körperklasse zu bevorzugen. Betrachten wir zwei Körper, die miteinander im Wärmegleichgewicht stehen; der Zustand dieses Körperpaares sei dann durch die Angabe der Volumina v_1 und v_2 und die Gesamtenergie u bestimmt. Wir können nun einen adiabatischen Prozeß in reversibler Weise durchführen, indem wir v_1 und v_2 langsam ändern, während die beiden Körper thermisch

　　　　　　　Leo Szilard,

verbunden bleiben. Wir können so etwa v_2 als Funktion von v_1 eine beliebig geschlossene Kurve in reversibler Weise durchlaufen lassen und berechnen, wie sich entlang dieser Kurve die Gesamtenergie u ändert.

Geht man von einer Reihe solcher zusammengesetzten Körper mit normaler Energiestatistik aus und läßt dann v_2 als Funktion von v_1 eine geschlossene Kurve durchlaufen, so darf dabei der Energiemittelwert der Reihe nicht abgenommen haben, wenn der zweite Hauptsatz zu Recht besteht. Läßt man v_2 als Funktion von v_1 eine bestimmte geschlossene Kurve durchlaufen und bezeichnet man die Änderung des Energieinhaltes von einem Exemplar dabei mit $\varDelta(u)$, so muß also

$$\int_0^\infty \varDelta(u)\,W^*(u)\,d\,u \geq 0 \qquad (33)$$

sein. Führen wir für die adiabatische Veränderung bei festem v_2 die Bezeichnung

$$\frac{d\,u}{d\,v_1} = -\,P_1(v_1,\,v_2,\,u)$$

ein, und analog bei festem v_1

$$\frac{d\,u}{d\,v_2} = -\,P_2(v_1,\,v_2,\,u),$$

so gilt, wenn die Kurve innerhalb einer genügend engen Umgebung des Ausgangspunktes verläuft, je nach Umlaufssinn

$$\frac{\varDelta}{f} = \frac{d\,P_1}{d\,v_2} - \frac{d\,P_2}{d\,v_1}$$

bzw.

$$\frac{\varDelta}{f} = -\left(\frac{d\,P_1}{d\,v_2} - \frac{d\,P_2}{d\,v_1}\right).$$

Dabei bezeichnet f die umschlossene Fläche und es ist etwa $\dfrac{d\,P_1}{d\,v_2}$ ausführlich geschrieben

$$\frac{d\,P_1}{d\,v_2} = \frac{\partial\,P_1}{\partial\,v_2} + \frac{\partial\,P_1}{\partial\,u}\left\{\frac{d\,u}{d\,v_2}\right\}_{v_1\,=\,\text{const}}$$

Da Nr. (33) sowohl für den einen als auch für den anderen Umlaufssinn gelten muß, so ist

$$\int_0^\infty \varDelta(u)\,W^*(u)\,d\,u = 0$$

oder, indem man

$$W^* = C(T)\,g(u)\,e^{\varphi(T)\,u}$$

berücksichtigt,

$$\int_0^\infty \varDelta(u)\,g(u)\,e^{\varphi(T)\,u}\,d\,u = 0,$$

und da dies für alle Werte T bzw. φ gelten muß, so gilt auch für alle u

$$\varDelta \equiv 0$$

und es ist also

$$\frac{d\,P_1}{d\,v_2} - \frac{d\,P_2}{d\,v_1} = 0. \qquad (34)$$

Über die Ausdehnung der phänomenologischen Thermodynamik usw.   785

[Man sieht, daß $d\,u = P_1\,d\,v_1 + P_2\,d\,v_2$ ein vollständiges Differential ist, d. h. es muß u eine Funktion von (v_1, v_2) sein.]

Es ist nun gemäß Nr. (15)

$$P_1\,(u) = \int_0^\infty p_1\,(u_1)\,V_{12}\,(u_1,\,u)\,d\,u_1,$$

$$P_2\,(u) = \int_0^\infty p_2\,(u_2)\,V_{21}\,(u_2,\,u)\,d\,u_2.$$

Machen wir nun von der Näherung auf S. 763 Gebrauch, so ergibt sich aus der Forderung, daß P_1 und P_2 Koeffizienten eines vollständigen Differentials seien [Gleichung (34)], nach einer einfachen Rechnung*) für ψ wiederum die Differentialgleichung

$$T\frac{d\,\psi}{d\,T} - 2\,\psi = 0.$$

13) Es gilt einerseits die Identität

$$T\frac{\partial}{\partial T}\frac{\int_0^\infty p\,g\,e^{-\frac{u}{k\,T}}\,du}{\int_0^\infty g\,e^{-\frac{u}{k\,T}}\,du} = \frac{\int_0^\infty p\,g\,e^{-\frac{u}{k\,T}}\,du}{\int_0^\infty g\,e^{-\frac{u}{k\,T}}\,du} + k\,T^2\frac{\partial}{\partial T}\frac{\int_0^\infty \frac{\partial}{\partial u}[p\,g]\,e^{-\frac{u}{k\,T}}\,du}{\int_0^\infty g\,e^{-\frac{u}{k\,T}}\,du},\quad (35)$$

wovon man sich überzeugen kann, indem man auf der rechten Seite

$$\int_0^\infty \frac{\partial}{\partial u}[p\,g]\,e^{-\frac{u}{k\,T}}\,du = \frac{1}{k\,T}\int_0^\infty p\,g\,e^{-\frac{u}{k\,T}}\,du$$

substituiert (daß dies gestattet ist, zeigt man durch partielle Integration, wobei man $[p\,g\,e^{-\frac{u}{k\,T}}]_0^\infty = 0$ berücksichtigt) und dann die Differentiation nach T ausführt. Ferner gilt die Identität

$$\frac{\partial}{\partial v}\frac{\int_0^\infty u\,g\,e^{-\frac{u}{k\,T}}\,du}{\int_0^\infty g\,e^{-\frac{u}{k\,T}}\,du} = k\,T^2\frac{\partial}{\partial T}\frac{\int_0^\infty \frac{\partial g}{\partial v}\,e^{-\frac{u}{k\,T}}\,du}{\int_0^\infty g\,e^{-\frac{u}{k\,T}}\,du},\quad (36)$$

indem man den Identitäten Nr. (35) und Nr. (36) entsprechend die Ausdrücke

$$T\frac{\partial}{\partial T}\frac{\int_0^\infty p\,g\,e^{-\frac{u}{k\,T}}\,du}{\int_0^\infty g\,e^{-\frac{u}{k\,T}}\,du}\quad \text{und}\quad \frac{\partial}{\partial v}\frac{\int_0^\infty u\,g\,e^{-\frac{u}{k\,T}}\,du}{\int_0^\infty g\,e^{-\frac{u}{k\,T}}\,du}$$

---

*) Vgl. L. Szilard, Dissert. Berlin 1922.

786   Leo Szilard,

in die Gleichung Nr. (24)

$$\frac{\int\limits_0^\infty p\,g\,e^{-\frac{u}{kT}}\,du}{\int\limits_0^\infty g\,e^{-\frac{u}{kT}}\,du} + \frac{\partial}{\partial v}\frac{\int\limits_0^\infty u\,g\,e^{-\frac{u}{kT}}\,du}{\int\limits_0^\infty g\,e^{-\frac{u}{kT}}\,du} = T\frac{\partial}{\partial T}\frac{\int\limits_0^\infty p\,g\,e^{-\frac{u}{kT}}\,du}{\int\limits_0^\infty g\,e^{-\frac{u}{kT}}\,du}$$

substituiert, erhält man

$$\frac{\partial}{\partial T}\frac{\int\limits_0^\infty \left\{\frac{\partial}{\partial u}[p\,g] - \frac{\partial}{\partial v}g\right\} e^{-\frac{u}{kT}}\,du}{\int\limits_0^\infty g\,e^{-\frac{u}{kT}}\,du} = 0.$$

[14]) Führen wir die Bezeichnungen

$$\int\limits_0^\infty W\log\frac{g}{W}\,du = J$$

und

$$\int\limits_0^\infty W*\log\frac{g}{W*}\,du = J*$$

ein, so ist

$$\frac{d}{dv}\left(k\int\limits_0^\infty W*\log\frac{g}{W*}\,du - k\int\limits_0^\infty W\log\frac{g}{W}\,du\right) = k\frac{dJ*}{dv} - k\frac{dJ}{dv};$$

es ist nun

$$\frac{dJ}{dv} = \int\limits_0^\infty \frac{\partial W}{\partial v}\log g\,du + \int\limits_0^\infty \frac{W}{g}\frac{\partial g}{\partial v}\,du - \int\limits_0^\infty \frac{\partial W}{\partial v}\log W\,du - \int\limits_0^\infty \frac{\partial W}{\partial v}\,du$$

und indem man hier $\frac{\partial W}{\partial v}$ aus der Differentialgleichung Nr. (14)

$$\frac{\partial W}{\partial v} = \frac{\partial}{\partial u}[p\,W]$$

und $\frac{\partial g}{\partial v}$ aus der Differentialgleichung für thermodynamische Körper, vgl. Nr. (25),

$$\frac{\partial g}{\partial v} = \frac{\partial}{\partial u}[p\,g] + K\,g$$

substituiert und hierauf partiell integriert, findet man

$$\frac{dJ}{dv} = +K. \tag{37}$$

Es ist ferner

$$\frac{dJ*}{dv} = \frac{\partial}{\partial v}J* + \frac{dT}{dv}\frac{\partial}{\partial T}J*, \tag{38}$$

Über die Ausdehnung der phänomenologischen Thermodynamik usw.   787

für J^* findet man, indem man

$$W^* = C(T)\, g(u)\, e^{-\frac{u}{kT}}$$

setzt und ausintegriert,

$$J^* = -\log C(T) + \frac{\bar{u}^*}{kT},$$

dabei ist

$$C = \frac{1}{\displaystyle\int_0^\infty g\, e^{-\frac{u}{kT}}\, du}\,.$$

Es ist also

$$\frac{\partial}{\partial v} J^* = -\frac{\dfrac{\partial C}{\partial v}}{C} + \frac{\dfrac{\partial \bar{u}^*}{\partial v}}{kT}\,.$$

Für das erste Glied auf der rechten Seite findet man

$$-\frac{\dfrac{\partial C}{\partial v}}{C} = C \int_0^\infty \frac{\partial g}{\partial v}\, e^{-\frac{u}{kT}}\, du,$$

und wenn man für $\dfrac{\partial g}{\partial v}$ gemäß der Gleichung

$$\frac{\partial g}{\partial v} = \frac{\partial}{\partial u} [p\, g] + K g$$

substituiert und sodann partiell integriert

$$-\frac{\dfrac{\partial C}{\partial v}}{C} = \frac{\bar{p}^*}{kT} + K$$

$\left(\text{dabei ist } \bar{p}^* \text{ die Abkürzung für das Integral } \displaystyle\int_0^\infty p\, C\, g\, e^{-\frac{u}{kT}}\, du\right)$, so daß wir erhalten

$$\frac{\partial}{\partial v} J^* = \frac{\bar{p}^*}{kT} + K + \frac{\dfrac{\partial \bar{u}^*}{\partial v}}{kT}\,. \tag{39}$$

Es ist ferner

$$\frac{\partial}{\partial T} J^* = -\frac{\dfrac{\partial C}{\partial T}}{C} - \frac{\bar{u}^*}{kT^2} + \frac{\dfrac{\partial \bar{u}^*}{\partial T}}{kT}$$

und da die Identität

$$\frac{\dfrac{\partial C}{\partial T}}{C} = -\frac{\bar{u}^*}{kT^2}$$

gilt, so ist

$$\frac{\partial}{\partial T} J^* = \frac{\dfrac{\partial \bar{u}^*}{\partial T}}{kT}\,. \tag{40}$$

788 Leo Szilard, Über die Ausdehnung der phänomenologischen Thermodynamik usw.

Schließlich ergibt sich $\dfrac{d\,T}{d\,v}$ aus der Bedingung

$$d\,\bar{u} = \frac{\partial\,\bar{u}^*}{\partial\,v}\,d\,v + \frac{\partial\,\bar{u}^*}{\partial\,T}\,d\,T$$

zu

$$\frac{d\,T}{d\,v} = \frac{\dfrac{d\,\bar{u}}{d\,v} - \dfrac{\partial\,\bar{u}^*}{\partial\,v}}{\dfrac{\partial\,\bar{u}^*}{\partial\,T}},$$

oder, indem wir hier

$$\frac{d\,\bar{u}}{d\,v} = -\int_0^\infty p\,(u)\,W\,(u)\,d\,u = -\bar{p}$$

schreiben, erhalten wir

$$\frac{d\,T}{.d\,v} = \frac{-\bar{p} - \dfrac{\partial\,\bar{u}^*}{\partial\,v}}{\dfrac{\partial\,\bar{u}^*}{\partial\,T}} \tag{41}$$

Wir erhalten somit

$$\frac{d\,T}{d\,v} \cdot \frac{\partial}{\partial\,T}\,J^* = -\frac{\bar{p}}{k\,T} - \frac{\dfrac{\partial\,\bar{u}^*}{\partial\,v}}{k\,T} \tag{42}$$

und es ist also gemäß (39) und (42)

$$\frac{d\,J^*}{d\,v} = \frac{\bar{p}^* - \bar{p}}{k\,T} + K \tag{43}$$

und somit

$$\frac{d}{d\,v}\Big(k\int_0^\infty W^* \log\frac{g}{W^*}\,d\,u - k\int_0^\infty W \log\frac{g}{W}\,d\,u\Big) = k\frac{d\,J^*}{d\,v} - k\frac{d\,J}{d\,v} = \frac{\bar{p}^* - \bar{p}}{T}.$$

On the Extension of Phenomenological Thermodynamics to Fluctuation Phenomena

By Leo Szilard in Berlin–Dahlem
(Translation for this volume of preceding paper)

It is shown that purely phenomenological thermodynamic considerations can lead to an understanding of the laws governing fluctuation phenomena, the usual derivation of which is generally based on other principles.

In a system in thermodynamic equilibrium, the parameters (e.g., the energy content of a subsystem) vary in time—they fluctuate. We want to show here that the second law [of thermodynamics] not only gives information about the mean value of these fluctuating parameters but also about the laws governing deviations from these mean values. Superficial considerations may give the impression that this is not possible, since the second law seems to fail just at the point where the fluctuations begin to be noticeable. This objection, however, is directed more toward the form rather than the basic ideas of thermodynamics. It is indeed possible to show that purely phenomenological thermodynamic considerations can lead to an understanding of those laws, the derivation of which is generally based on Boltzmann's principle.

First we have to sharpen the second law giving special consideration to the fluctuations. The point of departure for this is a well-known formulation of this classical postulate of thermodynamics. We consider a closed, isolated system consisting of:

a) a system of different bodies that undergo cyclic processes;
b) a weight that can be moved up and down and which serves as energy reservoir;
c) a number of infinitely large heat reservoirs, called support reservoirs, which serve for the absorption of the heat released in the cyclic processes.

The classical postulate then says: In a cyclic process in which the heat contents of all but one support reservoir remain the same, the heat content of this particular reservoir cannot decrease. If we denote the quantity of heat absorbed by the ith reservoir during the cycle by Q_i, then we have for the indicated reservoir

$$Q \geqslant 0$$

while for the others we have

$$Q_i = 0$$

If we do not disregard the fluctuation phenomena, we will have to make the following statement: if the system of bodies undergoes the same cycle repeatedly and if we consider the quantity of heat which one of the reservoirs, say the ith, absorbs, then because of the fluctuations we will find a different value for Q_i every time. By repeating this same process many times, we arrive at a statistic for the value of Q_i. This value is a matter of chance in each individual process. For the individual process, the probability of having Q_i lie between Q and $Q + dQ$ is given by some probability law[1]

$$W_i(Q)\, dQ$$

We now postulate: for all cyclic processes the mean value of the quantity of absorbed heat must satisfy the relation

$$\bar{Q} \geqslant \int_{-\infty}^{\infty} QW(Q)\, dQ \geqslant 0$$

while for the other reservoirs the mean of the absorbed heat quantity satisfies

$$\bar{Q}_i = \int_{-\infty}^{\infty} QW_i(Q)\, dQ = 0$$

If this postulate is violated for some particular cycle, then repeating it a large number of times would almost certainly result in an arbitrarily large quantity of heat taken from the reservoir under consideration. Hence the postulate merely expresses the conviction that exploitation of the fluctuation phenomena will not lead to the construction of a perpetuum mobile of the second kind.

On account of these fluctuations, it is in general impossible to perform exact cyclic processes. The cycles under consideration do not have to be exact in the spirit of the classical theory. This is easily seen in the following example:

When a body is in equilibrium with some infinite heat reservoir, its energy content will fluctuate about some mean value. If we separate it at some moment from the reservoir, a certain amount of energy will be trapped in the body, the magnitude of which is a matter of chance. Despite the fact that the energy content of the separated body is not exactly determined, we shall regard its state as the initial state in a cyclic process. Now we change one parameter of the body, e.g., its volume. With respect to this parameter, we let the body undergo a complete cycle and bring it once again into thermal contact with the previously used heat reservoir. We separate them again after a sufficiently long period of time. Now the final state of the process has been attained and we say that the body has undergone a cycle, although its energy content after the process will in general differ from the energy content in the initial state, since both values are a matter of chance. However, for the application of the second law, it is sufficient that this process can be repeated indefinitely, just as long as the absorbed quantity of heat follows the same probability law.

We can see the point even better if we consider instead of one body undergoing many cycles, many identical bodies undergoing the cycle at the same time. We shall call it again a "cyclic process" just as long as the statistic of the large number of bodies is the same before and after the process. The postulate of thermodynamics holds for both cases: (a) averaging the heat quantity absorbed by the reservoirs over many bodies, or (b) averaging the heat quantity absorbed by one reservoir over many experiments. Since the wording of the second law is simpler for the case in which we consider a large number of identical bodies, we shall use it in what follows.

[1] The claim that some probability law applies is an assumption on which this work is based.

In this context we shall not talk of a temperature as far as the bodies undergoing cyclic processes are concerned. We assign, however, to each infinite heat reservoir i a definite temperature t_i. This assignment is until now quite arbitrary as long as the warmer reservoir i is assigned the higher temperature t_i.[2]

Consider now only those reversible cycles which are special in the sense that only two heat reservoirs, say t_1 and t_2, absorb or reject heat. From the classical postulate we conclude that the heat quantities, absorbed by the two systems in this process, satisfy

$$\frac{Q_1}{Q_2} = -\frac{f(t_1)}{f(t_2)}$$

where $f(t)$ is independent of the choice among the many possible reversible processes, and is hence a universal function of the arbitrarily defined temperature t.

It follows analogously from our postulate that for such processes we have

$$\frac{\overline{Q_1}}{\overline{Q_2}} = -\frac{f(t_1)}{f(t_2)}$$

For the quantity of heat absorbed by the reservoirs in an arbitrary cycle, the classical postulate leads to

$$\sum \frac{Q_i}{T_i} \geqslant 0$$

where T_i is the thermodynamic temperature of the reservoir, which we define in terms of the universal function

$$T_i = f(t_i)$$

In similar fashion our postulate then leads to

$$\sum \frac{\overline{Q_i}}{T_i} \geqslant 0 \tag{1}$$

For the present it is sufficient to define the entropy change only for the infinite reservoirs. If an infinite reservoir at temperature T absorbs, in a particular case, the heat quantity Q, then the quotient

$$S = \frac{Q}{T}$$

will be called its entropy change. For several reservoirs we have accordingly

$$S = \sum \frac{Q_i}{T_i}$$

---

[2] If two infinite reservoirs are in thermal contact and if a net amount of heat is flowing from one to the other, then the former reservoir is the warmer.

The classical postulate demanded for all cyclic processes is

$$S \geqslant 0$$

However, we have to say the following: the repetition of one and the same cyclic process will lead to different sets of values $(Q_1, Q_2, \ldots)$ and hence to different values of S. For a particular cyclic process, however, one definite probability function

$$W(S)\,dS$$

gives the probability for obtaining an entropy increase between S and $S + dS$ for the reservoir after one cycle has been carried out. $W(S)$ can easily be determined from the function $W_i(Q)$ which holds for the absorbed heat quantity in the same cyclic process. For the mean value of S

$$\bar{S} = \int_0^\infty SW(S)\,dS$$

we find

$$\bar{S} = \sum \frac{\overline{Q_i}}{T_i}$$

Hence our postulate demands for all cyclic processes

$$\bar{S} \geqslant 0$$

This inequality may be interpreted as follows: Consider somebody playing a thermodynamical gamble with the help of cyclic processes and with the intention of decreasing the entropy of the heat reservoirs. Nature will deal with him like a well established casino, in which it is possible to make an occasional win but for which no system exists ensuring the gambler a profit.[3] In very similar fashion, Smoluchowski has expressed the conviction that a perpetuum mobile of the second kind cannot be constructed, even if fluctuation phenomena are used.

Using the example of the energy fluctuations, we will show how the postulate of thermodynamics, into which enter only certain mean values $\bar{Q}$, not only gives us information concerning the mean values of the fluctuating parameters but also insight into the laws governing the deviations from these mean values. Hence there exists the possibility of a unified treatment of fluctuation phenomena.

---

[3] It is plausible to express an apprehension against the present form of the second law, which originates from Maxwell:

If we had some demon at our disposal who could accurately guess the instantaneous values of the parameters and take the appropriate action, then it would certainly be possible to construct a perpetuum mobile of the second kind, if this demon were willing to help. We human beings cannot guess these parameters, but we can measure them and could take appropriate action. This raises the question whether we do not arrive in this way at a contradiction with the exact form of the second law.

We hope to give a satisfactory answer in a forthcoming paper and avoid in the present one this difficulty by separating in our "Gedanken"-experiments the action taken from the fluctuations. All our actions on the system are of such a kind that they could be done periodically by machines.

The Energy Fluctuations

Consider a body in thermal connection with an infinite heat reservoir. If this connection is severed at some instant of time, the body will retain a certain energy content, the amount of which is governed by a probability law which will also depend on the temperature of the reservoir. The probability for the energy content to lie between u and $u + du$ shall be given by

$$W^*(u; T)\, du$$

This same distribution function $W^*(u; T)$ also describes the energy statistic of a series of an infinite number of samples of the same body, provided each sample was in thermal contact with a reservoir of the same temperature for a sufficiently long period of time.

We shall call this energy statistic the normal distribution at temperature T, since we shall find that it is characteristic of the body concerned and is independent of any subsidiary conditions. For example, it is immaterial whether the thermal connection was via heat radiation or heat conduction. This is related to the stability of the normal distribution, a property which puts it above other distributions. If we are given a series with any other energy statistic $W(u)$ then it is possible to change its distribution by letting each sample of the series undergo a thermal process. This can be done without effecting any changes in other bodies.

If the series is given by its normal distribution, the case is different. We will find that in this case it is impossible to change the energy statistic, keeping the mean energy fixed, without some compensation. This stability is related to the fact that a series without normal distribution can be transferred into the normal one, keeping the mean energy fixed and forcing upon the support reservoirs a very definite entropy decrease.

Determination of the Equilibrium Statistic

Further on we shall discuss in what sense we can conclude, from the second law, that for each mean energy there exists a distribution for the series, which cannot be changed without compensation and which we therefore call the stable distribution. In the present section, however, we shall assume its existence and determine the laws which the stable distribution must necessarily obey.

It is easy to see that the distribution obtained by bringing each sample of the series into thermal equilibrium with an infinite heat reservoir must be a stable distribution. We previously called it the normal distribution. To see this we have to consider that this stable distribution, the mean energy of which corresponds to the temperature of the reservoir, can be transformed into the normal distribution by bringing the system into contact with the heat reservoir. The mean of the heat taken up by the reservoir vanishes. This represents a change of the stable distribution without compensation, unless the two distributions are identical. In order to learn more about the stable distributions we will make the following considerations.

We assume as given a set of samples 1) with arbitrary energy statistic $W_1(u)\,du$ and a set of samples 2) with the distribution $W_2(u)\,du$. We associate each sample of set 1) with a sample of set 2) and bring them into thermal contact. The probability for the energy of the composite sample to lie between u and $u + du$ is given by[4]

$$du\,W(u) = du \int_0^u W_1(\xi)W_2(u - \xi)\,d\xi \tag{2}$$

If after sufficiently long time, the two associated samples are separated, then in general the statistic of the two series 1) and 2) will have changed. Stability demands that the original distributions are retained after contact, provided the two sets were given in their normal statistic to begin with. From this demand alone we can obtain information about the equilibrium distribution. However we will make a second demand: Associate with each sample of set 1) a sample of set 2). From the knowledge of the energy content u_1 of body 1), we can in no way draw any conclusions about the energy content of the associated body 2), provided the equilibrium distribution of each was reached independently in some manner or another. The two values u_1 and u_2 are statistically independent. Now we demand that this statistical independence is maintained even after establishing contact.

If we bring a body of set 1) into thermal contact with a body of set 2), the energy of each body will fluctuate. If, after a sufficiently long period of time, such a connection is broken at some instant, the probability for the body 1) to have an energy-content falling between u_1 and $u_1 + du_1$ is given by some probability law

$$V_{12}(u_1; u)\,du_1$$

After sufficiently long contact, this law can only depend on the total energy content u of the two bodies. It cannot depend on the way this energy was originally distributed between the two bodies. Hence for a number of pairs in contact with energy content u, the relative number of those pairs of which body 1) has an energy content lying between u_1 and $u_1 + du_1$ after separation is given by $V_{12}(u_1; u)\,du_1$.[5]

Now we shall find out how the temperature independent function V_{12} can be calculated from the temperature dependent functions W_1^* and W_2^*. To do this we will consider two sets, which are given together with their equilibrium distribution, and bring them into thermal contact. After contact is broken, we will obtain again the original distributions W_1^* and W_2^*. According to statistical independence, the probability of finding the energy content of body 1) to lie between u_1 and $u_1 + du_1$ and that of corresponding body 2) to lie between u_2 and $u_2\,du_2$ is given by

$$W_1^*(u_1)W_2^*(u_2)\,du_1\,du_2$$

---

[4] See Remark 1. The remarks are collected at the end of this paper.
[5] Analogously we have for body 2)

$$V_{21}(u_2; u)\,du_2$$

and since $u_1 + u_2 = u$, we have

$$V_{21}(u - u_1; u) = V_{12}(u_1; u)$$

We can just as well characterize the state of the pair by the energy u_1 and the total energy $u_1 + u_2$. Since the corresponding Jacobian is equal to unity, we obtain for the probability q for the energy of body 1) to lie between u_1 and $u_1 + du_1$ and at the same time for the total energy to lie between u and $u + du$:

$$q = W_1^*(u_1)W_2^*(u - u_1)\,du_1\,du$$

This same probability q can be calculated in a different way. We select all pairs whose total energy lies between u and $u + du$. This selection will give us a subset. The probability for finding in this subset a pair of which body 1) has an energy content lying between u_1 and $u_1 + du_1$ is by definition $V_{12}(u_1, u)\,du_1$. On the other hand, the probability for picking out of the complete set a pair that belongs to this subset is given by $W^*(u)\,du$ (see equation 2). Hence the probability for a pair picked from the complete set to satisfy both conditions is given by:

$$q = V_{12}(u_1, u)W^*(u)\,du_1\,du$$

Comparison with the above found value yields

$$V_{12}(u_1, u) = \frac{W_1^*(u_1)W_2^*(u - u_1)}{W^*(u)}; \tag{4}$$

with

$$W^*(u) = \int_0^u W_1^*(\xi)W_2^*(u - \xi)\,d\xi$$

The functions W_1^*, W_2^* and W^* depend on the temperature, while the quotient $V_{12}(u_1, u)$ of course does not. From this it follows already that the distribution W_2^* has the form:

$$W_2^*(u, T) = C(T)g(u)\,e^{\varphi(T)u}$$

To see this we consider the logarithm. Denoting the logarithms of the quantities appearing in equation (4) by their respective small letters, we obtain

$$v_{12}(u_1, u) + w^*(u, T) = w_1^*(u_1, T) + w_2^*(u - u_1, T)$$

Taking the partial derivative with respect to u_1, keeping T fixed and evaluating at $u_1 = 0$, we find that $\partial w_2^*/\partial u$ is of the form

$$\frac{\partial}{\partial u}w_2^*(u, T) = f_1(u) + f_2(T)$$

Integrating with respect to u gives us for the form of w_2^*:

$$w_2^*(u, T) = f_3(u) + f_2(T)u + f_4(T)$$

and accordingly the normal distribution W_2^* must be of the form

$$W_2^*(u, T) = C(T)g(u)\,e^{\varphi(T)u} \tag{5}$$

The same result is found for the normal distribution W_1^* upon considering the quotient V_{21}.

Conversely if the distributions W_1^* and W_2^* are of the given form and introducing them into the quotient V_{12} (see equation 4) we see that temperature independence can be maintained if and only if φ is the same function of T for both bodies (up to a constant independent of T which can always be combined with the function g). Since this holds for two arbitrary bodies, the function φ is independent of the particular choice of bodies and hence a universal function of the temperature. The function g remains undetermined while $C(T)$ is determined by the normalization condition:

$$\int_0^\infty W^*(u)\, du = 1$$

We have

$$C(T) = \frac{1}{\displaystyle\int_0^\infty g(u)\, e^{\varphi(T)u}\, du}$$

The normal distribution for a body will be fixed by the weight function g.[6]

Actually we should also consider the possibility that specific energy values, especially the energy zero, and not only energy intervals have a finite probability (quantized energy distribution), all the more so since no choice of $g(u)$ will give us the T^3 law for the heat capacity, as should be the case for black-body-radiation which obeys the Stefan–Boltzmann law.[7] In spite of this we want to limit ourselves here to the continuous distribution so as not to complicate matters.

Approximate Calculation: The Deviation from the Normal Distribution

If we perform an approximate calculation, the above considerations can be carried out easily on the basis of the second law, without invoking the existence of a stable distribution as we had to do in the preceding section. For the sake of elegance of method, we apply in what follows this approximation procedure. We limit ourselves, however, to showing in this section how one could construct statistical thermodynamics on the basis of the second law alone. In the remaining sections of this paper, we will again proceed with mathematical rigor.

The approximation mentioned is based on the assumption that the fluctuations may be considered very small. Hence the distribution functions $W(u)$ will fall off rapidly on both sides of the maximum. In that case they can be characterized by their mean

---

[6] Conversely, if the normal distribution is given, g is not unambiguously defined by only up to a factor which does not depend on u.

[7] See Note 2.

value and their deviation:

$$\bar{u} = \int_0^\infty u W(u)\, du$$

$$\eta = \int_0^\infty (u - \bar{u})^2 W(u)\, du$$

Assuming that $f(u)$ is a slowly varying function, we have in particular:

$$\int_0^\infty f(u) W(u)\, du \cong f(\bar{u}) + \left\{ \frac{d^2 f}{du^2} \right\}_{u=\bar{u}} \frac{\eta}{2} \tag{6}$$

The latter is obtained by making a Taylor series expansion of $f(u)$ about the point $u = \bar{u}$ and by disregarding the integral of the remainder

$$J = \int_0^\infty \frac{d^3 f}{du^3} \frac{(u - \bar{u})^3}{3!} W(u)\, du$$

where $d^3 f/du^3$ is to be evaluated at points between u and $\bar{u}$.

  In the spirit of this approximation, we will now derive the stability of the normal distribution from the second law. We do this by showing that it is impossible to change the deviation of a set with normal statistic without any compensation. From the stability of the normal distribution we will be able to conclude that its value is given by

$$\eta^* = \psi(T) \frac{d\bar{u}^*}{dT} \tag{7}$$

In this equation, ψ represents a number which is the same for all bodies, i.e., it is a universal function of the temperature. We argue as follows:

  Assume we are given two sets of samples, which were in contact with a heat reservoir for a sufficiently long time. The second law requires, directly, that the mean energy of one set cannot be raised by simply lowering the mean energy of the second set, unless we make room for compensation. If this were possible, we would only have to cool one set reversibly and correspondingly heat the other set until the original mean energies were reached and thus force upon the used heat reservoirs an entropy decrease. (The reason for this is that upon heating the one set, the reservoirs reject the same amount of heat as they receive due to cooling the other set; however this happens at different temperatures.)

  Consider two sets of samples whose energy statistics are given by W_1 and W_2 and bring each member of one set into thermal contact with one member of the other set. Separating these after a sufficiently long thermal contact will result in having two sets with new energy statistics denoted by W_1' and W_2'. After separation, we have for the mean energy content of set one

$$\bar{u}_1' = \int_0^\infty u_1 W_1'(u_1)\, du_1$$

in our approximation[8]

$$\bar{u}_1' \cong \bar{u}_{12}(\bar{u}) + \left\{ \frac{d^2 \bar{u}_{12}}{du^2} \right\}_{u=\bar{u}} \frac{\eta_1 + \eta_2}{2}. \tag{8}$$

where

$$\bar{u}_{12}(u) = \int_0^\infty u_1 V_{12}(u_1\,; u)\, du_1.$$

If we do the same thing for two normal sets, then the mean energy must remain unchanged and we have

$$\bar{u}_1^* = \bar{u}_{12}(\bar{u}^*) + \left\{ \frac{d^2 \bar{u}_{12}}{du^2} \right\}_{u=\bar{u}^*} \frac{\eta_1^* + \eta_2^*}{2}. \tag{9}$$

[8] See Note 3.

From equation (8) we see that the mean energy after thermal contact is not fully determined by the mean energies before contact, but in our approximation it will also depend on the deviation. This has the following consequence: consider again two sets with normal statistics. If we could change, without any compensation, the deviation of the first set, the second law would certainly be violated, since establishing thermal contact between the two would result in a change of the mean energy. We obtain for this change in the mean energy

$$\bar{u}'_1 - \bar{u}^*_1 = \left\{\frac{d^2\bar{u}_{12}}{du^2}\right\}_{u=\bar{u}} \left(\frac{\eta_2 - \eta^*_2}{2} + \frac{\eta_1 - \eta^*_1}{2}\right),$$

by subtracting equation (9) from equation (8). A change in the mean energy follows if, for example, $\eta_1 = \eta^*_1$; $\eta_2 \neq \eta^*_2$. Hence we have shown by thermodynamic reasoning that the normal deviation is indeed stable in our approximation.

From the stability it follows that starting from two normal sets and establishing thermal contact by the method described above, after separation, we will regain the original normal deviations. For the deviation after separation

$$\eta'_1 = \int_0^\infty (u_1 - \bar{u}'_1)^2 W'_1(u_1)\,du_1,$$

$$\eta'_2 = \int_0^\infty (u_2 - \bar{u}'_2)^2 W'_2(u_2)\,du_2$$

we have in our approximation[9]

$$\eta'_1 = \eta_{12}(\bar{u}) + \left\{\frac{d\bar{u}_{12}}{du}\right\}_{u=\bar{u}}^2 (\eta_1 + \eta_2),$$

$$\eta'_2 = \eta_{21}(\bar{u}) + \left\{\frac{d\bar{u}_{21}}{du}\right\}_{u=\bar{u}}^2 (\eta_1 + \eta_2),$$

where

$$\eta_{12}(u) = \int_0^\infty [u_1 - \bar{u}_{12}(u)]^2 V_{12}(u_1, u)\,du_1,$$

If we start with two normal sets then we have

$$\eta^*_1 = \eta_{12}(\bar{u}) + \left\{\frac{d\bar{u}_{12}}{du}\right\}_{u=\bar{u}}^2 (\eta^*_1 + \eta^*_2),$$

$$\eta^*_2 = \eta_{21}(\bar{u}) + \left\{\frac{d\bar{u}_{21}}{du}\right\}_{u=\bar{u}}^2 (\eta^*_1 + \eta^*_2) \tag{10}$$

Taking account of the fact that

$$\eta_{12} = \eta_{21} \tag{11}$$

we obtain from equations (10)[10]

$$\frac{\eta^*_1}{\dfrac{d\bar{u}^*_1}{dT}} = \frac{\eta^*_2}{\dfrac{d\bar{u}^*_2}{dT}};$$

However this holds true for two arbitrary bodies, so that the quotient

$$\frac{\eta^*}{\dfrac{d\bar{u}^*}{dT}} = \psi(T) \tag{12}$$

must be the same for all bodies, i.e., a universal function of the temperature.

It is remarkable that the existence of such a universal function follows already from the phenomenological second law.

Let us now compare the results of the exact and approximate calculations. The former resulted in the normal distribution:

$$W^* = C(T)g(u)\,e^{\varphi(T)u};$$

[9] See Note 4.
[10] See Note 5.

forming the deviation

$$\eta^* = \int_0^\infty (u - \bar{u})^2 W^* \, du,$$

we obtain the identity

$$\frac{d\bar{u}^*}{dT} = \frac{d\varphi}{dT}\eta^*$$

If we now put

$$\frac{d\varphi}{dT} = \frac{1}{\psi(T)},$$ (13)

then we obtain, in agreement with equation (7):

$$\eta^* = \psi(T)\frac{d\bar{u}^*}{dT}$$

"Free" and "Normal" Adiabatic Processes. "Normal" Cyclic Processes

We will now consider a body whose state is determined by its volume and its energy content. One might think of a gas[11] contained in a cylindrical enclosure, closed off at one end by a movable piston. By moving the piston up and down the gas is compressed or expanded and hence we have a body whose state is completely specified by its volume and energy content.

If we now expand adiabatically such a body from the exact[12] volume v_1 to the exact volume v, then its energy content in the final state is completely determined by the energy content in the initial state. The expansion must be carried out sufficiently slowly, so that the pressure fluctuations may be neglected.

We will now consider a set consisting of an infinite number of such bodies. If we are given the probability law that describes the energy statistic of the set in the initial state $v = v_1$, then it is easy to calculate the probability law that describes the statistic after expansion, provided we know the energy of one sample during adiabatic expansion as a function of the initial energy u_1 and the volume v. Let us denote by $W(u, v)$ the probability law which holds after the expansion to volume v; we then have the partial differential equation[13]

$$\frac{\partial}{\partial u}[p(u, v)W(u, v)] - \frac{\partial}{\partial v}W(u, v) = 0$$ (14)

---

[11] We do not have to limit ourselves to an ideal gas, but just consider a body whose state is determined by its energy and volume.
[12] See Note 6.
[13] See Note 7.

where $p(u, v)$ has been defined with the help of the set of adiabatics $u = u(u_1, v)$ in such a way that along the adiabatic

$$du = -p\, dv$$

holds. Hence we have

$$p(u, v) = -\frac{\partial u(u_1, v)}{\partial v}$$

Then $p(u)$ is nothing else but the time average of the pressure for the case that the energy content of the body does not fluctuate but permanently maintains the exact value u.

We find from equation $(14)^{14}$

$$\frac{d\bar{u}}{dv} = -\int_0^\infty p(u, v)W(u, v)\, du \qquad (15)$$

or abbreviated

$$\frac{d\bar{u}}{dv} = -\bar{p}.$$

For the deviation we find

$$\frac{d\eta}{dv} = -2\int_0^\infty (u - \bar{u})p(u, v)W(u, v)\, du \qquad (16)$$

As can be seen from equation (15) the decrease in the mean energy during the adiabatic expansion is not determined solely by the respective mean energy $\bar{u}$ but is in general also dependent on the entire distribution $W(u)$.

If we start with a normal set, then during the adiabatic expansion the statistic will definitely cease to be normal. It will deviate more or less during this free adiabatic expansion from the normal statistic. However, it is possible to conduct the expansion in such a way that the statistic remains normal and that we are still presented with a process which, with some justification, we may call adiabatic. Such a normal adiabatic process can be obtained as follows: Starting from a normal state, the set is expanded "freely adiabatically" by an infinitesimal amount and is then brought into thermal contact with a reservoir of the proper temperature. The temperature of the reservoir is chosen in such a way that during thermal contact it will not absorb any net heat. This thermal contact has as its only effect that within the set the normal distribution, from which the distribution after the infinitesimal free adiabatic expansion deviates already by an infinitesimal amount, is reestablished exactly. Now the contact with the reservoir is broken and the whole thing repeated, and so on. By maintaining equilibrium with the help of reservoirs of the proper temperature along the entire expansion, we achieve an expansion during which the statistic remains normal. Since

[14] See Note 8.

the reservoirs do not absorb any net heat, we will also call this expansion adiabatic, in particular "normal" adiabatic to distinguish it from the "free" adiabatic expansion discussed earlier. Starting with a set with normal distribution, during the expansion the mean energy will run along different curves in the (u, v) diagram depending on whether we are expanding "freely" or "normally" adiabatically.

Conversely it is possible to heat up a set of bodies reversibly by bringing them into thermal contact with reservoirs of ever higher temperature. The mean energy in the (u, v) diagram runs along an ordinate. By combining heating at constant volume with normal adiabatic expansion in a suitable way, we can vary the mean energy along arbitrary curves in a reversible way while maintaining normal statistics. For these normal processes it is possible to apply the consequential methodology of classical thermodynamics. If the value of the mean energy runs through a closed curve, then we have for the participating reservoirs[15]

$$\bar{S} = 0$$

This means that for a body, fully determined by volume and energy content,

$$d\bar{S} = -\frac{\bar{p}^* \, dv + d\bar{u}^*}{T}$$

is a complete differential, which yields the relation

$$\bar{p}^* + \frac{\partial \bar{u}^*}{\partial v} = T \frac{\partial \bar{p}^*}{\partial T} \tag{17}$$

Here

$$\bar{p}^*(T) = \int_0^\infty p(u) W^*(u, T) \, du,$$

so that along the normal adiabatic expansion, according to equation (15), we have

$$d\bar{u} = -\bar{p}^* \, dv$$

---

[15] Remember analogous considerations in classical thermodynamics. There we had for all reversible cyclic processes

$$S = 0$$

This leads to the result, that

$$dS = -\frac{p \, dv + du}{T}$$

is a complete differential. From this we obtain the relation

$$p + \frac{\partial u}{\partial v} = T \frac{\partial p}{\partial T}$$

where p is the pressure and, along the adiabate, we have for the work done

$$du = -p \, dv$$

The Temperature Dependence of the Normal Distribution, and its Determination from Thermodynamics[16]

Since the quotient

$$\frac{\eta^*}{d\bar{u}^*/dT} = \psi(T)$$

is the same number for all bodies, it is fully determined as soon as we know η^* as a function of temperature for one body. If we are presented with a set with normal energy statistic, then we know how the statistic varies during the "free" adiabatic expansion. We did not gain very much from this since during this expansion we do not have in general the normal statistic corresponding to the instantaneous mean energy $\bar{u}$. However we shall see that for bodies for which p is of the form

$$p(u, v) = uf(v),$$

the statistic during the "free" adiabatic expansion must always remain normal if the second law is indeed to be true.[17] We note at first, that for such bodies the decrease of the mean energy (see equation (15))

$$\frac{d\bar{u}}{dv} = f(v) \int_0^\infty uW(u)\,du = f(v) \cdot \bar{u}$$

depends only on the mean energy itself and does not depend on the entire distribution $W(u)$. From this it follows immediately that starting from a set with normal energy distribution the mean during the free adiabatic expansion varies along the same curve in the (u, v) diagram as in the case of the normal expansion. We can see this by remembering that the normal adiabatic expansion is composed of many infinitesimally small free expansions. The repeated contact with the reservoir, which is employed to keep the distribution normal, does not change the mean energy $\bar{u}$. In the present case, $d\bar{u}/dv$ will not be influenced either by the repeated thermal contact, so that we obtain the same function $\bar{u}(v)$ as for the free expansion.

This property of the abovementioned class of bodies allows us to construct a contradiction with the second law if we assume that for these bodies the originally normal distribution does not remain normal during the free adiabatic expansion. To see this we consider a normal set and let each body expand freely adiabatically up to a certain volume. For the sake of argument we shall assume that the distribution will not be normal after the expansion. Now we can perturb another normal set by bringing each sample of it into thermal contact with a sample of the first, abnormal set. After sufficiently long time the samples are separated. Now we have achieved the fact that the second set did not remain normal,[18] even though by proper choice of

[16] See Note 12.
[17] The ideal gas and black body radiation in an enclosure belong in this category.
[18] See Note 9.

temperatures of the second set the mean energies did not change. However, the statistic of our expanding set was changed. This means that we will not regain the original mean energy if we compress adiabatically to the original volume. For the bodies in question, for which

$$p(u, v) = uf(v)$$

the mean energy during compression varies along the same curve as for the previous expansion, in spite of the modified statistic (of course we can bring its statistic back to normal by bringing it into thermal contact with a reservoir of the proper temperature). So the only net change that remains is the change of the normal statistic of the second set. Since this is impossible on account of the stability, we conclude that for these kind of bodies the statistic must remain normal during a "free" adiabatic expansion.

Hence, to calculate (T) all we have to do is to find out how the deviation η for these bodies varies during the "free" expansion, given that we start out with a normal statistic. The normal deviation must change in the same fashion during the normal expansion.

However, during the "free" adiabatic expansion the deviation varies according to equation (16)

$$\frac{d\eta}{dv} = -2 \int_0^\infty (u - \bar{u})p(u, v)W(u, v)\,du$$

For the class of bodies in question, for which

$$p(u, v) = uf(v),$$

we have accounted for the decrease in the deviation for a normal initial state[19]

$$\frac{d\eta}{dv} = -2\psi(T)\frac{\partial \bar{u}}{\partial T}\,f(v) \tag{18}$$

On the other hand, the decrease for the "normal" adiabatic expansion is

$$\frac{d\eta^*}{dv} = \frac{\partial}{\partial v}\eta^* + \frac{dT}{dv}\frac{\partial}{\partial T}\eta^*, \tag{19}$$

and for this class of bodies this goes over into[20]

$$\frac{d\eta^*}{dv} = -T\frac{d\psi}{dT}\frac{\partial \bar{u}}{\partial T}f(v) \tag{20}$$

The assumption that the deviation changes in the same fashion for the "free" as well as for the "normal" expansion yields by equations (18) and (20) a differential equation for ψ

$$T\frac{d\psi}{dT} - 2\psi = 0$$

[19] See Note 10.
[20] See Note 11.

Hence we have

$$\psi = k \cdot T^2 \tag{21}$$

And the formula for the normal deviation reads:

$$\eta^* = kT^2 \frac{\partial \bar{u}^*}{\partial T}$$

Here k is an integration constant whose value must be determined by experiment, since thermodynamics cannot give any information about it. For the quotient

$$\frac{\eta^*}{T^2 \dfrac{\partial u^*}{\partial T}} = k$$

however, thermodynamics demands that it be the same number for all bodies. Hence it is a universal constant. Upon comparison with microscopic theory we find that k is identical with the Boltzmann constant.

According to equation (13), we obtain for the function φ

$$\varphi = -\frac{1}{kT}$$

The Adiabatic Invariance of the Statistical Weight

After the function φ has been determined with the help of the second law, the normal energy-statistic takes the following form:

$$W^* = C(T)g(u)\,e^{-u/kT}, \tag{22}$$

where

$$C(T) = \frac{1}{\displaystyle\int_0^\infty g(u)\,e^{-u/kT}\,du} \tag{23}$$

The equation for the "thermodynamical" body (17)

$$\bar{p}^* + \frac{\partial \bar{u}^*}{\partial v} = T\frac{\partial \bar{p}^*}{\partial T}.$$

can now be written in detail

$$\frac{\displaystyle\int_0^\infty pg\,e^{-u/kT}\,du}{\displaystyle\int_0^\infty g\,e^{-u/kT}\,du} + \frac{\partial}{\partial v}\frac{\displaystyle\int_0^\infty ug\,e^{-u/kT}\,du}{\displaystyle\int_0^\infty g\,e^{-u/kT}\,du} = T\frac{\partial}{\partial T}\frac{\displaystyle\int_0^\infty pg\,e^{-u/kT}\,du}{\displaystyle\int_0^\infty g\,e^{-u/kT}\,du} \tag{24}$$

In this equation both p and g are functions of u and v. A simple rearrangement yields[21]

$$\frac{\partial}{\partial T} \frac{\displaystyle\int_0^{\infty} \left\{ \frac{\partial}{\partial u}[pg] - \frac{\partial}{dv}g \right\} e^{-u/kT}\, du}{\displaystyle\int_0^{\infty} g\, e^{-u/kT}\, du} = 0$$

and after integration with respect to T

$$\int_0^{\infty} \left\{ \frac{\partial}{\partial u}[pg] - \frac{\partial}{\partial v}g + K(v)g \right\} e^{-u/kT}\, du = 0$$

Here $K(v)$ is the constant of integration, which is a function only of v. Since this equation holds for all temperatures we also have

$$\frac{\partial}{\partial u}[pg] - \frac{\partial}{\partial v}g + K(v)g = 0 \tag{25}$$

For all thermodynamical bodies, for which (17) holds, the functions p and g obey the relation (25).

Since the function $g(u)$ is determined up to a factor depending on volume,[22] we have a certain amount of freedom to normalize g in some useful way. If we put

$$G(u) = \gamma(v)g(u)$$

and then introduce G into the equation (25) we see that by proper choice of $\gamma(v)$ the normalized function will satisfy the equation[23]

$$\frac{\partial}{\partial u}[pG] - \frac{\partial}{\partial v}G = 0 \tag{26}$$

This means that $G(u)$ satisfies the same partial differential equation which holds for any distribution $W(u)$ for the case of "free" adiabatic expansion (compare equation (14)). This equation was derived above as a sufficient condition for the relation

$$W(u_1, v_1)\, du_1 = W(u_2, v_2)\, du_2$$

to hold for the free adiabatic expansion that took v_1 to v_2 and changed u_1 to u_2. Accordingly we also have

$$G(u_1, v_1)\, du_1 = G(u_2, v_2)\, du_2$$

and hence $G\, du$ is an invariant under a "free" adiabatic expansion.

---

[21] See Note 13.

[22] See note on page 763 [Note 6].

[23] If we substitute G into (25) we obtain

$$\frac{1}{\gamma}\left\{ \frac{\partial}{\partial u}[pG] - \frac{\partial}{\partial v}G \right\} + G\left\{ K(v)\frac{1}{\gamma} - \frac{\partial}{\partial v}\left(\frac{1}{\gamma}\right) \right\} = 0$$

If $\gamma(v)$ is chosen in such a way that

$$\left\{ K(v)\frac{1}{\gamma} - \frac{\partial}{\partial v}\frac{1}{\gamma} \right\} = 0$$

then G will satisfy the equation (26).

The Entropy of a Distribution

In some cases it is possible to change an abnormal distribution into the normal distribution of the same mean energy with the help of very simple reversible processes. In this case it is possible to force upon the support reservoirs involved very definite entropy decreases, so that thermodynamic reasons make it plausible to associate with a certain distribution a certain entropy. It is easy to make quantitative statements about it.

To do this we consider anew a body whose state is determined by its energy content and its volume. Consider a set of such bodies with $v = v_a$ and in their normal state. Now we expand to volume v_b, once "normally" adiabatic and once "freely" adiabatic. The mean energy starts from the same point A at v_a but follows two different curves to v_b (Fig. 1). During the "normal" expansion, we follow the lower line to B^* while the "free" expansion leads us along the upper line to B. For the "free" expansion we will arrive at point B with a statistic which differs from the normal statistic associated with the mean energy $\bar{u}_B$.

Conversely, if we are presented with this "abnormal" set at B we can transform the statistic in a reversible way to the normal statistic associated with the mean energy $\bar{u}_B$. We have to "freely" adiabatically compress the set up to $v = v_a$. Then we expand "normally" adiabatically this already normal set to $v = v_b$ and then warm it up at constant volume. During the compression the mean energy runs along the upper curve to A and during the expansion it follows along the lower curve to B^*. Hence all that is left to do is to warm the set up, keeping the volume constant, until the mean energy goes from B^* to B. Then we will have transferred the distribution into the normal one without changing the mean energy u_B. The entropy decrease of the reservoirs takes place only during warming at constant volume so that the entropy change is given by the integral

$$\mathfrak{S} = \int_{B^*}^{B} \frac{\partial \bar{u}^*/\partial T}{T} \, dT$$

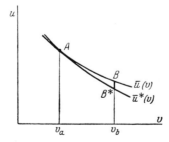

Fig. 1.

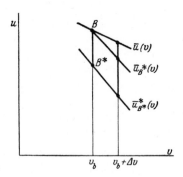

Fig. 2.

Now we want to calculate the differential quotient $d\mathfrak{S}/dv$. To do this we draw the line $\bar{u}_{B*}$ starting at B. The mean energy decreases along this line during the normal adiabatic expansion, provided we were starting out with a normal set at B. Since the entropy must remain constant during an adiabatic expansion and since the entropy difference between two normal sets at the same volume is given by the integral over $(\partial \bar{u}^*/\partial T)/T$, we have for this expansion

$$\Delta \int_{\bar{u}^* = \bar{u}^*_{B*}}^{\bar{u}^* = \bar{u}^*_B} dT \frac{\partial \bar{u}^*/\partial T}{T} = 0$$

and up to terms of higher order is given at the point B by

$$\Delta \mathfrak{S} = \frac{\Delta \bar{u} - \Delta \bar{u}^*}{T}$$

This means

$$\frac{d\mathfrak{S}}{dv} = \frac{1}{T}\left(\frac{d\bar{u}}{dv} - \frac{d\bar{u}^*}{dv} \right)$$

or more succinctly

$$\frac{d\mathfrak{S}}{dv} = \frac{\bar{p}^* - \bar{p}}{T} \qquad (27)$$

if we introduce the notation

$$\bar{p} = -\frac{d\bar{u}}{dv} = \int_0^\infty p(u)W(u)\,d(u)$$

$$\bar{p}^* = -\frac{d\bar{u}^*}{dv} = \int_0^\infty p(u)W^*(u;T)\,du$$

according to equation (15). As can be easily[24] checked out, the following relation holds:

$$\frac{\bar{p}^* - \bar{p}}{T} = \frac{d}{dv}\left(k \int_0^\infty W^* \log \frac{g}{W^*} du - k \int_0^\infty W \log \frac{g}{W} du\right)$$

According to (27) and the initial condition $\mathfrak{S} = 0$ for $v = v_a$, $\mathfrak{S}$ is given by

$$\mathfrak{S} = k \int_0^\infty W^* \log \frac{g}{W^*} du - k \int_0^\infty W \log \frac{g}{W} du \qquad (28)$$

If we want to associate with an energy statistic a certain entropy in such a fashion that the entropy difference $S_1 - S_2$ between two statistics is equal to the entropy decrease forced upon the reservoirs during the transfer of distribution 2 into distribution 1 for each sample, we must write:

$$S - S^* = -\mathfrak{S}$$

and hence

$$S - S^* = k \int_0^\infty W \log \frac{g}{W} du - k \int_0^\infty W^* \log \frac{g}{W^*} du$$

It is easy to show in general that

$$S_1 - S_2 = k \int_0^\infty W_1 \log \frac{g}{W_1} du - k \int_0^\infty W_2 \log \frac{g}{W_2} du \qquad (29)$$

if for volume 1 equal to volume 2 the two distributions are made equal[25] by a combination of "free" and "normal" adiabatic expansion and heating at constant volume. The restriction of equal volume can be dropped if we use for g the normalized weight function G (29). Hence on purely thermodynamic grounds, it seems appropriate to associate with a distribution the entropy

$$S = k \int_0^\infty W \log \frac{G}{W} du \qquad (30)$$

These examples were intended to show how the second law must be handled to obtain the laws governing the fluctuation phenomena. It is obviously possible to construct statistical thermodynamics on the basis of the second law from purely thermodynamic considerations. It is not easy to proceed on this path,[26] and it seems more reasonable to us to follow a path indicated by Einstein, if we want to obtain the statistical laws of thermodynamics. However, here we have had another goal. We wanted

---

[24] See Note 14.

[25] For two distributions that cannot be transferred into one another in this way, it will be harder to prove formula (29) with the tools presented here.

[26] Since we will have the opportunity to go into this in more detail elsewhere, we just want to draw attention to a paper by Einstein, which has not received due attention (Verh. d. Deutsch. Phys. Ges. **12**, 820, 1914).

to show that the second law does not lose anything of its exact character even in view of fluctuations and does not deteriorate to an approximate law but evolves in some higher harmony thus determining the laws of the fluctuations as well.

The present paper is closely connected with my dissertation, which will not appear in print.[27] I want to use the occasion to thank Prof. v. Laue for his friendly support and encouragement.

Remarks

1) If the energy content of the sample from set 1) is u_1 and of the sample from set 2) is u_2, then the energy content of the two together is

$$u = u_1 + u_2$$

If we have for the quantity u_1 the probability law W_1 and for the quantity u_2 the probability law W_2, then for the sum u we have the probability law

$$W(u) = \int_0^u W_1(\xi) W_2(u - \xi) \, d\xi$$

provided u_1 and u_2 are statistically independent.

2) We write (see equation (19)):

$$\frac{1}{\displaystyle\int_0^\infty g(u) \, e^{-u/kT} \, du} = C(T) \tag{31}$$

then the following identity holds:

$$\frac{d}{dT} \log C(T) = -\frac{\bar{u}^*}{kT^2}$$

and by integration we obtain

$$C(T) = e^{\int_\tau^A \frac{\bar{u}^*(T)}{kT^2} dT},$$

where A is the constant of integration. For any choice of $g(u)$, $C(T)$ goes to infinity for T going to 0, hence the integral in the exponent goes to infinity. We see from this that no choice of g will give us the T^4 dependence for $\bar{u}^*$ (as must be the case for black body radiation in an enclosure), because in that case the above integral would not diverge for $T = 0$.

3) If we bring into contact in the described fashion two sets with the distributions W_1 and W_2 then, for the set created, each sample of which consists of two bodies, we have

$$W(u) = \int_0^u W_1(\xi) W_2(u - \xi) \, d\xi$$

[27] Dissertation, Berlin 1922.

Now we want to find the new distributions of the two sets after separation provided the contact was sufficiently long. If all the composite samples had the same energy content u, then the two distributions after separation would be given by $V_{12}(u_1, u)\, du_1$ and $V_{21}(u_2, u)\, du_2$. However, the composite samples do not have the same energy content. The probability for the energy of one such sample to lie between u and $u + du$ is governed by a probability law $W(u)\, du$. Hence the energy distribution of set 1 after separation will be given by the integral

$$W_1'(u_1) = \int_0^\infty V_{12}(u_1, u) W(u)\, du \tag{32}$$

For the new mean energy

$$\bar{u}_1' = \int_0^\infty u_1 W_1'(u_1)\, du_1$$

we find using equation (32) and the notation

$$\bar{u}_{12}(u) = \int_0^u u_1 V_{12}(u_1, u)\, du_1$$

the following

$$\bar{u}_1' = \int_0^u \bar{u}_{12}(u) W(u)\, du$$

and in our approximation (see equation (6))

$$\bar{u}_1' = \bar{u}_{12}(\bar{u}) + \left\{ \frac{d^2 \bar{u}_{12}}{du^2} \right\}_{u=\bar{u}} \cdot \frac{\eta}{2}$$

where η is given by

$$\eta = \int_0^\infty (u - \bar{u})^2 W(u)\, du$$

According to (3) we have:

$$\eta = \eta_1 + \eta_2$$

and in this way we obtain equation (8).

4) For the deviation of the new distribution

$$\eta_1' = \int_0^\infty (u_1 - \bar{u}_1')^2 W_1'(u_1)\, du_1$$

we obtain, using W_1' from (32) and the notation

$$\eta_{12} = \int_0^\infty [u_1 - \bar{u}_{12}(u)]^2 V_{12}(u_1, u)\, du_1$$

in our approximation (see equation (6))

$$\eta_1' = \eta_{12}(\bar{u}) + \left\{\frac{d\bar{u}_{12}}{du}\right\}_{u=\bar{u}}^2 \left(\frac{\eta_1 + \eta_2}{2}\right)$$

where we neglected terms involving products of deviations.

5) In (10) we substitute $\{d\bar{u}_{12}/du\}_{u=\bar{u}*}$ for

$$\left\{\frac{d\bar{u}_{12}}{du}\right\}_{u=\bar{u}*} = \frac{\dfrac{d\bar{u}_1^*}{dT}}{\dfrac{d\bar{u}_1^*}{dT} + \dfrac{d\bar{u}_2^*}{dT}}$$

We can do this in our approximation, since according to (9):

$$\bar{u}_{12}[\bar{u}^*(T)] \cong \bar{u}_1^*(T)$$

hence

$$\left\{\frac{d\bar{u}_{12}}{du}\right\}_{u=\bar{u}*(T)} = \frac{d\bar{u}_1^*}{dT}\frac{dT}{d\bar{u}^*} = \frac{\dfrac{d\bar{u}_1^*}{dT}}{\dfrac{d\bar{u}^*}{dT}}$$

and since

$$\bar{u}^*(T) = \bar{u}_1^*(T) + \bar{u}_2^*(T)$$

we have

$$\left\{\frac{d\bar{u}_{12}}{du}\right\}_{u=\bar{u}*} = \frac{\dfrac{d\bar{u}_1^*}{dT}}{\dfrac{d\bar{u}_1^*}{dT} + \dfrac{d\bar{u}_2^*}{dT}}$$

Quite analogously we find

$$\left\{\frac{d\bar{u}_{21}}{du}\right\}_{u=\bar{u}*} = \frac{\dfrac{d\bar{u}_2^*}{dT}}{\dfrac{d\bar{u}_1^*}{dT} + \dfrac{d\bar{u}_2^*}{dT}}$$

Now equation (10) represents a homogeneous system of equations for η_1^*, η_2^*, and $\eta_{12} = \eta_{21}$, and we find from this

$$\frac{\eta_1^*}{\eta_2^*} = \frac{\dfrac{d\bar{u}_1^*}{dT}}{\dfrac{d\bar{u}_2^*}{dT}}$$

6) Speaking exactly, we are not allowed to start from an exact volume v but should start from a state in which the piston can move freely a finite if very small interval. Because we should note that by acting on the piston with a constant force which balances the pressure on the average, the free piston will not remain at rest but will perform a sort of Brownian motion, and its position will fluctuate about the equilibrium position. If we want to limit the extent of the fluctuations, by approaching the piston from both sides with stops, then we have to do work. This process can be compared with the compression of a gas with only one molecule. The work expended will diverge when the interval of maximum fluctuations is reduced to zero. However, we can fix once and for all a small interval representing a volume Δ inside of which the piston can move freely up and down. The work that has to be expended to compress from a volume between v_1 and $v_1 + \Delta$ to a volume lying between v_2 and $v_2 + \Delta$ is then simply the work of compression for the gas in the cylindrical container, provided we choose Δ small enough.

7) The ordinate at v_1 (Fig. 3) is projected onto the ordinate at v by the set of adiabatics

$$u = u(u_1, v)$$

Given that the probability of finding a point on the left ordinate between u_1 and $u_1 + du_1$ is

$$W(u_1, v_1)\, du_1$$

then the probability for its projection to lie on the correct ordinate between u and $u + du$ is given by

$$W(u, v)\, du = W(u_1, v_1)\, du_1$$

Hence we have

$$W[u(u_1, v_1), v] \frac{\partial u(u_1, v)}{\partial u_1} = W(u_1, v_1)$$

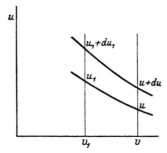

Fig. 3.

If we partially differentiate this equation with respect to v keeping u_1 and v_1 fixed, and evaluate at $v = v_1$, we obtain

$$\frac{\partial u(u_1, v)}{\partial v} = -p(u, v)$$

where we used the notation

$$\frac{\partial}{\partial u}(pW) - \frac{\partial}{\partial v}W = 0$$

8) We have

$$\bar{u}(v) = \int_0^\infty uW(u, v)\, du$$

and because of the necessary assumption about continuity:

$$\frac{d\bar{u}}{dv} = \int_0^\infty u\frac{\partial}{\partial v}W\, du$$

Substituting from the partial differential equation (14) $\partial W/\partial v$

$$\frac{d\bar{u}}{dv} = \int_0^\infty u\frac{\partial}{\partial u}[pW]\, du$$

we obtain from this, by partial integration, setting $(pW)_0^\infty = 0$

$$\frac{d\bar{u}}{dv} = -\int_0^\infty pW\, du$$

In similar fashion we obtain from the partial differential equation (14)

$$\eta = \int_0^\infty (u - \bar{u})^2 W(u)\, du$$

$$\frac{d\eta}{dv} = -2\int_0^\infty (u - \bar{u})pW\, du$$

9) We will first show that one arrives at an abnormal composite distribution if one brings into thermal contact each sample of a set with normal statistic $W_1^*(u)$ with one sample of a second set with abnormal energy statistic $W_2(u)$. The statistic of the composite set is given by

$$W(u) = \int_0^u W_1^*(n - \xi)W_2(\xi)\, d\xi$$

On the other hand, combining two normal sets in this way the statistics of the composite set will be normal, given by

$$W^*(u) = \int_0^u W_1^*(u - \xi)W_2^*(\xi)\, d\xi$$

Subtracting these two equations we get

$$W^*(u) - W(u) = \int_0^u W_1^*(u - \xi)\{W_2^*(\xi) - W_2(\xi)\}\, d\xi$$

If $W = W^*$ then we would have for all u

$$\int_0^u W_1^*(u - \xi)\{W_2^*(\xi) - W_2(\xi)\}\, du \equiv 0$$

which will be impossible if

$$W_2 \neq W_2^*$$

In order to show that thermal contact between a normal and an abnormal set will indeed affect the originally normal set we have yet to show that by splitting such an abnormal composite set we obtain again two abnormal sets. This is the case since upon separation we obtain the statistic (see equation (32))

$$W_1'(u_1) = \int_0^\infty \frac{g_1(u_1)g_2(u - u_1)}{g_{12}(u)} W(u)\, du$$

But since on the other hand

$$W_1^*(u_1) = \int_0^\infty \frac{g_1(u_1)g_2(u - u_1)}{g_{12}(u)} W^*(u)\, du$$

for the case that $W_1' = W_1^*$, we would have to have for all u_1

$$\int_0^\infty g_1(u_1)g_2(u - u_1)\frac{W^*(u) - W(u)}{g_{12}(u)}\, du \equiv 0$$

This is impossible if

$$W \neq W^*$$

and the functions under consideration vanish sufficiently fast for $u = \infty$.

10) For the class of bodies in question $[p = uf(v)]$ equation (16) yields

$$\frac{d\eta}{dv} = -2f(v)\int_0^\infty u(u - \bar{u})W(u)\, du$$

Since the following identity holds true

$$\int_0^\infty u(u - \bar{u})W(u)\, du = \int_0^\infty (u - \bar{u})^2 W(u)\, du = \eta$$

we can also write

$$\frac{d\eta}{dv} = -2\eta f(v)$$

and for the normal initial state we have to set

$$\eta = \eta^* = \psi(T)\frac{\partial \bar{u}^*}{\partial T}$$

For the decrease in deviation in the case of free expansion we therefore obtain

$$\frac{d\eta}{dv} = -2\psi(T)\frac{\partial \bar{u}^*}{dT}\, f(v)$$

11) If we put

$$\eta^* = \psi(T)\frac{\partial \bar{u}^*}{\partial T}$$

we obtain

$$\frac{\partial}{\partial v}\eta^* = \psi\frac{\partial^2 \bar{u}}{\partial v\,\partial T}$$

$$\frac{\partial}{\partial T}\eta^* = \frac{\partial \psi}{\partial T}\frac{\partial \bar{u}^*}{\partial T} + \psi\frac{\partial^2 \bar{u}^*}{\partial T^2}$$

For all bodies, the decrease in temperature for adiabatic expansion is given by

$$\frac{dT}{dv} = -\frac{T\dfrac{\partial \bar{p}^*}{\partial T}}{\dfrac{\partial \bar{u}^*}{\partial T}}$$

so that according to equation (19)

$$\frac{d\eta^*}{dv} = \psi\frac{\partial^2 \bar{u}^*}{\partial v\,\partial T} - \frac{T\dfrac{\partial \bar{p}^*}{\partial T}}{\dfrac{\partial \bar{u}^*}{\partial T}}\left(\frac{d\psi}{dT}\frac{\partial \bar{u}^*}{\partial T} + \psi\frac{\partial^2 \bar{u}^*}{\partial T^2}\right)$$

For the class of bodies in question, for which

$$p(u, v) = uf(v)$$

we have

$$\frac{\partial \bar{p}^*}{\partial T} = \frac{\partial \bar{u}^*}{\partial T}f(v)$$

and since, in general for these bodies (see equation (17)),

$$\frac{\partial^2 \bar{u}^*}{\partial v\,\partial T} = T\frac{\partial^2 \bar{p}^*}{\partial T^2}$$

we have in our case

$$\frac{\partial^2 \bar{u}^*}{\partial v\,\partial T} = T\frac{\partial^2 \bar{u}^*}{\partial T^2}f(v)$$

This will yield

$$\frac{d\eta^*}{dv} = -T\frac{d\psi}{dT}\frac{\partial \bar{u}^*}{\partial T}f(v)$$

12) In the text the class of bodies for which

$$p(u,v) = uf(v)$$

played a special role. Of course it is possible to determine the function $\varphi(T)$ without treating any class of bodies preferentially. Consider two bodies which are in thermal equilibrium with each other. Assume that the state of these bodies is given by the volumes v_1 and v_2 and the total energy u. We can now perform an adiabatic process in a reversible way by slowly varying v_1 and v_2, while maintaining thermal contact between the two bodies. In this way we can let v_2 follow a closed curve as a function of v_1 in a reversible way, and we may calculate how the total energy u varies along this curve.

If we start with a set of such composite bodies with normal energy statistic and let v_2 vary along a closed curve as a function of v_1, the mean energy of the set should not decrease if the second law is indeed true. If we let v_2 vary along a particular closed curve as a function of v_1, and if we denote the change in energy of one sample by $\Delta(u)$, then

$$\int_0^\infty \Delta(u)W^*(u)\,du \geqslant 0 \tag{33}$$

If we introduce for the adiabatic change at constant v_2 the notation

$$\frac{du}{dv_1} = -P_1(v_1, v_2, u)$$

and analogously for constant v_1

$$\frac{du}{dv_2} = -P_2(v_1, v_2, u)$$

then we have, depending on whether we go around clockwise or counterclockwise,

$$\frac{\Delta}{f} = \frac{dP_1}{dv_2} - \frac{dP_2}{dv_1}$$

$$\frac{\Delta}{f} = -\left(\frac{dP_1}{dv_2} - \frac{dP_2}{dv_1}\right)$$

provided we stay close enough to the initial point. Here f denotes the area enclosed, and writing dP_1/dv_2 in more detail

$$\frac{dP_1}{dv_2} = \frac{\partial P_1}{\partial v_2} + \frac{\partial P_1}{\partial u}\left\{\frac{du}{dv_2}\right\}_{v_1 = \text{const}}$$

Since (33) must hold for the clockwise as well as counterclockwise case, we have

$$\int_0^\infty \Delta(u)W^*(u)\,du = 0$$

or taking into consideration that

$$W^* = C(T)g(u)\,e^{\varphi(T)u}$$

we get

$$\int_0^\infty \Delta(u)g(u)\,e^{\varphi(T)u}\,du = 0$$

and since this must hold for all values of T and φ, we will have for all u

$$\Delta \equiv 0$$

and hence

$$\frac{dP_1}{dv_2} - \frac{dP_2}{dv_1} = 0 \qquad (34)$$

[We see that $du = P_1 dv_1 + P_2 dv_2$ is a complete differential, i.e., u must be a function of (v_1, v_2).]

Now according to (15)

$$P_1(u) = \int_0^\infty p_1(u_1)V_{12}(u_1, u)\,du_1$$

$$P_2(u) = \int_0^\infty p_2(u_2)V_{21}(u_2, u)\,du_2$$

Making use of the approximation on page 763 and the demand that P_1 and P_2 are coefficients in a complete differential (34), we obtain from a simple calculation[28] that ψ satisfies the differential equation

$$T\frac{d\psi}{dT} - 2\psi = 0$$

13) On the one hand the following identity holds true:

$$T\frac{\partial}{\partial T}\frac{\int_0^\infty pg\,e^{-u/kT}\,du}{\int_0^\infty g\,e^{-u/kT}\,du} = \frac{\int_0^\infty pg\,e^{-u/kT}\,du}{\int_0^\infty g\,e^{-u/kT}\,du} + kT^2\frac{\partial}{\partial T}\frac{\int_0^\infty \frac{\partial}{\partial u}[pg]\,e^{-u/kT}\,du}{\int_0^\infty g\,e^{-u/kT}\,du} \qquad (35)$$

[28] See L. Szilard, Dissertation Berlin 1922.

One can convince oneself of this by substituting on the right-hand side

$$\int_0^\infty \frac{\partial}{\partial u}[pg] \, e^{-u/kT} \, du = \frac{1}{kT} \int_0^\infty pg \, e^{-u/kT} \, du$$

(the legitimacy of this can be seen by partial integration, taking into account that $[pg \, e^{-u/kT}]_0^\infty = 0$) if we differentiate with respect to T.

On the other hand we have the identity

$$\frac{\partial}{\partial v} \frac{\displaystyle\int_0^\infty ug \, e^{-u/kT} \, du}{\displaystyle\int_0^\infty g \, e^{-u/kT} \, du} = kT^2 \frac{\partial}{\partial T} \frac{\displaystyle\int_0^\infty \frac{\partial g}{\partial v} e^{-u/kT} \, du}{\displaystyle\int_0^\infty g \, e^{-u/kT} \, du} \qquad (36)$$

In the spirit of these identities we substitute the expressions

$$T\frac{\partial}{\partial T} \frac{\displaystyle\int_0^\infty pg \, e^{-u/kT} \, du}{\displaystyle\int_0^\infty g \, e^{-u/kT} \, du} \qquad \text{and} \qquad \frac{\partial}{\partial v} \frac{\displaystyle\int_0^\infty ug \, e^{-u/kT} \, du}{\displaystyle\int_0^\infty g \, e^{-u/kT} \, du}$$

into equation (24)

$$\frac{\displaystyle\int_0^\infty pg \, e^{-u/kT} \, du}{\displaystyle\int_0^\infty g \, e^{-u/kT} \, du} + \frac{\partial}{\partial v} \frac{\displaystyle\int_0^\infty ug \, e^{-u/kT} \, du}{\displaystyle\int_0^\infty g \, e^{-u/kT} \, du} = T\frac{\partial}{\partial T} \frac{\displaystyle\int_0^\infty pg \, e^{-u/kT} \, du}{\displaystyle\int_0^\infty g \, e^{-u/kT} \, du}$$

and obtain

$$\frac{\partial}{\partial T} \frac{\displaystyle\int_0^\infty \left\{ \frac{\partial}{\partial u}[pg] - \frac{\partial}{\partial v}g \right\} e^{-u/kT} \, du}{\displaystyle\int_0^\infty g \, e^{-u/kT} \, du} = 0$$

14) Introducing the notation

$$\int_0^\infty W \log \frac{g}{W} \, du = J$$

and

$$\int_0^\infty W^* \log \frac{g}{W^*} \, du = J^*$$

we obtain

$$\frac{d}{dv}\left(k \int_0^\infty W^* \log \frac{g}{W^*} \, du - k \int_0^\infty W \log \frac{g}{W} \, du \right) = k\frac{dJ^*}{dv} - k\frac{dJ}{dv}$$

Now we have

$$\frac{dJ}{dv} = \int_0^\infty \frac{\partial W}{\partial v} \log g \, du + \int_0^\infty \frac{W}{g} \frac{\partial g}{\partial v} du - \int_0^\infty \frac{\partial W}{\partial v} \log W \, du - \int_0^\infty \frac{\partial W}{\partial v} du$$

and by substituting $\partial W/\partial v$ from the differential equation (14)

$$\frac{\partial W}{\partial v} = \frac{\partial}{\partial u}[pW]$$

and for $\partial g/\partial v$ from the differential equation for thermodynamic bodies (see equation (25))

$$\frac{\partial g}{\partial v} = \frac{\partial}{\partial u}[pg] + Kg$$

and after partial integration we find that

$$\frac{dJ}{dv} = +K \tag{37}$$

We have also

$$\frac{dJ^*}{dv} = \frac{\partial}{\partial v}J^* + \frac{dT}{dv}\frac{\partial}{\partial T}J^* \tag{38}$$

and we find for J^* by putting

$$W^* = C(T)g(u)\, e^{-u/kT}$$

and integrating

$$J^* = -\log C(T) + \frac{\bar{u}^*}{kT}$$

where

$$C = \frac{1}{\displaystyle\int_0^\infty g\, e^{-u/kT}\, du}$$

Hence we have

$$\frac{\partial}{\partial v}J^* = -\frac{\dfrac{\partial C}{\partial v}}{C} + \frac{\dfrac{\partial \bar{u}^*}{\partial v}}{kT}$$

For the first term on the right-hand side, we find

$$-\frac{\dfrac{\partial C}{\partial v}}{C} = C\int_0^\infty \frac{\partial g}{\partial v}\, e^{-u/kT}\, du$$

and if we substitute for $\partial g / \partial v$ from the equation

$$\frac{\partial g}{\partial v} = \frac{\partial}{\partial u}[pg] + Kg$$

and then integrate partially

$$-\frac{\dfrac{\partial C}{\partial v}}{C} = \frac{\bar{p}^*}{kT} + K \qquad (39)$$

(where $\bar{p}^*$ is the abbreviation for the integral $\int_0^\infty pCg\, e^{-u/kT}\, du$) we obtain

$$\frac{\partial}{\partial v}J^* = \frac{\bar{p}^*}{kT} + K + \frac{\dfrac{\partial \bar{u}^*}{\partial v}}{kT} \qquad (39)$$

We have further

$$\frac{\partial}{\partial T}J^* = -\frac{\dfrac{\partial C}{\partial T}}{C} - \frac{\bar{u}^*}{kT^2} + \frac{\dfrac{\partial \bar{u}^*}{\partial T}}{kT}$$

and since the identity

$$\frac{\dfrac{\partial C}{\partial T}}{C} = -\frac{\bar{u}^*}{kT^2}$$

holds true, we get

$$\frac{\partial}{\partial T}J^* = \frac{\dfrac{\partial \bar{u}^*}{\partial T}}{kT} \qquad (40)$$

Finally we get for dT/dv from the condition

$$d\bar{u} = \frac{\partial \bar{u}^*}{\partial v}\, dv + \frac{\partial \bar{u}^*}{\partial T}\, dT$$

the result

$$\frac{dT}{dv} = \frac{\dfrac{d\bar{u}}{dv} - \dfrac{\partial \bar{u}^*}{\partial v}}{\dfrac{\partial \bar{u}^*}{\partial T}}$$

or, by writing here for

$$\frac{d\bar{u}}{dv} = -\int_0^\infty p(u)\, W(u)\, du = -\bar{p}$$

we obtain

$$\frac{dT}{dv} = \frac{-\bar{p} - \dfrac{\partial \bar{u}^*}{\partial v}}{\dfrac{\partial \bar{u}^*}{\partial T}} \tag{41}$$

With this we obtain

$$\frac{dT}{dv} \cdot \frac{\partial}{\partial T} J^* = -\frac{\bar{p}}{kT} - \frac{\dfrac{\partial \bar{u}^*}{\partial v}}{kT} \tag{42}$$

and according to (39) and (42)

$$\frac{dJ^*}{dv} = \frac{\bar{p}^* - \bar{p}}{kT} + K \tag{43}$$

and hence

$$\frac{d}{dv}\left(k \int_0^\infty W^* \log \frac{g}{W^*}\, du - k \int_0^\infty W \log \frac{g}{W}\, du \right) = k\frac{dJ^*}{dv} - k\frac{dJ}{dv} = \frac{\bar{p}^* - \bar{p}}{T}$$

840

Über die Entropieverminderung in einem thermodynamischen System bei Eingriffen intelligenter Wesen.

Von **L. Szilard** in Berlin.

Mit 1 Abbildung. (Eingegangen am 18. Januar 1928.)

Es wird untersucht, durch welche Umstände es bedingt ist, daß man scheinbar ein Perpetuum mobile zweiter Art konstruieren kann, wenn man ein Intellekt besitzendes Wesen Eingriffe an einem thermodynamischen System vornehmen läßt. Indem solche Wesen Messungen vornehmen, erzeugen sie ein Verhalten des Systems, welches es deutlich von einem sich selbst überlassenen mechanischen System unterscheidet. Wir zeigen, daß bereits eine Art Erinnerungsvermögen, welches ein System, in dem sich Messungen ereignen, auszeichnet, Anlaß zu einer dauernden Entropieverminderung bieten kann und so zu einem Verstoß gegen den zweiten Hauptsatz führen würde, wenn nicht die Messungen selbst ihrerseits notwendig unter Entropieerzeugung vor sich gehen würden. Zunächst wird ganz universell diese Entropieerzeugung aus der Forderung errechnet, daß sie im Sinne des zweiten Hauptsatzes eine volle Kompensation darstellt [Gleichung (1)]. Es wird dann auch an Hand einer unbelebten Vorrichtung, die aber (unter dauernder Entropieerzeugung) in der Lage ist, Messungen vorzunehmen, die entstehende Entropiemenge berechnet und gefunden, daß sie gerade so groß ist, wie es für die volle Kompensation notwendig ist: die wirkliche Entropieerzeugung bei der Messung braucht also nicht größer zu sein, als es Gleichung (1) verlangt.

Es gibt einen schon historisch gewordenen Einwand gegen die allgemeine Gültigkeit des zweiten Hauptsatzes der Thermodynamik, welcher in der Tat einen recht bedrohlichen Eindruck macht. Es ist dies der Einwand des Maxwellschen Dämons, der in verschiedener Umkleidung auch heute noch immer wieder auftaucht, und vielleicht nicht ganz mit Unrecht insofern, als hinter der präzis gestellten Frage sich quantitative Zusammenhänge zu verbergen scheinen, die bisher nicht aufgeklärt worden sind. Den Einwand in seiner ursprünglichen Formulierung, die mit einem Dämon operiert, welcher die raschen Moleküle abfängt und die langsamen passieren läßt, kann man allerdings mit der Entgegnung abtun, daß wir Menschen den Wert der thermisch schwankenden Parameter ja prinzipiell nicht jeweils erraten können; aber es läßt sich nicht leugnen, daß wir den Wert eines solchen schwankenden Parameters sehr wohl messen könnten und dann sicherlich Arbeit auf Kosten der Wärme gewinnen könnten, indem wir unsere Eingriffe dann je nach dem Resultat der Messung passend einrichten. Freilich bleibt es zunächst dahingestellt, ob wir nicht einen Fehler begehen, wenn wir den eingreifenden Menschen selbst nicht mit zum System rechnen und seine Lebensvorgänge nicht mitberücksichtigen.

L. Szilard, Über die Entropieverminderung usw.   841

Abgesehen von diesem ungeklärten Punkt ist heute erkannt, daß in einem sich selbst überlassenen System trotz der Schwankungserscheinungen kein Perpetuum mobile zweiter Art, d. h. genauer, „keine automatische Wärme niedrigster Temperatur verbrauchende Maschine von fortdauernder endlicher Arbeitsleistung" wirksam sein kann. Es wäre dies eine Maschine, deren Wirksamkeit, auch wenn man lange Zeiten ins Auge faßt, das Resultat hätte, daß ein Gewicht auf Kosten des Wärmeinhalts eines Wärmereservoirs gehoben wird. Etwas anders ausgedrückt kann man auch sagen, daß, wenn wir die Schwankungserscheinungen benutzen wollen, um Arbeit auf Kosten der Wärme zu gewinnen, wir in derselben Lage sind wie bei einem Glücksspiel, bei welchem wir ab und zu gewisse Beträge gewinnen können, bei dem aber der Erwartungswert (mathematische Hoffnung) des Gewinns null oder negativ ist. Dasselbe gilt für ein System, das zwar nicht ganz sich selbst überlassen ist, bei dem aber die Eingriffe von außen her streng periodisch erfolgen, etwa durch periodisch sich bewegende Maschinen. Diese Erkenntnis betrachten wir als gesichert* und wollen uns hier nur mit den Schwierigkeiten befassen, die auftreten, wenn etwa intelligente Wesen Eingriffe im System vornehmen und die dabei auftretenden quantitativen Zusammenhänge zu erkennen trachten.

Smoluchowski schreibt**: „Soweit unsere jetzigen Kenntnisse reichen, gibt es also trotz molekularer Schwankungen kein automatisches, dauernd wirkendes Perpetuum mobile, aber wohl könnte eine solche Vorrichtung regelmäßig funktionieren, falls sie durch intelligente Wesen in passender Weise betätigt würde. ...

Ein Perpetuum mobile ist also möglich, falls man nach der üblichen Methode der Physik den experimentierenden Menschen als eine Art »Deus ex machina« auffaßt, welcher von dem momentanen Zustand der Natur fortwährend genau unterrichtet ist und die makroskopischen Naturvorgänge in beliebigen Momenten ohne Arbeitsleistung in Gang setzen oder unterbrechen kann. Er brauchte somit durchaus nicht wie ein Maxwellscher Dämon die Fähigkeit zu besitzen, einzelne Moleküle abzufangen, würde sich aber doch schon in obigen Punkten durchaus von wirklichen Lebewesen unterscheiden. Denn die Hervorbringung irgend eines physikalischen Effekts durch Betätigung des sensorischen wie auch des motorischen Nervensystems derselben ist immer mit einer Energieentwertung verbunden, abgesehen davon, daß ihre ganze Existenz an fortwährende Dissipation derselben gebunden ist.

---

* Vgl. z. B. L. Szilard, ZS. f. Phys. **32**, 753, 1925.
** Vorträge über die kinetische Theorie der Materie u. Elektrizität, S. 89. Leipzig 1914.

842   L. Szilard,

Ob also bei Berücksichtigung dieser Umstände wirkliche Lebewesen
dauernd, oder wenigstens in regelmäßiger Weise Arbeit auf Kosten der
Wärme niederster Temperatur erzeugen könnten, erscheint wohl recht
zweifelhaft, wiewohl unsere Unkenntnis der Lebensvorgänge eine definitive
Antwort ausschließt. Doch führen die zuletzt berührten Fragen schon
über den Rahmen der eigentlichen Physik hinaus …"

Es scheint nun, daß uns die Unkenntnis der Lebensvorgänge nicht
zu stören braucht, um das zu erkennen, auf was es unserer Ansicht nach
hier ankommt. Denn wir dürfen ja sicher sein, daß man die intelligenten
Wesen, sofern es sich um ihre Eingriffe in ein thermodynamisches System
handelt, durch unbelebte Vorrichtungen ersetzen kann, deren „Lebensvorgänge" sich verfolgen ließen, so daß man feststellen könnte, ob dabei
in der Tat eine Kompensation für die Entropieverminderung auftritt,
welche durch die Eingriffe der Vorrichtung im System hervorgerufen wird.

Wir wollen zunächst zu erkennen trachten, durch welchen Umstand
bei dem Eingreifen intelligenter Wesen in ein thermodynamisches System
die in diesem hervorgebrachte Entropieverminderung bedingt wird, und
werden dabei sehen, daß es auf Kopplungen besonderer Art zwischen verschiedenen Parametern des Systems ankommt. Wir wollen eine besonders
einfache Art dieser bedrohlichen Kopplungen betrachten. Wir werden
kurz von einer „Messung" reden, wenn es uns gelingt, den Wert eines
Parameters y (z. B. der Lagenkoordinate des Zeigers eines Meßinstrumentes)
in einem Augenblick mit dem Momentanwert eines schwankenden Parameters x des Systems zu koppeln, so daß man aus dem Werte von y
Rückschlüsse auf den Wert, den x zum Zeitpunkt der „Messung" gehabt
hat, ziehen kann. Dabei seien x und y nach der Messung wieder ent-
koppelt, so daß sich x verändern kann, während y noch seinen Wert eine
Zeitlang beibehält. Solche Messungen sind keine harmlosen Eingriffe;
ein System, in welchem solche Messungen vorkommen, weist ja eine Art
Erinnerungsvermögen auf, in dem Sinne, daß man an dem Zustandspara-
meter y erkennen kann, was für einen Wert ein anderer Zustandspara-
meter x zu einem früheren Zeitpunkt gehabt hat, und wir werden sehen,
daß schon vermöge einer solchen Erinnerung der zweite Hauptsatz verletzt
wäre, wenn der Vorgang der Messung sich ohne Kompensation abspielen
würde. Daß der zweite Hauptsatz durch diese Entropieverminderungen
nicht so stark bedroht ist, wie man es ursprünglich meinen könnte, werden
wir dann sehen, wenn wir zunächst erkennen, daß die durch die Eingriffe
bedingte Entropieverminderung jedenfalls schon vollständig kompensiert
wäre, falls die Vornahme einer solchen Messung ganz universell beispiels-

Über die Entropieverminderung in einem thermodynamischen System usw. 843

weise jedesmal mit einer Entropieerzeugung von dem Betrage $k.\log 2$ einherginge. Es wird dann gelingen, ein etwas allgemeineres Entropiegesetz aufzufinden, welches sich ganz universell auf alle Messungen bezieht. Zum Schluß wird dann eine sehr einfache (natürlich unbelebte) Vorrichtung betrachtet, welche in der Lage ist, dauernd „Messungen" vorzunehmen, und deren „Lebensvorgänge" wir leicht verfolgen können. Durch direktes Ausrechnen findet man dann in der Tat eine dauernde Entropieerzeugung von dem Betrage, wie ihn das erwähnte, allgemeinere, aus der Gültigkeit des zweiten Hauptsatzes hergeleitete Entropiegesetz fordert.

Das erste Beispiel, welches wir als typisch jetzt näher betrachten wollen, ist nun das folgende: Ein etwa stehender Hohlzylinder, der nach unten und oben durch einen Boden, bzw. einen Deckel abgesperrt ist, läßt sich in zwei im allgemeinen ungleich große Hälften vom Volumen V_1 bzw. V_2 teilen dadurch, daß man in irgend einer festgesetzten Höhenlage von der Seite her eine Zwischenwand einschiebt. Diese Zwischenwand bildet einen Stempel, der sich im Zylinder nach oben und unten verschieben läßt. Ein unendlich großes Wärmereservoir von irgend einer Temperatur T sorgt dafür, daß ein etwa im Zylinder vorhandenes Gas bei der Bewegung des Stempels eine isotherme Expansion durchmacht. Dieses Gas soll nun aus einem einzigen Molekül bestehen, welches, solange der Stempel nicht in den Zylinder eingeschoben wird, sich vermöge seiner thermischen Bewegung in dem ganzen Zylinder herumtummelt.

Wir denken uns nun etwa einen Menschen, der zu irgend einem Zeitpunkt den Stempel in den Zylinder hineinschiebt und gleichzeitig irgendwie feststellt, ob nun das Molekül in der oberen oder unteren Hälfte des Zylinders, im Volumen V_1 oder V_2, gefangen ist. Findet er dann etwa, daß das erstere der Fall ist, so wird er den Stempel langsam nach unten schieben, bis der Boden des Zylinders erreicht ist. Während dieser langsamen Bewegung des Stempels bleibt das Molekül selbstverständlich immer oberhalb des Stempels, nicht aber in der oberen Hälfte des Zylinders, prallt vielmehr viele Male gegen den bereits in der unteren Hälfte des Zylinders sich bewegenden Stempel. Das Molekül überträgt auf diese Weise eine gewisse Menge Arbeit auf den Stempel. Es ist dies die Arbeit, die einer isothermen Expansion eines aus einem einzigen Molekül bestehenden idealen Gases vom Volumen V_1 auf das Volumen $V_1 + V_2$ entspricht. Nach einiger Zeit, wenn der Stempel den Boden des Gefäßes erreicht hat, steht dem Molekül wieder das ganze Volumen $V_1 + V_2$ zur Verfügung, und der Stempel wird alsdann herausgezogen. Der Vorgang kann beliebig oft wiederholt werden, wobei der eingreifende

56*

Mensch den Stempel, je nachdem das Molekül in der oberen oder unteren
Hälfte des Stempels gefangen ist, eben nach unten oder nach oben verschiebt, d. h. genauer, etwa mit einem Gewicht, das gehoben werden soll,
durch eine mechanische Kraftübertragung jeweils so koppelt, daß das
Gewicht durch den Stempel jeweils nach oben verschoben wird. Auf
diese Weise wächst die potentielle Energie des Gewichts sicher dauernd.
(Die Kraftübersetzung auf das Gewicht wird zweckmäßig so gestaltet,
daß die von dem Gewicht auf den Stempel übertragene Kraft bei jeder
Stellung des Stempels dem mittleren Druck des Gases gerade das Gleichgewicht hält.) Es ist klar, daß auf solche Weise dauernd Arbeit auf
Kosten der Wärme gewonnen wird, sofern die Lebensvorgänge des eingreifenden Menschen nicht mit in Rechnung gestellt werden.

Um zu erkennen, was der Mensch hier eigentlich für das System
leistet, stellt man sich am besten vor, daß die Bewegung des Stempels
maschinell erfolgt und daß die ganze Tätigkeit des Menschen sich darin
erschöpft, daß er die Höhenlage des Moleküls feststellt und daß er einen
Hebel, der die Bewegung des Stempels steuert, nach rechts oder nach
links legt, je nachdem, ob die festgestellte Höhenlage eine Abwärts- oder
eine Aufwärtsbewegung des Stempels erfordert. Das heißt, der Eingriff
des Menschen besteht lediglich in der Kopplung von zwei Lagenkoordi-
naten, nämlich einer Koordinate x, welche die Höhenlage des Moleküls
festlegt, mit einer anderen Koordinate y, welche die Lage des Hebels
definiert und welche also dafür bestimmend ist, ob dem Stempel eine
aufwärts oder abwärts gerichtete Geschwindigkeit erteilt wird. Die
Masse des Stempels denkt man sich dabei am besten groß und die er-
teilte Geschwindigkeit hinreichend hoch, so daß die thermische Agitation,
die der Stempel bei der betreffenden Temperatur hat, daneben zu ver-
nachlässigen ist.

Wir wollen bei den Vorgängen des hier betrachteten typischen Bei-
spiels zwei Zeitabschnitte unterscheiden, und zwar:

1. den Zeitabschnitt der Messung, bei welchem der Stempel bereits
in die Zylindermitte eingeschoben und das Molekül entweder in der oberen
oder unteren Zylinderhälfte eingeschlossen ist, so daß, wenn wir den
Koordinatenanfangspunkt passend wählen, seine x-Koordinate auf das
Intervall $x > 0$ oder $x < 0$ beschränkt ist;

2. den Zeitabschnitt der Ausnutzung der Messung, währenddessen die
Abwärts- oder Aufwärtsbewegung des Stempels erfolgt, den „Zeitabschnitt
der Entropieverminderung". Während dieses Zeitabschnitts wird die x-Ko-
ordinate des Moleküls keineswegs auf das ursprüngliche Intervall $x > 0$

Über die Entropieverminderung in einem thermodynamischen System usw.   845

bzw. $x < 0$ beschränkt bleiben.   Vielmehr muß, wenn das Molekül im Zeitabschnitt der Messung sich in der oberen Hälfte des Zylinders befunden hat, so daß $x > 0$ war, das Molekül auf den sich jetzt entsprechend in der unteren Hälfte des Zylinders nach abwärts bewegenden Stempel aufprallen, wenn es auf diesen Arbeit übertragen soll.   Das heißt, es muß die x-Koordinate des Moleküls in das Intervall $x < 0$ eintreten. Dagegen wird der Hebel seine der Abwärtsbewegung entsprechende Rechtsstellung während des ganzen Zeitabschnitts beibehalten.   Ist diese Rechtslage des Hebels etwa durch $y = 1$ festgelegt (und entsprechend die Linkslage $y = -1$), so sehen wir, daß während des Zeitabschnitts der Messung der Lage $x > 0$ $y = 1$ zugeordnet wird, daß aber nachher $y = 1$ bleibt, obschon x in das andere Intervall $x < 0$ eintritt.   Man sieht, daß bei der Ausnutzung der Messung die Kopplung der beiden Parameter x und y wieder verloren geht.

Wir wollen ganz allgemein davon reden, daß ein Parameter y den Wert eines etwa nach einem Wahrscheinlichkeitsgesetz schwankenden Parameters x „mißt", wenn der Wert von y sich danach richtet, welchen Wert der Parameter x zu einem bestimmten Zeitpunkt annimmt.   Die Vornahme einer Messung liegt der Entropieverminderung bei den Eingriffen intelligenter Wesen zugrunde.

Es ist nun naheliegend, anzunehmen, daß die Vornahme einer Messung prinzipiell mit einer ganz bestimmten mittleren Entropieerzeugung verbunden ist, und daß dadurch der Einklang mit dem zweiten Hauptsatz wieder hergestellt wird; größer dürfte die bei der Messung entstehende Entropiemenge freilich immer sein, nicht aber kleiner.   Genauer gesprochen, müssen hier zwei Entropiegrößen unterschieden werden, die eine, $\bar{S}_1$, wird erzeugt, wenn bei dem Meßprozeß y den Wert 1 annimmt, und die andere, $\bar{S}_2$, wenn y den Wert -1 erhält.   Wir können nicht erwarten, ganz allgemein über $\bar{S}_1$ oder $\bar{S}_2$ allein etwas zu erfahren, wohl aber werden wir sehen, daß aus der Annahme, daß diese bei der „Messung" erzeugte Entropiemenge die durch die Ausnutzung erzielte Entropieverminderung des Systems im Sinne des zweiten Hauptsatzes kompensieren muß, ganz allgemein die Beziehung

$$e^{-\frac{\bar{S}_1}{k}} + e^{-\frac{\bar{S}_2}{k}} \leqq 1 \qquad (1)$$

folgt.

Man sieht aus dieser Formel, daß wir wohl eine der Größen, also etwa $\bar{S}_1$, beliebig klein machen können, daß aber dann dafür die andere Größe, $\bar{S}_2$, entsprechend groß wird.   Es kann ferner auffallen, daß es gar

846   L. Szilard,

nicht auf die Größe des betrachteten Intervalls ankommt. Immerhin kann man leicht einsehen, daß es auch gar nicht anders sein kann.

Wir können auch umgekehrt sagen, sobald die bei der Messung entstehenden Entropien S_1 und S_2 der Ungleichung (1) genügen, dürfen wir beruhigt sein, daß dadurch die nachher bei der Messung erzielbare Entropieverminderung schon voll kompensiert wird.

Bevor wir nun auf den Beweis der Ungleichung (1) eingehen, können wir an dem bisher betrachteten mechanischen Beispiel zusehen, wie sich dies alles zusammenreimt. Für die bei der Messung erzeugten Entropien $\bar{S}_1$ und $\bar{S}_2$ machen wir als Spezialfall den Ansatz:

$$\bar{S}_1 = \bar{S}_2 = k \log 2.$$

Dieser Ansatz gehorcht der Ungleichung (1), und der mittlere Wert der bei einer Messung erzeugten Entropiemenge ist (in diesem Spezialfall natürlich unabhängig von dem Verhältnis der Häufigkeit w_1, w_2 der beiden Ereignisse):

$$\bar{S} = k \log 2.$$

Bei dem betrachteten Beispiel erzielt man bei der isothermen Expansion Entropieverminderung*:

$$-s_1 = -k \log \frac{V_1}{V_1 + V_2}; \quad -\bar{s}_2 = -k \log \frac{V_2}{V_1 + V_2},$$

je nachdem, ob beim Einschieben des Stempels das Molekül im Volumen V_1 oder V_2 angetroffen wurde. (Die Entropieverminderung ist gleich dem Quotienten der Wärmemenge, die bei der isothermen Expansion dem Wärmereservoir entzogen wird, und der Temperatur des betreffenden Wärmereservoirs.) Da nun im vorliegenden Falle die Häufigkeit w_1, w_2 der beiden Ereignisse sich verhält wie die Volumina V_1, V_2, so ist der mittlere Betrag der erzeugten Entropie (negative Zahl)

$$\bar{s} = w_1 (+\bar{s}_1) + w_2 \cdot (+\bar{s}_2) = \frac{V_1}{V_1 + V_2} k \log \frac{V_1}{V_1 + V_2} + \frac{V_2}{V_1 + V_2} k \log \frac{V_2}{V_1 + V_2}.$$

Wie man sieht, ist nun in der Tat

$$\frac{V_1}{V_1 + V_2} k \log \frac{V_1}{V_1 + V_2} + \frac{V_2}{V_1 + V_2} k \log \frac{V_2}{V_1 + V_2} + k \log 2 \gtreqless 0$$

und also

$$\bar{S} + \bar{s} \gtreqless 0.$$

---

* Die erzeugte Entropie ist mit $\bar{s}_1$, $\bar{s}_2$ bezeichnet.

Über die Entropieverminderung in einem thermodynamischen System usw.   847

Es wäre also. tatsächlich im vorliegenden Spezialfall eine volle Kompensation für die bei der Ausnutzung der Messung erzielbare Entropieverminderung vorhanden.

Wir wollen nun darauf verzichten, weiter spezielle Beispiele zu
betrachten, und gleich mit Hilfe einer allgemeinen Betrachtung über die
hier obwaltenden Verhältnisse Klarheit zu schaffen und Formel (1) ab-
zuleiten versuchen. Wir wollen hierzu das ganze System, in welchem
die Koordinate x irgendwelchen thermischen Schwankungen unterworfen
ist und durch y in der soeben erörterten Weise gemessen werden kann,
in sehr vielen Exemplaren vorgegeben denken, die alle in einem gemeinsamen Kasten eingeschlossen sind. Jedes einzelne dieser Exemplare sei
frei beweglich, so daß die einzelnen Systeme als die Moleküle eines
idealen Gases betrachtet werden können, die infolge der thermischen
Agitation in dem gemeinsamen Kasten voneinander unabhängig umher-
irren und auf die Wände des Kastens einen bestimmten, der Temperatur
der fortschreitenden Bewegung entsprechenden Druck ausüben. Wir
werden nun zwei dieser Moleküle als chemisch verschieden und durch
semipermeable Wände prinzipiell trennbar ansehen, falls die x-Koordinate
für das eine Molekül sich in einem hervorgehobenen Intervall befindet
und für das andere Molekül außerhalb des Intervalls. Ebenso werden
wir sie als chemisch verschieden ansehen, wenn sie sich nur darin unterscheiden, daß der Wert der y-Koordinate für das eine 1 ist und für das
andere — 1 beträgt.

Dem Kasten, in welchem diese „Moleküle" eingeschlossen sind, wollen
wir nun die Form eines Hohlzylinders geben, in welchem vier Stempel
angebracht sind (siehe Fig. 1). A und A' sind fest, die
beiden anderen B und B' beweglich, und zwar so, daß
der Abstand BB' konstant gleich dem Abstand AA'
gehalten wird, wie dies in der Figur durch die beiden
Klammern angedeutet ist. A', der Boden, und B, der
Deckel des ganzen Gefäßes, sind für alle „Moleküle"
undurchdringlich, dagegen A und B' semipermeabel,
und zwar A nur permeabel für jene „Moleküle", für

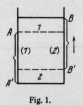

Fig. 1.

welche der Parameter x im hervorgehobenen Intervall $x_1 x_2$ liegt, B' nur
permeabel für die übrigen.

Anfänglich liegt der Stempel B bei A und also B' bei A', und alle
„Moleküle" befinden sich in dem Zwischenraum. Bei einem bestimmten
Bruchteil der Moleküle befindet sich x in dem hervorgehobenen Intervall.
Die Wahrscheinlichkeit dafür, daß dies bei einem herausgegriffenen

848   L. Szilard,

Molekül der Fall ist, sei mit w_1 bezeichnet, dafür, daß x außerhalb des
Intervalls fällt, mit w_2. Es gilt dann:

$$w_1 + w_2 = 1.$$

Die Verteilung der y-Parameter möge auf die beiden Werte 1 und −1
in irgend einem Verhältnis, aber jedenfalls von den x-Werten unabhängig
sein. Wir denken uns nun etwa durch ein intelligentes Wesen einen
Eingriff vorgenommen, durch welchen y der Wert 1 erteilt wird in allen
„Molekülen", für welche in dem betreffenden Zeitpunkt x in das hervorgehobene Intervall fällt, und entsprechend im umgekehrten Falle der
Wert − 1. Tritt dann infolge der thermischen Schwankung bei irgend
einem „Molekül" der Parameter x aus dem hervorgehobenen Intervall
heraus oder, wie wir uns hier auch ausdrücken können, erleidet das
„Molekül" bezüglich x eine monomolekulare chemische Reaktion (durch
welche es aus der Modifikation, welche durch den semipermeablen
Stempel A durchgelassen wird, in eine Modifikation überführt wird, für
welche dieser Stempel undurchlässig ist), so behält der Parameter y zunächst seinen Wert 1 unverändert bei, so daß sich das „Molekül" vermöge
des Parameters y während des ganzen späteren Prozesses daran „er-
innert", daß x ursprünglich in das hervorgehobene Intervall fiel. Wir
werden gleich sehen, welche Rolle diese Erinnerung spielen kann. Nach
dem soeben besprochenen Eingriff verschieben wir nun die Stempel, so
daß wir ohne Arbeitsleistung die beiden Molekülsorten voneinander
trennen. Es liegen dann zwei Gefäße vor, von denen das eine nur die
eine Modifikation, das andere nur die andere Modifikation enthält. Jede
Modifikation allein nimmt jetzt dasselbe Volumen ein wie früher das
Gemisch. In einem dieser Gefäße für sich betrachtet, herrscht jetzt bezüglich der beiden „Modifikationen in x" nicht die Gleichgewichtsverteilung. Das Mengenverhältnis der beiden Modifikationen ist natürlich $w_1 : w_2$ geblieben. Lassen wir bei konstant gehaltenem Volumen und
Temperatur in beiden Gefäßen für sich diese Gleichgewichtsverteilung
entstehen, so hat dabei die Entropie des Systems bestimmt zugenommen.
Denn die Gesamtwärmetönung ist dabei 0, da doch das Verhältnis der
beiden „Modifikationen in x" $w_1 : w_2$ sich dabei nicht ändert. Wenn wir
die Einstellung der Gleichgewichtsverteilung in beiden Gefäßen für sich
auf reversiblem Wege leiten, so wird dabei die Entropie der übrigen
Welt um denselben Betrag abnehmen. Die Entropie nimmt also um eine
negative Größe zu, und zwar ist der Wert der Entropiezunahme pro
Molekül:

$$\bar{s} = k\,(w_1 \log w_1 + w_2 \log w_2). \tag{2}$$

Über die Entropieverminderung in einem thermodynamischen System usw.   849

(Die Entropiekonstanten, die wir beiden „Modifikationen in x" zuzuordnen haben, kommen hier nicht explizite vor, weil der Prozeß die Gesamtanzahl der Moleküle, die der einen oder der anderen Modifikation angehören, ungeändert läßt.)

Wir können freilich jetzt durch einfache Rückverschiebung der Stempel die beiden Gase nicht mehr ohne weiteres ohne Arbeitsleistung wieder auf das ursprüngliche Volumen bringen. Denn es sind jetzt etwa in dem Gefäß, welches durch die Stempel BB' begrenzt ist, auch solche Moleküle da, deren x-Koordinate außerhalb des hervorgehobenen Intervalls fällt und für die infolgedessen der Stempel A nicht mehr permeabel ist. So sieht man, daß die gefundene Entropieverminderung (2) nicht einen Widerspruch zum zweiten Hauptsatz bedeutet. Solange wir nicht die Tatsache benutzen, daß die Moleküle in dem Gefäß BB' vermöge der Koordinate y sich noch daran „erinnern", daß die x-Koordinate für die Moleküle dieses Gefäßes ursprünglich im hervorgehobenen Intervall lag, solange ist auch für die errechnete Entropieverminderung bestimmt die volle Kompensation da in der Tatsache, daß die Partialdrucke in den beiden Gefäßen jetzt kleiner sind als im ursprünglichen Gemisch.

Wir können aber nun die Tatsache, daß alle Moleküle im Gefäß BB' die y-Koordinate 1 haben und im anderen Gefäß entsprechend — 1, benutzen, um alle Moleküle wieder auf das ursprüngliche Volumen zu bringen. Hierzu brauchen wir bloß die semipermeable Wand A durch eine Wand A^* zu ersetzen, welche semipermeabel ist nicht mit Rücksicht auf x, sondern mit Rücksicht auf y, und zwar so, daß sie permeabel ist für die Moleküle mit der y-Koordinate 1 und impermeabel für die anderen, und umgekehrt B' ersetzen durch einen Stempel B'^*, der impermeabel ist für die Moleküle mit $y = -1$ und permeabel ist für die anderen. Dann lassen sich die beiden Gefäße wieder ohne Arbeitsleistung ineinander verschieben. Die Verteilung der y-Koordinate auf 1 und — 1 ist nunmehr von den x-Werten statistisch unabhängig geworden, und wir können außerdem noch die ursprüngliche Verteilung auf 1 und — 1 herstellen: wir durchliefen so einen vollständigen Kreisprozeß. Die einzige Veränderung, die wir dann zu registrieren haben, ist die erzielte Entropieverminderung von dem Betrage (2),

$$\bar{s} = k(w_1 \log w_1 + w_2 \log w_2).$$

Wenn wir also nicht zugeben wollen, daß der zweite Hauptsatz verletzt ist, so müssen wir schließen, daß der Eingriff, der die

850   L. Szilard,

Kopplung zwischen y und x bewirkt, die Messung von x durch y,
unlöslich mit einer Entropieerzeugung verknüpft ist. Wird
ein bestimmter Weg zur Erzielung dieser Kopplung eingeschlagen, und
wird die dabei notwendig erzeugte Entropiemenge mit S_1 und S_2 bezeichnet, wobei dann S_1 die mittlere Entropievergrößerung angibt, die
eintritt, wenn y den Wert 1 erhält, und entsprechend S_2 die, welche
eintritt, wenn y den Wert — 1 erhält, so wird die bei einem solchen im
Mittel erzeugte Entropiemenge gegeben sein durch

$$w_1 S_1 + w_2 S_2 = \bar{S}.$$

Damit der zweite Hauptsatz zu Recht besteht, muß diese Entropiemenge
größer sein als die durch die Verwertung der Messung erzielbare Entropieverminderung $\bar{s}$ nach (2). Es muß daher gelten:

$$\bar{S} + \bar{s} \gtreqqless 0,$$
$$w_1 S_1 + w_2 S_2 + k(w_1 \log w_1 + w_2 \log w_2) \gtreqqless 0. \qquad (3)$$

Diese Ungleichheit muß für beliebige Werte von w_1 und w_2 gelten*;
natürlich darf die Nebenbedingung $w_1 + w_2 = 1$ nicht verletzt sein. Im
besonderen fragen wir danach, bei welchem w_1 und w_2 für gegebene
S-Werte der Ausdruck ein Minimum wird. Auch für dieses Wertepaar
w_1 und w_2 muß die Ungleichheit (3) noch gelten. Unter der erwähnten
Nebenbedingung tritt das Minimum ein, wenn

$$\frac{S_1}{k} + \log w_1 = \frac{S_2}{k} + \log w_2 \qquad (4)$$

gilt.   Dann ist aber

$$e^{-\frac{S_1}{k}} + e^{-\frac{S_2}{k}} \leqq 1.$$

Dies kann man sehr einfach sehen, wenn man die Bezeichnung

$$\frac{S_1}{k} + \log w_1 = \frac{S_2}{k} + \log w_2 = \lambda$$

einführt.   Es ist dann

$$w_1 = e^\lambda . e^{-\frac{S_1}{k}}; \quad w_2 = e^\lambda . e^{-\frac{S_2}{k}}. \qquad (5)$$

Setzt man diese Werte in die Ungleichheit (3) ein, so ergibt sich

$$\lambda e^\lambda \left(e^{-\frac{S_1}{k}} + e^{-\frac{S_2}{k}} \right) \gtreqqless 0.$$

Es gilt also auch

$$\lambda \gtreqqless 0.$$

---

* Die Entropievergrößerung kann ja nur von der Art der Messung und seinem
Resultate, nicht aber davon abhängen, wie viele Systeme der einen oder anderen
Art vorhanden sind.

Über die Entropieverminderung in einem thermodynamischen System usw.   851

Setzt man aber die Werte für w_1 und w_2 aus (5) in die Gleichung $w_1 + w_2 = 1$ ein, so erhält man

$$e^{-\frac{S_1}{k}} + e^{-\frac{S_2}{k}} = e^{-\lambda}.$$

Und wegen $\lambda \geqq 0$ gilt also

$$e^{-\frac{S_1}{k}} + e^{-\frac{S_2}{k}} \leqq 1. \tag{6}$$

Diese Formel muß also ganz allgemein gelten, wenn die Thermodynamik nicht verletzt sein soll.

Solange wir die Eingriffe durch intelligente Wesen vornehmen lassen, ist freilich eine direkte Nachprüfung ausgeschlossen. Wir können jedoch versuchen, einfache unbelebte Vorrichtungen anzugeben, welche eine solche Kopplung bewirken, und zusehen, ob dabei in der Tat eine Entropie erzeugt wird und von welchem Betrage. Wenn wir erst einmal erkannt haben, daß es lediglich auf eine bestimmte charakteristische Art der Kopplung, eine „Messung", ankommt, so brauchen wir nicht irgendwelche komplizierten Modelle zu konstruieren, welche die Eingriffe der Lebewesen weitgehend nachahmen, sondern können uns mit dem Zustandebringen dieser besonderen, mit Erinnerung behafteten Kopplung begnügen.

Bei dem Beispiel, das wir hier anführen wollen, wird die Lagenkoordinate eines hin und her schwankenden Zeigers durch den Energieinhalt eines Körpers K „gemessen". Der Zeiger soll zunächst rein mechanisch bewirken, daß der Körper K, durch dessen Energieinhalt die Lage des Zeigers gemessen werden soll, je nach Lage des Zeigers mit dem einen von zwei Zwischenstücken, A bzw. B, wärmeleitend verbunden ist, und zwar so, daß der Körper mit A verbunden ist, solange die Koordinate, welche die Lage des Zeigers festlegt, in ein bestimmtes hervorgehobenes, im übrigen aber beliebig großes oder kleines Intervall a fällt, und sonst, im Intervall b, mit B. Beide Zwischenstücke seien bis zu einem gewissen Zeitpunkt, dem Zeitpunkt der „Messung", mit einem Wärmereservoir von der Temperatur T_0 in thermischem Kontakt. In diesem Zeitpunkt wird etwa durch eine periodisch funktionierende mechanische Vorrichtung das Zwischenstück A in reversibler Weise auf die Temperatur T_A abgekühlt, d. h. nach sukzessiver Berührung mit Wärmereservoiren von Zwischentemperaturen mit einem Wärmereservoir von der Temperatur T_A in Berührung gebracht. Gleichzeitig wird das Zwischenstück B in derselben Weise auf die Temperatur T_B erwärmt. Alsdann werden die Zwischenstücke wieder von den betreffenden Wärmereservoiren thermisch isoliert.

Wir setzen voraus, daß die Lage des Zeigers sich so lamgsam verändert, daß alle hier skizzierten Operationen noch bei ein und derselben Zeigerlage vorgegangen sind. Fiel die Lagenkoordinate des Zeigers in das hervorgehobene Intervall, so war während der erwähnten Operation der Körper mit dem Zwischenstück A verbunden und ist infolgedessen jetzt auf die Temperatur T_A abgekühlt. Im umgekehrten Falle ist der Körper jetzt auf die Temperatur T_B erwärmt. Sein Energieinhalt wird also, je nachdem wie die Zeigerlage zu dem Zeitpunkt der „Messung" war, der Temperatur T_A entsprechend klein oder der Temperatur T_B entsprechend groß sein und seinen Wert beibehalten, auch wenn der Zeiger nun im Laufe der Zeit aus dem hervorgehobenen Intervall austritt bzw. in dasselbe eintritt. Nach einiger Zeit, während der Zeiger weiter seine Schwankungen ausführt, läßt sich dann aus dem Energieinhalt des Körpers K kein Schluß mehr auf die augenblickliche Lage des Zeigers ziehen, wohl aber ein sicherer Schluß auf die Lage des Zeigers zum Zeitpunkt der Messung. Dann ist die Messung vollbracht.

Die erwähnte periodisch funktionierende mechanische Vorrichtung, die uns hier die Eingriffe vornimmt, soll dann nach der vollbrachten Messung die jetzt thermisch isolierten Zwischenstücke A und B mit dem Wärmereservoir T_0 direkt in Verbindung bringen. Dies hat den Zweck, den Körper K, der jetzt mit einem der beiden Zwischenstücke ebenfalls in Verbindung steht, wieder in den ursprünglichen Zustand zu bringen, in welchem er vor der „Messung" war. Die direkte Verbindung der Zwischenstücke und damit des auf T_A abgekühlten bzw. T_B erwärmten Körpers K mit dem Reservoir T_0 bewirkt notwendig eine Entropievergrößerung. Dies läßt sich aber gar nicht vermeiden, denn es hätte jetzt keinen Sinn, etwa das Zwischenstück A durch sukzessive Berührung mit Reservoiren von Zwischentemperaturen in reversibler Weise auf die Temperatur T_0 erwärmen zu wollen und B ebenso abzukühlen; denn nach vollbrachter Messung wissen wir nicht, mit welchem der beiden Zwischenstücke der Körper K jetzt im Kontakt steht, und ebensowenig wissen wir, ob er zuletzt mit T_A oder T_B in Verbindung gestanden hat. Wir wissen also auch nicht, ob wir Zwischentemperaturen zwischen T_A und T_0 oder T_0 und T_B anwenden sollen.

Der Betrag der auf diese Weise im Mittel pro Messung erzeugten Entropiemengen $\bar{S}_1$ und $\bar{S}_2$ läßt sich angeben, falls die Wärmekapazität als Funktion der Temperatur $\bar{u}(T)$ für den Körper K bekannt ist, denn

Über die Entropieverminderung in einem thermodynamischen System usw.   853

die Entropie ist ja aus der Wärmekapazität zu berechnen; von der Wärmekapazität der Zwischenstücke wird natürlich abstrahiert. War die Lagenkoordinate des Zeigers zum Zeitpunkt der „Messung" in dem hervorgehobenen Intervall und entsprechend der Körper mit dem Zwischenstück A in Verbindung, so wurde den Wärmereservoiren bei der sukzessiven Abkühlung die Entropie

$$\int_{T_A}^{T_0} \frac{1}{T} \frac{d\bar{u}}{dT}\, dT$$

zugeführt. Dafür wurde aber nachher bei der direkten Berührung mit dem Wärmereservoir T_0 diesem Wärmereservoir die Entropie

$$\frac{\bar{u}(T_0) - \bar{u}(T_A)}{T_0}$$

entzogen.   Es wurde also in summa die Entropie mit dem Betrage

$$S_A = \frac{\bar{u}(T_A) - \bar{u}(T_0)}{T_0} + \int_{T_A}^{T_0} \frac{1}{T} \frac{d\bar{u}}{dT}\, dT \qquad (7)$$

vermehrt. Analog wird die Entropie, wenn der Körper zum Zeitpunkt der „Messung" mit dem Zwischenstück B in Verbindung war, den Zuwachs

$$S_B = \frac{\bar{u}(T_B) - \bar{u}(T_0)}{T_0} + \int_{T_B}^{T_0} \frac{1}{T} \frac{d\bar{u}}{dT}\, dT \qquad (8)$$

erfahren.

Wir wollen nun diese Ausdrücke auswerten für den besonders einfachen Fall, daß der Körper, dessen wir uns bedienen, nur zwei Energiezustände, einen unteren und einen oberen Zustand, besitzt. Ist ein solcher Körper mit einem Wärmereservoir von irgend einer Temperatnr T in thermischer Berührung, dann ist die Wahrscheinlichkeit dafür, daß er sich im unteren bzw. oberen Zustand befindet, gegeben durch

$$\left.\begin{aligned} p(T) &= \frac{1}{1 + g\, e^{-\frac{u}{kT}}} \\[2ex] q(T) &= \frac{g\, e^{-\frac{u}{kT}}}{1 + g\, e^{-\frac{u}{kT}}} \cdot \end{aligned}\right\} \qquad (9)$$

bzw.

854   L. Szilard,

Dabei bedeutet u den Energieunterschied der beiden Zustände und g das statistische Gewicht. Die Energie des unteren Zustandes können wir, ohne daß dies eine Einschränkung bedeuten würde, gleich Null setzen. Es ergibt sich dann*

bzw.
$$S_A = q(T_A)\,k \log \frac{q(T_A)}{q(T_0)}\frac{p(T_0)}{p(T_A)} + k \log \frac{p(T_A)}{p(T_0)} \left.\vphantom{\frac{\frac{q}{q}}{\frac{p}{p}}}\right\}$$
$$S_B = p(T_B)\,k \log \frac{q(T_0)}{q(T_B)}\frac{p(T_B)}{p(T_0)} + k \log \frac{q(T_B)}{q(T_0)}. \left.\vphantom{\frac{\frac{q}{q}}{\frac{p}{p}}}\right\} \qquad (10)$$

Dabei bedeuten q und p die durch Formel (9) gegebenen Funktionen von T, die hier für die betreffenden Argumente T_0, T_A und T_B zu nehmen sind

Wollen wir nun erreichen, daß ein sicherer Schluß von dem Energieinhalt des Körpers K auf die Lagenkoordinate des Zeigers möglich sei, wie dies bei der hier verwendeten begrifflichen Abgrenzung für die „Messung" zu fordern ist, so müssen wir erreichen, daß der Körper sicher in den unteren Zustand gerät, wenn er mit T_A in Berührung kommt, und sicher in den oberen Zustand, wenn er mit T_B in Berührung kommt, was der Aussage entspricht, daß

$$p(T_A) = 1, \quad q(T_A) = 0; \qquad p(T_B) = 0, \quad q(T_B) = 1$$

gilt. Dies läßt sich zwar nicht erreichen, aber beliebig genau approximieren dadurch, daß man T_A gegen den absoluten Nullpunkt und das statistische Gewicht g gegen Unendlich gehen läßt. [Bei dem Grenzübergang wird auch T_0 verändert so, daß $p(T_0)$ und $q(T_0)$ festgehalten werden.] Aus den Formeln (10) wird dann

$$S_A = -k \log p(T_0); \quad S_B = -k \log q(T_0), \qquad (11)$$

und wenn wir nun den Ausdruck $e^{-\frac{S_A}{k}} + e^{-\frac{S_B}{k}}$ bilden, so finden wir

$$e^{-\frac{S_A}{k}} + e^{-\frac{S_B}{k}} = 1.$$

Es konnte also der durch unsere vorangehende Überlegung gerade noch zugelassene Grenzfall realisiert werden; die Verwendung von semipermeablen Wänden nach Fig. 1 erlaubt eine volle Ausnützung der Messung: Ungleichung (1) läßt sich sicher nicht verschärfen.

Wie wir an diesem Beispiel gesehen haben, kann eine einfache unbelebte Vorrichtung mit Bezug auf das, was wesentlich ist, dasselbe

---

* Siehe den Anhang am Schluß der Arbeit.

Über die Entropieverminderung in einem thermodynamischen System usw.   855

leisten, was die Eingriffe intelligenter Wesen leisten würden. Wir haben an diesem Beispiel die „Lebensvorgänge" einer unbelebten Vorrichtung verfolgt und sehen können, daß durch sie genau die von der Thermodynamik geforderte Entropiemenge erzeugt wird.

Anhang. In dem betrachteten Falle, wenn die Häufigkeit der beiden Zustände von der Temperatur nach den Formeln

$$p(T) = \frac{1}{1 + g\,e^{-\frac{u}{kT}}}\,; \quad q(T) = \frac{g\,e^{-\frac{u}{kT}}}{1 + g\,e^{-\frac{u}{kT}}}$$

abhängt und somit die mittlere Energie des Körpers durch

$$\bar{u}(T) = u\,q(T) = \frac{u\,g\,e^{-\frac{u}{kT}}}{1 + g\,e^{-\frac{u}{kT}}}$$

gegeben ist, gilt die Identität

$$\frac{1}{T}\frac{d\bar{u}}{dT} = \frac{d}{dT}\left\{\frac{\bar{u}(T)}{T} + k\log\left(1 + e^{-\frac{u}{kT}}\right)\right\}.$$

Damit können wir den Ausdruck

$$S_A = \frac{\bar{u}(T_A) - \bar{u}(T_0)}{T_0} + \int_{T_A}^{T_0} \frac{1}{T}\frac{d\bar{u}}{dT}\,dT$$

auch schreiben:

$$S_A = \frac{\bar{u}(T_A) - \bar{u}(T_0)}{T_0} + \left\{\frac{\bar{u}(T)}{T} + k\log\left(1 + g\,e^{-\frac{u}{kT}}\right)\right\}_{T_A}^{T_0},$$

und indem wir die Grenzen einsetzen, erhalten wir

$$S_A = \bar{u}(T_A)\left(\frac{1}{T_0} - \frac{1}{T_A}\right) + k\log\frac{1 + g\,e^{-\frac{u}{kT_0}}}{1 + g\,e^{-\frac{u}{kT_A}}}.$$

Schreiben wir nun in diesem letzten Ausdruck zunächst gemäß (9)

$$1 + g\,e^{-\frac{u}{kT}} = \frac{1}{p(T)}$$

für T_A und T_0, so erhalten wir

$$S_A = \bar{u}(T_A)\left(\frac{1}{T_0} - \frac{1}{T_A}\right) + k\log\frac{p(T_A)}{p(T_0)},$$

und wenn wir dann noch gemäß (12)

$$\bar{u}(T_A) = u\,q(T_A)$$

schreiben, so wird daraus

$$S_A = q(T_A)\left(\frac{u}{T_0} - \frac{u}{T_A}\right) + k\log\frac{p(T_A)}{p(T_0)}.$$

856   L. Szilard, Über die Entropieverminderung usw.

Schreiben wir schließlich noch gemäß (9)

$$\frac{u}{T} = - k \log \frac{q(T)}{g\,p(T)}$$

für T_A und T_0, so erhalten wir

$$S_A = q(T_A)\, k \log \frac{p(T_0)}{q(T_0)} \frac{q(T_A)}{p(T_A)} + k \log \frac{p(T_A)}{p(T_0)}.$$

Die entsprechende Formel für S_B erhalten wir, indem wir hier den Index A mit B vertauschen. Wir erhalten dann:

$$S_B = q(T_B)\, k \log \frac{p(T_0)}{q(T_0)} \frac{q(T_B)}{p(T_B)} + k \log \frac{p(T_B)}{p(T_0)}.$$

Erstere ist die Formel, die im Text für S_A angeführt ist.

Wir können die Formel für S_B auch noch auf etwas andere Gestalt bringen, indem wir

$$q(T_B) = 1 - p(T_B)$$

schreiben, ausmultiplizieren und das erste und letzte Glied zusammenziehen. Es wird dann:

$$S_B = p(T_B)\, k \log \frac{q(T_0)}{p(T_0)} \frac{p(T_B)}{q(T_B)} + k \log \frac{q(T_B)}{q(T_0)}.$$

Dies ist die Formel, die im Text für S_B angeführt ist.

Reprinted from BEHAVIORAL SCIENCE
Vol. 9, No. 4, October, 1964
Printed in U.S.A.

In memory of Leo Szilard, who passed away on May 30, 1964, we present an English translation of his classical paper *Über die Entropieverminderung in einem thermodynamischen System bei Eingriffen intelligenter Wesen*, which appeared in the *Zeitschrift für Physik*, 1929, 53, 840–856. The publication in this journal of this translation was approved by Dr. Szilard before he died, but he never saw the copy. At Mrs. Szilard's request, Dr. Carl Eckart revised the translation.

This is one of the earliest, if not the earliest paper, in which the relations of physical entropy to information (in the sense of modern mathematical theory of communication) were rigorously demonstrated and in which Maxwell's famous demon was successfully exorcised: a milestone in the integration of physical and cognitive concepts.

ON THE DECREASE OF ENTROPY IN A THERMODYNAMIC SYSTEM BY THE INTERVENTION OF INTELLIGENT BEINGS

by Leo Szilard

Translated by Anatol Rapoport and Mechthilde Knoller from the original article "Über die Entropieverminderung in einem thermodynamischen System bei Eingriffen intelligenter Wesen." Zeitschrift für Physik, 1929, 53, 840–856.

ᔆᴕ

The objective of the investigation is to find the conditions which apparently allow the construction of a perpetual-motion machine of the second kind, if one permits an intelligent being to intervene in a thermodynamic system. When such beings make measurements, they make the system behave in a manner distinctly different from the way a mechanical system behaves when left to itself. We show that it is a sort of a memory faculty, manifested by a system where measurements occur, that might cause a permanent decrease of entropy and thus a violation of the Second Law of Thermodynamics, were it not for the fact that the measurements themselves are necessarily accompanied by a production of entropy. At first we calculate this production of entropy quite generally from the postulate that full compensation is made in the sense of the Second Law (Equation [1]). Second, by using an inanimate device able to make measurements—however under continual entropy production—we shall calculate the

Requests for reprints of this paper should be sent to:
Mrs. Leo Szilard
2380 Torrey Pines Road
La Jolla, California 92038

resulting quantity of entropy. We find that it is exactly as great as is necessary for full compensation. The actual production of entropy in connection with the measurement, therefore, need not be greater than Equation (1) requires.

ᔆᴕ

THERE is an objection, already historical, against the universal validity of the Second Law of Thermodynamics, which indeed looks rather ominous. The objection is embodied in the notion of Maxwell's demon, who in a different form appears even nowadays again and again; perhaps not unreasonably, inasmuch as behind the precisely formulated question quantitative connections seem to be hidden which to date have not been clarified. The objection in its original formulation concerns a demon who catches the fast molecules and lets the slow ones pass. To be sure, the objection can be met with the reply that man cannot in principle foresee the value of a thermally fluctuating parameter. However, one cannot deny that we can very well measure the value of such a fluctuating parameter and therefore could certainly gain energy at the expense of heat by arranging our interven-

302   LEO SZILARD

tion according to the results of the measurements. Presently, of course, we do not know whether we commit an error by not including the intervening man into the system and by disregarding his biological phenomena.

Apart from this unresolved matter, it is known today that in a system left to itself no "perpetuum mobile" (perpetual motion machine) of the second kind (more exactly, no "automatic machine of continual finite work-yield which uses heat at the lowest temperature") can operate in spite of the fluctuation phenomena. A perpetuum mobile would have to be a machine which in the long run could lift a weight at the expense of the heat content of a reservoir. In other words, if we want to use the fluctuation phenomena in order to gain energy at the expense of heat, we are in the same position as playing a game of chance, in which we may win certain amounts now and then, although the expectation value of the winnings is zero or negative. The same applies to a system where the intervention from outside is performed strictly periodically, say by periodically moving machines. We consider this as established (Szilard, 1925) and intend here only to consider the difficulties that occur when intelligent beings intervene in a system. We shall try to discover the quantitative relations having to do with this intervention.

Smoluchowski (1914, p. 89) writes: "As far as we know today, there is no automatic, permanently effective perpetual motion machine, in spite of the molecular fluctuations, but such a device might, perhaps, function regularly if it were appropriately operated by intelligent beings...."

A perpetual motion machine therefore is possible if—according to the general method of physics—we view the experimenting man as a sort of *deus ex machina*, one who is continuously and exactly informed of the existing state of nature and who is able to start or interrupt the macroscopic course of nature at any moment without expenditure of work. Therefore he would definitely not have to possess the ability to catch single molecules like Maxwell's demon, although he would definitely be different from real living beings in possessing the above abilities. In eliciting any physical effect by action of the sensory

as well as the motor nervous systems a degradation of energy is always involved, quite apart from the fact that the very existence of a nervous system is dependent on continual dissipation of energy.

Whether—considering these circumstances—real living beings could continually or at least regularly produce energy at the expense of heat of the lowest temperature appears very doubtful, even though our ignorance of the biological phenomena does not allow a definite answer. However, the latter questions lead beyond the scope of physics in the strict sense.

It appears that the ignorance of the biological phenomena need not prevent us from understanding that which seems to us to be the essential thing. We may be sure that intelligent living beings—insofar as we are dealing with their intervention in a thermodynamic system—can be replaced by non-living devices whose "biological phenomena" one could follow and determine whether in fact a compensation of the entropy decrease takes place as a result of the intervention by such a device in a system.

In the first place, we wish to learn what circumstance conditions the decrease of entropy which takes place when intelligent living beings intervene in a thermodynamic system. We shall see that this depends on a certain type of coupling between different parameters of the system. We shall consider an unusually simple type of these ominous couplings.[1] For brevity we shall talk about a "measurement," if we succeed in coupling the value of a parameter y (for instance the position co-ordinate of a pointer of a measuring instrument) at one moment with the simultaneous value of a fluctuating parameter x of the system, in such a way that, from the value y, we can draw conclusions about the value that x had at the moment of the "measurement." Then let x and y be uncoupled after the measurement, so that x can change, while y retains its value for some time. Such measurements are not harmless interventions. A system in which such measurements occur shows a sort of memory

[1] The author evidently uses the word "ominous" in the sense that the possibility of realizing the proposed arrangement threatens the validity of the Second Law.—*Translator*

faculty, in the sense that one can recognize by the state parameter y what value another state parameter x had at an earlier moment, and we shall see that simply because of such a memory the Second Law would be violated, if the measurement could take place without compensation. We shall realize that the Second Law is not threatened as much by this entropy decrease as one would think, as soon as we see that the entropy decrease resulting from the intervention would be compensated completely in any event if the execution of such a measurement were, for instance, always accompanied by production of $k \log 2$ units of entropy. In that case it will be possible to find a more general entropy law, which applies universally to all measurements. Finally we shall consider a very simple (of course, not living) device, that is able to make measurements continually and whose "biological phenomena" we can easily follow. By direct calculation, one finds in fact a continual entropy production of the magnitude required by the above-mentioned more general entropy law derived from the validity of the Second Law.

The first example, which we are going to consider more closely as a typical one, is the following. A standing hollow cylinder, closed at both ends, can be separated into two possibly unequal sections of volumes V_1 and V_2 respectively by inserting a partition from the side at an arbitrarily fixed height. This partition forms a piston that can be moved up and down in the cylinder. An infinitely large heat reservoir of a given temperature T insures that any gas present in the cylinder undergoes isothermal expansion as the piston moves. This gas shall consist of a single molecule which, as long as the piston is not inserted into the cylinder, tumbles about in the whole cylinder by virtue of its thermal motion.

Imagine, specifically, a man who at a given time inserts the piston into the cylinder and somehow notes whether the molecule is caught in the upper or lower part of the cylinder, that is, in volume V_1 or V_2. If he should find that the former is the case, then he would move the piston slowly downward until it reaches the bottom of the cylinder. During this slow movement of the piston the molecule stays, of course, above the piston.

However, it is no longer constrained to the upper part of the cylinder but bounces many times against the piston which is already moving in the lower part of the cylinder. In this way the molecule does a certain amount of work on the piston. This is the work that corresponds to the isothermal expansion of an ideal gas—consisting of one single molecule—from volume V_1 to the volume $V_1 + V_2$. After some time, when the piston has reached the bottom of the container, the molecule has again the full volume $V_1 + V_2$ to move about in, and the piston is then removed. The procedure can be repeated as many times as desired. The man moves the piston up or down depending on whether the molecule is trapped in the upper or lower half of the piston. In more detail, this motion may be caused by a weight, that is to be raised, through a mechanism that transmits the force from the piston to the weight, in such a way that the latter is always displaced upwards. In this way the potential energy of the weight certainly increases constantly. (The transmission of force to the weight is best arranged so that the force exerted by the weight on the piston at any position of the latter equals the average pressure of the gas.) It is clear that in this manner energy is constantly gained at the expense of heat, insofar as the biological phenomena of the intervening man are ignored in the calculation.

In order to understand the essence of the man's effect on the system, one best imagines that the movement of the piston is performed mechanically and that the man's activity consists only in determining the altitude of the molecule and in pushing a lever (which steers the piston) to the right or left, depending on whether the molecule's height requires a down- or upward movement. This means that the intervention of the human being consists only in the coupling of two position co-ordinates, namely a co-ordinate x, which determines the altitude of the molecule, with another co-ordinate y, which determines the position of the lever and therefore also whether an upward or downward motion is imparted to the piston. It is best to imagine the mass of the piston as large and its speed sufficiently great, so that the thermal agita-

304   Leo Szilard

tion of the piston at the temperature in question can be neglected.

In the typical example presented here, we wish to distinguish two periods, namely:

1. The period of *measurement* when the piston has just been inserted in the middle of the cylinder and the molecule is trapped either in the upper or lower part; so that if we choose the origin of co-ordinates appropriately, the x-co-ordinate of the molecule is restricted to either the interval $x > 0$ or $x < 0$;

2. The period of *utilization of the measurement*, "the period of decrease of entropy," during which the piston is moving up or down. During this period the x-co-ordinate of the molecule is certainly not restricted to the original interval $x > 0$ or $x < 0$. Rather, if the molecule was in the upper half of the cylinder during the period of measurement, i.e., when $x > 0$, the molecule must bounce on the downward-moving piston in the lower part of the cylinder, if it is to transmit energy to the piston; that is, the co-ordinate x has to enter the interval $x < 0$. The lever, on the contrary, retains during the whole period its position toward the right, corresponding to downward motion. If the position of the lever toward the right is designated by $y = 1$ (and correspondingly the position toward the left by $y = -1$) we see that during the period of measurement, the position $x > 0$ corresponds to $y = 1$; but afterwards $y = 1$ stays on, even though x passes into the other interval $x < 0$. We see that in the utilization of the measurement the coupling of the two parameters x and y disappears.

We shall say, quite generally, that a parameter y "measures" a parameter x (which varies according to a probability law), if the value of y is directed by the value of parameter x at a given moment. A measurement procedure underlies the entropy decrease effected by the intervention of intelligent beings.

One may reasonably assume that a measurement procedure is fundamentally associated with a certain definite average entropy production, and that this restores concordance with the Second Law. The amount of entropy generated by the measurement may, of course, always be greater than this funda-

mental amount, but not smaller. To put it precisely: we have to distinguish here between two entropy values. One of them, $\bar{S}_1$, is produced when during the measurement y assumes the value 1, and the other, $\bar{S}_2$, when y assumes the value -1. We cannot expect to get general information about $\bar{S}_1$ or $\bar{S}_2$ separately, but we shall see that *if* the amount of entropy produced by the "measurement" is to compensate the entropy decrease affected by utilization, the relation must always hold good.

$$e^{-\bar{S}_1/k} + e^{-\bar{S}_2/k} \leqq 1 \qquad (1)$$

One sees from this formula that one can make one of the values, for instance $\bar{S}_1$, as small as one wishes, but then the other value $\bar{S}_2$ becomes correspondingly greater. Furthermore, one can notice that the magnitude of the interval under consideration is of no consequence. One can also easily understand that it cannot be otherwise.

Conversely, as long as the entropies $\bar{S}_1$ and $\bar{S}_2$, produced by the measurements, satisfy the inequality (1), we can be sure that the expected decrease of entropy caused by the later utilization of the measurement will be fully compensated.

Before we proceed with the proof of inequality (1), let us see in the light of the above mechanical example, how all this fits together. For the entropies $\bar{S}_1$ and $\bar{S}_2$ produced by the measurements, we make the following Ansatz:

$$\bar{S}_1 = \bar{S}_2 = k \log 2 \qquad (2)$$

This ansatz satisfies inequality (1) and the mean value of the quantity of entropy produced by a measurement is (of course in this special case independent of the frequencies w_1, w_2 of the two events):

$$\bar{S} = k \log 2 \qquad (3)$$

In this example one achieves a decrease of entropy by the isothermal expansion:[2]

$$- \bar{s}_1 = -k \log \frac{V_1}{V_1 + V_2} \, ;$$
$$- \bar{s}_2 = -k \log \frac{V_2}{V_1 + V_2} , \qquad (4)$$

[2] The entropy generated is denoted by $\bar{s}_1, \bar{s}_2$.

depending on whether the molecule was found in volume V_1 or V_2 when the piston was inserted. (The decrease of entropy equals the ratio of the quantity of heat taken from the heat reservoir during the isothermal expansion, to the temperature of the heat reservoir in question). Since in the above case the frequencies w_1, w_2 are in the ratio of the volumes V_1, V_2, the mean value of the entropy generated is (a negative number):

$$\bar{s} = w_1 \cdot (+\bar{s}_1) + w_2 \cdot (+\bar{s}_2) =$$

$$\frac{V_1}{V_1 + V_2} k \log \frac{V_1}{V_1 + V_2} + \qquad (5)$$

$$\frac{V_2}{V_1 + V_2} k \log \frac{V_1}{V_1 + V_2}$$

As one can see, we have, indeed

$$\frac{V_1}{V_1 + V_2} k \log \frac{V_1}{V_1 + V_2} + \frac{V_2}{V_1 + V_2}$$

$$\cdot k \log \frac{V_2}{V_1 + V_2} + k \log 2 \geqq 0 \qquad (6)$$

and therefore:

$$\bar{S} + \bar{s} \geqq 0. \qquad (7)$$

In the special case considered, we would actually have a full compensation for the decrease of entropy achieved by the utilization of the measurement.

We shall not examine more special cases, but instead try to clarify the matter by a general argument, and to derive formula (1). We shall therefore imagine the whole system—in which the co-ordinate x, exposed to some kind of thermal fluctuations, can be measured by the parameter y in the way just explained—as a multitude of particles, all enclosed in one box. Every one of these particles can move freely, so that they may be considered as the molecules of an ideal gas, which, because of thermal agitation, wander about in the common box independently of each other and exert a certain pressure on the walls of the box—the pressure being determined by the temperature. We shall now consider two of these molecules as chemically different and, in principle, separable by semipermeable walls, if the co-ordinate x for one molecule is in a preassigned interval while the corresponding co-ordinate of the other molecule falls outside that interval. We

also shall look upon them as chemically different, if they differ only in that the y co-ordinate is $+1$ for one and -1 for the other.

We should like to give the box in which the "molecules" are stored the form of a hollow cylinder containing four pistons. Pistons A and A' are fixed while the other two are movable, so that the distance BB' always equals the distance AA', as is indicated in Figure 1 by the two brackets. A', the bottom, and B, the cover of the container, are impermeable for all "molecules," while A and B' are semipermeable; namely, A is permeable only for those "molecules" for which the parameter x is in the preassigned interval, i.e., (x_1, x_2), B' is only permeable for the rest.

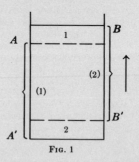

Fig. 1

In the beginning the piston B is at A and therefore B' at A', and all "molecules" are in the space between. A certain fraction of the molecules have their co-ordinate x in the preassigned interval. We shall designate by w_1 the probability that this is the case for a randomly selected molecule and by w_2 the probability that x is outside the interval. Then $w_1 + w_2 = 1$.

Let the distribution of the parameter y be over the values $+1$ and -1 in any proportion but in any event independent of the x-values. We imagine an intervention by an intelligent being, who imparts to y the value 1 for all "molecules" whose x at that moment is in the selected interval. Otherwise the value -1 is assigned. If then, because of thermal fluctuation, for any "molecule," the parameter x should come out of the preassigned interval or, as we also may put it, if the "molecule" suffers a monomolecular chemical reaction with regard to x (by which it is transformed from a species that can pass the semipermeable piston A into a species for which the piston is impermeable), then the parameter y retains its value 1 for the time being, so that the "molecule," because of the value of the parameter y, "remembers" during the whole following process that x originally was in the preassigned interval. We shall see immediately what part this memory may play. After the intervention just discussed, we move the piston, so that we separate the two kinds of molecules without doing work. This results in two containers, of which the first contains only the one modification and the second only the other. Each modification now occupies the same volume as the mixture did previously. In one of these containers, if considered by itself, there is now no equilibrium with regard to the two "modifications in x." Of course the ratio of the two modifications has remained $w_1 : w_2$. If we allow this equilibrium to be achieved in both containers independently and at constant volume and temperature, then the entropy of the system certainly has increased. For the total heat release is 0, since the ratio of the two "modifications in x" $w_1 : w_2$ does not change. If we accomplish the equilibrium distribution in both containers in a reversible fashion then the entropy of the rest of the world will decrease by the same amount. Therefore the entropy increases by a negative value, and, the value of the entropy increase per molecule is exactly:

$$\bar{s} = k(w_1 \log w_1 + w_2 \log w_2). \quad (9)$$

(The entropy constants that we must assign to the two "modifications in x" do not occur here explicitly, as the process leaves the total number of molecules belonging to the one or the other species unchanged.)

Now of course we cannot bring the two gases back to the original volume without expenditure of work by simply moving the piston back, as there are now in the container—which is bounded by the pistons BB'—also molecules whose x-co-ordinate lies outside of the preassigned interval and for which the piston A is not permeable any longer. Thus one can see that the calculated decrease of entropy (Equation [9]) does not mean a contradiction of the Second Law. *As* *long as we do not use the fact that the molecules* *in the container BB', by virtue of their co-* *ordinate y, "remember" that the x-co-ordinate* *for the molecules of this container originally* *was in the preassigned interval, full compensa-* *tion exists for the calculated decrease of entropy,* by virtue of the fact that the partial pressures in the two containers are smaller than in the original mixture.

But now we can use the fact that all mole- *cules in the container BB' have the y-co-ordi-* *nate 1, and in the other accordingly -1, to* *bring all molecules back again to the original* *volume.* To accomplish this we only need to replace the semipermeable wall A by a wall A^*, which is semipermeable not with regard to x but with regard to y, namely so that it is permeable for the molecules with the y-co-ordinate 1 and impermeable for the others. Correspondingly we replace B' by a piston B'^*, which is impermeable for the molecules with $y = -1$ and permeable for the others. Then both containers can be put into each other again without expenditure of energy. The distribution of the y-co-ordinate with regard to 1 and -1 now has become statistically independent of the x-values and besides we are able to re-establish the original distribution over 1 and -1. Thus we would have gone through a complete cycle. The only change that we have to register is the resulting decrease of entropy given by (9):

$$\bar{s} = k(w_1 \log w_1 + w_2 \log w_2). \quad (10)$$

If we do not wish to admit that the Second Law has been violated, we must conclude *that the intervention which establishes the* *coupling between y and x, the measurement of* *x by y, must be accompanied by a production* *of entropy.* If a definite way of achieving this coupling is adopted and if the quantity of entropy that is inevitably produced is designated by S_1 and S_2, where S_1 stands for the mean increase in entropy that occurs when y acquires the value 1, and accordingly S_2 for the increase that occurs when y acquires the value -1, we arrive at the equation:

$$w_1 S_1 + w_2 S_2 = \bar{S} \quad (11)$$

In order for the Second Law to remain in force, this quantity of entropy must be greater than the decrease of entropy $\bar{s}$, which according to (9) is produced by the utilization of the measurement. Therefore the following unequality must be valid:

$$\bar{S} + \bar{s} \geqq 0$$

$$w_1 S_1 + w_2 S_2 \qquad (12)$$

$$+ k(w_1 \log w_1 + w_2 \log w_2) \geqq 0$$

This equation must be valid for any values of w_1 and w_2,[3] and of course the constraint $w_2 + w_2 = 1$ cannot be violated. We ask, in particular, for which w_1 and w_2 and given S-values the expression becomes a minimum. For the two minimizing values w_1 and w_2 the inequality (12) must still be valid. Under the above constraint, the minimum occurs when the following equation holds:

$$\frac{S_1}{k} + \log w_1 = \frac{S_2}{k} + \log w_2 \qquad (13)$$

But then:

$$e^{-S_1/k} + e^{-S_2/k} \leqq 1. \qquad (14)$$

This is easily seen if one introduces the notation

$$\frac{S_1}{k} + \log w_1 = \frac{S_2}{k} + \log w_2 = \lambda; \quad (15)$$

then:

$$w_1 = e^\lambda \cdot e^{-S_1/k}; \quad w_2 = e^\lambda \cdot e^{-S_2/k}. \quad (16)$$

If one substitutes these values into the inequality (12) one gets:

$$\lambda e^\lambda (e^{-S_1/k} + e^{-S_2/k}) \geqq 0. \qquad (17)$$

Therefore the following also holds:

$$\lambda \geqq 0. \qquad (18)$$

If one puts the values w_1 and w_2 from (16) into the equation $w_1 + w_2 = 1$, one gets

$$e^{-S_1/k} + e^{-S_2/k} = e^{-\lambda}. \qquad (19)$$

And because $\lambda \geqq 0$, the following holds:

$$e^{-S_1/k} + e^{-S_2/k} \leqq 1. \qquad (20)$$

This equation must be universally valid, if thermodynamics is not to be violated.

As long as we allow intelligent beings to perform the intervention, a direct test is

[3] The increase in entropy can depend only on the types of measurement and their results but not on how many systems of one or the other type were present.

not possible. But we can try to describe simple nonliving devices that effect such coupling, and see if indeed entropy is generated and in what quantity. Having already recognized that the only important factor is a certain characteristic type of coupling, a "measurement," we need not construct any complicated models which imitate the intervention of living beings in detail. We can be satisfied with the construction of this particular type of coupling which is accompanied by memory.

In our next example, the position co-ordinate of an oscillating pointer is "measured" by the energy content of a body K. The pointer is supposed to connect, in a purely mechanical way, the body K—by whose energy content the position of the pointer is to be measured—by heat conduction with one of two intermediate pieces, A or B. The body is connected with A as long as the co-ordinate—which determines the position of the pointer—falls into a certain preassigned, but otherwise arbitrarily large or small interval a, and otherwise if the co-ordinate is in the interval b, with B. Up to a certain moment, namely the moment of the "measurement," both intermediate pieces will be thermally connected with a heat reservoir at temperature T_0. At this moment the insertion A will be cooled reversibly to the temperature T_A, e.g., by a periodically functioning mechanical device. That is, after successive contacts with heat reservoirs of intermediate temperatures, A will be brought into contact with a heat reservoir of the temperature T_A. At the same time the insertion B will be heated in the same way to temperature T_B. Then the intermediate pieces will again be isolated from the corresponding heat reservoirs.

We assume that the position of the pointer changes so slowly that all the operations that we have sketched take place while the position of the pointer remains unchanged. If the position co-ordinate of the pointer fell in the preassigned interval, then the body was connected with the insertion A during the above-mentioned operation, and consequently is now cooled to temperature T_A.

In the opposite case, the body is now heated to temperature T_B. Its energy content becomes—according to the position of

308   LEO SZILARD

the pointer at the time of "measurement"—small at temperature T_A or great at temperature T_B and will retain its value, even if the pointer eventually leaves the preassigned interval or enters into it. After some time, while the pointer is still oscillating, one can no longer draw any definite conclusion from the energy content of the body K with regard to the momentary position of the pointer but one can draw a definite conclusion with regard to the position of the pointer at the time of the measurement. Then the measurement is completed.

After the measurement has been accomplished, the above-mentioned periodically functioning mechanical device should connect the thermally isolated insertions A and B with the heat reservoir T_0. This has the purpose of bringing the body K—which is now also connected with one of the two intermediate pieces—back into its original state. The direct connection of the intermediate pieces and hence of the body K—which has been either cooled to T_A or heated to T_B—to the reservoir T_0 consequently causes an increase of entropy. This cannot possibly be avoided, because it would make no sense to heat the insertion A reversibly to the temperature T_0 by successive contacts with the reservoirs of intermediate temperatures and to cool B in the same manner. After the measurement we do not know with which of the two insertions the body K is in contact at that moment; nor do we know whether it had been in connection with T_A or T_B in the end. Therefore neither do we know whether we should use intermediate temperatures between T_A and T_0 or between T_0 and T_B.

The mean value of the quantity of entropy S_1 and S_2, per measurement, can be calculated, if the heat capacity as a function of the temperature $\bar{u}(T)$ is known for the body K, since the entropy can be calculated from the heat capacity. We have, of course, neglected the heat capacities of the intermediate pieces. If the position co-ordinate of the pointer was in the preassigned interval at the time of the "measurement," and accordingly the body in connection with insertion A, then the entropy conveyed to the heat reservoirs during successive cooling was

$$\int_{T_A}^{T_0} \frac{1}{T} \frac{d\bar{u}}{dT}.\qquad(21)$$

However, following this, the entropy withdrawn from the reservoir T_0 by direct contact with it was

$$\frac{\bar{u}(T_0) - \bar{u}(T_A)}{T_0}.\qquad(22)$$

All in all the entropy was increased by the amount

$$S_A = \frac{\bar{u}(T_A) - \bar{u}(T_0)}{T_0} + \int_{T_A}^{T_0} \frac{1}{T} \frac{d\bar{u}}{dT} dT.\qquad(23)$$

Analogously, the entropy will increase by the following amount, if the body was in contact with the intermediate piece B at the time of the "measurement":

$$S_B = \frac{\bar{u}(T_B) - \bar{u}(T_0)}{T_0} + \int_{T_B}^{T_0} \frac{1}{T} \frac{d\bar{u}}{dT} dT.\qquad(24)$$

We shall now evaluate these expressions for the very simple case, where the body which we use has only two energy states, a lower and a higher state. If such a body is in thermal contact with a heat reservoir at any temperature T, the probability that it is in the lower or upper state is given by respectively:

$$\left.\begin{aligned} p(T) &= \frac{1}{1 + ge^{-u/kT}} \\ q(T) &= \frac{ge^{-u/kT}}{1 + ge^{-u/kT}} \end{aligned}\right\}\qquad(25)$$

Here u stands for the difference of energy of the two states and g for the statistical weight. We can set the energy of the lower state equal to zero without loss of generality. Therefore:[4]

$$\left.\begin{aligned} S_A &= q(T_A)\, k \log \frac{q(T_A)\, p(T_0)}{q(T_0)\, p(T_A)} \\ &\qquad + k \log \frac{p(T_A)}{p(T_0)} \\ S_B &= p(T_B)\, k \log \frac{q(T_0)\, p(T_B)}{q(T_B)\, p(T_0)} \\ &\qquad + k \log \frac{q(T_B)}{q(T_0)} \end{aligned}\right\}.\qquad(26)$$

Here q and p are the functions of T given

[4] See the Appendix.

ON THE DECREASE OF ENTROPY   309

by equation (25), which are here to be taken for the arguments T_0, T_A, or T_B.

If (as is necessitated by the above concept of a "measurement") we wish to draw a dependable conclusion from the energy content of the body K as to the position co-ordinate of the pointer, we have to see to it that the body surely gets into the lower energy state when it gets into contact with T_B. In other words:

$$p(T_A) = 1, q(T_A) = 0; \qquad (27)$$
$$p(T_B) = 0, q(T_B) = 1.$$

This of course cannot be achieved, but may be arbitrarily approximated by allowing T_A to approach absolute zero and the statistical weight g to approach infinity. (In this limiting process, T_0 is also changed, in such a way that $p(T_0)$ and $q(T_0)$ remain constant.) The equation (26) then becomes:

$$S_A = -k \log p(T_0);$$
$$S_B = -k \log q(T_0) \qquad (28)$$

and if we form the expression $e^{-S_A/k} + e^{-S_B/k}$, we find:

$$e^{-S_A/k} + e^{-S_B/k} = 1. \qquad (29)$$

Our foregoing considerations have thus just realized the smallest permissible limiting care. The use of semipermeable walls according to Figure 1 allows a complete utilization of the measurement: inequality (1) certainly cannot be sharpened.

As we have seen in this example, a simple inanimate device can achieve the same essential result as would be achieved by the intervention of intelligent beings. We have examined the "biological phenomena" of a nonliving device and have seen that it generates exactly that quantity of entropy which is required by thermodynamics.

APPENDIX

In the case considered, when the frequency of the two states depends on the temperature according to the equations:

$$p(T) = \frac{1}{1 + ge^{-u/kT}} ; q(T) = \frac{ge^{-u/kT}}{1 + ge^{-u/kT}} \qquad (30)$$

and the mean energy of the body is given by:

$$\bar{u}(T) = uq(T) = \frac{uge^{-u/kT}}{1 + ge^{-u/kT}}, \qquad (31)$$

the following identity is valid:

$$\frac{1}{T}\frac{d\bar{u}}{dT} = \frac{d}{dT}\left\{\frac{\bar{u}(T)}{T} + k \log\left(1 + e^{-u/kT}\right)\right\}. \qquad (32)$$

Therefore we can also write the equation:

$$B_A = \frac{\bar{u}(T_A) - \bar{u}(T_0)}{T_0} + \int_{T_A}^{T_0} \frac{1}{T}\frac{d\bar{u}}{dT}dT \qquad (33)$$

as

$$S_A = \frac{\bar{u}(T_A) - \bar{u}(T_0)}{T_0}$$
$$+ \left\{\frac{\bar{u}(T)}{T} + k \log(1 + ge^{-u kT})\right\}_{T_A}^{T_0}, \qquad (34)$$

and by substituting the limits we obtain:

$$S_A = \bar{u}(T_A)\left(\frac{1}{T_0} - \frac{1}{T_A}\right) + k \log \frac{1 + ge^{-u/kT_0}}{1 + ge^{-u/kT_A}}. \qquad (35)$$

If we write the latter equation according to (25):

$$1 + ge^{-u/kT} = \frac{1}{p(T)} \qquad (36)$$

for T_A and T_0, then we obtain:

$$S_A = \bar{u}(T_A)\left(\frac{1}{T_0} - \frac{1}{T_A}\right) + k \log \frac{p(T_A)}{p(T_0)} \qquad (37)$$

and if we then write according to (31):

$$\bar{u}(T_A) = uq(T_A) \qquad (38)$$

we obtain:

$$S_A = q(T_A)\left(\frac{u}{T_0} - \frac{u}{T_A}\right) + k \log \frac{p(T_A)}{p(T_0)}. \qquad (39)$$

If we finally write according to (25):

$$\frac{u}{T} = -k \log \frac{q(T)}{gp(T)} \qquad (40)$$

for T_A and T_0, then we obtain:

$$S_A = q(T_A) k \log \frac{p(T_0)}{q(T_0)} \frac{q(T_A)}{p(T_A)}$$
$$+ k \log \frac{p(T_A)}{p(T_0)}. \qquad (41)$$

310   LEO SZILARD

We obtain the corresponding equation for S_B, if we replace the index A with B. Then we obtain:

$$S_B = q(T_B)\, k \log \frac{p(T_0)}{q(T_0)}\frac{q((T_B)}{p((T_B)} + k \log \frac{p(T_B)}{p(T_0)}. \quad (42)$$

Formula (41) is identical with (26), given, for S_A, in the text.

We can bring the formula for S_B into a somewhat different form, if we write:

$$q(T_B) = 1 - p(T_B), \quad (43)$$

expand and collect terms, then we get

$$S_B = p(T_B)\, k \log \frac{q(T_0)}{p(T_0)}\frac{p(T_B)}{q(T_B)} + k \log \frac{q(T_B)}{q(T_0)}. \quad (44)$$

This is the formula given in the text for S_B.

REFERENCES

Smoluchowski, F. *Vorträge über die kinetische Theorie der Materie u. Elektrizitat.* Leipzig: 1914.

Szilard, L. Zeitschrift fur Physik, 1925, 32, 753.

Experimental Work on X-Ray Effects in Crystals

Ein einfacher Versuch zur Auffindung eines selektiven Effektes bei der Zerstreuung von Röntgenstrahlen.

Von **H. Mark** und **L. Szilard** in Berlin-Dahlem.

Mit einer Abbildung. (Eingegangen am 23. Juni 1925.)

Es wird eine Methode angegeben, die eine sprunghafte Änderung im Beugungs-
vermögen für Röntgenstrahlen mit großer Empfindlichkeit nachzuweisen gestattet.
Sie beruht darauf, daß von der Oktaederebene des Rb Br in erster Ordnung zwar
keine Röntgenstrahlung reflektiert wird, welche das Brom- und Rubidiumion in
gleicher Weise zerstreuen, die Reflexion aber auftritt, sobald die Strahlung vom
Bromion selektiv anders gebeugt wird als vom Rubidiumion. Eine solche selektive
Beugung wurde z. B. für Strontium-$K\alpha$-Strahlung gefunden, welche härter ist als
die Absorptionskante des Broms, aber weicher als die des Rb, so daß sie wohl
vom Br, nicht aber vom Rb selektiv absorbiert wird.

Bestrahlt man irgend eine Substanz mit Röntgenlicht, so gehen von
ihr sekundäre Röntgenstrahlen aus, welche man im Rahmen der klassischen Theorie als Kugelwellen ansehen kann und welche sich dadurch
charakterisieren lassen, daß man angibt, wie sich ihre Amplitude zur
Amplitude des Primärstrahles verhält und welchen Phasenunterschied
die sekundären Kugelwellen gegen die Primärwellen besitzen. Stellt man
die Frage, wie sich Phasendifferenz und Amplitudenverhältnis verändern,
wenn man die Frequenz des auffallenden Primärstrahles variiert, so ist es
wohl sicher, daß man im allgemeinen beide als langsam veränderliche
Funktionen der Wellenlänge ansehen darf. Es könnte aber sein, daß bei
bestimmten Wellenlängen, welche für die zerstreuende Substanz charakteristisch sind, fast sprunghafte Änderungen auftreten. Wenn man das
Bild, welches sich die klassische Theorie vom Zerstreuungsvorgang macht,
als richtig ansieht, so muß man z. B. erwarten, daß bei kontinuierlicher
Verkleinerung der Primärwellenlänge eine derartige sprunghafte Änderung dann eintritt, wenn man aus dem langwelligen Bereich diesseits
der Absorptionskante, innerhalb dessen die Strahlung von der Substanz
nur sehr wenig absorbiert wird, durch Überschreiten der Absorptionskante
in denjenigen Frequenzbereich eintritt, in welchem die Strahlung sehr
stark absorbiert wird. Dies ist um so mehr zu erwarten, als der von der
klassischen Theorie geforderte Zusammenhang zwischen Amplitude und
Phase der sekundären Kugelwellen einerseits, und dem Absorptionskoeffi-
zienten andererseits für das optische Gebiet sich in dem bekannten anomalen
Gange des Brechungsindex in der Nähe der Absorptionslinie widerspiegelt.

Um eine derartige, etwa vorhandene sprunghafte Änderung im
Gebiet der Röntgenfrequenzen wirklich nachzuweisen, kann man die

H. Mark und L. Szilard, Ein einfacher Versuch zur Auffindung usw.   689

Reflexionen von Röntgenstrahlen an einem Rubidiumbromidkristall untersuchen; Rubidium und Brom stehen im periodischen System nahe bei-
sammen, sie sind nur durch ein Edelgas voneinander getrennt. Da bei
der Bildung des Salzes das Rubidiumion ein Elektron abgegeben und
das Bromion eines aufgenommen hat, ist außerdem die Elektronenzahl
der beiden Ionen die gleiche. Es ist daher nach der bisherigen Erfahrung [1]) — von selektiven Effekten abgesehen — zu erwarten, daß
Röntgenstrahlen von beiden Ionen in gleicher Weise zerstreut werden.
Belichtet man z. B. die (111)-Ebene (Oktaederebene) von Rubidiumbromid
unter dem Glanzwinkel in erster Ordnung mit Kupfer K-Strahlen, so ist
zu erwarten, daß keine Reflexion eintritt; denn die (111)-Ebenen sind
abwechselnd gleich dicht mit Brom und Rubidiumionen belegt, und wenn
diese Ebenen die Röntgenstrahlen in gleicher Weise zerstreuen, so heben
sich ihre Wirkungen gegenseitig gerade auf. Tatsächlich fanden wir
diese Reflexion bei Verwendung von Kupfer-, Eisen-, Kobalt- oder Zink-
strahlung vollkommen ausgelöscht. Belichtet man aber mit einer
Wellenlänge, welche von Bromionen anders zerstreut wird,
als von Rubidiumionen, so muß die Reflexion auftreten. Dies
liefert eine empfindliche Methode, um einen Sprung in Phase oder Amplitude der sekundären Kugelwellen nachzuweisen.

Wir haben nun an mehreren wohlausgebildeten Rubidiumbromid-
kristallen von etwa 6 mm Kantenlänge die Oktaederebene angeschliffen
und dann Strontium-$K\alpha$-Strahlung daran in verschiedenen Ordnungen
reflektiert. Diese Strahlung ist ein wenig härter als die Absorptionskante des Broms, aber weicher als die betreffende Kante des Rubidiums;
ihre Wellenlänge liegt in dem schmalen Bereich zwischen den beiden
Kanten und die Strahlung wird daher vom Brom selektiv absorbiert,
nicht aber vom Rubidium. Wenn sie daher vom Brom auch anders zerstreut wird als vom Rubidium, so müßten nunmehr die sonst ausgelöschten
Reflexionen der ungeraden Ordnungen von (111) hier auftreten. Sie
treten in der Tat mit recht merklicher Intensität auf, dagegen verhält
sich der Kristall der Emissionslinie des Broms gegen über wie gegen
Kupferstrahlung. Die Emissionslinie des Broms wird also vom Bromion
nicht merklich selektiv zerstreut, wie man dies nach einer Arbeit von
R. Clark und W. Duane [2]) hätte erwarten können.

---

[1]) W. H. und W. L. Bragg, Proc. Roy. Soc. **88**, 428, 1913; **89**, 248, 1914,
und W. P. Davey, Phys. Rev. **21**, 143, 1923.
[2]) R. Clark und W. Duane, Journ. Optical Soc. **8**, 90, 1922.

690   H. Mark und L. Szilard,

Zur Herstellung der verwendeten Rubidiumbromidkristalle haben
wir reinstes Rb Br aus fast gesättigter wässeriger Lösung durch lang-
sames Abdunsten auskristallisiert; nach mehreren Wochen bildeten sich
hierbei neben unbrauchbaren Kristalldrusen auch ganz wasserklare Würfel
bis zu 6 mm Kantenlänge. Die Analyse mehrerer solcher Kristalle ergab
einen Bromgehalt von 48,10, 48,35, 48,45 Proz., während sich 48,33 Proz.
berechnet.   Da es uns nicht gelang, die Oktaederebene als Wachstums-
fläche zu erhalten, haben wir sie durch Anschleifen für die spektro-
skopischen Aufnahmen freigelegt.   Da der Kristall sehr hygroskopisch

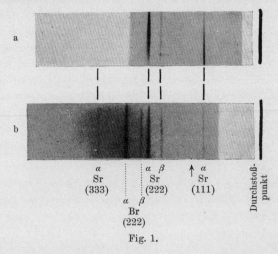

Fig. 1.

ist, muß man dabei schnell verfahren und ihn sofort mit einem wasser-
schützenden Lack überziehen, weil er sonst in kurzer Zeit unbrauchbar wird.

Die Fig. 1a zeigt eine Spektralaufnahme von Strontiumstrahlung
an der (111)-Ebene in einer Seemannschen Kamera.   Man erkennt die
erste und zweite Ordnung der Reflexion, welche der Lage nach genau
mit der theoretisch geforderten übereinstimmen (vgl. Tabelle 1).   Das
Intensitätsverhältnis ist jedoch zugunsten der ersten Ordnung verschoben,
da bei dieser Aufnahme — welche im wesentlichen die Auffindung der
ersten Ordnung bezweckte — die Wackelvorrichtung so eingestellt war,
daß der Kristall in der Reflexionsstellung erster Ordnung etwa fünf- bis
achtmal solange verweilte, als unter dem Glanzwinkel der zweiten Ordnung.

Die Fig. 1b zeigt eine Spektralaufnahme eines anderen Rb Br-Kri-
stalls in derselben Kamera.   Diesmal wurde eine Antikathode aus Sr Br$_2$
verwendet; die Drehung des Kristalls erfolgte gleichmäßig.   Man kann

Ein einfacher Versuch zur Auffindung eines selektiven Effektes usw.   691

Tabelle 1.

Kristall Nr.	Verwendete Strahlung	Abstand der Linie vom Null-punkt in Milli-meter $1^0 = 2{,}54$ mm	$\vartheta/2$ gefund.	sin $\vartheta/2$ gefund.	sin $\vartheta/2$ berechn.	In-dizierung	
1 Br = 48,10 Proz.	Cu K $\lambda_\beta = 1{,}389$ $\lambda_\alpha = 1{,}541$	51,5 57,2 112,2 128,0	$20^0 20'$ 22 35 44 10 40 20	0,347 0,384 0,696 0,769	0,348 0,385 0,696 0,770	222 222 444 444	β α β α
3 Br = 48,45 Proz.	Zn K $\lambda_\beta = 1{,}294$ $\lambda_\alpha = 1{,}435$	48,0 53,8 102,3 116,5	$18^0 55'$ 21 10 40 15 45 50	0,323 0,360 0,646 0,717	0,323 0,359 0,646 0,718	222 222 444 444	β α β α
2 Br = 48,35 Proz.	Cu K + Sr K $\lambda_\beta = 0{,}779$ (Sr) $\lambda_\alpha = 0{,}875$ (Sr)	16,1 28,4 32,4 49,0 51,5 57,3 58,2 66,2	$6^0 20'$ 11 10 12 45 19 15 20 20 22 35 22 55 26 5	0,110 0,194 0,220 0,329 0,347 0,385 0,389 0,439	$0{,}109_5$ 0,194 0,219 0,329 0,348 0,385 0,388 0,438	111 Sr 222 Sr 222 Sr 333 Sr 222 Cu 222 Cu 444 Sr 444 Sr	α β α α β α β α
1 Br = 48,10 Proz.	Sr K $\lambda_\beta = 0{,}779$ $\lambda_\alpha = 0{,}875$	16,1 28,3 32,4 49,1 58,2 66,2	$6^0 20'$ 11 10 12 45 19 20 22 55 26 5	0,110 0,193 0,219 0,330 0,388 0,439	0,1095 0,194 0,219 0,329 0,388 0,438	111 222 222 333 444 444	α β α α β α
2 Br = 48,35 Proz.	Br K + Sr K $\lambda_\beta = 0{,}929$ $\lambda_\alpha = 1{,}040$	16,0 28,2 32,0 33,8 38,0 47,8	$6^0 20'$ 11 6 12 36 13 20 15 0 18 50	0,110 0,192 0,217 0,231 0,259 0,323	$0{,}109_5$ 0,194 0,219 0,232 0,260 0,329	111 Sr 222 Sr 222 Sr 222 Br 222 Br 333 Sr	α β α β α α

auf diesem Diagramm zunächst die erste Ordnung der Strontiumlinie er-kennen (Sr α 111); etwa 2 mm links von ihr müßte die erste Ordnung der Bromemissionslinie liegen (durch einen Pfeil angedeutet); sie ist jedoch nicht[1]) vorhanden, obwohl, wie man an den Reflexionen der zweiten Ordnungen erkennen kann, die Intensitäten der Brom- und Strontiumstrahlung im Primärstrahl von derselben Größe waren.

Die Tabelle 1 enthält die Vermessung der beiden Aufnahmen sowie noch einiger mit anderen Strahlungen hergestellter Diagramme.

Kaiser Wilhelm-Institut für Faserstoffchemie und Institut für theoretische Physik der Universität.

---

[1]) Wenigstens nicht mit einer merklichen Intensität.

743

Die Polarisierung von Röntgenstrahlen durch Reflexion an Kristallen.

Von **H. Mark** und **L. Szilard** in Berlin-Dahlem.

Mit fünf Abbildungen. (Eingegangen am 27. November 1925.)

Es wird in einigen Fällen gezeigt, daß das an einem Kristall unter 90⁰ nach der Braggschen Beziehung reflektierte Röntgenlicht praktisch vollständig polarisiert ist und daß bei Belichtung eines Kristalls mit linear polarisiertem Röntgenlicht keine Reflexion eintritt, falls die Richtung des reflektierten Strahles in die Schwingungsrichtung des polarisierten einfallenden Strahles fällt. Es lassen sich so Kristalle als Polarisatoren und Analysatoren für Röntgenstrahlen verwenden. Dabei erfolgt gleichzeitig eine spektrale Auflösung, so daß die Polarisationsverhältnisse des Fluoreszenzlichtes und des abgebeugten Lichtes getrennt zutage treten. Es zeigte sich so bei der Verwendung von Eisen als Streustrahler, daß die Fluoreszenzstrahlung im Gegensatz zum abgebeugten Licht auch unter einem Winkel von 90⁰ zum primären Strahl unpolarisiert ist. Die Polarisationsverhältnisse beim abgebeugten Licht lagen auch bei den untersuchten schweren Elementen so, als ob die Zerstreuung durch isotrop gebundene Elektronen erfolgen würde.

Die grundlegenden Versuche über die Polarisationserscheinungen an Röntgenstrahlen sind im Jahre 1905 von Barkla durchgeführt worden[1]). Er zerstreute ein Bündel von Röntgenstrahlen an einem Kohleblock und fand, daß die unter 90⁰ zum Primärstrahl abgehenden sekundären — gestreuten — Strahlen sich insofern bemerkenswert verhalten, als die durch sie an einem zweiten Kohlestück ausgelösten Tertiärstrahlen sich bezüglich ihrer Intensität ungleichförmig um den Sekundärstrahl herum verteilen. Seine Beobachtung deutete er durch die Annahme, daß die unter 90⁰ vom ersten Streustrahler ausgehenden gestreuten Strahlen vollständig polarisiert sind und daß dieses polarisierte Licht bei der Zerstreuung durch den zweiten Kohleblock in der Schwingungsrichtung des elektrischen Vektors keine und senkrecht hierzu die maximale Intensität ergibt. Das Ergebnis seiner Versuche läßt sich zusammenfassen in die Aussage: Bei der diffusen Zerstreuung von Röntgenstrahlen durch Kohle und andere leichte Elemente verlaufen die Erscheinungen so, als ob die Zerstreuung durch isotrop gebundene Elektronen nach den Gesetzen der klassischen Elektrodynamik erfolgte.

Es schien uns nun der Mühe wert, die Laueschen Kristallreflexionen mit Bezug auf die Polarisation zu untersuchen. Da die Reflexion am Kristall als eine Beugungserscheinung aufgefaßt werden muß, hat man

---

[1]) Vgl. den zusammenfassenden Bericht von Ch. G. Barkla, Jahrb. d. Radioakt. u. Elektronik **5**, 246, 1908.

*

744   H. Mark und L. Szilard,

auf Grund der Barklaschen Versuche mit diffus zerstreutem Lichte zu
erwarten, daß das an einer Kristallfläche nach der Braggschen Beziehung
unter dem Glanzwinkel reflektierte Röntgenlicht vollständig polarisiert
ist, falls der reflektierte Strahl senkrecht zum Primärstrahl steht. Ebenso
war zu erwarten, daß ein Kristall oder Kristallpulver linear polarisierte
Röntgenstrahlen überhaupt nicht reflektiert, wenn der reflektierte Strahl
in die Richtung des elektrischen Vektors des polarisierten Primärstrahles
fallen würde.

    Um zu sehen, ob dies tatsächlich zutrifft, wurde ein Versuch an-
gestellt, welcher durch Fig. 1 schematisch dargestellt ist. Die mono-

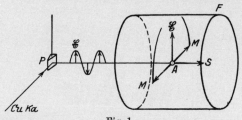

chromatische K-Strah-
lung einer Kupferanti-
kathode wird an einem
Steinsalzkristall (Polarisator P) so reflektiert,
daß der Ablenkungs-
winkel nahezu 90⁰ be-
trägt. Der reflektierte
Strahl S fällt dann auf

Fig. 1.
Schematische Darstellung der Versuchsanordnung.

ein Kristallpulver A und erzeugt Debye-Scherrerkegel. Uns interessieren
nun jene Kegel, deren Erzeugende ungefähr einen Winkel von 90⁰ mit dem
Primärstrahl einschließen. Wenn nämlich die am Polarisator reflektierte
Strahlung vollständig polarisiert ist, dann schwingt der elektrische Vektor
des auf A treffenden Strahles S parallel der Richtung $\mathfrak{E}$, und die vom
analysierenden Kristallpulver A unter 90⁰ ausgehenden Debye-Scherrerkegel dürfen in der Richtung $\mathfrak{E}$ keine Intensität enthalten; senkrecht
hierzu müssen sie hingegen die Maximalintensität besitzen.

    Legt man um das Kristallpulver einen zylindrischen Film F, dessen
Achse mit dem Strahl S zusammenfällt, so wird dieser durch die Debye-
Scherrerkegel in Kreisen geschnitten, welche nach dem Ausbreiten des
Zylinders in gerade Linien übergehen. Die Schwärzung entlang dieser
geraden Linien wird nun keineswegs gleichförmig sein, wie dies bei
einem Debye-Scherrerkreis der Fall wäre, welcher in der gebräuchlichen
Weise mit unpolarisiertem· Licht erzeugt ist; sie wird vielmehr einen
charakteristischen Verlauf zeigen.

    Die Fig. 2 zeigt ein in dieser Weise aufgenommenes Bild. Statt
an einem Kristallpulver wurde an einer Aluminiumfolie als Analysator
reflektiert, welche um die Strahlachse gedreht wurde, um eine eventuell
vorhandene Abweichung von der regellosen Verteilung der Kriställchen

Die Polarisation von Röntgenstrahlen durch Reflexion an Kristallen. 745

auszugleichen. Die beiden geraden Linien sind durch Aufrollen des Films aus den Debye-Scherrerkreisen hervorgegangen, welche den Reflexionen an zwei verschiedenen Netzebenen entsprechen [(222) und (311)]; man sieht deutlich die zu erwartenden Maxima und Minima der Schwärzung.

Bei der Fig. 2 verschwindet im Minimum die Schwärzung nicht vollkommen, was wohl damit zusammenhängt, daß die beiden Reflexionen an den betreffenden Stellen nicht genau unter 90⁰ erfolgten. Bei der als

Cu $K\alpha$-Strahlung an NaCl (422) polarisiert. $\vartheta = 86^0$.

· (311) 78⁰
· (222) 82⁰ 45′

Fig. 2.

Zn $K\alpha$-Strahlung an Al polarisiert. $\vartheta = 90^0$.

Al (400) $\vartheta = 90^0$.

Fig. 3.

Cu $K\alpha$-Strahlung an NaCl (422) polarisiert. ϑ 86⁰.

86⁰ (222) .
80⁰ (311) ·

Pt-Folie.

96⁰ (222) :
90⁰ 30′ (311)

Cu-Folie.
Fig. 4.

Fig. 3 wiedergegebenen Aufnahme war dies jedoch der Fall; eine Schwärzung ist hier in der Tat im Minimum nicht bemerkbar.

In Fig. 4 sind zwei Aufnahmen wiedergegeben, welche mit polarisierter Kupfer-K-Strahlung an Kupfer- und Platinfolien als Analysator erhalten wurden. Man sieht auch hier einen Schwärzungsverlauf, welcher einer praktisch vollständigen Polarisation entspricht.

Auch bei den schwereren Elementen erfolgt also offenbar die Beugung so, als ob sie klassisch durch isotrop gebundene Elektronen bewirkt würde. Barkla fand bei seinen Versuchen an schwereren Elementen allerdings einen scheinbar geringeren Polarisationsgrad. Man muß aber

*

746   H. Mark und L. Szilard,

bedenken, daß bei Polarisationsversuchen mit diffus zerstreuten Röntgenstrahlen die Fluoreszenzstrahlung sich dem abgebeugten Lichte über-
lagert und so einen zu geringen Polarisationsgrad vortäuschen kann.
Dies war auch die Deutung, die Barkla seinen Beobachtungen bei
schwereren Elementen gab.

    Während die bisherigen Untersuchungen stets mit diffus zerstreuten
Strahlen vorgenommen wurden, ergibt sich bei der Verwendung eines
Kristallpulvers oder eines Einkristalls als Analysator der Vorteil, daß
dann zugleich eine spektrale Auflösung erfolgt und daß man daher unter-
scheiden kann, wie die Polarisationsverhältnisse bei den verschiedenen
Wellenlängen liegen, welche im einfallenden Röntgenlicht vorhanden sind.

90°

· Cu $K\beta$
· Cu $K\alpha$
· Fe $K\beta$
· Fe $K\alpha$

Fig. 5.

Cu K-Strahlung wurde an Eisen unter
84° gestreut (211) und der Sekundärstrahl an Gips (060) spektroskopiert.
Hierbei wird die Fluoreszenzstrahlung
Fe $K\alpha$ unter einem Winkel von 101°
reflektiert; sie ist unpolarisiert; die
gestreute Cu $K\alpha$-Strahlung wird unter
75° 40′ reflektiert und zeigt normalen
Polarisationseffekt.

So erkennt man an Hand von
Fig. 5 im Falle des Eisens, daß die
unter 90° abgebeugte Strahlung fast
vollständig polarisiert ist, während
die Fluoreszenzstrahlung unter 90°
nicht merklich polarisiert ist. Die
Aufnahme erfolgte so, daß die Strahlung einer Kupferantikathode an
einem Eisenblock — Polarisator —
zerstreut wurde und ein senkrecht
zum Primärstrahl ausgehendes Bündel
auf einen gedrehten Kristall — Analysator — traf, welcher die Rolle
des bei den früheren Versuchen verwendeten Kristallpulvers spielte. Bei

der Verwendung eines Einkristalls statt eines Kristallpulvers als Analysator
steht eine geringere Mannigfaltigkeit von Kristallagen zur Verfügung, und
es muß daher das auf den Analysator fallende Bündel einen bestimmten
Öffnungswinkel haben, wenn die in ihm vorhandenen verschiedenen Wellenlängen zur Reflexion kommen sollen.   Im Falle unseres Versuchs enthält
das Bündel hauptsächlich die abgebeugte Kupfer-K-Strahlung und die
Fluoreszenzstrahlung des Eisens, welche von der Kupfer-K-Strahlung stark
angeregt wird. Der Glanzwinkel der Kupfer-K-Strahlung lag ungefähr in
demselben Betrag unter 90°, um welchen der Glanzwinkel für die Eisen-
strahlung 90° überschritt, so daß die Intensitätsverteilung entlang der
beiden Interferenzkreise bei gleichem Polarisationsgrad für beide Strahlungen dieselbe sein müßte.   Wie die Aufnahme zeigt, erscheint jedoch
der Eisenkreis als praktisch unpolarisiert, während der Kupferkreis deut-

Die Polarisation von Röntgenstrahlen durch Reflexion an Kristallen. 747

lich den Polarisationseffekt zeigt, so daß man aus der Aufnahme auf das Fehlen der Polarisation bei der Fluoreszenzstrahlung schließen und die diesbezüglichen Angaben von Barkla bestätigen kann.

Der Notgemeinschaft der deutschen Wissenschaften danken wir bestens für die gewährte Unterstützung, und insbesondere möchten wir nicht verfehlen, Herrn Dr. Stuchtey für seine stets bereitwillige Hilfeleistung zu danken.

Berlin-Dahlem, Kaiser Wilhelm-Institut für Faserstoffchemie und Institut für theoretische Physik an der Universität Berlin.

Nuclear Physics

Introduction by Maurice Goldhaber*

In this essay, I shall deal with Leo Szilard's publications in the field of nuclear physics which stem from his brief but historically important "English period." Since I was partly responsible for fanning his interest in nuclear physics at that time, I am glad to have been asked to introduce his work in this field.

Leo Szilard had wide interests, few fields of human endeavor escaping his inquiring mind. When confronted with a problem, he tried to isolate its essential core, and this was especially so with scientific problems. In contemplating the life of a person with such a fertile mind, it is of interest to see what problems he found important enough to pursue to completion. This is best shown in his published papers. His published scientific papers are few in number. In each one he tried to make an important point, to prove something of general interest and consequence.

Let us consider here Szilard's contributions to nuclear physics and chemistry. His first experiments in these fields were carried out in London. He had come there early in 1933 from Berlin, where I had met him while he was Privatdozent and I was a student. I was struck by the way he would talk to young people without stuffiness or condescension. I came to England in the spring of that year and entered Cambridge University in the fall as a research student. I saw him often during his years in England, mostly in London and occasionally in Cambridge or Oxford. We usually discussed progress in nuclear physics and the "state of the world." It is not widely appreciated that Szilard, who started out to be a theoretical physicist, had great powers of observation as his two "accidental" discoveries discussed below illustrate: one of the first examples of an artificially produced nuclear isomer (in indium) and, independent of Fermi and Amaldi, the first observation of "neutron resonances."

"Chemical Separation of the Radioactive Element from its Bombarded Isotope in the Fermi Effect," by Leo Szilard and T. A. Chalmers, *Nature*, **134**, 462 (Sept. 22, 1934).

The discovery described in this paper has become known as the "Szilard-Chalmers effect." Its genesis is typical of Szilard's way of thinking and acting. He was at that time in London. He was contemplating (he told me once) the possibility that some of the neutron-produced radioactive isotopes discovered by Fermi and his collaborators might make medically useful compounds. However, Szilard realized immediately that chemical compounds exposed to neutrons would be disrupted by neutron capture. Characteristically, he turned this apparent defeat around, and it led him to a brilliantly simple method of isotope separation. At that time any isotope separation was still a major feat. He therefore proposed to a young physicist, T. A. Chalmers, who was working at St. Bartholomew's Hospital, that they should collaborate on investigating the possibility of concentrating I^{128} produced by an (n, γ) reaction in an iodine compound. They started with ethyl iodide in solution and precipitated the "free" radio-active iodine, which does not exchange with the iodine in the ethyl iodide, as silver iodide. This method is now classical. Its mechanism has been widely studied and there are many variations in the means of production of the free radioactive isotope from molecules [e.g., by (γ, n) reactions, by conversion electrons accompanying isomeric transitions, etc.] There is some retention of the radioactive isotope in the starting compound under certain conditions, as well as exchange of the recoil atoms; this allows promising medical applications and thus comes back to Szilard's original hope.

* Director, Brookhaven National Laboratory, Upton, Long Island, New York.

"Detection of Neutrons Liberated from Beryllium by Gamma Rays: A New Technique for Inducing Radioactivity," by Leo Szilard and T. A. Chalmers, *Nature*, **134**, 494–495 (Sept. 29, 1934).

Soon after the observation of photoprotons from the photodisintegration of the deuteron was reported, Szilard and Chalmers used their newly developed method of concentrating radioactive iodine as a sensitive way of detecting photoneutrons from beryllium disintegrated by radium γ rays. These photoneutrons have turned out to be a very useful source of fairly low-energy neutrons (a few hundred keV). They are produced by γ rays above the (γ, n) threshold of Be^9, which has the lowest known threshold for stable isotopes. A Ra-γ-Be photoneutron source was often used by Szilard in his later researches, particularly in fission experiments.

"Liberation of Neutrons from Beryllium by X-Rays: Radioactivity Induced by Means of Electron Tubes," by A. Brasch, F. Lange, A. Waly, T. E. Banks, T. A. Chalmers, Leo Szilard, and F. L. Hopwood, *Nature*, **134**, 880 (Dec. 8, 1934).

This paper well illustrates Szilard's mode of research—ahead of his time: If γ rays, at that time available in very limited sources, can disintegrate beryllium, why not use powerful high-energy x-ray sources? The only source of this kind well known to him was at Berlin, but distance was no barrier to him. Nowadays researchers take working at a distant machine for granted. He got in touch with his former colleagues there and arranged to have bromoform exposed to photoneutrons produced from beryllium by x rays with energies exceeding 1.5 MeV. The bromoform was flown to London where the radioactive Br was separated by a "Szilard-Chalmers" separation. A weak activity decaying with a six-hour period was observed. (We now know that slow neutrons induce in Br activities of 4.5 h and 35.3 h half-life besides shorter lived activities, another early example of nuclear isomers.) The question of whether a sharp energy threshold exists is raised in this paper. Szilard has recounted the fact that at that time (because of a mistake in the measurement of the Be^9 mass) it was not certain that a sharp threshold would exist and that Be^9 might be metastable. This led him to his interesting first speculations on nuclear chain reactions before the discovery of fission. The threshold for the photodisintegration of Be^9 is now known to be 1.66 MeV.

"Radioactivity Induced by Neutrons," by Leo Szilard and T. A. Chalmers, *Nature*, **135**, 98 (Jan. 19, 1935).

Here Szilard and Chalmers report an interesting observation in indium exposed to neutrons. They found a new period which they estimated to have a $3\frac{1}{2}$ h half-life (now known to be 4.4 h). This period had been overlooked by Fermi's group (who had found half-lives of 13 sec and 54 min), and Szilard and Chalmers correctly concluded that "one of the two indium isotopes is activated with more than one period." They realized that this could not be caused by one of the processes already established by Fermi and his collaborators, because the new activity was not "water sensitive." Since In seemed too heavy an element to undergo (n, p) or (n, α) reactions with Rn-α-Be neutrons, their "Letter" ended characteristically with the remark that their findings "seem to deserve further investigation, for which adequate instruments of observation are not at present at our disposal."

Thus ended the short but exceedingly fruitful collaboration between Szilard and Chalmers. Decades later, Chalmers told me how fondly he remembered those days.

"Absorption of Residual Neutrons," by Leo Szilard, *Nature*, **136**, 950–951 (Dec. 14, 1935).

In 1935, at the invitation of Professor Lindemann (later Lord Cherwell), Szilard moved to Oxford, where he enjoyed the hospitality of the Clarendon Laboratory. The work on "Absorption of Residual Neutrons" appears to be the only experiment that Szilard carried out without any collaborator. He reports the discovery that slow neutrons, filtered by cadmium and detected through the 54 min activity induced in indium, are selectively absorbed by indium. He observed a similar effect in iodine and correctly concluded "that some elements have fairly sharp regions of strong absorption in an energy region for which cadmium is transparent." To carry out these experiments cleanly, he used a "paraffin wax tube of 13 cm inner diameter and 20 to 40 cm length" to lead the neutrons from a radon–beryllium source to the detector. I have found this invention quite useful for research in which the fast neutron background had to be reduced, and "descendants" of this invention are now used at many reactors. The discovery of "residual neutrons" (resonance neutrons) was made independently also by Amaldi and Fermi. These observations proved of great importance in stimulating the development of the Breit-Wigner formula and Niels Bohr's compound model of the nucleus.

"Gamma Rays Excited by Capture of Neutrons," by J. H. E. Griffiths and Leo Szilard, *Nature*, **139**, 323–324 (Feb. 20, 1937).

By making use of a paraffin tube filled with lead to reduce both fast neutron and γ-ray background from a radon–beryllium source, Griffiths and Szilard made some interesting observations on neutron-capture γ rays. This is still a very active field of research today. They found that, per neutron captured, the effect produced in a γ-ray counter does not differ much for a great variety of elements. One deviation from the "normal" effect they correctly interpreted as due to an undiscovered long-lived activity in chlorine (now known to be the 3×10^5-year activity in Cl^{36} produced with an (n, γ) cross section of 44 barns in Cl^{35}). Other deviations they ascribed to strongly absorbing impurities. They remarked, "The values obtained indicate that more than seven quanta are emitted from cadmium per captured neutron, but this result requires confirmation by an independent method which is now being attempted." I do not remember hearing about this method, and Szilard appears to have become interested in other things. The average number of quanta per neutron captured is now known to be about three to four.

"Radioactivity Induced by Nuclear Excitation: I. Excitation by Neutrons," by M. Goldhaber, R. D. Hill, and Leo Szilard, *Physical Review*, **55**, 47–49 (Jan. 1, 1939).

While Szilard was at Oxford, he and I often discussed the puzzling "extra" activity in indium which Szilard and Chalmers had observed. Sometime during 1937, we decided to collaborate on a systematic investigation of this activity, and R. D. Hill, who had come to the Cavendish Laboratory from Australia as a research student, joined us in these experiments. In December 1937 Szilard left England for the United States. Before leaving he came up to Cambridge for a farewell dinner with me at Magdalene College. At dinner he met I. A. Richards (of *Basic English* fame) and learned that Richards was going to the United States on a visit. Szilard at that time was rather pessimistic about the future of Europe and quipped to Richards that he had better buy a one-way ticket. (Later Richards became a Harvard Professor.) After Szilard left, Hill and I continued to investigate the indium activity until the spring of 1938. At that time I visited the United States for the first time to see this country and science here for myself. Szilard met me at the boat and introduced me to New York. During the Washington meeting of the American Physical Society, Professor F. W. Loomis, Chairman of the Physics Department of the University of Illinois, offered me a position as Assistant Professor and Szilard strongly advised me to look into this carefully. After a visit to Urbana-Champaign, I accepted the position, returned to England and immigrated to this country in time for the academic year in the fall of 1938. I arrived at Illinois shortly before the Munich crisis. Szilard spent the time of the crisis mostly in my new apartment in Champaign, and we listened to the news on a radio which I had bought for this purpose. Hill had come along with me to Illinois, and together with Szilard we wrote up our results on the indium activity.

We could show that this activity was due to a new process: the excitation in In^{115} by inelastic neutron scattering into an isomeric state. It was later shown that this isomeric state can be excited by many types of radiation (γ, p, α, e, etc.), and many other examples of such an isomeric activity have since been found. In^{115m} was probably the first well-understood isomer, and thus the observation of Szilard and Chalmers turned out to be very important for the development of systematic isomer research.

(*Reprinted from* NATURE, *Vol.* 134, *page* 462, *September* 22, 1934.)

CHEMICAL SEPARATION OF THE RADIOACTIVE ELEMENT FROM ITS BOMBARDED ISOTOPE IN THE FERMI EFFECT

BY

DR. LEO SZILARD

AND

T. A. CHALMERS

Physics Department, Medical College,
St. Bartholomew's Hospital, London,
E.C.1.

FOLLOWING the pioneer experiment of Fermi, it has been found by Fermi, Amaldi, D'Agostino, Rasetti and Segrè that many elements up to the atomic number 30, when bombarded by neutrons from a radon-beryllium source, are transmuted into a radioactive element which is chemically different from the bombarded element. In several cases of this type, they succeeded in separating chemically the active substance from the bulk of the bombarded element, and there is no inherent difficulty in getting any desirable concentration of the radioactive element.

They have not observed such chemical changes in elements above the atomic number 30, though many of these heavier elements show strong Fermi effects. For some of these, for example, arsenic, bromine, iodine, iridium and gold, they could show that the activity is carried by the bombarded element, which in the circumstances leads to the conclusion that the radioactive element is an isotope of the bombarded element.

In order to separate the radioactive isotope of the bombarded element from the bulk of the bombarded element, one has to find a new principle of separation. We have attempted to apply the following principle. If we irradiate by a neutron source a chemical compound of the element in which we are interested, we might expect those atoms of the element which are struck by a neutron to be removed from the compound. Whether the atoms freed in this way will interchange with their isotopes bound in the irradiated chemical compound will depend on the nature of the chemical compound with which we have to deal. If we work under conditions in which such an interchange does not take place, we obtain the radioactive isotope 'free', and by separating the 'free' element from the compound we can obtain any desirable concentration of the radioactive isotope.

We have applied this principle to iodine. Ethyl iodide has been irradiated and a trace of free iodine added to protect the radioactive isotope. By reduction and precipitation as silver iodide in water, it was easy to concentrate the activity so as to get from the precipitate ten times as many impulses of the Geiger-Müller β-ray counter as directly from the irradiated ethyl iodide[1]. Apparently a large fraction of the active substance could be extracted from the ethyl iodide. The quantity of the active element obtainable in the precipitate will naturally depend on the quantity of the compound subjected to irradiation.

This principle of isotopic separation has also been applied to some other elements which, like iodine, are transmuted into their own isotopes, and further experiments mostly with organic compounds are in progress.

[1] *Proc. Roy. Soc.*, A, **146**, 483; 1934.

Printed in Great Britain by FISHER, KNIGHT & Co., LTD., St. Albans

(Reprinted from NATURE, *Vol.* 134, *page* 494, *September* 29, 1934.)

DETECTION OF NEUTRONS LIBERATED FROM BERYLLIUM BY GAMMA RAYS: A NEW TECHNIQUE FOR INDUCING RADIOACTIVITY

BY

DR. LEO SZILARD

AND

T. A. CHALMERS

Physics Department, Medical College, St. Bartholomew's Hospital, London, E.C.1.

WE have observed that a radiation emitted from beryllium under the influence of radium gamma rays excites induced radioactivity in iodine, and we conclude that neutrons are liberated from beryllium by gamma rays.

Chadwick and Goldhaber were the first to observe a nuclear disintegration due to the action of gamma rays. In their pioneer experiment[1], they used a small ionisation chamber filled with heavy hydrogen and observed that protons were ejected from the heavy hydrogen under the influence of gamma rays from thorium C. Their method can be used for the detection of the gamma ray disintegrations of other elements, as such a disintegration would generally be accompanied by the ejection of *charged nuclei* which their method is designed to detect. On the other hand, apart from the unique case of heavy hydrogen, their method does not appear to give direct evidence on *neutron* radiations, which may in certain cases accompany gamma ray disintegrations.

It appeared to us of interest to search for such neutron radiations, and we thought that the Fermi effect might conveniently be used as an indicator of their presence. For certain reasons, we chose to use as indicators elements which, like iodine, are transmuted in the Fermi effect into their own radioactive isotopes.

In order to make our test more sensitive, we applied in this work the new principle of isotopic separation which we recently described[2]. In the present experiment we have used iodine as indicator, and separated radio-iodine from the bombarded iodine.

In one experiment we surrounded 150 mgm. of radium (in sealed containers of 1·0 mm. platinum filtration) with 25 gm. of beryllium, which was further surrounded by 100 c.c. ethyl iodide. The silver iodide precipitate obtained after irradiation from the ethyl iodide showed an activity decaying with a half period of 30 minutes. In spite of the inefficient geometrical arrangement of the beryllium in this experiment, we obtained from the active precipitate 200 impulses of the Geiger-Müller beta ray counter per minute. In the control experiment omitting the beryllium, we obtained less than 12 impulses per minute. The effect observed is sufficiently strong to be easily detected without separating chemically the radioactive element.

Our observations show that it will be possible to make experiments on induced radioactivity by using the gamma rays of sealed radium containers, which are available in many hospitals for therapeutic purposes. Further, it will be possible to have very much stronger sources of neutrons and to produce thereby larger quantities of radioactive elements by using X-rays from high-voltage electron tubes.

[1] NATURE, **134**, 237, Aug. 18, 1934.
[2] NATURE, **134**, 462, Sept. 22, 1934.

Printed in Great Britain by FISHER, KNIGHT & Co., LTD., St. Albans

(*Reprinted from* NATURE, *Vol.* 134, *page* 880, *December* 8, 1934.)

Liberation of Neutrons from Beryllium by X-Rays: Radioactivity Induced by Means of Electron Tubes

IT has been recently reported[1] that neutrons are liberated from beryllium by γ-rays of radium and that these are able to induce radioactivity in iodine. Following up this work, we have attempted to liberate neutrons from beryllium by means of hard X-rays, produced by high-voltage electron tubes. An electron tube, which could conveniently be operated by a high-voltage impulse generator at several million volts[2], is at present in use in the High Tension Laboratory of the A.E.G. in Berlin, and has served in the present experiment for the production of X-rays.

X-rays from a tungsten anticathode generated at a voltage above $1\cdot5 \times 10^6$ v. were allowed to fall on beryllium. An organic bromine compound (bromoform) was exposed to the radiation of the beryllium and this compound was then sent by air from Berlin to London. Here, at St. Bartholomew's Hospital, after an isotopic separation[3] of the radio-bromine from the ordinary bromine, a weak activity decaying with the six-hour period of radio-bromine was observed.

Afterwards, at a higher voltage, but still below 2×10^6 v., very much stronger activities were induced in bromine and were observed both in Berlin and London. Strong activities were observed in Berlin both in bromine and iodine (30 minutes half-life period) in co-operation with K. Philipp and O. Erbacher of the Kaiser Wilhelm Institute for Chemistry, the activity and its decay being easily measured by means of an electroscope. Recently, Fermi, Amaldi, Pontecorvo, Rasetti and Segrè discovered[4] that by surrounding the irradiated material with substances containing hydrogen the efficiency of activation of certain elements by neutron bombardment is greatly increased. Use was made of this effect in these experiments.

A very sharp increase of the induced activity with increasing voltage is to be expected if there is a more or less sharply defined upper limit of the wave-length at which the liberation of neutrons from beryllium begins. If there is such a critical wave-length, and if the voltage applied to the tube only slightly exceeds the corresponding critical voltage, a small fraction only of the total X-ray energy will be present in the form of radiation of sufficiently short wave-length; this fraction will then increase sharply with the excess voltage.

We wish to thank Prof. L. Meitner for her kind assistance in the Berlin experiments.

	A. BRASCH.
Berlin.	F. LANGE.
	A. WALY.

Medical College,	T. E. BANKS.
St. Bartholomew's Hospital,	T. A. CHALMERS.
London, E.C.1.	LEO SZILARD.
Nov. 26.	F. L. HOPWOOD.

[1] Szilard and Chalmers, NATURE, **134** 494, Sept. 29, 1934.
[2] Brasch and Lange, *Z. Phy.*, **70**, H. 1/2.
[3] Szilard and Chalmers, NATURE, **134**, 462, Sept. 22, 1934.
[4] Fermi, Amaldi, Pontecorvo, Rasetti and Segrè, *La Ricerca Scientifica*, **2**, Nos. 7–8.

Printed in Great Britain by FISHER, KNIGHT & CO., LTD., St. Albans

(*Reprinted from* NATURE, *Vol.* 135, *page* 98, *January* 19, 1935.)

Radioactivity Induced by Neutrons

AMALDI, d'Agostino and Segrè[1] report that, using neutrons from a radon - alpha particle - beryllium source, they have induced an activity in indium of a very short half-life period (13 sec.) and also one of half-life period of about one hour (54 min.).

Our own unpublished observations on indium show the one hour period and a longer period of several hours (estimated at $3\frac{1}{2}$ h.). If indium is irradiated in air these two periods show strong initial intensities of the same order of magnitude, but if it is irradiated in water, the one hour period is so strongly reinforced that it overshadows the long period and may thereby prevent its detection. Thus three periods appear to exist for indium, and the two shorter ones of these are reported[2] to be strongly water-sensitive.

Indium has two known isotopes[3] (mass numbers 113 and 115, the ratio of their abundance being less than one to ten). It has an odd atomic number and since, apart from the isolated case of hydrogen, there is no precedent for such an element having more than two isotopes, we tentatively assume that no further stable indium isotope is involved. Accordingly we conclude that one of the two indium isotopes is activated with more than one period.

The question arises whether the observed periods can be interpreted on the basis of the primary processes which have so far been recognised in the Fermi effect. These recognised processes are : (*a*) capture of the neutron by the nucleus (all cases so far investigated were reported to be water-sensitive) ; (*b*) ejection of a heavy positively charged particle—a proton or an alpha particle—from the nucleus (all cases so far investigated were reported not to be water-sensitive). Some isotopes of lighter elements are known to be activated with two or three periods, the ejection of a proton or an alpha particle being quite a common process for elements lighter than zinc (atomic number 30). No such processes have so far been observed for elements heavier than zinc.

In the circumstances the Fermi effect of indium (atomic number 49) seems to deserve further investigation, for which adequate instruments of observation are not at present at our disposal.

LEO SZILARD.
T. A. CHALMERS.
Physics Department, Medical College,
St. Bartholomew's Hospital, E.C.1.
Dec. 18.

[1] Amaldi, d'Agostino, Segrè, *Ricerca Scientifica*, V, **2**, No. 9–10 November 1934.
[2] Amaldi, d'Agostino, Fermi, Pontecorvo, Rasetti, Segrè, *Ricerca Scientifica*, V, **2**, No. 11–12, December 1934.
[3] Wehrli, *Helvetica Physica Acta*, **7**, 6, 611; 1934.

Printed in Great Britain by FISHER, KNIGHT & CO., LTD., St. Albans

(*Reprinted from* NATURE, *Vol.* 136, *page* 950, *December* 14, 1935.)

Absorption of Residual Neutrons

AMALDI, D'Agostino, Fermi, Pontecorvo, Rasetti and Segrè have discovered that certain elements strongly absorb neutrons which have been slowed down by paraffin wax[1]. They report, for example, that thin sheets of cadmium or indium of 0·013 gm./cm.^{2} and 0·3 gm./cm.^{2} thickness respectively cut down the intensity of a beam of slow neutrons to half its value and find for iodine a half-value thickness of 4 gm./cm.^{2}.

Thick sheets of a strongly absorbing element, such as cadmium, will, however, still transmit an appreciable fraction of the incident heterogeneous beam, and in these circumstances it appeared to be of interest to investigate the absorption of such residual neutrons in some elements.

In one set of experiments, I filtered slow neutrons by a sheet of cadmium, 1·6 mm. (1·4 gm./cm.^{2}) thick, and determined the absorption of the residual neutrons in several elements, using radioactivity induced in indium (54 min. period) as an indicator of the neutron intensity. The residual neutrons from the thick cadmium filter are scarcely absorbed by cadmium itself—a 0·5 mm. thick cadmium absorber will absorb perhaps less than 10 per cent of the residual neutrons. Yet I find that these residual neutrons are strongly absorbed by some elements—for example, a thin indium absorber of less than 0·3 gm./cm.^{2} (0·4 mm.) thickness absorbs more than two thirds of the intensity.

Cadmium absorbs the bulk of the unfiltered beam much more strongly than indium, but as we see, it is transparent for some component of the unfiltered beam which in its turn is strongly absorbed by indium. This fact is in contradiction to the current conclusions drawn from the theory of radiative capture.

Fermi *et al.*, Bethe, Perrin and Elsasser attempted to explain the observed large absorbing cross-sections of elements such as cadmium without assuming long-range forces of the nucleus. They have assumed that the range of these forces is small compared with the wave-length λ of the slow neutron and that the neutron is captured in a deep energy level of the nucleus, the excess energy of several million volts being carried away by an emitted photon. They have demonstrated that large effective absorbing cross-sections result if resonance occurs, and it can be shown that the degree of this resonance will not change appreciably with the neutron energy between thermal energies and some 10,000 electron volts. Consequently, if an element has an absorbing cross-section larger than another element for one particular neutron energy, it should not have a smaller absorbing cross-section for any other energy within this energy range.

The present observation on indium contradicts this conclusion. It cannot be argued in defence of the theory that the observed effect might be due to neutrons of energies higher than some 10,000 volts, since the observed very large absorbing cross-section of the indium atom for the residual neutrons would then have to exceed λ^2/π—the limit set by the theory.

In our experiments, use was made of the fact that slow neutrons will diffuse through a paraffin wax tube in much the same way as a gas will diffuse through a tube, if the mean free path in the gas is large compared with the diameter of the tube. The neutrons were led from a radon-beryllium source of about 200 mC. through a paraffin wax tube of 13 cm. inner diameter and 20–40 cm. in length to filter, absorber and indicator sheets and passed only once through these sheets.

Control experiments show that the observed absorption is not due to reflection (back scattering) from the indium absorber. Reversal of the position of cadmium filter and indium absorber produces no change in the transmitted intensity, and this fact indicates that the observed highly absorbable residual neutrons are not produced in our cadmium filter but form part of the unfiltered beam. We are thus led to the conclusion that we have to deal in these experiments with types of absorption spectra for which the present form of the theory cannot account.

There were earlier observations[2], especially those reported some time ago by Moon and Tillman, which did not seem to fit in with the theory. Tillman and Moon showed that the absorption of slow neutrons in an element appears to be different if different elements are used as indicators, and that it often appears to be comparatively high, if the same element is used as absorber and indicator.

In the present experiments, the residual neutrons from cadmium show such selective absorption effects with some combinations of indium, silver and iodine, and show them much more markedly than the unfiltered beam. Moreover, some elements show, if one and the same element is used as absorber and indicator, a larger absorption for the residual neutrons than for the unfiltered neutrons; for example, less than 1 gm./cm.^{2} of iodine absorbs more than half of the residual neutrons if iodine is used as indicator.

It would therefore seem that some elements have fairly sharp regions of strong absorption in an energy region for which cadmium is transparent.

It would be interesting to know the energy values which correspond to these absorbing regions. An attempt is now being made to determine them by studying the absorption in boron and lithium of the 'highly absorbable' components of the residual beam.

The observed strong absorption of residual neutrons makes it possible to construct efficient slits or shutters for the purpose of stopping out a well defined beam.

LEO SZILARD.

Clarendon Laboratory,
    Oxford.
    Nov. 19.

[1] *Proc. Roy. Soc.*, A, **149**, 522 ; 1935.
[2] Artsimovitch, T. Kourtschatov, Miccowskii and Palibin, *C.R.*, **200**, 2159 ; 1935. Bjerge and Westcott, *Proc. Roy. Soc.*, A, **150**, 709; 1935. Amaldi, D'Agostino, Fermi, Pontecorvo and Segrè, *Ric. Sci.*, (vi), **1**, No. 11–12. Moon and Tillman, NATURE, **135**, 904 ; 1935. Tillman and Moon, NATURE, **136**, 66, July 13, 1935. Ridenour and Yost, *Phys. Rev.*, **48**, 383 ; 1935.

Printed in Great Britain by FISHER, KNIGHT & CO., LTD., St. Albans.

[By J. H. E. Griffiths and Leo Szilard,
Nature, **139**, 323–324 (1937) (Letter).]

Gamma Rays excited by Capture of Neutrons

LEA's observations and Fermi's pioneer work has shown that many elements may capture a neutron, and emit gamma rays in the process. This phenomenon has been further investigated especially by Rasetti[1], and more recently by Kikuchi[2] and Fleischmann[3], who used polonium-beryllium, deuteron-deuterium and radon-beryllium sources respectively.

We have used radon-beryllium sources, with an experimental arrangement designed to reduce the high background effects of the Geiger counter due to the gamma rays emitted by this type of neutron source, and also to concentrate slow neutrons on the counter. As indicated in the diagram (Fig. 1), a cylindrical block of lead 18 cm. in diameter and 44 cm. in length is sunk into a thick-walled paraffin wax tube closed at the bottom. The radon-beryllium neutron source is placed on the axis of the lead block at a distance of 29 cm. from the top. The neutrons emitted are slowed down by the paraffin wax, diffuse through the lead core, and slow neutrons emerge at the top of the lead block. The efficiency of this arrangement is based on the fact that the mean free path for scattering of slow neutrons is much larger in lead than in paraffin. Gamma ray effects were measured by a thin-walled magnesium counter covered by a thick-walled lead tube and placed in the neutron stream; beta rays and secondary electrons were measured with the same counter uncovered.

The capture radiation from a thick sheet of cadmium (0·4 gm./cm.²) placed below the counter produced 1,700 impulses per minute above the background; the background amounted to 440 impulses per minute with a radon-beryllium source of about 500 mc. The capture radiation of cadmium (and other strongly absorbing elements) is thus a sensitive, and in many respects a very convenient, indicator for slow neutron intensities in our arrangement.

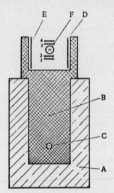

Fig. 1.

A, PARAFFIN WAX TUBE; B, LEAD CORE; C, RADON-BERYLLIUM SOURCE; D AND E, BORON SCREENS; F, GEIGER-MÜLLER COUNTER; 1, 2, 3, 4, POSITIONS OF THE SAMPLES.

We have used it as an indicator to determine the absorption of slow neutrons in a number of elements. A sample of the element was placed below the counter, a sheet of cadmium inserted between this sample and the counter, and the gamma ray effect measured. This, compared with the gamma ray effects obtained in the absence of the cadmium, and in the absence of the sample, gives the absorption of the sample. The results are shown in Table 1. Our values are not corrected for a slight divergence of the neutron stream, which may account for the fact that most of them are slightly above those of Dunning and Pegram[4], who used the disintegration of lithium as an indicator.

TABLE 1.

Element	Compound	gm./cm.²	Per cent transmitted	Cross-section (× 10⁻²⁴ cm.²)
Cl	CCl₄	1·30	34	52
Co	Co₃O₄	2·87	44	39
Rh	Rh	0·68	51	170
Ag	Ag	1·07	56	97
Cd	Cd	0·0285	50	4,500
Ir	Ir	0·61	46	410
Au	Au	3·14	37	103
Hg	HgO	0·70	50	355

Further, we have compared the gamma ray effect per captured neutron as measured by the counter for a number of strong absorbers of slow neutrons, and the results are given in Table 2. The value for cadmium is arbitrarily fixed at 100.

TABLE 2.

Cl	Co	Rh	Ag	Cd	In	Sm	Gd	Ir	Au	Hg
95	86	82	75	100	94	94	94	74	80	105

The same samples of elements were used to measure both the absorption cross-section and the gamma ray effect, and the values given in column four of Table 1 were then used to compute the relative gamma ray effect per captured neutron. The value for indium includes gamma rays from the induced activity of the 54-minute period. Otherwise the figures include only gamma rays from induced activities of periods shorter than half-hour; longer periods were not appreciably excited in our experiments.

The eleven elements investigated give almost equal effects although their atomic weights range from 35 to 200.

A similar comparison of the gamma rays emitted by elements which are weak absorbers of slow neutrons would, in our opinion, require better data on the elastic scattering of such elements than is at present available. So far, with no element heavier than chlorine have we found evidence to show that the gamma ray effect is appreciably different from 'normal'.

About equal effects might be expected if each element emitted one gamma ray quantum per captured neutron, but other observations do not support this simple interpretation of our result, since they seem to indicate that a larger number of quanta are emitted from the investigated elements.

We obtained some indication of this number by comparing the number of fast electrons ejected backwards from lead by the gamma rays from cadmium which were excited by a slow neutron beam, with the number of beta rays from a very thin silver foil made radioactive by the same neutron beam. Half-value thicknesses in lead were determined for the electrons from lead and for the gamma rays from cadmium. The values obtained indicate that more than seven quanta are emitted from cadmium per captured neutron, but this result requires confirmation by an independent method which is now being attempted.

The fact that chlorine shows a 'normal' gamma ray value indicates that its intense neutron absorption is due to radiative capture that leads to a chlorine isotope of long half-life period which has not yet been detected. Samples of yttrium prepared free from gadolinium by Dr. J. K. Marsh, Old Chemistry Laboratory, Oxford, showed only very weak capture radiation and, in agreement with a measurement of Hevesy[5], a small absorbing cross-section. Thus the widespread view[6], based on earlier measurements, that yttrium has a large capture cross-section for slow neutrons and therefore must transform into an undetected radioactive isotope, does not seem to be justified. Barium also shows a much smaller capture cross-section (< 10 × 10⁻²⁴ cm.²) than given by earlier measurements[4]. In agreement with Kikuchi, we find that boron shows a gamma ray effect of about a twentieth of that of cadmium.

J. H. E. GRIFFITHS.
LEO SZILARD.
Clarendon Laboratory,
University Museum,
Oxford.
Jan. 20.

¹ Rasetti, Z. Phys., 97, 64 (1935).
² Kikuchi, Husimi and Aoki, Proc. Phys.-Math. Soc. Japan, 18, 188 (1936).
³ Fleischmann, Z. Phys., 100, 307 (1936).
⁴ Dunning, Pegram, Fink and Mitchell, Phys. Rev., 48, 265 (1935).
⁵ Hevesy and Levi, NATURE, 137, 185 (1936).
⁶ Cf. Feather, "Nuclear Physics", p. 183 (1936).

JANUARY 1, 1939   PHYSICAL REVIEW   VOLUME 55
Printed in U. S. A.

Radioactivity Induced by Nuclear Excitation*

I. Excitation by Neutrons

M. GOLDHABER,† R. D. HILL,‡ *Cavendish Laboratory, Cambridge, England*

AND

LEO SZILARD, *Clarendon Laboratory, Oxford, England*

(Received November 5, 1938)

It is shown that the 4.1-hr. period of indium can be produced by nuclear excitation of indium and is to be attributed to an excited metastable state, In^{115},* of the stable In^{115}. This result is obtained by studying the radioactivity produced in indium by neutrons of different energy distributions and by studying the chain reactions produced in cadmium by fast neutrons. It is found that In^{115}* can be produced from In^{115} by 2.5-Mev neutrons, but not noticeably by strong sources of photoneutrons which have energies of a few hundred thousand electron volts. A radioactive Cd^{115} of 2.5-day half-life time is found to transform with emission of negative electrons into In^{115}*. Cd^{115} was produced by neutron loss from Cd^{116} and by neutron capture from Cd^{114}.

WHEN indium is bombarded with fast neutrons from a Rn-Be source, a radioactive indium isotope, of 4.1-hr. half-life time, is produced which emits negative electrons.[1] So far, it has been assumed, that when an element is transformed by neutron bombardment into its own radioactive isotope, the radioactive isotope is generated either through neutron capture or through neutron loss from the bombarded element. The present experiments, however, indicate that this radioactive In isotope is generated through a new type of nuclear process—nuclear excitation leading to a metastable isomer of the stable In^{115}. For this metastable isomer we introduce the symbol In^{115}*.

As it was known that more than two radioactive isotopes of indium could be generated by neutron bombardment from the two stable isotopes of indium, a further investigation of their generation appeared to be of interest. Two of these radioactive isotopes, of 13-sec. and 54-min. half-life time, have been assigned to mass number 116. They are isomers produced from In^{115} by

radiative capture of the neutron. In order to learn more about the generation of the 4.1-hr. period, which did not appear to be water-sensitive, we made the following experiments.

§1. GENERATION OF In^{115}* FROM INDIUM

A thick indium foil was bombarded with Rn-α-Be neutrons, and showed the 4.1-hr. period more strongly than the 54-min. capture period. A similar indium foil irradiated with photoneutrons from a strong radium-beryllium source (0.5 g Ra-260 g Be) did not show the 4.1-hr. period, though the photoneutron source was so strong that it produced the 54-min. capture period with an initial activity of 1200 impulses per minute recorded by a Geiger-Müller counter. Control experiments showed that this large activity was caused by the primary photoneutrons and not by slower scattered neutrons which would obviously produce a large activity of the 54-min. period by resonance capture. If the 4.1-hr. period, which is strongly produced by the fast Rn-α-Be neutrons, belonged to an isomer of In^{116}, it would be difficult to understand why this period is not appreciably produced from In^{115} by capture of the photoneutrons which have energies of a few hundred thousand electron volts. We conclude therefore that this period must not be attributed to In^{116}.

We made further experiments with neutrons from a radon-boron source. It was found that

* Some of these experiments have been reported at the Washington Meeting of the American Physical Society, April 1938, and at the British Association Meeting in Cambridge, August 1938. (See Nature **142**, 521 (1938)).
† Now at the University of Illinois.
‡ 1851 Exhibition Scholar, now at the University of Illinois.
[1] Szilard and Chalmers, Nature **135**, 99 (1935); Amaldi, d'Agostino, Fermi, Pontecorvo, Rasetti and Segrè, Proc. Roy. Soc. **A149**, 522 (1935); Lawson and Cork, Phys. Rev. **52**, 531 (1937).

48   GOLDHABER, HILL AND SZILARD

these neutrons excite the 4.1-hr. period more strongly than either the 25-min. period of I, or the 54-min. period of In, which are both caused by neutron capture; and also much more strongly than other investigated periods which are known to require fast neutrons for their excitation, such as the 10-min. and 15-hr. periods of Al, the 2.3-min. period of Si, and the 2.3-min. period of P. The 170-min. period of P which does not require very fast neutrons for its excitation is, however, more strongly activated than the 4.1-hr. period of In. These results are in accord with the fact that a radon-boron source is deficient in very fast neutrons compared with a Rn-α-Be source. The strong activation of the 4.1-hr. period in the present experiment cannot be attributed to a neutron loss reaction from either In^{113} or In^{115} for such a reaction requires very fast neutrons.

Nor can the strong activation of this period by radon-boron neutrons be attributed to neutron capture from In^{113}, since the relative abundance of this isotope is only five percent. The capture cross section of In^{113}, corrected for the half-value thickness of the β-rays, would have to be 100 times larger than the capture cross section of iodine. In the absence of slow neutrons we cannot expect such large differences of capture cross sections among the strongly absorbing elements of odd atomic number of such similar atomic weight. This is borne out by an unpublished survey of the relative capture cross sections of elements for Ra+Be photoneutrons which was made by J. H. E. Griffiths, Oxford, in collaboration with one of us.

Finally indium was irradiated with neutrons of about 2.5-Mev energy, produced by bombarding heavy ice with deuterons of 250 kev.

TABLE I. *Stable and some radioactive isotopes of cadmium, indium and tin in the mass range 112 to 117. For stable isotopes the relative abundance in percent is given in heavy type. For radioactive isotopes the half-life periods are given in italics.*

M	$_{48}Cd$		$_{49}In$	$_{50}Sn$
117	*4h*	→	*2.3h*	9.1
116	7.3		*13s*	15.5
			54m	
115	*2.5d*	→	*4.1h*	0.4
			95.5	
114	28.0			0.8
113	12.3		4.5	
112	24.2			1.1

The 4.1-hr. period was found to be strongly activated. Obviously this cannot be caused by a neutron loss reaction, since the available energy is certainly less than the binding energy of the neutron in an In isotope.

The experiments described rule out the possibility that this radioactive indium isotope is generated from a stable indium isotope by neutron capture, or neutron loss, and therefore exclude an indium isotope of mass number 112, 114 or 116 as the carrier of the 4.1-hr. period. Moreover, the fact that photoneutrons from a Ra-Be source do not excite the 4.1-hr. period, while neutrons from the D+D reaction do so, suggests that its carrier might arise through a process which has a real or apparent energy threshold.

These results forced us to assume that the 4.1-hr. period is generated by a new type of process. In order to explain this period we assumed that the In^{115} nucleus may be excited by neutron impact and left in a metastable excited state, In^{115*}. This metastable In^{115*} may be radioactive and transform either by β-disintegration into its stable isobar Sn^{115}, or by emission of γ-rays and internal conversion electrons into the stable In^{115}.

If this assumption is correct it should be possible to produce this radioactive indium isotope by other methods of nuclear excitation from indium, e.g. by proton impact. S. W. Barnes and P. W. Aradine, Rochester, will report in the following paper on the nuclear excitation of indium by protons.

It is conceivable that the existence of a metastable state of In^{115} is connected with the high nuclear spin (9/2) of In^{115}. If the metastable In^{115*} has a small spin it will be difficult to reach the metastable state from the ground state by a direct transition. As a rule, a higher excited state of indium will have to be produced, from which the metastable state can be reached by one or more spontaneous transitions. The apparent energy threshold of In^{115*} may therefore be considerably above the level of the metastable state.

§2. Generation of In^{115*} from Cadmium

Other methods of producing In^{115*} can be devised, and we have isolated In^{115*} from cadmium bombarded by neutrons from Li+D, thus confirming our assignment of the 4.1-hr. period.

As can be seen from Table I, cadmium has an isotope of mass number 116 having a relative abundance of 7.3 percent. By bombarding cadmium with fast neutrons of the Li+D reaction, one could hope to produce a radioactive Cd^{115} by the neutron loss reaction Cd^{116} $(n, 2n)$. Such a radioactive isotope could be expected to emit negative electrons and it might transform partly into the stable In^{115} and partly into the metastable In^{115*}.

We bombarded metallic cadmium for several hours with neutrons from Li+D using deuterons of 950 kev and currents averaging $30\mu a$. The irradiated cadmium was dissolved in nitric acid and a small quantity of indium was added. Indium was precipitated by adding ammonia. The indium which was separated from the cadmium solution two days after the irradiation showed a single activity decaying with a 4.1-hr. period. By successive separations of indium from the cadmium solution it was found that this radioactive indium isotope grew from a parent substance of 2.5-day half-life period. This indium isotope was found to emit negative electrons.

We have compared the half-life period and the electron absorption curve of this radioactive indium isotope with the half-life period and the electron absorption curve of the radioactive indium isotope produced by fast neutrons from indium. The result confirms our assumption that the two are identical and that the carrier of the 4.1-hr. period is In^{115*}. The range of the electrons in Al was found to be 0.15 g/cm². This corresponds to an energy of 550 ± 100 kev.

Our experiments show that In^{115*} grows from the parent Cd^{115} which decays with a half-life time of 2.5 days. We find by magnetic analysis that Cd^{115} emits negative electrons, thus transforming into In^{115*} and perhaps also into the stable In^{115}.[2]

To investigate whether Cd^{115} can also be obtained by slow neutrons we irradiated with neutrons a cadmium sheet which was enclosed between two similar cadmium sheets and placed inside a paraffin block. We found that the two outer sheets showed a stronger activity of the 2.5-day period than the inner sheet. This indicates that Cd^{115} is produced by slow neutron capture from Cd^{114}. The activity produced in this way is small compared with the activity produced by the neutron loss reaction from Cd^{116}.

The properties of In^{115*} are of great interest both from the point of view of the nature of isomeric nuclei and of isobaric pairs. A detailed investigation of the radiations emitted by In^{115*} is now being carried out. An account of some preliminary experiments has been given at the British Association Meeting.[3]

We wish to thank all those who have cooperated in this work: Dr. F. M. Brewer, Old Chemistry Laboratory, Oxford, who in the early stages of this work identified the carrier of the 4.1-hr. period as an isotope of indium; Mr. P. I. Dee, Cavendish Laboratory, Cambridge, who made it possible for us to use strong Li+D neutron sources; Drs. E. T. Booth, C. H. Collie and C. Hurst, Clarendon Laboratory, Oxford, who carried out the irradiations with D+D neutrons. Our thanks are also due to Professor F. A. Lindemann, F.R.S. and to Dr. J. D. Cockcroft, F.R.S. for their kind interest in this work. We are much indebted to the Anaconda Wire and Cable Company who very generously enabled us to work with large quantities of indium.

---

[2] A radioactive Cd isotope of similar half-life time (58 hr.) has been previously reported to exist by Cork and Thornton (Phys. Rev. **51**, 608 (1937)), but assigned by them to mass number 117. They also reported a radioactive isotope of In of 2.3-hr. half-life time which they ascribed to In^{117} and which appeared to grow from a 58-hr. Cd parent. We find this indium isotope only when we precipitate indium from the cadmium solution in a time less than two days after irradiation. By successive separations we find that it grows from a 4-hr. Cd parent, a radioactive isotope which has been previously reported by Cork and Thornton and at the time assumed to be Cd^{115}.

[3] See Nature **142**, 521 (1938).

Reprinted from THE PHYSICAL REVIEW, Vol. 55, No. 8, 799–800, April 15, 1939
Printed in U. S. A.

Instantaneous Emission of Fast Neutrons in the Interaction of Slow Neutrons with Uranium*

Recently it became known[1] that uranium can be split by neutrons into two elements of about equal atomic weight. In this fission of uranium the two elements produced have a large neutron excess; moreover they are probably produced in an excited nuclear state. One might therefore expect that these excited fragments instantaneously emit neutrons and that perhaps the number emitted is even larger than one per fission.

One might also expect a delayed emission of neutrons—as was first pointed out by Fermi—if some of the fragments go through one or more beta-transformations before they emit a neutron. Delayed emission of neutrons caused by the action of both slow and fast neutrons on uranium has recently been reported by Roberts, Meyer, and Wang,[2] who find a period of about 12 seconds.

In order to see if there is an instantaneous emission of neutrons from the fission of uranium we have performed the following experiment. We exposed uranium oxide to neutrons which were slowed down by paraffin wax, using as a source of neutrons a block of beryllium from which photoneutrons were liberated by the gamma-rays of radium. A helium-filled ionization chamber connected to a linear amplifier served as a detector for fast neutrons. The ionization pulses of the chamber were observed visually by means of a cathode-ray oscillograph and were recorded by the usual counting arrangement.

Figure 1 shows a diagram of the experimental arrangement. The ionization chamber is covered by a cadmium sheet cap G which prevents the thermal neutrons from penetrating to the helium ionization chamber. A cadmium sheet shield H, 0.5 mm thick, is used to cover the cylindrical box E which contains 2300 g of uranium oxide. The uranium oxide is screened from the thermal neutrons by this shield and can be exposed to them simply by removing the shield.

We observed about 50 pulses per minute from the helium chamber when we exposed the uranium oxide to the thermal neutrons in the absence of the cadmium shield H, but obtained only 5 pulses per minute when the uranium was screened from the thermal neutrons by the cadmium shield. The difference of about 45 pulses per minute we have to attribute to fast neutrons emitted from uranium under the action of thermal neutrons. It is reasonable to assume that this emission of fast neutrons is connected with the fission of uranium.

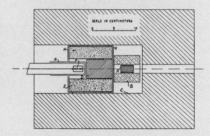

FIG. 1. Arrangement for the observation of the emission of fast neutrons from uranium. *A*, Radium. *B*, Beryllium block. *C*, Paraffin wax. *D*, Lead block. *E*, Box filled with uranium oxide. *F*, Ionization chamber. *G*, Cadmium sheet cap. *H*, Cadmium sheet shield.

Control experiments were carried out in which uranium was replaced by lead. The effect of the presence and absence of the cadmium shield H and the cadmium cap G was tested.

In order to estimate the number of fast neutrons emitted per fission under the action of thermal neutrons we used an ionization chamber lined with a thick layer of uranium oxide having an area of 25 cm². This uranium chamber was put in place of the helium chamber without otherwise materially changing the experimental arrangement. Under these conditions the uranium chamber gave about 45 fissions per minute. Assuming the range of the fission fragments to be about 0.005 g per cm² in uranium oxide, the observed 45 fissions per minute should occur in a surface layer, weighing 0.13 g, of the thick uranium oxide lining. Accordingly, about 800,000 fissions per minute should occur in the 2300 g of uranium oxide which was used in our experiment. By taking into account the solid angle, the size of the helium chamber and the pressure used, and by assuming that the "fission neutrons" have an average collision cross section in helium of 3.5×10^{-24} cm² we find the number of neutrons emitted per fission to be about two.

This number is of course only a rough estimate; the main cause of uncertainty is the considerable variation of the cross section of helium with the neutron energy in the region around one million volts.[3] A hydrogen-filled ionization chamber is now being used in order to obtain a more accurate estimate. It seems to be established, however, that the order of magnitude is one neutron per fission.

Anderson, Fermi and Hanstein have independently, and by a different method, carried out experiments on the neutron emission connected with the fission of uranium. Our observations are consistent with their results, and we wish to thank them for communicating their results to us before publication.

While from our observations we can only say that the time delay involved in this "instantaneous" neutron emission appears to be less than one second, we should expect, for theoretical reasons, this emission to take place within less than 10^{-14} second.

We have also looked for a delayed emission of fast neutrons by performing the following experiment. The uranium oxide was irradiated for some length of time in the arrangement shown in Fig. 1. Then the radium was quickly removed from the beryllium block and the cathode-ray oscillograph screen was watched for a period of 15 seconds for an indication of a delayed emission of fast neutrons. After the radium is removed there is no gamma-ray background to set a lower limit for the observable helium recoil energy; the only slight background remaining is due to electrical fluctuations of the amplifier. In 50 experiments, corresponding to a total observation time of

more than 12 minutes, we observed only two pulses which may or may not have been due to a delayed emission of fast neutrons. This is to be compared with the emission of 45 fast neutrons per minute, the number observed while the radium is inside the beryllium block. We conclude that, if slow neutrons falling on uranium cause a delayed emission of neutrons which are sufficiently fast for us to observe, their number must be very much smaller than the number of neutrons which we have observed in the instantaneous emission.

We are indebted to Dr. S. Seely for his assistance in carrying out some of these experiments. We wish to thank the Department of Physics of Columbia University for the hospitality and the facilities extended to us, and also wish to thank the Association for Scientific Collaboration for enabling us to use one gram of radium in these experiments.

LEO SZILARD
WALTER H. ZINN

Pupin Physics Laboratories,
Columbia University,
New York, New York,
March 16, 1939.

* Publication assisted by the Ernest Kempton Adams Fund for Physical Research of Columbia University.
[1] O. Hahn and F. Strassmann, Naturwiss. 27, 11 (1939); L. Meitner and R. Frisch, Nature (February, 1939).
[2] R. B. Roberts, R. C. Meyer, and P. Wang, Phys. Rev. 55, 510 (1939).
[3] H. Staub and W. E. Stephens, Phys. Rev. 55, 131 (1939).

AUGUST 1, 1939   PHYSICAL REVIEW   VOLUME 56
                              Printed in U. S. A.

Neutron Production and Absorption in Uranium*

H. L. ANDERSON, E. FERMI AND LEO SZILARD
Columbia University, New York, New York
(Received July 3, 1939)

IT has been found[1–3] that there is an abundant emission of neutrons from uranium under the action of slow neutrons, and it is of interest to ascertain whether and to what extent the number of neutrons emitted exceeds the number absorbed.

This question can be investigated by placing a photo-neutron source in the center of a large water tank and comparing, with and without uranium in the water, the number of thermal neutrons present in the water. In the previous experiments of this type[1,3] it was attempted to have as closely as possible a spherically symmetrical distribution of neutrons. The number of thermal neutrons present in the water was determined by measuring along one radius the neutron density ρ as a function of the distance r from the center, and then calculating $\int r^2 \rho dr$. A difference in favor of uranium of about five percent was reported by von Halban, Joliot and Kovarski.[4]

Since one has to measure a small difference, slight deviations from a spherically symmetrical distribution might give misleading results. The present experiments which are based on the same general principle do not require such symmetry. In order to measure the number of thermal neutrons in the water we filled the tank with a ten-percent solution of $MnSO_4$. The activity induced in manganese is proportional to the number of thermal neutrons present. A physical averaging was performed by stirring the solution before measuring the activity of a sample with an ionization chamber. To obtain an effect of sufficient magnitude, about 200 kg of U_3O_8 was used.

The experimental arrangement is shown in Fig. 1. A photo-neutron source, consisting of about 2 g of radium and 250 g of beryllium was placed in the center of the tank. The geometry was such that practically all neutrons emitted by the source and by the uranium oxide were slowed down and absorbed within the tank. Each irradiation extended over several half-life periods of radiomanganese and the observed activity of the solution was about four times the background of the ionization chamber. Alternating measurements were taken with the cans filled with uranium oxide, and with empty cans of the same dimensions. The activity proved to be about ten percent higher with uranium oxide than without it. This result shows that in our arrangement more neutrons are emitted by uranium than are absorbed by uranium.

In order to find the average number of fast neutrons emitted by uranium for each thermal neutron absorbed by uranium, we have to determine what fraction of the total number of neutrons emitted by the photo-neutron source is, in our experiment, absorbed in the thermal region by uranium. The number of photo-neutrons

* Publication assisted by the Ernest Kempton Adams Fund for Physical Research of Columbia University.

[1] v. Halban, Joliot and Kovarski, Nature **143**, 470 (1939).
[2] L. Szilard and W. H. Zinn, Phys. Rev. **55**, 799 (1939).
[3] Anderson, Fermi and Hanstein, Phys. Rev. **55**, 797 (1939).
[4] v. Halban, Joliot and Kovarski, Nature **143**, 680 (1939).

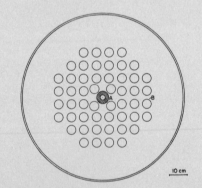

FIG. 1. Horizontal section through center of cylindrical tank which is filled with 540 liters of 10-percent $MnSO_4$ solution. A, Photo-neutron source composed of 2.3 grams of radium and 250 grams of beryllium. B, One of 52 cylindrical cans 5 cm in diameter and 60 cm in height, which are either empty or filled with uranium oxide.

285   NEUTRON PRODUCTION AND ABSORPTION

emitted by the source is indicated by the activity of the solution in the tank when the irradiation is carried out with empty cans surrounding the source. We obtained a measure of this number by taking into account that in our solution about 20 percent of the neutrons are captured by manganese and the rest by hydrogen. In order to obtain, in the same units, a measure of the number of neutrons absorbed by uranium we proceeded in the following way: A mixture of sand and manganese powder, having the same thermal neutron absorption as uranium oxide replaced the uranium oxide in $\frac{1}{4}$ of the cans which were distributed uniformly among the other uranium oxide-filled cans. After irradiation, all this powder was mixed together, a ten-percent $MnSO_4$ solution was prepared from a sample, and its activity was measured with our ionization chamber.

In this way we found that about 50 percent of the neutrons emitted by the source are absorbed as thermal neutrons by uranium in our arrangement. It follows that, if uranium absorbed only thermal neutrons, the observed ten-percent increase in activity obtained with uranium present would correspond to an average emission of about 1.2 neutrons per thermal neutron absorbed by uranium. This number should be increased, to perhaps 1.5, by taking into account the neutrons which, in our particular arrangement, are absorbed at resonance in the nonthermal region by uranium, without causing neutron emission.

From this result we may conclude that a nuclear chain reaction could be maintained in a system in which neutrons are slowed down without much absorption until they reach thermal energies and are then mostly absorbed by uranium rather than by another element. It remains an open question, however, whether this holds for a system in which hydrogen is used for slowing down the neutrons.

In such a system the absorption of neutrons takes place in three different ways: The neutrons are absorbed at thermal energies, both by hydrogen and uranium, and they are also absorbed by uranium at resonance before they are slowed down to thermal energies. Our result is independent of the ratio of the concentrations of hydrogen and uranium, insofar as it shows that, for thermal neutrons, the ratio of the cross section for neutron production and neutron absorption in uranium is greater than one, and probably about 1.5. What fraction of the neutrons will reach thermal energies without being absorbed will, however, depend on the ratio of the average concentrations of hydrogen and uranium. Since there is an appreciable absorption even far from the center of the resonance band, it follows that the fraction of neutrons absorbed by uranium at resonance will increase with decreasing hydrogen concentration. This has to be taken into account in discussing the possibility of a nuclear chain reaction in a system composed essentially of uranium and hydrogen. A chain reaction would require that more neutrons be produced by uranium than absorbed by uranium and hydrogen together. In our experiment the ratio of the average concentration of hydrogen to uranium atoms was 17 to 1, and in the experiment of von Halban, Joliot and Kovarski this ratio was 70 to 1. At such concentrations the absorption of hydrogen in the thermal region will prevent a chain reaction. By reducing the concentration of hydrogen one would obtain the following effect: On the one hand a larger fraction of those neutrons which, reach thermal energies will be absorbed by uranium; on the other hand fewer neutrons reach the thermal region due to an increased absorption by uranium at resonance. Of these two counteracting factors the first is more important for high hydrogen concentrations and the second is more important for low hydrogen concentrations. Starting with high hydrogen concentrations, the ratio of neutron production to total neutron absorption will thus first rise, then pass through a maximum, and, as the hydrogen concentration is decreased, thereafter decrease. We attempted to estimate the quantities involved from the information available about resonance absorption in uranium[5–7] and from the observed net gain of 0.2 in the number of neutrons in our experiment. The effect of the absorption at resonance turns out to be so

[5] Meitner, Hahn and Strassman, Zeits. f. Physik **106**, 249 (1937).
[6] v. Halban, Kovarski and Savitch, Comptes rendus **208**, 1396 (1939).
[7] H. L. Anderson and E. Fermi, Phys. Rev. **55**, 1106 (1939).

ANDERSON, FERMI AND SZILARD   286

large that even at the optimum concentration of hydrogen it is at present quite uncertain whether neutron production will exceed the total neutron absorption. More information concerning the resonance absorption of uranium as well as more accurate measurement of some of the values which enter into our calculation are required before we can conclude whether a chain reaction is possible in mixtures of uranium and water.

We wish to thank Dr. D. W. Stewart, of the Department of Chemistry, and Mr. S. E. Krewer, for advice and assistance in carrying out some of these experiments. We are much indebted to the Eldorado Radium Corporation for enabling us to work with large quantities of uranium oxide in our experiments, and to the Association for Scientific Collaboration for the use of the photo-neutron source and other facilities.

OCTOBER 1, 1939   PHYSICAL REVIEW   VOLUME 56
                              Printed in U. S. A.

Emission of Neutrons by Uranium*

W. H. Zinn, *City College, The College of the City of New York, New York*

Leo Szilard, *Pupin Physics Laboratories, Columbia University, New York, New York*
(Received August 14, 1939)

Fast neutrons emitted by uranium under the action of thermal neutrons were studied by using a radium-beryllium photoneutron source. The background due to the primary neutrons can be neglected since only a few of the photo-neutrons are sufficiently fast to be counted. Data are obtained concerning the energy spectrum of the uranium fission neutrons by recording photographically by means of a linear amplifier and cathode-ray oscillograph the pulses due to helium atoms projected in an ionization chamber. Visual inspection of the record gives an upper limit of the spectrum of 3.5 Mev. The number of neutrons emitted is estimated by analyzing the pulse distribution of hydrogen atoms projected by uranium neutrons in an ionization chamber filled with hydrogen and argon. The number found is brought into relationship with the number of fissions, observed under comparable conditions, in an ionization chamber lined with a thin film of uranium oxide containing a known amount of uranium. In this way it is found that about 2.3 neutrons are emitted per fission. The method used would permit a greater accuracy in the determination of this number than the actual accuracy obtained in the present experiments. This number, together with the fission cross section and the cross section for radiative capture gives the number of neutrons produced for each thermal neutron absorbed in uranium.

W E reported[1] some time ago that fast neutrons are emitted—apparently instantaneously—from uranium under the action of thermal neutrons and we found, as a rough estimate, an average of two neutrons per fission process. This result was obtained by counting the helium recoil nuclei which the fission neutrons project in a helium-filled ionization chamber. The emission of neutrons in the fission of uranium was independently discovered by von Halban, Joliot and Kowarski[2] as well as by Anderson, Fermi and Hanstein,[3] who observed an increase of the thermal neutron density in water in the presence of uranium. Others[4] have investigated the same phenomenon.

Before this "instantaneous" emission had been observed, Roberts, Meyer and Wang[5] discovered a delayed emission of neutrons from uranium which apparently follows a beta-transformation of a half-life period of twelve seconds. We had found that the instantaneous emission was very much stronger than the delayed emission and we assumed that it corresponds to a direct ejection of neutrons from the uranium fragments, without being preceded by a beta-transformation, and that accordingly the time delay involved is far too small to be measured by the usual

* Publication assisted by the Ernest Kempton Adams Fund for Physical Research of Columbia University.

[1] L. Szilard and W. H. Zinn, Phys. Rev. **55**, 799 (1939).
[2] H. von Halban, F. Joliot and L. Kowarski, Nature **143**, 470 (1939).
[3] H. L. Anderson, E. Fermi and H. B. Hanstein, Phys. Rev. **55**, 797 (1939).
[4] G. P. Thomson, J. L. Michiels and G. Parry, Nature **143**, 760 (1939); G. von Droste and H. Reddeman, Nat. Wiss. **20/21**, 371 (1939).
[5] R. B. Roberts, R. C. Meyer and P. Wang, Phys. Rev. **55**, 510 (1939).

620   W. H. ZINN AND L. SZILARD

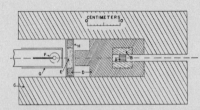

Fig. 1. Arrangement for the observation of the emission of fast neutrons from uranium. *A*—radium; *B*—beryllium block; *C*—paraffin wax; *D*—lead; *E*—uranium cell; *F*—spherical ionization chamber; *G*—cadmium sheet cap; *H*—cadmium sheet shield.

methods. This assumption was based on the arguments that it would be very difficult to explain the great abundance of the instantaneous neutron emission without assuming direct ejection and that no hard beta-rays were observed which should be expected to be present if the neutron emission followed a very short-lived beta-transformation. From direct experimental evidence, however, we could not exclude a delay smaller than one-tenth of a second. Gibbs and Thomson[6] have now shown by direct experiments that the delay is smaller than one-thousandth of a second and this appears to leave little doubt as to a direct ejection of neutrons.

In the present experiments helium recoils were used for investigating the energy distribution of the fission neutrons, but hydrogen recoils were

6 D. F. Gibbs and G. P. Thomson, Nature **144**, 202 (1939).

used for estimating the number emitted per fission.

The experimental arrangement is shown in Fig. 1. The source of thermal neutrons was about one gram of radium, *A*, placed in the center of a beryllium block, *B*, and surrounded by a paraffin cylinder, *C*. Fast neutrons emitted under the action of the thermal neutrons by about 430 grams of uranium metal enclosed in the cell *E*, were detected by the spherical ionization chamber, *F*. The pulses from the chamber were fed into a linear amplifier and were made visible by means of a cathode-ray oscillograph. A camera with a moving film was used to obtain a photographic record of the pulses appearing on the oscillograph screen.

Two such records are shown in Fig. 2; one was obtained in the absence and the other in the presence of the cadmium sheet shield, *H*. The shield *H* completely surrounds the uranium and shuts it off from most of the thermal neutrons, leaving a background of pulses which is partly due to particularly fast photoneutrons from the source, and partly due to fission neutrons from the uranium emitted under the action of the few thermal neutrons which pass through the cadmium shield. This background amounts to less than one pulse per minute.

The ionization chamber, which was filled with 10 atmospheres of hydrogen and 8 atmospheres of argon, contained a small amount of nitrogen. By removing the cadmium cap, *G*, from the chamber the nitrogen atoms in the chamber can be exposed to the action of thermal neutrons and

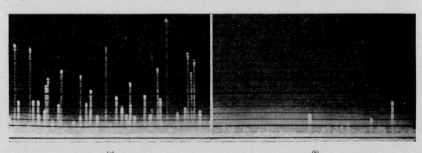

(a)   (b)

Fig. 2. (*a*) Oscillograph record of the fast neutrons from uranium. Cadmium sheet shield, *H* of Fig. 1, absent. Thermal neutrons falling on the uranium in the cell *E*. Ionization chamber filled with 10 atmospheres of hydrogen and 8 atmospheres of argon. (*b*) Record obtained with the cadmium shield *H* shutting off the thermal neutrons from the cell *E*.

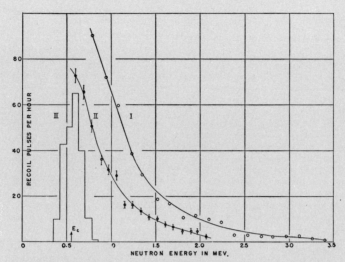

FIG. 3. Curve I: Pulse distribution due to helium recoils. Ionization chamber filled with 10 atmospheres of helium and 10 atmospheres of argon. Curve II: Pulse distribution $P(E)$ due to hydrogen recoils. Ionization chamber filled with 10 atmospheres of hydrogen and 8 atmospheres of argon. Curve III: Pulse distribution due to protons emitted, under the action of thermal neutrons, by a small amount of nitrogen in the chamber filled with 10 atmospheres of helium and 10 atmospheres of argon.

will then emit protons of about 0.6 Mev energy.[7] The pulses due to these protons were recorded and their distribution is shown in curve III of Fig. 3. This curve shows a sharp maximum which should correspond to an energy of about 0.6 Mev, and therefore this curve was used for calibrating the ionization chamber.

In order to find from the observed number of hydrogen recoils the number of neutrons which pass through the chamber it is necessary to know something about the energy distribution of the fission neutrons. This knowledge is required for two reasons. First, the scattering cross section of hydrogen is a function of the neutron energy; secondly, the observed pulse distribution of the hydrogen recoils is cut off at a certain energy E_c, which in this case was 0.55 Mev, in order to avoid the counting of pulses in the region which is affected by the gamma-ray background. Neutrons which have an energy below this cut-off

[7] J. Chadwick and M. Goldhaber, Proc. Camb. Phil. Soc. **31**, 612 (1935); T. W. Bonner and W. M. Brubaker, Phys. Rev. **49**, 778 (1936); M. S. Livingston and H. A. Bethe, Rev. Mod. Phys. **9**, 344 (1937).

energy, E_c, do not contribute to the recorded pulse distribution and their number has to be determined from the shape of the neutron spectrum, provided this spectrum is known.

If the time required for the collection of ions in the chamber were short compared with the time constant of the amplifier, the size of the pulses recorded by the oscillograph might be considered a fair measure of the energy which the recoil proton loses in the chamber. Even so, the size of the pulses cannot be considered a measure of the initial energy of the recoil protons if these lose only part of their energy in the chamber and are stopped by the walls. Therefore, if $R(E)dE$ is the number of recoil protons having an initial energy between E and $E+dE$, and if $P(E)dE$ is the number of recoil protons which lose in the chamber an amount of energy between E and $E+dE$, these two functions will be rather different in the high energy region where the range of the recoil protons cannot be neglected in comparison with the linear dimensions of the chamber. For the hydrogen-argon filled chamber

622   W. H. ZINN AND L. SZILARD

which was used the two functions can be expected to coincide very nearly in the region of the cut-off energy E_c, and can be expected to differ widely for energies above 1.5 Mev.

For this reason helium recoils (which have about $\frac{1}{10}$ the range of recoil protons for equal neutron energies) had to be used instead of hydrogen recoils in order to find the upper end of the energy spectrum of the fission neutrons. Curve I, in Fig. 3, shows the pulse distribution of helium recoils obtained with 10 atmospheres of helium and 10 atmospheres of argon in the chamber. This curve shows that the spectrum of the fission neutrons extends to about 3.5 Mev. Though the existence of a small number of high energy neutrons such as reported by von Halban, Joliot and Kowarski,[8] is not inconsistent with our result, the number of neutrons having energies above 4 Mev appears to be too small to have much bearing on our estimate of the total number of fission neutrons.

Since the calibration of the chamber, which we performed by means of protons, is not entirely satisfactory for correlating the size of the pulses due to helium recoils with the energy of the helium recoils, the helium-argon filled chamber was also calibrated by means of D+D neutrons of 2.5 Mev energy. The two calibrations coincided within the limits of the experimental error.

An estimate of the number of fission neutrons should be based on a count of hydrogen recoils rather than helium since the scattering cross section of helium has a sharp maximum[9] for neutrons of about 1.0 Mev energy, and helium is therefore not suitable for the purposes of a quantitative estimate. It can be shown that, if the neutron-proton scattering is spherically symmetrical in the system of the center of gravity, the number of neutrons $N(E)dE$ which pass through the chamber, and which have an energy between E and $E+dE$, is given by:

$$N(E) = -\frac{E}{H\sigma(E)}\frac{dR(E)}{dE},$$

where $\sigma(E)$ is the scattering cross section of the proton and H is the number of hydrogen atoms in

[8] H. von Halban, F. Joliot and L. Kowarski, Nature 143, 939 (1939).
[9] H. Staub and W. E. Stephens, Phys. Rev. 55, 131 (1939).

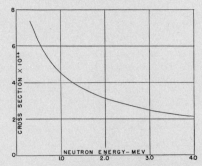

Fig. 4. Neutron-proton cross section as a function of neutron energy according to current theory.

the ionization chamber. From this we derive, for the total number of neutrons, N, passing through the chamber:

$$N = \int_0^\infty N(E)dE = \frac{\alpha}{\sigma_N H}\left[E_c R(E_c) + \int_{E_c}^\infty R(E)dE\right],$$

where α is the ratio of the total number of neutrons to the number of neutrons which have an energy in excess of E_c, and σ_N is an average scattering cross section of the proton, the value of which has to be determined from the energy spectrum of the neutrons. Since we have:

$$\int_0^\infty R(E)dE = \int_0^\infty P(E)dE$$

and since, for the reasons stated above, we have with good approximation:

$$R(E) \approx P(E) \quad \text{for} \quad E \leqq E_c,$$

we can express N in terms of $P(E)$ instead of $R(E)$. We then have:

$$N = \frac{\alpha}{\sigma_N H}\left[E_c P(E_c) + \int_{E_c}^\infty P(E)dE\right].$$

Let it now be assumed for the sake of argument that all the neutrons are emitted from a moving uranium fragment which has a mass number of about 120 and a kinetic energy of about 100 Mev. If all the neutrons were emitted from such a moving fragment with a single energy E_0, the energy distribution of the neutrons in the laboratory reference system would stretch from:

$$E_{min} = (0.9 - E_0^{\frac{1}{2}})^2 \text{ Mev}$$

to:

$$E_{max} = (0.9 + E_0^{\frac{1}{2}})^2 \text{ Mev}.$$

It is easy to see that the neutrons should be uniformly distributed in this energy interval if their distribution is spherically symmetrical in the center of mass system. One obtains accordingly:

and
$$\alpha = (E_{max} - E_{min})/(E_{max} - E_c)$$
$$\frac{1}{\sigma_{N}} = \frac{1}{E_{max} - E_c} \int_{E_c}^{E_{max}} \frac{dE}{\sigma(E)}.$$

Using for $\sigma(E)$ the curve shown in Fig. 4 which has been theoretically derived,[10] the following values of α/σ_{N} are obtained for various values of E_{max}.

E_{max}	2.0 Mev	3.0 Mev	4.0 Mev
α/σ_{Av}	$0.316 \times 10^{+24}$	$0.353 \times 10^{+24}$	$0.388 \times 10^{+24}$

The variation of α/σ_{N} with E_{max} is so slight because of the manner in which both α and σ_{N} decrease with rising E_{max}.

The value of the expression:

$$E_c P(E_c) + \int_{E_c}^{\infty} P(E) dE$$

was found from the observed pulse distribution (curve II of Fig. 3) to be 13.7 pulses per minute. Since the number H of hydrogen atoms in the chamber was $H = 6.9 \times 10^{+21}$ the number of neutrons passing through the chamber is

$$N = 1.98 \times 10^{-21} (\alpha/\sigma_{N}) \text{ per minute}.$$

If the cadmium cap G is removed the number of thermal neutrons reaching the uranium cell is increased and the number of fast neutrons passing through the ionization chamber is increased by the same factor. This factor was found to be 1.22 by filling the ionization chamber with pure hydrogen and then counting the hydrogen recoils giving rise to pulses above a certain arbitrarily set level, both in the presence and absence of the cadmium cap. Thus the number of neutrons N^* which pass through the

___
[10] J. Schwinger and E. Teller, Phys. Rev. **52**, 286 (1937).

chamber in the absence of the cadmium cap, G, is

$$N^* = 2.415 \times 10^{-21} (\alpha/\sigma_{N}) \text{ per minute}.$$

From N^* the number, K, of neutrons emitted per minute by the uranium was calculated by taking into account the geometrical factors, including the variation of the thermal neutron density within the uranium cell. K is thus found to be:

$$K = 8.25 \times 10^{-19} (\alpha/\sigma_{N}) \text{ per minute}.$$

In order to obtain the number of neutrons emitted per fission it is necessary to compare K with the number of fissions, L, which occur in the uranium under the conditions of this experiment. For this purpose both the ionization chamber and the uranium cell were removed and a parallel plate ionization chamber lined with a thick layer of uranium oxide was placed in the position previously occupied by the uranium cell. The number of fissions produced in this chamber was observed and found to be 19 per minute. The chamber was then calibrated by comparing the number of fissions obtained from the thick uranium oxide layer with the number of fissions obtained in the same chamber at the same thermal neutron intensity from a thin layer of uranium oxide containing 1.4 mg of uranium. The calibration was carried out by using a particularly strong neutron source, so as to obtain a sufficiently large number of counts from the thin layer. The ratio of the fission counts from the thick layer and from the thin layer was found to be 29.2, from which it is concluded that 196,000 fissions per minute should take place in the uranium cell containing 427.7 grams of uranium. This would be the number of fissions if the density of the thermal neutrons were not reduced in the uranium cell due to the absorption of such neutrons in uranium. We estimate that the average density of thermal neutrons within the cell is reduced by a factor of 0.715. The number of fissions L actually taking place within the cell is therefore

$$L = 140,000 \text{ per minute}.$$

In order to estimate the reduction of the average thermal neutron density within the uranium cell leading to the factor of 0.715, we first explored the anisotropy of the thermal neutron radiation near the uranium cell by means

624   W. H. ZINN AND L. SZILARD

of a rhodium indicator, and then calculated the thermal neutron density within the uranium by assuming the distribution of thermal neutrons to be the same as would result from the superposition of two parallel thermal neutron beams, one directed away from the source and the other towards it, and having an intensity ratio of 3 to 1. We assume exponential absorption for these two beams within the uranium and an exponent corresponding to a half-value thickness in uranium of 14 g per cm^2.

The number of neutrons emitted per fission is

$$K/L = 5.9 \times 10^{-24} (\alpha/\sigma_{N}).$$

This number should be increased by perhaps 10 percent in order to correct for the fact that $P(E)$ does not exactly coincide with $R(E)$ even for $E \leq E_c$. The magnitude of this correction was estimated by comparing for D+D neutrons of 2.5 Mev energy the observed pulse distribution $P(E)$ with the calculated distribution $R(E)$ in the region of the low recoil energies. Making this correction one finds for ρ, the number of neutrons per fission

$$\rho = 6.5 \times 10^{-24} (\alpha/\sigma_{N}).$$

Using for α/σ_{N} the value 0.353×10^{-24} which corresponds to $E_{max} = 3$ Mev rather than to the actually observed upper limit of the fission neutron spectrum, one finds

$$\rho = 2.3.$$

Since the fission neutrons hardly will be emitted with a single energy E_0, too high a value for α/σ_{N} would be obtained if the observed value of the upper limit of the energy spectrum were used for E_{max}. In any case the error introduced by the uncertainty of the actual energy distribution of the fission neutrons should be small since one finds for

$$E_{max} = 2 \text{ Mev} \quad \rho = 2.0$$

and for

$$E_{max} = 4 \text{ Mev} \quad \rho = 2.5.$$

More serious, however, may be a number of experimental inaccuracies which might conceivably add up to give a considerable error.

The interest in the number of neutrons emitted per fission arose out of its obvious importance from the point of view of the possibility of nuclear chain reactions. At present we have the following set of values: number of neutrons per fission, 2.3; fission cross section,[11] 2.0×10^{-24} cm^2; cross section for radiative capture[12] 1.3 or 1.2×10^{-24} cm^2. According to these values, the number of neutrons emitted by uranium per thermal neutron absorbed should be 1.4, which agrees with the value of 1.5 recently obtained by another method by Anderson, Fermi and Szilard.[13] Too much significance should not be attributed to this agreement, since the values given above are subject to fairly wide experimental errors.

If required the present experiments could be repeated with greater accuracy since the method used is quite capable of being applied with greater precision. Moreover, it gives the number of neutrons per fission independently of the value of the fission cross section which enters into the method used by von Halban, Joliot and Kowarski. These authors report[14] a value of 3.5 ± 0.7 neutrons per fission.

It should be mentioned that it appears to be essential for the method presented here to work with a low background count. The background is due to the primary neutrons and can be kept small by using a photoneutron source. We did not find it possible to obtain quantitative results by using neutrons from radon-beryllium sources or from the D+D reaction on account of the high background count due to the primary neutrons.

We are indebted to Dr. G. N. Glasoe for suggesting the method of obtaining the photographic records and for much valuable advice in this connection, and to Dr. E. T. Booth for determining by means of an alpha-particle count the uranium content of the thin uranium sheet which we used for purposes of calibration. Also, we wish to thank the Department of Physics of Columbia University for the laboratory facilities placed at our disposal and the Association for Scientific Collaboration for enabling us to obtain the radium used in this experiment.

[11] H. Anderson, E. Booth, J. Dunning, E. Fermi, G. Glasoe and F. Slack, Phys. Rev. 55, 511 (1939).
[12] H. v. Halban, L. Kowarski and P. Savitch, Comptes rendus 208, 1396 (1939); H. L. Anderson and E. Fermi, Phys. Rev. 55, 1106 (1939).
[13] H. L. Anderson, E. Fermi and L. Szilard, Phys. Rev. 56, 284 (1939).
[14] H. von Halban, F. Joliot and L. Kowarski, Nature 143, 680 (1939).

Part III

Documents Relating to the Manhattan Project
(1940–1945)

Introduction by Bernard T. Feld*

Preparation

The genius of Leo Szilard had many facets: a power of foresight, bordering on the visionary, that enabled him to comprehend the long-range consequences of a scientific discovery or of a social situation long before they were imagined by even the most astute; an ingenuity of approach and execution that led him to discover the most unexpected, but direct, solutions to difficult problems; a tenacity that prevented him from abandoning difficult problems but forced him to return and return again, devising new approaches at each attempt, and yet permitting him to leave the details of the solution in other, competent hands once the appropriate approach had been found and the path charted; and an overriding sense of responsibility, toward his work, toward science, but most of all toward his fellow men.

All these combined to make Leo Szilard an indispensable factor in the successful achievement of the first man-made nuclear chain reaction and in the vast wartime enterprise known as the Manhattan Project, which culminated in the first man-made nuclear explosion.

It was no accident that Leo Szilard was on the spot, at Columbia University, immediately after the announcement of the discovery of nuclear fission in January 1939, ready to plunge into the experimental investigations of its ramifications. He had sensed, with that uncanny instinct which never deserted him, that this was where the action would be; and he had been scheming and dreaming about the controlled release of nuclear energy ever since a remark by Lord Rutherford in the fall of 1933—that "anyone who expects a source of power from transformation of these atoms is talking moonshine"—had intrigued and goaded him into taking up the study of nuclear physics. Characteristically, he immediately realized that, of all the available nuclear particles, it was the neutron which held the key; the neutron had the ability to penetrate matter without the energy-consuming electromagnetic interactions which prevented all but a minute fraction of the other, charged nuclear projectiles from inducing nuclear reactions. The problem was to discover a neutron-induced reaction in which at least one neutron is emitted for each neutron absorbed.

While Szilard was a refugee in England (1933–1938) with no academic base and practically no resources, he embarked on a series of fundamental researches that, even though they did not succeed in uncovering the type of reaction he had envisaged, were to give him that profound understanding of nuclear phenomena which was indispensable to his exploitation of the fission reaction when it was finally discovered.

The scientific aspects of Szilard's work in England, prior to his emigration to New York at the beginning of 1938, are described in Part II of this volume. It is sufficient to observe that if his late entry into the field of nuclear physics did not permit him to participate in its most fundamental discoveries, it did not take him long to overcome the handicaps of a newcomer; by 1938 Szilard could be counted among the elite of the small international community of nuclear physicists. Still, he was always profoundly aware of his place in the historical scheme, with no illusions and a delicate balance of judgment with which he was always able to view history. As usual, this judgment is best described in his own words, and it is appropriate to reproduce at the end of this essay "Creative Intelligence and Society," an address delivered by Szilard at the University of Chicago in 1946. There is, to my knowledge, no more succinct or penetrating summary of the history of nuclear physics from its inception until the first chain reaction, nor of his own role in this historical chain.

* Professor of Physics, Massachusetts Institute of Technology, Cambridge, Massachusetts.

Genesis

Although much has been written about the genesis of the wartime atomic energy project, its prehistory—the period between the arrival of Bohr in the United States in January 1939 with fresh knowledge of the discovery of fission, and the setting up of the official Manhattan Project in 1941—remains somewhat obscure and confused. In part, this is a natural consequence of the fact that in science, when the time is ripe, similar ideas occur simultaneously at many sources, and the rapid interchange of ideas and information serves to confuse their origins, even in the minds of the most important contributors. Furthermore, most scientists are too concerned with their current research to take the time to unravel and record their recollections of the past.

Fortunately, Szilard has left us with a clear record of the steps leading to the conception of the chain-reacting system utilizing a lattice of natural uranium embedded in a matrix of graphite. These steps are outlined in a letter written to E. P. Wigner in 1956, which is reproduced at the end of this essay, together with the documents to which it refers.

Having established, in an experiment with Walter H. Zinn, the emission of neutrons in the fission process,[1] the problem was how to achieve a self-sustaining system—one in which the neutrons emitted in a fission would be capable of inducing at least one more fission reaction. There are two requirements for such a system. First, the fast neutrons emitted in the fission process must be slowed down to thermal energies, at which the probability of producing a fission in uranium is much increased. Second, it is necessary to avoid the parasitic absorption of these neutrons, both by uranium during the slowing-down process and by the light elements that are required to reduce rapidly the neutrons to thermal energies before they undergo such absorption or escape from a necessarily finite system.

After careful consideration of the possibility of utilizing mixtures of uranium and water for this purpose [Anderson, Fermi, and Szilard, *Phys. Rev.*, **56**, 284 (1939); also Halban, Joliot, Kowarski, and Perrin, *J. Phys. Radium*, **10**, 428 (1939)], Fermi and Szilard eliminated water from consideration as the slowing-down matrix owing to the strong neutron absorption properties of hydrogen. (In Szilard's case, at least, consideration of the engineering difficulties involved in extracting large amounts of power from a water–uranium system were also involved; from the start, he had in mind the necessity of producing large amounts of plutonium if the chain reaction was to be of any wartime importance.) Heavy water was also soon discarded owing to the unavailability of large quantities. Carbon, in the form of graphite, was the next choice, despite its considerably poorer energy-moderating properties than the hydrogens, both because of its availability in large quantities and of its small neutron absorption cross section (although it was not known for some time just how small the cross section is; as it later turned out, the upper limit known at that time could be accounted for entirely on the basis of impurities).

The problem of avoiding the parasitic (resonance) absorption of neutrons by uranium during the slowing-down process was more difficult. Szilard, and also Fermi,[2] hit upon the solution of separating the uranium from the graphite, so that the moderating process would take place in the absence of uranium. It remained to work out the necessary proportions of the constituents and their optimum geometrical distribution, on the basis of neutron reaction properties very incompletely known and in the absence of a satisfactory theoretical framework for treating such heterogeneous mixtures.

[1] The same fact was simultaneously and independently ascertained by Anderson, Hanstein, and Fermi, also at Columbia University [*Phys. Rev.*, **55**, 797 (1939)] and, slightly before, by Halban, Joliot, and Kowarski in Paris [*Nature*, **143**, 479 (1939)].
[2] Although Fermi was apparently thinking in terms of alternate layers of uranium and graphite, rather than the lattice structure devised by Szilard. (See letter from Fermi to Szilard, July 9, 1939, in the next section.)

By the end of 1939, Szilard had worked out a rough theory of the nuclear chain reaction in a graphite–uranium lattice, which served to convince him of its feasibility. This he submitted first as a letter and then as a paper to *The Physical Review* in February 1940. However, being by then convinced of the favorable prospects for its success and concerned that its publication would aid Germany in its prior achievement, its publication was indefinitely deferred at his request. Early in 1946, the editors of *The Physical Review* and Szilard agreed that the original version should be published with some footnotes and a "note added May 1946" for the purpose of clarifying some aspects of the paper in the light of new information. However, the authorities of the Manhattan Project at first refused and then delayed its declassification until December 1946. For reasons unknown, it was never published. This classic paper is here published for the first time.[3]

In acknowledgment of their invention, Fermi and Szilard were jointly awarded a patent on the nuclear reactor in 1955 by the United States Government.

Attainment

With the broad outlines of the solution generally agreed upon, Szilard was content to leave to Fermi the main responsibility for the experiments that were to prove the feasibility of the graphite–uranium chain reaction and explore its properties. Although he was deeply involved in the conception of the "exponential" experiment for measuring the neutron absorption properties of graphite and for testing the multiplicative properties of subcritical uranium–graphite systems, and he continued to take an active interest in the details of the experiments, Szilard's main efforts, from 1940 on, were directed toward planning for the earliest possible use of the reactor for the large-scale production of plutonium.

The fact that enough plutonium was produced to be used militarily before the end of the war (albeit just barely) is mainly owed to the acute and incisive foresight exhibited by Szilard in the earliest stages of the project.

This may be the appropriate point for some remarks on the question of indispensability: In the history of science very few men have been indispensable; particularly in applied science, no man is indispensable, and the achievement of the nuclear chain reaction definitely falls into this category. Many men in many countries had the same or similar ideas, and some advanced them relatively far under much more adverse circumstances.[4] The remarkable exploit involved in the plutonium project was the short time span between the discovery of fission at the end of 1938 and the production by means of a high-power-producing nuclear reactor of kilogram quantities of an entirely new element, plutonium, by 1944—all this requiring the invention and reduction to operation of the nuclear reactor and the development of a completely new branch of engineering in a field where many of the scientific fundamentals were either not yet known or imperfectly understood.

The scientists of the Manhattan Project thought they were involved in a race against the Germans; they were certainly racing against time. With hindsight, one may denigrate the importance of their accomplishment or regret, as Szilard was afterwards frequently to regret, that they succeeded so well; but retrospect cannot dim the luster of their achievement. In this accomplishment no man was more important than Leo

---

[3] This paper received only very limited circulation in the Manhattan Project; its existence was quite widely known, as A-55, but under the confusing and wholly misguided system of compartmentalization it was not possible, even for members of the project, to obtain most reports without establishing a "need to know" their contents. Early (pre-Chicago project) reports on the chain reaction were classified under the heading A; later, a more elaborate system was established: C for Chicago, CP for physics, CE for engineering, CF for fast neutron work, etc. But the system served more to confuse than to clarify.

[4] See the recent revelations on the exploits of Kurchatov and co-workers in the U.S.S.R., *Bull. At. Scientists*, December 1967.

Szilard, and only a few—notably Enrico Fermi and Eugene P. Wigner for the plutonium-production aspect of the project—can be placed at the same level.

As noted, Szilard's major contributions came in the earliest stages, in 1940–1941, beginning even before the Columbia group moved to Chicago and was incorporated into the many-faceted Manhattan Project. It was he who first recognized the importance of securing copious quantities of *pure* graphite, and who initiated the steps necessary to assure this. More important, realizing at a very early stage the necessity of securing pure uranium in its high-density metallic form—uranium metal had not previously been industrially produced, and its metallurgy was only vaguely comprehended—Szilard initiated and personally supervised a program aimed at the industrial production of uranium metal of high purity. A brief report (A-24), reproduced later in this section, shows how important was his contribution to the success of this aspect of the project.

Szilard's first act on moving to Chicago was to set up a Technological Division, of which he was the first head, concerned with the anticipation and solution of the engineering and metallurgical problems involved in the achievement of a high-power nuclear reactor. To this group he recruited a number of outstanding physicists and engineers who undertook, at his instigation, important studies of such problems as uranium metal fabrication, coating and cladding, heat transfer, as well as studies of alternative (to water) means of removing the fission-heat energy in an operating nuclear reactor.

Szilard's usefulness at this stage stemmed as much from his abilities as an engineer as from his understanding of nuclear physics. Like Wigner, whose early training in Budapest was in engineering—a budding scientist in Hungary usually turned to engineering as the most effective entrée into science—Szilard was a superb engineer. It is therefore no coincidence that the most farsighted and imaginative, as well as solid, contributions to the solution of the project's unprecedented engineering problems came from Szilard's Technological Division and Wigner's Theoretical Division.

The papers reproduced later in this section, all heretofore unpublished, represent two aspects of Szilard's contributions in this early stage of preparation for the large-scale release of nuclear fission energy in a chain reaction aimed at plutonium production. The first set are concerned with the measurement of basic nuclear constants and their application to the design of the nuclear reactor; the second represent engineering studies, some exploratory and some detailed, on reactor design.

Consolidation

Once the decision had been taken to construct a plutonium production reactor at Hanford, Washington, based on the graphite–uranium system with water cooling, Szilard's direct involvement slowly came to an end. The assignment of providing liaison between the project physicists and the engineers of the Du Pont Company, responsible for the construction and operation of the Hanford reactors, was given to Wigner's Theoretical Division; most of Szilard's Technological Division was incorporated into the engineering team, while some of its key members drifted off to other, more immediately pressing aspects of the project. A small group of physicists, under my immediate supervision, continued to work under his direction on some of the unsolved problems of nuclear physics only indirectly associated with the reactor.

However, even though relieved of direct responsibility for the success of the project, Leo Szilard could not divest himself of his deep involvement in its success. He assumed the role of contingency planner, devising alternative approaches in the event of the failure of the water-cooled system, which was encountering unforeseen problems and delays. His favored alternatives were a helium-cooled reactor and cooling by a eutectic mixture of liquid Bi and Pb. Szilard devoted great efforts toward the detailed design of such reactors; for example, he and I developed a detailed design for a magnetic pump for liquid bismuth, based on a principle which had been invented by him and Einstein in 1926 for the purpose of circulating liquid refrigerants in a system without moving parts.

In this period, Szilard was intrigued with the problem of devising the most efficient system for a nuclear power reactor. In addition to He and liquid Bi cooling (he also foresaw the possibility of using the electromagnetic pump for the circulation of liquid Na) Szilard foretold and investigated the engineering problems associated with most of the reactor systems that have since been seriously considered or successfully utilized for power purposes—a reactor cooled by the boiling of water; a beryllium moderated reactor; CO_2 cooling. In addition, he placed very high priority on the problem of breeding—of designing a reactor that would produce at least one nucleus of a fissile element (U^{233} from Th and Pu^{239} from U^{238} neutron absorption) for each nucleus consumed in fission. The idea of a fast neutron breeder reactor was first suggested by Szilard in a "Memorandum on the Production of 94 and the Production of Power by Means of the Fast Neutron Reaction" sent to Dr. A. H. Compton on January 8, 1943.

But more and more, convinced that we had won the race against the Germans, Szilard was concerned with the problem of using, or not using, the atomic bomb in a fashion most calculated to ensure peace and stability in the postwar world. In part, his concern for more efficient power reactors was motivated by a belief that it would strengthen the hand of the United States Government, in seeking a postwar system of stringent controls over nuclear weapons, to have accumulated, in the early postwar period, a substantial store of fissile material.

And thus, to an increasing extent, in the last years of the war, Szilard's efforts were devoted to the organization of a strong body of opinion among his colleagues in the project for exerting pressures on the government against the hasty and ill-considered use of the first atomic bombs. Although these efforts failed, their failure only spurred him toward greater efforts at political action in the postwar years. But that is another aspect of history.[5]

L'Envoi

I could not forego this opportunity to write some very personal recollections of the unique experience I had in working in close scientific collaboration with Leo Szilard during World War II.

At our first meeting, sometime early in 1941, in response to his telephoned invitation to lunch, I succumbed, with only token resistance, to his irresistibly exciting invitation to join him in the attempt to produce a nuclear chain reaction.

At the time, I was in my second year of graduate studies at Columbia University and was just becoming involved in research in Rabi's molecular beam laboratory. But I had heard of, and had been intrigued by, the discovery of fission—it was in the air at Columbia—and, more important, I was convinced that the United States would soon (should already have) become involved in the war and I was chafing for a chance to make some contribution. Fortunately, I had already hit upon a subject that I thought might be suitable for a (theoretical) thesis, and Szilard convinced me that I could continue work on this subject in my spare time and helped to arrange this with Willis Lamb, who agreed to be my thesis supervisor.[6]

I never knew why Szilard chose me as his assistant; I suppose he needed someone who would pursue some of his myriad ideas, who would follow his instructions without too much independence, who had enough competence in theoretical techniques to carry out a program of computation without too much supervision, and who

---

[5] See, for example, Alice Smith, *A Peril and a Hope*, Chicago: University of Chicago Press, 1965.
[6] By fortunate chance, this worked out as anticipated, and I was able to submit my thesis and receive my degree immediately after the war.

understood enough—but not too much—to pick up a partly developed idea and carry it to completion. Whatever his original expectations, I seemed to suit him well, for we worked in close collaboration until I left Chicago for Los Alamos in 1944.

Szilard was an ideal boss. He supervised and inspired but never held me back. He encouraged independence and enthusiasm. When I became excited about the early experiments of Fermi and his group at Columbia on the uranium–graphite lattice (the first exponential pile experiments), he encouraged me to join that group and to participate in the crucial experiments that proved the feasibility of the chain reaction. By the time we moved to Chicago, I was leader of a small group who carried on measurements on fundamental nuclear processes, mostly inspired by Szilard and followed by him with close interest.

The day I came to work for Szilard he installed me in his cluttered room in the King's Crown Hotel, adjacent to Columbia, presented me with a 20-inch slide rule (which, until today, remains my favorite computer), outlined a program for improved computation of the properties of the carbon–uranium lattice which was to occupy me for months, and informed me that he had to leave town for a week or so. At that time Szilard was continuously traveling, to Washington to assure continuing support for the project, to the laboratories and factories capable of producing pure graphite and uranium metal to goad and cajole their adoption of programs for the large-scale production of these vital materials even in the absence of governmental assurance of support of these programs. This was the critical period of preparation for the chain reaction, and Szilard's finger was in every pie. But he did not need to be around continually in order to supply ideas and inspiration enough to occupy all the efforts of more than one eager assistant.

Characteristically, Szilard spoke very little of his earlier life and prewar activities. Only little by little, in bits and pieces, did I learn of his work in Berlin with von Laue and Einstein—mostly during an automobile trip from New York to Princeton on an occasion when he wanted to consult with Einstein and on which he thoughtfully took the opportunity to introduce me to the great man. His foresight in leaving Germany immediately after Hitler's seizure of power he explained with the comment that it only had required a short airplane trip from Berlin to London, passing over that impossible jumble of frontiers and unstable political entities, to convince him that the Europe of 1932 was no place in which a sane man could stake his future. My interest in Chinese poetry—I was reading a book of translations by Arthur Waley—occasioned some fragmentary anecdotes concerning his acquaintance with Waley during the formation in England of a committee to aid the exodus and placement in British universities of young German-Jewish scholars, a project, as I learned later, for whose conception and execution Szilard was primarily responsible. And despite our very close collaboration during the period 1941–1944, I never heard directly from him of his difficulties with the narrow-minded and unimaginative higher authorities of the Manhattan Project, who could neither understand nor appreciate his independence and unconventionality. His method of teaching and guidance was by leadership and example and never by resort to preachment or authority.

After the success of the plutonium project was assured, and I departed with his blessing to work on the next stage of the project, we corresponded only occasionally on aspects of the physics problems with which we had been mutually concerned. But I saw him from time to time, when project business brought me to Chicago, and I could not help but be infected with his deep concerns over the consequences of the forces he had been so instrumental in releasing.

After the war his scientific interests turned to biology, but on our occasional meetings he exhibited the same keen and penetrating interest in the problems of particle physics, in which I had become engaged, as he had in the problems of nuclear physics on which we had worked together. He always maintained a deeply human, if apparently impersonal, interest in me as an individual.

More and more after 1946 our mutual concerns converged on the problems of controlling the nuclear genie. And so, in the last years of his life, I again became his assistant in the enterprise of building a more livable world, the ideal that guided all of Leo Szilard's enterprises from his earliest days until his death.

**Creative Intelligence and Society: The Case of
Atomic Research, The Background in Fundamental Science**

(Public Lecture by Leo Szilard, University of Chicago, July 31, 1946)

As you probably know, it all started one day around the turn of the century when Becquerel in Paris noticed that uranium minerals, placed in a drawer near some photographic plates, blackened those plates.

Uranium minerals affected photographic plates even if they were wrapped in black paper, yes, even if they were placed in a wooden or metal case.

This effect of uranium minerals was due to penetrating rays which they emitted and which were similar to x rays.

Immediately the question arose, where does the energy for these x-rays come from. But it took a few years before the answer to this question was found.

Madame Curie was at that time a graduate student with Becquerel, and she had a suspicion that uranium minerals contained some element other than uranium which was more active than uranium.

And as you know, years later she isolated radium from such minerals.

Radium, which was an element until then unknown, is chemically rather similar to barium.

It slowly changes over into another radioactive element which is chemically different from it.

This other element is not similar to barium but is similar to neon, and like neon, it is a gas.

It is called radon or radium emanation in order to indicate that it emanates from radium.

We refer to this by saying that radium undergoes a spontaneous transmutation. Its half lifetime is about 1,500 years which means that out of one gram of radium during a period of 1,500 years, $\frac{1}{2}$ gram will transmute into radium emanation.

Transmuting one chemical element into another chemical element was, as you know, the unsolved problem of the alchemists.

But Madame Curie, who isolated radium, could not pride herself to be a successful alchemist.

She did not produce radium.

She merely separated it chemically from a mineral in which it was previously contained and she was in no position either to accelerate or even to slow down the transmutation of radium into radium emanation.

So, in spite of this new discovery, God remained the first and only successful alchemist.

As uranium and radium were further investigated it gradually became apparent that we had to deal with a whole family of naturally occurring radioactive elements.

Any one member of this family either emits beta rays, that is, negatively charged electrons, or it emits alpha rays, that is, positively charged, fast moving helium atoms.

Up to the 30's, it was much easier to observe the single alpha particles of the alpha rays than it was to observe the single electrons of the beta rays.

In air such an alpha particle travels in a straight line over a length of a few centimeters and then it stops.

The range of alpha particles in air is a few centimeters.

Because the alpha particle carries a positive charge, it ionizes the air in its path, and therefore it is possible to make visible the track of the alpha particle.

Its track in air can be made visible to the naked eye and can also easily be photographed.

When an atom of a radioactive element emits an alpha particle, it undergoes a spontaneous transmutation and goes over into a chemically different element.

A radium atom for instance, which is similar to barium, emits an alpha particle and becomes an atom of radium emanation which is similar to neon.

The alpha particle which is emitted carries away the energy liberated in this transformation from one chemical element into another chemical element, and the energy liberated in such a process is of the order of several million electron-volts.

You have to contrast this with the energy which is liberated when carbon is burned, that is, when a carbon atom combines chemically with two oxygen atoms.

The energy liberated when carbon is burned amounts to a few electron volts.

So you see that as far as energies are concerned, the two types of processes differ by a factor of a million.

Becquerel's original observation that uranium minerals emit rays that blacken the photographic plate was the very first manifestation of the enormous energies liberated in nuclear transformations.

Until that observation was made, physicists had no inkling that there are any sources of energy other than the chemical processes of which the burning of carbon is the most common example.

Much of the investigation of the properties of radioactive elements and the behavior of alpha particles was done by Rutherford.

Rutherford noticed that an alpha particle does not always travel in air in a straight line along the whole course of its path but that occasionally its track is sharply deflected.

As long as an alpha particle merely ionizes the air, its track remains straight but when it hits the nucleus of a nitrogen or oxygen atom, the alpha particle gets sharply deflected.

But sometimes the track of the alpha particle in air is neither a straight line nor a line with a break, but a line which shows a fork.

Pictures of this sort were first obtained by Blackett in Rutherford's laboratory and were interpreted on the basis of some early experiments of Rutherford, by assuming that an alpha particle smashed a nitrogen nucleus and knocked a proton out of it while the alpha particle itself was captured by the nucleus.

According to Rutherford the long track of the proton and the short track of the rest of the nucleus formed the fork which was observed.

It was evident that in an individual process of this type, energy could be liberated, but it was also easy to see that processes of this type could *not* lead to large scale liberation of nuclear energy.

Because for every track which showed a fork there were thousands of tracks which did *not* and this means that in most cases the alpha particle does not produce energy but loses its own energy by ionizing the air through which it travels.

Whenever a charged particle moves through matter, it will ionize and in the vast majority of the cases it will lose its energy and stop before it has had a chance to make a collision with a nucleus and to smash that nucleus.

Looking back at the steps which led us from the discovery of radium to the large scale liberation of atomic energy by means of a chain reaction, we can now see that the alpha particles were able to give us the answer to almost all of our questions.

In order to discover all the things which we had to know, we merely had to observe the tracks of these alpha particles in air and to look sharp and see what happened when alpha particles were allowed to bombard various elements.

Bothe in Germany bombarded, with alpha particles, some light elements such as boron and beryllium and he noticed a penetrating radiation emanating during the bombardment from these light elements.

He thought he had to deal with penetrating x-rays or gamma rays, and this was an interesting discovery, but nothing to get excited about.

But now Joliot performed a curious experiment with these penetrating rays.

For some reason or other he placed a sheet of paraffin into their path and looked to see if anything was happening to the paraffin.

You might say that this was a rather pointless experiment and there really was very little justification for it.

However, this experiment changed the course of history.

For Joliot found that hydrogen atoms were knocked out of the paraffin by the penetrating radiation to which it was exposed.

Joliot further found that these hydrogen atoms or, more correctly, protons, had curiously enough, energies of several million volts and inevitably the question arose, by what mechanism could x-rays passing through paraffin transmit several million volts of energy to the protons in the paraffin.

There was no such mechanism known and it was therefore pretty evident that Joliot had found a phenomenon which could not be explained on the assumption that the penetrating radiations with which he had to deal were x-rays.

Chadwick, in Cambridge, looked therefore at the tracks of these protons in air and saw that the radiation which was responsible for them must consist of particles which carry no charge since their track in air could not be made visible.

Chadwick found further to his great surprise that the mass of these particles must be pretty close to the mass of the proton itself.

This meant that he had to deal with some new particle which carried neither a positive nor a negative charge, and he decided therefore to call the particle, neutron.

Now this was a very interesting discovery but at first these neutrons proved to be rather elusive.

You could not make visible and see the track of the neutron.

You could produce neutrons but you could not collect them in any bottle because, having no charge, the neutrons would pass through the wall of the bottle.

In those early days experiments with neutrons were rather clumsy to perform; they required some instrument like a Wilson cloud chamber which is suitable for some experiments but not so suitable for others.

Progress of neutron physics would have probably remained slow if it had not been for another discovery made by Joliot and his wife, Irene Curie.

I am speaking of the discovery that it is possible to make elements artificially radio-active.

Now the thought that when we shoot alpha particles at some element and smash nuclei of that element that we might then produce unstable or radioactive elements did not escape earlier investigators.

If you look at the last edition of the classical textbook on radioactivity by Rutherford, Chadwick, and Ellis, printed in the 1930's, you will find a paragraph devoted to this topic.

Rutherford and his school made some such experiments and they summed up their results in one short paragraph.

Their experiments showed that the disintegration of elements by alpha particles leads to stable and not to radioactive elements.

"This conclusion appears inescapable"—so they stated—"unless the elements produced should happen to be beta ray emitters."

But the joke is that the artificial radioactive elements which Joliot later produced from ordinary elements by bombarding them with alpha particles all happen to be beta ray emitters.

Beta rays cannot be easily observed in the same manner as alpha particles can, but they can be detected by means of a so-called Geiger–Muller counter tube.

It may be that the discovery of artificial radioactivity had to wait until the Geiger–Muller counter became available.

Each time an electron passes through a Geiger–Muller tube, it causes an electric discharge which can be made visible and can be registered.

Such a Geiger–Muller tube will not only register the electrons from the beta rays of radioactive elements, but will also register the electrons of the cosmic rays which pass through it.

In the counter tube of the usual size, you obtain something like 20 discharges per minute from the cosmic ray particles passing through the counter and this count of about 20 per minute is called the background of the counter.

E. O. Lawrence of Berkeley, California, relates that sometime in the 30's, he thought of using such a Geiger–Muller tube for various experiments in the laboratory where his cyclotron was installed.

He had no previous experience with these counter tubes and he put one of his students to work on it.

Of course, during the time when the cyclotron was in operation its radiations interfered with the counter, but whenever the cyclotron was switched off, they could turn on the counter tube and begin to experiment with it.

To his dismay, Lawrence found however that he could not get the background of his counter tube to be constant.

On some days the background count was down to 20 per minute, as it should be, but on other days it was 100–200 or even 1,000 counts per minute.

Clearly, the background of the counter was as the physicist would put it, irreproducible.

Since he had no previous experience with it, Lawrence concluded that the Geiger–Muller counter was an unreliable instrument, not fit to be used in his laboratory.

Today, we can easily understand just what took place in these Berkeley experiments.

During the operation of the cyclotron the copper parts of the cyclotron became radioactive.

When the cyclotron was switched off, these copper parts continued to give off radiations which affected the Geiger–Muller counter and were registered by the counter in the form of 100 or on some days, 1,000 counts per minute.

So what may be considered the greatest discovery of this century was passed by at Berkeley as a mere nuisance.

In physics it is really difficult to know how to behave.

Perhaps Lawrence did make the mistake of not trusting his eyes.

But in physics, while you may look like an idiot if you do not trust your eyes, you may also look like a fool if you do.

In physics, there are no recipes which you can follow, and all we can say is that in the present case, Joliot succeeded where Lawrence failed.

In 1933 Joliot was again bombarding some light elements such as boron with alpha particles and he noticed that, apart from the mysterious penetrating radiation which he now knew were neutrons, boron also emitted positive electrons under the effect of this boron bombardment.

Joliot registered these positive electrons by means of a Geiger–Muller counter and one day, toward the end of 1933, he noticed that when he cut off the alpha particles from the boron, the Geiger–Muller counter still continued to register electrons.

These electrons did not keep on coming however for very long.

Their number fell off fairly rapidly and they vanished after a few minutes.

At first Joliot thought he had to deal with some curious after effect in the Geiger–Muller counter.

He did not conclude, as Lawrence did, that the counter is an unreliable instrument.

He merely concluded that there was something wrong with his particular counter.

So he had his counter tested, and having convinced himself that there was nothing wrong with it, he saw at last that he had to deal with a new phenomenon.

Boron, if bombarded with alpha particles, undergoes a nuclear transmutation.

It turns into a radioactive nitrogen which emits positive electrons and in doing so, transmutes into stable carbon.

A number of other light elements can be made radioactive in the same way, that is, simply by bombarding them with alpha particles.

Now, the discovery of artificial radioactivity had been predicted as early as 1914.

It had been predicted not by any physicist, but by H. G. Wells. Wells put this discovery into the year of 1933, the year in which it actually happened.

His book, called *The World Set Free*, was published before the First World War, and goes far beyond predicting Joliot's discovery.

It also predicts the large-scale liberation and industrial use of atomic energy, the manufacture of atomic bombs, and a world war in 1956 in which Chicago, Paris, London, and other cities are destroyed at the very outbreak of the war.

According to Wells, these cities are transformed into rubble, or to be quite precise, into radioactive rubble.

Finally, the book describes how after this catastrophe, a world government is established and begins to operate.

For all we know, every single one of these predictions may yet come true.

I happened to read Wells' book in 1932 and at that time I did not have the notion that it had much to do with reality.

I remember very clearly that the first thought that liberation of atomic energy might in fact be possible came to me in October 1933, as I waited for the change of a traffic light in Southampton Row in London.

The thought did not come entirely out of the clear sky.

A week or two earlier there had been the annual meeting of the British Association and Lord Rutherford was reported to have said at that meeting that whoever talks of the large scale liberation of atomic energy is talking moonshine.

I was wondering whether Rutherford was right when it occurred to me that neutrons, in contrast to alpha particles, do not ionize the substance through which they pass.

Consequently, neutrons need not stop until they hit a nucleus with which they may react.

If we could now find an element which captures neutrons, and in the process of doing so, emits further neutrons, we might have something like a chain reaction.

This looked like a rather attractive possibility, but the question was, is there such an element which captures a neutron and emits other neutrons in its place and if there is such an element, how would we recognize it?

I did not see how to go about looking for such elements, until a few months later, when I learned of Joliot's discovery that alpha particles can make elements radioactive.

If alpha particles can do this it was reasonable to think that neutrons could do it also and indeed this was demonstrated shortly afterwards by Fermi.

Fermi found that if you have a neutron radiation and if you expose for instance iodine to this radiation, the neutrons will be captured by the iodine atoms and you obtain a radioactive iodine which will emit beta rays and which decays to half of its amount in about twenty-five minutes.

Many other elements behave quite similarly.

Now it became possible to devise new types of experiments with neutrons, in which the presence of the invisible neutrons could be detected by their ability to make iodine, and other elements, radioactive.

This new technique appeared to me to be so promising that I decided to start experimenting with neutrons even though prior to this time I had done no work whatsoever in the field of nuclear physics.

By 1935, I began to take very seriously the thought that some elements might emit more than one neutron for every neutron which they capture, and that they might therefore be capable of sustaining a chain reaction.

I particularly suspected the elements which showed a peculiar phenomenon called isomerism and among them were indium, bromine, and uranium. But there was no telling which of the elements would really perform this service.

So it appeared that rather than to try to be too clever, one ought to examine patiently all of the 92 elements in order to detect which of them might show such a phenomenon.

None of my colleagues among the physicists seemed to show much enthusiasm for such a project and I thought that I might perhaps have better luck with some of the chemists.

To the chemist, the word "chain reaction" had at least a familiar sound.

For chain reactions, of a sort, can be set up in some chemical mixtures, for instance, in a mixture of chlorine and hydrogen gas.

So, one day, I paid a visit to a distinguished chemist who had shown signs of vision and courage in the past and told him of these thoughts, suggesting that we organize a survey, going through the whole periodic system of the elements.

The cost of the survey was estimated at $8,000.

This proposal was favorably received, but somehow the funds did not materialize and the survey did not get under way.

Though some preparations were made and some apparatus was actually built.

Of course, I had no conception of uranium breaking up into two about equally heavy fragments; that is, I had no conception of what is these days called the fission process.

Later on, as my knowledge of nuclear physics increased, my faith in the possibility of a chain reaction gradually decreased, and just about reached the vanishing point when fission was actually discovered in Europe.

Apparently between the years of 1935 and 1938 I went through the process of becoming an expert, that is, a man who knows what cannot be done.

I have no apology to offer and my only consolation is that I was in very good company.

For fission really ought to have been discovered as early as 1934.

Uranium was not overlooked by Fermi who was the first one to make elements radioactive by bombarding them with neutrons.

He bombarded uranium with neutrons and found that quite a number of radioactive elements are produced from it.

Since uranium was the heaviest element and since these radioactive elements could not be identified with any of the other known radioactive elements which are lighter than uranium, Fermi concluded that he produced transuranic elements.

For the discovery of these transuranic elements, he was awarded the Nobel Prize in 1938.

The Swedish Academy has always been very anxious to avoid awarding the Nobel Prize for advances which later on might turn out to have been in error and therefore in general it does not like to award the Nobel Prize for results which are derived by means of theory rather than by means of experiments.

This is probably the reason why the Swedish Academy preferred to give Fermi the Nobel Prize for the discovery of transuranic elements rather than for his beautiful theory of beta ray emission.

But unfortunately, truth in science is a rather elusive creature, and the principle of "safety first" is not a reliable guide for action in any field of human endeavor.

Fermi's transuranic elements were further investigated and were incorporated in a more and more elaborate pattern of new radioactive elements by Hahn and Meitner in Germany.

This went on until the end of 1938 when suddenly the soap bubble burst and Hahn himself discovered that he had been wrong all along.

Uranium splits into two approximately equally heavy fragments in a number of different ways and these fragments which are radioactive are responsible for the radioactivities which were previously thought to belong to transuranic elements.

It seems to me we ought to thank God that the fission of uranium was not discovered, as it should have been, in 1934 or in 1935.

It is almost certain that if this discovery had been made at that time, with Germany planning for war and England and America being in the frame of mind in which they were, the Germans would have found a way to make a chain reaction and would have won the war within a few weeks after they started it.

Perhaps those of us who missed this discovery 12 years ago, ought to be considered as candidates for the next award of the Nobel Prize for Peace.

The news of fission reached us here in the United States in January 1939.

I personally heard about fission first through Mr. Wigner, whom I visited in Princeton and immediately I was convinced that neutrons would be emitted in the process.

So the question arose, how to demonstrate this fact by means of experiments? As it turned out, Fermi in New York and Joliot in Paris, had also thought of this possibility and were also devising experiments for the same purpose.

Three different experiments got thus under way and were completed practically at the same time, in the first week of March, 1939.

About two neutrons were found to be emitted from every uranium atom which undergoes fission and the big question was: does it mean that we can make a chain reaction in a large mass containing uranium?

The mere fact that two neutrons are emitted in fission is not sufficient for answering this question.

As Bohr immediately realized, uranium 238, which makes up 99 percent of uranium, does not undergo fission if it absorbs a neutron.

Only uranium 235 which accounts for less than 1 percent of the uranium, will split when it absorbs a neutron and will emit additional neutrons.

So the question arose, will the uranium 238 which forms the bulk of uranium, so strongly compete for neutrons with uranium 235 that it will make a chain reaction in natural uranium impossible?

Since uranium 235 likes to react with slow neutrons, conditions for a chain reaction could obviously be made more favorable by mixing uranium with something that will slow down the neutrons.

Now the classical agent for slowing down neutrons is hydrogen.

The neutrons will collide with hydrogen atoms and on the average, their velocity will drop down to half in each such collision.

As a practical matter, hydrogen is used not in the form of hydrogen gas, but in the form of water.

And so, in May of 1939, Fermi, Anderson, and I began to experiment at Columbia University with mixtures of uranium and water.

Joliot, Halban and Kowarski did the same in Paris.

As I said before, a neutron which is produced in such a system will not necessarily cause fission in U235.

It may die, by being absorbed in U238 which makes up 99 percent of the natural uranium.

It may also die after it has been slowed down in such a system by being absorbed in the water rather than in uranium.

So the question, whether a uranium–water mixture can maintain a self-sustaining chain reaction, had to be decided by rather elaborate measurement of the balance of all these different absorption processes.

The last experiment which was performed by us on such a system in the United States was completed in June of 1939 and was not wholly decisive.

But I personally rather lost faith in this system at the time.

I felt that even if a chain reaction could be maintained in a uranium–water mixture, we would have to deal with very disagreeable chemical processes in such a system, if we attempted to use it for the liberation of atomic energy on a large scale.

The radiations emitted from uranium would decompose water into hydrogen and oxygen and this explosive gas mixture would have to be removed.

So, from an engineering point of view, the water–uranium system appeared to be a rather messy one.

It was this consideration which first led me to contemplate the possibility of using graphite, rather than water, for slowing down the neutrons. Then gradually, from July 1939 until February 1940, I became more and more convinced that it should be possible to set up a chain reaction in a graphite–uranium system.

The first experiments with graphite were started at Columbia University in April of 1940, and would have very quickly led to the establishment of a chain reaction if it had not been for the slowness of securing the needed materials in the required degree of purity.

The fact is, however, that the chain reaction was not actually produced until the 2nd of December, 1942, when it was first in operation here at the campus of the University of Chicago.

Perhaps this is a good time to pause and to ask ourselves what were the fundamental discoveries which led to the liberation of atomic energy.

In answer to this question I would list only three discoveries:

1) The discovery of Becquerel, that uranium minerals emit x-rays, which led to the study of radioactivity.

2) Joliot's discovery, that the penetrating radiation emitted from boron, under bombardment with alpha particles, is capable of knocking out protons from paraffin wax.

It was this discovery which led to the study of the neutron.

3) And finally, Joliot's discovery that boron, if bombarded by alpha particles, emits positive electrons, because this led to the discovery of artificial radioactivity.

It may very well be that if Becquerel had not lived, another hundred years might have had to pass before radium was discovered.

It may very well be, that if Joliot had not fooled around with paraffin wax, another fifty years might be needed before the neutron was discovered.

And, it may very well be, that if Joliot had not noticed the positive electrons coming from the boron, the discovery of artificial radioactivity might have been delayed for a further twenty-five years.

It is curious indeed that all these fundamental discoveries were made in France.

Naturally, there were many other important discoveries along the road, many of them fully deserving the Nobel Prize, but in a way, all these other discoveries were inevitable consequences of these three fundamental steps which I have listed.

Returning now to the state of our knowledge as it was in February 1940, even though it appeared almost certain that we can set up a chain reaction in a uranium–graphite system, there was still another step missing before we became aware of the full importance of this line of development.

This missing step was made by Turner in Princeton.

Turner, who knew of the work which Fermi and I were getting under way in the Spring of 1940, sent us a manuscript in which he pointed out the importance of a secondary reaction which would accompany any such chain reaction.

As I mentioned repeatedly, a fraction of the neutrons in the chain reaction will be captured by uranium 238.

This leads to a radioactive element which after going through two beta transformations, goes over into a long-lived element nowadays called plutonium.

Fermi and I considered this absorption of neutrons by U238 merely as a nuisance.

Turner pointed out that the resulting element, plutonium, will be capable of fission, just as U235 is capable of fission.

There are however significant differences between plutonium and U235.

U235 is chemically not different from uranium and it takes therefore a very laborious process to separate it from uranium.

Plutonium, being chemically different from uranium, can be separated by ordinary chemical methods.

Another significant difference is that the amount of U235 is strictly limited, its abundance being less than 1 percent in natural uranium.

Plutonium, since it is produced from uranium 238, which forms more than 99 percent of natural uranium, is, at least in theory, not so severely limited in quantity.

With this remark of Turner, a whole landscape of the future of atomic energy arose before our eyes in the Spring of 1940 and from then on the struggle with ideas ceased and the struggle with the inertia of Man began.

Such further ideas as were necessary for the completion of this work concerned either technical details or concerned the detonation of the bomb.

The former class would not be of interest to you, and the latter one is in an area which is considered highly secret.

The first use of plutonium, as you know, was in the form of a bomb which destroyed a city.

The next use of plutonium might be the same again.

With the production of plutonium carried out on an industrial scale during the war, the dream of the alchemists came true and now we can change, at will, one element into another.

That is more than Mme. Curie could do.

But while the first successful alchemist was undoubtedly God, I sometimes wonder whether the second successful alchemist may not have been the Devil himself.

Correspondence and Memoranda

Szilard Letter to Eugene P. Wigner
(February 1956)

1155 E. 57th Street
Chicago 37, Illinois

February 1, 1956

Professor E. P. Wigner
Department of Physics
Princeton University
Princeton, New Jersey

Dear Wigner:

You expressed some interest in seeing certain documents which relate to my work in the early development of atomic energy.

Enclosed you will find the following:

1) A letter to the Editor of the *Physical Review* (Szilard and Zinn, *Phys. Rev.*, **55**, p. 799, 1939) that describes the discovery of the neutron emission in the fission of uranium and states that about two neutrons are emitted per fission.[1] The same discovery was made about the same time, but by a different method, independently by Halban, Joliot, and Kowarski, as well as by Anderson and Fermi.

2) An article in the *Physical Review* (Anderson, Fermi, and Szilard, *Phys. Rev.*, **56**, p. 284, 1939) in which the possibility of maintaining a chain-reaction in a uranium–water system is explored.[2]

3) Three letters written by me to Fermi in July 1939, dated July 3rd, 5th and 8th, proposing the use of the uranium–carbon system.[3]

4) Fermi's answer, dated July 9, which crossed my July 8th letter to him and also my answer to Fermi's letter.[4]

5) During the second week of July, I saw that by using a lattice of uranium spheres embedded in graphite, one would have a great advantage over using alternate layers of uranium and carbon. I also saw that if we have a lattice of uranium spheres in graphite a further advantage can be gained by using uranium metal instead of using uranium oxide. It was these facts which led me to believe from then on that there is a good chance of maintaining a chain-reaction in a uranium–graphite system.

---

[1] Reproduced in Part II.
[2] Reproduced in Part II.
[3] Reproduced in Part III.
[4] Reproduced in Part III.

Early in July, I told you of my pre-occupation with the uranium–graphite system. The two of us paid a call on Einstein and you left then for the West Coast. Teller came to New York to spend the summer there, and I kept him informed throughout July on my thoughts concerning the uranium sphere–lattice system. My appraisal for the chances of a chain-reaction in such a system was communicated to Einstein when Teller and I called on him later in July.

Einstein's letter to President Roosevelt and my memorandum which was attached to it are enclosed.[5]

When Fermi returned to New York in the fall of 1939, I showed him my computations on the uranium sphere lattice in graphite and he showed me his of a homogeneous mixture of uranium and carbon. In accordance with this the memorandum which I submitted to Dr. Briggs in October 1939 contains the following passage:

"The properties of a system composed of uranium and graphite have been calculated independently, for a homogeneous mixture, by Fermi, and, for a lattice of spheres of uranium oxide, or uranium metal, embedded in graphite, by myself. The results of these two independent calculations are in reasonable agreement and show that the two arrangements have different properties. For instance, in the case of using a lattice of spheres a great advantage could be obtained by using uranium metal instead of uranium oxide, whereas in the case of the homogeneous mixture the use of uranium metal would be of no great advantage."

So that you do not gain the wrong impression from this passage, I should stress that Fermi knew that it is better to keep the uranium and carbon in separate layers rather than to use a homogeneous mixture. This is shown by the passage in his letter of July 9, 1939, which reads as follows:

"Since however the amount of uranium that can be used, especially in a homogeneous mixture is exceedingly small, even a very small absorption by carbon either at thermal energy or even before might be sufficient for preventing the chain-reaction; perhaps the use of thick layers of carbon separated by layers of uranium might allow to use a somewhat larger percentage of uranium."

In the case of the uranium–water system, Fermi had computed as early as June 1939 that some advantage can be gained by keeping the uranium and water in separate layers. I have no knowledge, however, that Fermi has recognized the advantage of having a lattice of uranium spheres or rods of small diameter embedded in graphite, or the advantage of using the metal in place of the oxide in such a lattice—until I presented my results to him.

The full text of the memorandum which I submitted to Dr. Briggs in October 1939 is enclosed.[6]

6) Report A-55 of the Uranium Committee which is a copy of a paper that I submitted to the *Physical Review* in February 1940.[7] This paper was accepted for publication, but publication was deferred at my request because of the nature of the paper.

[5] Reproduced in Part III.
[6] Reproduced in Part III.
[7] Reproduced in Part III.

On the basis of measurements which Joliot and his co-workers have performed on a uranium–water system, it is concluded in this paper that one should be able to maintain a chain-reaction in a uranium–graphite system if a lattice of uranium metal spheres is embedded in graphite. The paper gives a rough theory of such a system. It explains that the chain-reaction can be controlled by shifting an absorber between positions differing in neutron density and it states that the delayed neutron emission of uranium permits it to move such absorbers quite slowly.

7) A page from *Chemical Engineering News* showing that the "first nuclear reactor patent (No. 2,708,656) was issued to Leo Szilard and the late Enrico Fermi on May 17, 1955 by the United States Patent Office." [8]

[Signed] Leo Szilard

[8] An article from the New York Times is substituted for this report and is reproduced in the Appendix to Part III.

Hotel King's Crown
420 West 116th Street
New York City

July 3rd, 1939

Dear Fermi:

This is to keep you informed of the trend of my ideas concerning chain reactions. It seems to me now that there is a good chance that carbon might be an excellent element to use in place of hydrogen, and there is a strong temptation to gamble on this chance. The capture cross-section of carbon is not known: the only experimental evidence available asserts an upper limit of 0.01 times 10^{-24} cm^2. If the cross-section were 0.01 carbon would be no better than hydrogen, but the cross-section is perhaps much smaller, and it might be for instance 0.001. If it were so carbon not only could be used in place of hydrogen, but would have great advantages, even if a chain reaction were possible with hydrogen also. The concentration of uranium oxide in carbon could be kept very low, so that one could have about 2 gm of carbon per cc. This compares favorably with $\frac{1}{2}$ gm of water per cc at the most and means that the mean square of the displacement of a neutron for slowing down to thermal velocities would be only 1.5 times as large in the carbon–uranium-oxide mixture than in the water–uranium-oxide mixture. If capture by carbon can be neglected, the concentration of uranium oxide is determined by the consideration that the average displacement of a thermal neutron for capture by uranium in the mixture must not become too large. With this as a limiting factor about $\frac{1}{10}$ of the weight of the mixture would have to be uranium, and that means that one would need only a few tons of uranium oxide if our present data about uranium are correct.

I personally would be in favor of trying a large scale experiment with a carbon–uranium-oxide mixture if we can get hold of the material.

I intend to plunge in the meantime into an experiment designed for measuring small capture cross-sections for thermal neutrons. This is the proposed experiment: A sphere of carbon of 20 cm radius or larger is surrounded by water and a neutron source is placed in the center of the sphere. The slow neutron density is measured inside the carbon sphere by an indium or rhodium indicator at two points, one close to the surface, and one close to the center. The slow neutron density at these two points is measured once with, and once without, an absorbing layer of boron (or cadmium), covering the surface of the sphere. It is easy to calculate, from the observed ratio of the differences (of the observed neutron density with and without absorber at the surface of the sphere) obtained for the two points and the scattering cross-section, the ratio of the capture cross-section to the scattering cross-section for thermal neutrons. I calculate that a ratio of the neutron densities of the order of magnitude of 75 to 100 would for instance be obtained for two points in a sphere of carbon of about 20 cm radius if the capture cross-section of carbon were 0.005. It seems that very small capture cross-sections can conveniently be measured by this method.

If carbon should fail, our next best guess might be heavy water, and I have therefore taken steps to find out if it is physically possible to obtain a few tons of heavy water. Heavy hydrogen is supposed to have a capture cross-section below 0.003, and the scattering cross-section ought to be 3 or 4 times 10^{-24} for neutrons above the 1 volt region. (It is 6 to 7 times 10^{-24} for the thermal region.) Since heavy hydrogen slows down about as efficiently per collision as ordinary hydrogen, and since hydrogen has a capture cross-section of 0.27 and a scattering cross-section of 20, heavy hydrogen is more favorable.

Yours,
[Signed] Leo Szilard

Hotel King's Crown
420 West 116th Street
New York City

July 5th, 1939

Dear Fermi:

I think the letter I wrote you on July 3rd contains a mistake insofar as the ratio of the thermal neutron density at the center of the sphere and at the surface of the sphere is not 75 to 100, but 95 to 100 for the values given in that letter. The thermal neutron density within the sphere obeys the equation

$$D \frac{d^2(r\rho)}{dr^2} - A(r\rho) = 0$$

with

$$D = \frac{w\lambda_{sc}}{3}; \qquad A = \frac{w}{\frac{\sigma_{sc}}{\sigma_c}\lambda_s}$$

it is:

$$\rho(r) = C \frac{e^{ar} - e^{-ar}}{r}$$

where

$$a = \frac{\sqrt{3}}{\lambda_{sc}} \sqrt{\frac{\sigma_c}{\sigma_{sc}}}$$

For small ar we have

$$\rho(r) = 2aC\left(1 + \frac{a^2 r^2}{6}\right)$$

and the ratio of the densities on the surface and in the center is given by

$$\frac{\rho(r)}{\rho(0)} = \frac{e^{ar} - e^{-ar}}{2ar} \cong 1 + \frac{a^2 r^2}{6}$$

For $r = 20\,\text{cm}$, $\lambda_{sc} = 2\,\text{cm}$, and $\dfrac{\sigma_c}{\sigma_{sc}} = \dfrac{1}{1000}$ we have

$$\frac{\rho(20)}{\rho(0)} \cong 1.05$$

As you see, the method is beginning to get somewhat awkward in the case of carbon for smaller capture cross-sections than 0.005. It seems that it will be possible to get sufficiently pure carbon at a reasonable price. Carbon would also have an advantage over hydrogen insofar as there is no change in the scattering cross-section in the transition from the resonance region to the thermal region. Consequently, if layers of uranium oxide of finite thickness are used, the diffusion of the thermal neutrons produced in the carbon to the uranium layer is not adversely affected as in the case of hydrogen by such a change. Whether this point is of any importance depends of course on the absolute value of the carbon cross-section. Pending reliable information about carbon we ought perhaps to consider heavy water as the "favorite," and I shall let you know as soon as I can how many tons could be obtained within reasonable time.

> With kind regards to all,
> Yours,
> [Signed] Leo Szilard

> Hotel King's Crown
> 420 West 116th Street
> New York City
>
> July 8th, 1939

Dear Fermi:

Sorry to bombard you with so many letters about carbon. This is just to tell you that I have reached the conclusion that it would be the wisest policy to start a large scale experiment with carbon right away without waiting for the outcome of the absorption measurement which was discussed in my last two letters. The two experiments might be done simultaneously. The following can be said in favor of this procedure:

A chain reaction with carbon is so much more convenient and so much more important from the point of view of applications than a chain reaction with heavy water or helium that we must know in the shortest possible time whether we can make it go. This can be decided with certainty in a relatively short time by a large scale experiment, and therefore this experiment ought to be performed. If we waited for the absorption measurements we would lose three months, and in case the result is positive we would still not know with a 100 % certainty the answer with respect to the question of the chain reaction.

I thought that perhaps 50 tons of carbon and 5 tons of uranium should be used as a start. The value of the carbon would only be about $10.000. Since the carbon and the uranium oxide would not be mixed but built up in layers, or in any case used in some canned form, there will be no waste of material or waste of labor involved in unmixing after the experiment is over. Since the uranium layers may be separated by carbon layers of 20 to 30 cm thickness, or even more, we have to deal with a comparatively simple structure. Much simpler than would be the case for alternating water and uranium layers.

I told Professor Pegram yesterday how I felt about the situation, and he seemed to be not unwilling to take the necessary action. I wonder whether you think it wise to proceed as outlined in this letter.

> With kindest regards,
> Yours,
> [Signed] Leo Szilard

> University of Michigan
> Ann Arbor
> Department of Physics
>
> July 9, 1939

Dear Szilard,

Thank you for your letter. I was also considering the possibility of using carbon for slowing down the neutrons; in the obviously optimistic hypothesis that carbon should have no absorption at all for neutrons, and assuming for the resonance absorption band of uranium the usual data (which also I rather suspect to be optimistic) one finds from an elementary calculation that the ratio of the concentrations (ratio of the number of atoms) of uranium and carbon should be about one thousandth in order to avoid too much resonance absorption. According to my estimates a possible recipe might be about 39,000 kg of carbon mixed with 600 kg of uranium. If it were really so the amounts of materials would certainly not be too large.

Since however the amount of uranium that can be used, especially in a homogeneous mixture, is exceedingly small, even a very small absorption by carbon either at thermal energy or even before might be sufficient for preventing the chain reaction; *perhaps the use of thick layers of carbon separated by layers of uranium might allow use of a somewhat larger percentage of uranium.*

I have been thinking about the experiment that you propose for measuring the small absorption cross section in carbon. It seems to me that you have probably over estimated the difference between *rand* [rim] and center activity in the carbon sphere; moreover *I don't see how you can take into account the contribution of those neutrons that become thermal due to impacts against carbon.* Their number should probably not be very large, but might disturb very considerably the measurement of a small difference.

I had discarded heavy water as too expensive; but if you can easily get several tons of it, it might work very nicely.

The cyclotron here will start working again next week and I hope to be able to get reliable information on the so called resonance absorption of uranium. I shall inform you of the results.

Yours sincerely
[Signed] Enrico Fermi

P.S. I have received your second letter. If heavy water is too expensive, as I believe, it would be important to find some way of knowing something of the carbon absorption. It seems to me that the use of very thick layers of C might do the trick.

Yours,
[Signed] Enrico Fermi

Hotel King's Crown
420 West 116th Street
New York City

July 11th, 1939

Dear Fermi:

Many thanks for your letter of July 9th. It obviously crossed with my third letter about carbon which probably reached you on Monday. Today, being in a hurry, I confine myself to discuss one point which you mentioned. You write with reference to the carbon sphere experiment that it might be difficult to take into account the distribution of those neutrons which become thermal due to impacts against carbon, and I wish to say the following in this connection.

The number of such neutrons which become thermal within the carbon is quite large, but their number is taken fully into account by the proposed method.

The density of the thermal neutrons within the carbon obeys the equation

$$D\frac{d^2(\rho r)}{dr^2} - A(r\rho) + f(r) = 0$$

where $f(r)$ stands for the number of thermal neutrons produced in unit time and unit volume at any point within the carbon sphere of the radius, R.

Let $\rho_1(r)$ be a solution of this equation for the boundary condition $\rho_1(R) = \delta$ where δ is the thermal neutron density at the boundary surface of the carbon sphere in water under the conditions of the experiment.

Let further be $\rho_2(r)$ a solution of the same equation for the boundary condition $\rho_2(R) = 0$ which is realized by covering the surface of the carbon sphere with a thermal neutron absorber. The equation

$$D\frac{d^2(r\rho)}{dr^2} - A(r\rho) = 0$$

will then be obeyed by $\rho = \rho_1 - \rho_2$, and ρ will satisfy the boundary condition $\rho(R) = \delta$. Therefore we have

$$\frac{\rho(r)}{\rho(0)} = \frac{\rho_1(r) - \rho_2(r)}{\rho_1(0) - \rho_2(0)} = \frac{e^{ar} - e^{-ar}}{2ar}$$

So much for the "theory." Practical difficulties are of course present.

I may write you again in the next few days and wish today only to add this: Since Anderson did not get an acknowledgment from *Physical Review* about our note I asked Pegram today to enquire about it. It turned out that the note was too long for a Letter to the Editor and that it will appear as a short paper in the issue of August 1st.

Yours
[Signed] Leo Szilard

Einstein Letter to President Franklin D. Roosevelt
(August 2, 1939)

Albert Einstein
Old Grove Road
Nassau Point
Peconic, Long Island

August 2nd, 1939

F. D. Roosevelt
President of the United States
White House
Washington, D.C.

Sir:

Some recent work by E. Fermi and L. Szilard, which has been communicated to me in manuscript, leads me to expect that the element uranium may be turned into a new and important source of energy in the immediate future. Certain aspects of the situation which has arisen seem to call for watchfulness and, if necessary, quick action on the part of the Administration. I believe therefore that it is my duty to bring to your attention the following facts and recommendations:

In the course of the last four months it has been made probable—through the work of Joliot in France as well as Fermi and Szilard in America—that it may become possible to set up a nuclear chain reaction in a large mass of uranium by which vast amounts of power and large quantities of new radium-like elements would be generated. Now it appears almost certain that this could be achieved in the immediate future.

This new phenomenon would also lead to the construction of bombs, and it is conceivable—though much less certain—that extremely powerful bombs of a new type may thus be constructed. A single bomb of this type, carried by boat and exploded in a port, might very well destroy the whole port together with some of the surrounding territory. However, such bombs might very well prove to be too heavy for transportation by air.

The United States has only very poor ores of uranium in moderate quantities. There is some good ore in Canada and the former Czechoslovakia, while the most important source of uranium is Belgian Congo.

In view of this situation you may think it desirable to have some permanent contact maintained between the Administration and the group of physicists working on chain reactions in America. One possible way of achieving this might be for you to entrust with this task a person who has your confidence and who could perhaps serve in an inofficial capacity. His task might comprise the following:

a) to approach Government Departments, keep them informed of the further development, and put forward recommendations for Government action, giving particular attention to the problem of securing a supply of uranium ore for the United States.

b) to speed up the experimental work, which is at present being carried on within the limits of the budgets of University laboratories, by providing funds, if such funds be required, through his contacts with private persons who are willing to make contributions for this cause, and perhaps also by obtaining the cooperation of industrial laboratories which have the necessary equipment.

I understand that Germany has actually stopped the sale of uranium from the Czechoslovakian mines which she has taken over. That she should have taken such early action might perhaps be understood on the ground that the son of the German Under-Secretary of State, von Weizsäcker, is attached to the Kaiser-Wilhelm-Institut in Berlin where some of the American work on uranium is now being repeated.

Yours very truly,
[Signed] Albert Einstein

Szilard Memorandum
Attached to Albert Einstein's Letter to the President
(August 15, 1939)

MEMORANDUM

August 15, 1939

Much experimentation on atomic disintegration was done during the past five years, but up to this year the problem of liberating nuclear energy could not be attacked with any reasonable hope for success. Early this year it became known that the element uranium can be split by neutrons. It appeared conceivable that in this nuclear process uranium itself may emit neutrons, and a few of us envisaged the possibility of liberating nuclear energy by means of a chain reaction of neutrons in uranium.

Experiments were thereupon performed, which led to striking results. One has to conclude that a nuclear chain reaction could be maintained under certain well defined conditions in a large mass of uranium. It still remains to prove this conclusion by actually setting up such a chain reaction in a large-scale experiment.

This new development in physics means that a new source of power is now being created. Large amounts of energy would be liberated, and large quantities of new radioactive elements would be produced in such a chain reaction.

In medical applications of radium we have to deal with quantities of grams; the new radioactive elements could be produced in the chain reaction in quantities corresponding to tons of radium equivalents. While the practical application would include the medical field, it would not be limited to it.

A radioactive element gives a continuous release of energy for a certain period of time. The amount of energy which is released per unit weight of material may be very large, and therefore such elements might be used—if available in large quantities— as fuel for driving boats or airplanes. It should be pointed out, however, that the physiological action of the radiations emitted by these new radioactive elements makes it necessary to protect those who have to stay close to a large quantity of such an element, for instance the driver of the airplane. It may therefore be necessary to carry large quantities of lead, and this necessity might impede a development along this line, or at least limit the field of application.

Large quantities of energy would be liberated in a chain reaction, which might be utilized for purposes of power production in the form of a stationary power plant.

In view of this development it may be a question of national importance to secure an adequate supply of uranium. The United States has only very poor ores of uranium in moderate quantities; there is a good ore of uranium in Canada where the total deposit is estimated to be about 3000 tons; there may be about 1500 tons of uranium in Czechoslovakia, which is now controlled by Germany; there is an unknown amount of uranium in Russia, but the most important source of uranium, consisting of an unknown but probably very large amount of good ore, is Belgian Congo.

It is suggested therefore to explore the possibility of bringing over from Belgium or Belgian Congo a large stock of pitchblend, which is the ore of both radium and uranium, and to keep this stock here for possible future use. Perhaps a large quantity of this ore might be obtained as a token reparation payment from the Belgian Government. In taking action along this line it would not be necessary officially to disclose that the uranium content of the ore is the point of interest; action might be taken on the ground that it is of value to secure a stock of the ore on account of its radium content for possible future extraction of the radium for medical purposes.

Since it is unlikely that an earnest attempt to secure a supply of uranium will be made before the possibility of a chain reaction has been visibly demonstrated, it appears necessary to do this as quickly as possible by performing a large-scale experiment. The previous experiments have prepared the ground to the extent that it is now possible clearly to define the conditions under which such a large-scale experiment would have to be carried out. Still two or three different setups may have to be tried out, or alternatively preliminary experiments have to be carried out with several tons of material if we want to decide in advance in favor of one setup or another. These experiments cannot be carried out within the limited budget which was provided for laboratory experiments in the past, and it has now become necessary either to strengthen—financially and otherwise—the organizations which concerned themselves with this work up to now, or to create some new organization for the purpose. Public-spirited private persons who are likely to be interested in supporting this enterprise should be approached without delay, or alternatively the collaboration of the chemical or the electrical industry should be sought.

The investigations were hitherto limited to chain reactions based on the action of *slow* neutrons. The neutrons emitted from the splitting of uranium are fast, but they are slowed down in a mixture of uranium and a light element. Fast neutrons lose their energy in colliding with atoms of a light element in much the same way as a billiard ball loses velocity in a collision with another ball. At present it is an open question whether such a chain reaction can also be made to work with *fast* neutrons which are not slowed down.

There is reason to believe that, if fast neutrons could be used, it would be easy to construct extremely dangerous bombs. The destructive power of these bombs can only be roughly estimated, but there is no doubt that it would go far beyond all military conceptions. It appears likely that such bombs would be too heavy to be transported by airplane, but still they could be transported by boat and exploded in port with disastrous results.

Although at present it is uncertain whether a fast neutron reaction can be made to work, from now on this possibility will have to be constantly kept in mind in view of its far-reaching military consequences. Experiments have been devised for settling this important point, and it is solely a question of organization to ensure that such experiments shall be actually carried out.

Should the experiments show that a chain reaction will work with *fast* neutrons, it would then be highly advisable to arrange among scientists for withholding publications on this subject. An attempt to arrange for withholding publications on this subject has already been made early in March but was abandoned in spite of favorable response in this country and in England on account of the negative attitude of certain French laboratories. The experience gained in March would make it possible to revive this attempt whenever it should be necessary.

Leo Szilard

Szilard Memorandum to Lyman J. Briggs
(October 26, 1939)

THE POSSIBILITY OF A LARGE-SCALE
EXPERIMENT IN THE IMMEDIATE FUTURE

At present it appears quite possible that a nuclear chain reaction could be set up in a system composed of uranium oxide (or uranium metal) and graphite. The graphite would have to be piled up in a space of perhaps $4 \times 4 \times 4$ metres and might weigh about 100 metric tons. Perhaps 10 to 20 tons of uranium oxide would have to be used, embedded in some such pile of graphite.

The probable success or failure of such a large-scale experiment cannot be forecast at present with any degree of assurance. The properties of a system composed of uranium and graphite have been calculated independently, for a homogeneous mixture, by Fermi, and, for a lattice of spheres of uranium oxide, or uranium metal, embedded in graphite, by myself. The results of these two independent calculations are in reasonable agreement and show that the two arrangements have different properties. For instance, in the case of using a lattice of spheres a great advantage could be obtained by using uranium metal instead of uranium oxide, whereas in the case of the homogeneous mixture the use of uranium metal would be of no great advantage. In spite of these calculations, we cannot foretell with certainty whether or not a nuclear chain reaction can be maintained in such a system because the absorption cross section of carbon for slow neutrons is not sufficiently known.

In order to remove this uncertainty, Fermi and I have devised two different experiments by means of which the absorption cross section of carbon, which is very small, could be measured. It is assumed that one of these experiments, or both of them, will be started at Columbia University as soon as the facilities required can be obtained.

If the absorption of carbon should turn out to be comparatively large, we could conclude that the large-scale experiment is bound to fail, and in this case it need not be started. If the absorption of carbon should prove to be exceedingly small, the large-scale experiment would appear to be very promising, and it can be assumed that everybody will then be in favor of starting it without delay.

Unfortunately, we must be also prepared to find an intermediate value for the carbon absorption. In this case a large-scale experiment will have to be performed in order to find out whether or not a nuclear chain reaction can be achieved with a combination of uranium and graphite. So we may have to make the experiment and risk its possible failure.

It should be borne in mind that a negative result of the large-scale experiment could also be of value by showing with certainty that a chain reaction *cannot* be achieved with simple means in the near future. Otherwise there remains an ever-present potential threat arising out of experiments on uranium, which are carried out in certain other countries. Therefore, in my personal opinion, a large-scale experiment ought to be performed unless the possibility of its success can be excluded with reasonable assurance on the basis of experiments which are designed to determine the absorption of carbon, or other similar experiments which can be carried out on a moderately small scale.

RECOMMENDATIONS CONCERNING
LARGE-SCALE EXPERIMENTS

No expenses need be incurred in connection with large-scale experiments until the absorption of carbon has been measured. On the other hand, steps ought to be taken now in order to prepare the ground for a large-scale experiment, so that this can be started without delay at the proper time. For instance, the possibility of converting uranium oxide into uranium metal ought to be explored. An attempt ought to be made to obtain a promise on the part of certain industrial corporations to supply at the proper time the quantities of the materials, which are required. If possible, these materials ought to be loaned without any financial consideration. Barring an accident in the case of a successful large-scale experiment, most of the materials used would remain unaffected and could be returned after the experiment is completed.

100 metric tons of graphite represent a value of about $33,000—at the rate of 15¢ per pound. If a purer brand of graphite has to be used, which rates at 24¢ per lb, the value involved would be $53,000.

20 metric tons of uranium oxide represent a value of $100,000—at the rate of $2.50 per lb. If it need not be converted into uranium metal but can be used in the form of oxide in the large-scale experiment, this material could be kept pure and could be returned undamaged. It would be desirable to have up to 50 tons of uranium oxide readily available for experiments in the United States.

STATEMENT CONCERNING THE POTENTIAL ASSISTANCE
OF THE UNION MINIÈRE DU HAUT KATANGA

It would be of particular value to enlist the assistance of this Belgian corporation which is to some extent controlled by the Belgian Government. It appears to be the only corporation which could supply at short notice 20 metric tons of uranium oxide, and probably even 50 tons. I understand that the Managing Director, Mr. E. Sengier, is on a short visit in America.

From conversations which Professor G. B. Pegram of Columbia University had with a representative of the Eldorado Gold Mines, Ltd., it appears that this Canadian corporation might be able to supply uranium oxide for our purposes at the rate of 1 ton per week. If the uranium oxide were to be bought rather than obtained as a gift or a loan, it might be secured from Canada probably just as easily as from Belgium. On the other hand, the Canadian corporation is rather small and can hardly be asked to give away large quantities of material without financial compensation.

So far, radium up to about 2.5 gm was used in our experiments, and we had to pay a high rent to a subsidiary of the Union Minière, the only corporation from which large quantities of radium can be readily rented in this country. An attempt ought to be made to obtain radium for the purposes of such experiments rent-free from the Union Minière in the future.

Carnotites containing uranium are mined in the U.S.A. by the U.S. Vanadium Corporation which is owned by the Union Carbon and Carbide Corporation. A conversation which I recently had with William F. Barrett, Vice-President of this corporation, did not encourage the hope of obtaining large quantities of uranium oxide from this firm, but the issue could perhaps be reopened.

STATEMENT ABOUT URANIUM ORE

As far as I was able to find out, pitchblend, which is an ore rich in uranium, is mined in Czechoslovakia, Canada and Belgian Congo. The total content of uranium in the deposit in Czechoslovakia is estimated to be between 1000 and 1500 tons. The Canadian deposit visibly contains a total of 3000 tons. The amount of pitchblend in the Belgian Congo is not known, but it is believed to be very much larger. In the United States uranium occurs chiefly in the form of carnotites, which is an ore poor in uranium, and is mined for the sake of its vanadium content. The total deposit is estimated to contain 3000 tons of uranium oxide. (Perhaps there are in the United States larger quantities of ore containing a very small amount of uranium which are not included in the above estimate.)

RECOMMENDATION CONCERNING URANIUM ORE

Steps to secure a stock of uranium ores for the government can hardly be recommended at the present time if such steps would involve financial commitments on the part of the government. It might, however, be advisable to begin to study the question in what manner the government could secure such a stock at a later date if required.

For instance, the question has been raised whether it might not be possible to obtain for the government a large quantity of pitchblend from Belgium as a token reparation payment. Such a transaction would not cause alarm abroad if it were arranged before the world learns of the results of some successful large-scale experiment. The transaction could be justified without reference to the uranium content of the ore. Pitchblend is also the ore of radium, and action could be taken on the ground of securing the ore for the sake of its radium content, with a view of extracting the radium at some future date for medical purposes. Action taken on this ground alone might in fact be entirely justified.

Szilard Letter to John T. Tate
(February 6, 1940)
With Initial Version of "Divergent Chain Reaction in Systems
Composed of Uranium and Carbon"

February 6, 1940

John P. Tate, Editor
Physical Review
University of Minnesota
Minneapolis, Minn.

Dear Dr. Tate:

Enclosed you will find a manuscript which I am sending you with the request that
you have it printed in the *Physical Review* as a "letter."

Since this manuscript deals with a matter in which the government has shown a
certain amount of interest from the point of view of national defense it is felt that
inquiries should be made in Washington as a matter of courtesy before the letter is
actually printed. Would you, therefore, perhaps be kind enough to ask the Lancaster
Press not to print this manuscript until they have a telegram from me releasing the
matter for publication. I trust this way of proceeding will not cause any undue
inconvenience.

Yours very truly,
[Signed] Leo Szilard

DIVERGENT CHAIN REACTION IN SYSTEMS
COMPOSED OF URANIUM AND CARBON

Following the discovery[1,2,3] of an abundant emission of fast neutrons from uranium
the balance of neutron absorption and production has been studied by means of
different types of experiments.[4,5,6] Carrying out one type of experiment Anderson,
Fermi, and Szilard reported[5] that on the average about $\mu = 1.5$ fast neutrons are
emitted by uranium for every thermal neutron absorbed by uranium. A more accurate
value for μ can be obtained in principle from a second type of experiment such as the

[1] Halban, Joliot, and Kowarski, *Nature*, **143**, 479 (1939).
[2] Anderson, Fermi, and Hanstein, *Phys. Rev.*, **55**, 797 (1939).
[3] Szilard and Zinn, *Phys. Rev.*, **55**, 799 (1939).
[4] Halban, Joliot, and Kowarski, *Nature*, **143**, 680 (1939).
[5] Anderson, Fermi, and Szilard, *Phys. Rev.*, **56**, 284 (1939).
[6] Halban, Joliot, Kowarski, and Perrin, *Journ. de Phys.*, **10**, 428–429 (1939).

one performed by Halban, Joliot, Kowarski, and Perrin for which we have

$$\mu = \frac{1}{1-p} + \frac{n\sigma_H}{\sigma_U}\left(\frac{1}{1-p} - J\right) \tag{1}$$

p is the fraction of neutrons absorbed by uranium at resonance in a uranium-oxide–water mixture containing n atoms of hydrogen for one atom of uranium, σ_H/σ_U is the ratio of the capture cross-section of hydrogen and the absorption cross-section of uranium comprising both radiative capture and fission. The value of J is determined by the values directly measured in an experiment of this second type. For the experiment which has been actually carried out by Halban, Joliot, Kowarski, and Perrin[6] I find for J from the published data the value of $J = 1.8$ for $n = 3$. Taking into account that for $n = 3$, p has a value of 0.5, we have

$$\mu = 2 + 0.2\frac{3\sigma_H}{\sigma_U}$$

The value obtained for μ if calculated from this equation is scarcely affected by the wide limits of error of the present experimental values of σ_U. A very accurate value for μ could thus be obtained by measuring the value of p accurately and an attempt is being made in this direction. In the following we shall use as the best value available at present $\mu = 2$.

As stated before in a homogeneous mixture of uranium and water a considerable fraction of the neutrons is absorbed by uranium at resonance and it remains quite uncertain whether in any such mixture the number of neutrons produced can exceed the number of neutrons absorbed, i.e. whether a chain reaction in the true sense of the word can be maintained in such mixtures.[5]

From several points of view it would be preferable to use carbon in the form of graphite rather than hydrogen in the form of water for slowing down the neutrons. The capture cross-section of carbon for thermal neutrons σ_C is small; an upper limit of $\sigma_C > 0.01 \times 10^{-24}\,\mathrm{cm}^2$ has been reported for room temperatures by Frisch, Halban, and Koch. It is not possible though to state at present whether or not a chain reaction can be maintained in homogeneous mixtures of uranium or uranium oxide and carbon.

The purpose of this communication is to report that a method for obtaining a chain reaction in a system composed of uranium and carbon has been devised and worked out in detail. This method is based on the fact that in a lattice of uranium spheres embedded in graphite the ratio of the number of thermal neutrons and the number of resonance neutrons which are absorbed by the uranium spheres can be made very small if only the radius R of the uranium spheres is chosen rather small. In spite of our present insufficient knowledge of the value of the capture cross-section of carbon it can be predicted that by using such a lattice of uranium spheres instead of a homogeneous mixture a chain reaction can be maintained which will be divergent for a lattice of small uranium spheres embedded in a large sphere of graphite.

We find for a lattice of uranium spheres in carbon that a chain reaction can be maintained provided we have

$$\frac{4.4\mu}{(\mu - 1.1)^2} < \varepsilon \tag{2}$$

ε is defined in the following way: a single uranium sphere is embedded in an infinite mass of graphite. Neutrons are generated in the graphite and the numbers of thermal neutrons and resonance neutrons produced per c.c. and sec. are equal and have the same value throughout the whole infinite mass of graphite. ε is the ratio of the number of thermal neutrons and resonance neutrons absorbed under these conditions by the uranium sphere.

In order to obtain a conservative estimate for ε we calculate its value under the assumption that every neutron which reaches the uranium sphere by diffusion while its energy is between $0.2E$ and $2E$ is absorbed by the uranium at resonance, E being the energy at which the resonance absorption of uranium has its maximum. Using only such values of nuclear cross-sections in our computations which appear to be well established we find thus for a sphere of uranium metal of less than 8 cm. radius at 900° Centigrade, $\varepsilon > 14$. This value is higher than the value given by Formula (2) which is 11 for $\mu = 2$.

In these circumstances we can expect a chain reaction to take place in a sufficiently large mass of graphite which contains for instance a close-packed hexagonal or cubic lattice of uranium spheres. The capture cross-section of carbon may of course be much smaller than the upper limit quoted above and consequently there is hope that only moderately large masses of graphite and uranium or uranium oxide will be needed to reach the point of divergence at which nuclear transmutation can be maintained at an intensity which is limited only by the necessity of avoiding overheating. An experiment requiring several tons of graphite has been devised for the purpose of measuring the exact value of σ_C.

Large quantities of radio-active elements will be produced directly from the splitting uranium atoms and indirectly by the intense neutron emission. The necessity of protecting human beings from deadly irradiation emanating from the chain reaction will undoubtedly limit the scope of practical applications and perhaps will slow down the industrial development of this field but it is difficult to imagine that practical applications should not follow in due course of time the present turn of events in physics.

It can be shown that a chain reaction maintained in a lattice of uranium spheres embedded in graphite increases in intensity if the temperature of the carbon rises; this represents a type of thermal instability which can not easily be remedied. A dynamic method for stabilizing the chain reaction has therefore been devised and encounters less difficulties than might be supposed. The problem of leading away the heat produced (most of which is liberated within the uranium spheres) without blocking the chain reaction has also been investigated and is capable of a satisfactory solution.

In so far as the production of power for practical purposes is concerned the crucial question which will determine the scope of applications is now whether the rare isotope of uranium 235 or the abundant isotope 238 is the active agent in the thermal neutron reaction. If the rare isotope is the active agent ten tons of uranium may become exhausted by the chain reaction after having supplied as much power as can be obtained from about fifty thousand tons of coal. In case of the other alternative, ten tons of uranium could supply more power than five million tons of coal without being used up. Though N. Bohr put forward interesting arguments[7] in support of the view that it is the rare isotope which is split up by thermal neutrons this question will have to be decided by direct observations performed on small samples of the separated isotopes. Only after some such observation shall we be able to express a well balanced opinion upon the immediate future of "atomic engineering."

[Signed] Leo Szilard

Pupin Physics Laboratories
Columbia University
New York, New York
February 6, 1940

[7] N. Bohr, *Phys. Rev.*, **56**, 426 (1939).

Szilard Letter to Gregory Breit
(July 30, 1941)
With "Background of the Paper 'Divergent Chain Reaction
in Systems Composed of Uranium and Carbon.'"

July 30, 1941

Dr. Gregory Breit
National Bureau of Standards
Washington, D.C.

Dear Breit:

I have now a photostat copy made of the paper which was sent to the *Physical Review* on February 14, 1940, and am enclosing it for your personal use. This photostat is a positive made from the negative which was made from the manuscript before it was sent. You will find two pages marked "21." The one which is marked "21 repl" was written to replace the original page 21 after the manuscript was sent off, and this replacement page bears the postmark of February 21, 1940. The same holds for the summary.

Please note that I am sending you this photostat copy for your personal use only. It seems to me desirable that you should have such an original document at your disposal.

At the request of Tate I have attempted to shorten this paper and the mimeographed copies which I am going to send you represent such a shortened version. Otherwise there is no change either in the content or in the emphasis, as you will see if you compare it with the original. The somewhat long introduction will be dropped at the suggestion of Tate. On the other hand, I am anxious to include a rather full statement of the background of the paper, i.e. a full discussion of all unpublished information which was available at the time the paper was written, so that if credit should be due to others the reader should be in the position to pass judgment on this point for himself.

I am quite anxious that no version of the paper should be circulated which does not have attached to it a full statement of the background. It seems to me that since the normal channels of publication are blocked, we should be even more careful than ordinarily to give credit for all unpublished statements which might possibly have contributed to or accelerated the development of our own ideas.

Yours sincerely,
[Signed] Leo Szilard

cc:
1—Pegram   1—Szilard
1—Fermi   2—Mitchell

August 1941

BACKGROUND OF THE PAPER "DIVERGENT CHAIN REACTION
IN SYSTEMS COMPOSED OF URANIUM AND CARBON"
OF FEBRUARY 14, 1940

In May 1939, Anderson, Fermi, and I carried out an experiment* in which neutrons were slowed down by water and reacted with uranium. This experiment showed that more neutrons are emitted than absorbed by uranium under the particular conditions in which the experiment was performed.

Our system contained 17 H atoms per U atom and the thermal neutron absorption of hydrogen rules out the possibility of a chain reaction at such concentrations. When we started this experiment we had hoped, however, that we might perhaps be able to draw from it the conclusions that a chain reaction could be maintained at a lower hydrogen–uranium ratio.

That the possibility of a chain reaction can not be proved in such a way had been first realized by G. Placzek. He pointed out that with increasing uranium concentrations the fraction of neutrons captured by uranium at resonance increases so rapidly that a chain reaction might be very well impossible at any hydrogen–uranium ratio. Placzek put forward the suggestion of using helium in place of hydrogen for slowing down the neutrons. He emphasized that the absence of a thermal neutron absorption in helium would make it possible to use very high helium–uranium ratios with the result that a high fraction of the neutrons would be slowed down to thermal energies without being captured by uranium at resonance.

In the experiment of Anderson, Fermi, and Szilard* tubes filled with uranium oxide were immersed in a water tank. Adjacent tubes were separated by a few centimeters of water in order to keep these distances within the range of thermal neutrons in this medium. For the interpretation of this experiment it was necessary to calculate the fraction of the neutrons absorbed at resonance in these uranium-oxide filled tubes. Fermi carried out such a calculation and noticed that the arrangement which we had used was more favorable with respect to resonance absorption than a homogeneous mixture of uranium oxide and water. Assuming a sharp absorption line in uranium it is indeed quite easy to see a posteriori that most of the resonance absorption will take place in a thin outside layer of the uranium oxide, whereas the inside of the cylinder, which will not appreciably contribute to the resonance neutron absorption, will still materially contribute to the useful thermal neutron absorption. Thus the use of uranium oxide layers of finite thickness has an advantage over the use of infinitely thin uranium oxide layers which would be equivalent to a homogeneous mixture of uranium oxide and water. Fermi calculated in detail the balance of neutron absorption and emission for systems built up from alternating layers of uranium oxide and water. These calculations of Fermi showed that there was a much better chance for

---

* Paper reproduced in Part II.

obtaining a chain reaction in such an inhomogeneous system of water and uranium oxide than in a homogeneous system of these two substances. For the purpose of such calculations it might be sufficient to consider only the lowest resonance absorption line of uranium but the width and intensity of this line enter into these calculations as major determining factors, and these quantities were known only within rather wide limits of experimental error. For this reason it was difficult to say whether or not a chain reaction can be maintained in the heterogeneous uranium-oxide–water systems which were considered by Fermi. (June 1939)

Meanwhile it occurred to me that from the point of view of practical applications it would be very much better to use carbon in place of hydrogen for slowing down the neutrons. Led by this point of view, I became interested in this possibility in spite of the fact that carbon, on account of its larger atomic weight and smaller scattering cross-section, is very much less efficient in slowing down neutrons than hydrogen. (June 1939)

The rate at which a nuclear chain reaction can be maintained will obviously be determined by the rate at which the heat developed in the reaction can be dissipated. Since carbon can withstand high temperatures it is much more efficient in this respect than water or paraffin. Led by these considerations, I made inquiries about carbon and found that it can be obtained in the form of graphite bricks, very pure, and at a moderate price. A structure composed of alternate layers of graphite bricks and uranium oxide appeared to be the simplest from a purely practical point of view and I attempted to form an estimate of the chances of a chain reaction in such a structure. As soon as this problem was seriously considered it became evident that these chances were appreciable even if one assumed that the uranium layers would absorb all the neutrons which reach them while the energy of the neutrons is between 10 and 100 volts. (July 1939)

In this respect, a system in which the neutrons are slowed down by carbon is very different from a system in which the neutrons are slowed down by water. Of the two, carbon is very much superior and this is mainly due to the fact that the range of thermal neutrons in carbon is considerably larger than the range of the resonance neutrons of uranium in carbon, whereas the ranges of these two categories of neutrons are about equal in water.

Further simple considerations showed that the lattice of uranium spheres embedded in graphite appeared even more favorable from the point of view of a chain reaction than the system of plane uranium layers which was initially considered. The efficiency of such a system was calculated and it was found that if small spheres of uranium metal were embedded in graphite there would be quite a good chance of obtaining a chain reaction in an experiment performed on a large but entirely practicable scale. There appeared to be an appreciable chance for success even if uranium oxide had to be used in place of uranium metal. In the circumstances, the conclusion was reached that it would be better to perform an experiment on a large scale rather than to wait for measurements to be performed for the purpose of determining all the nuclear constants involved. (July 9, 1939)

Note for Memorandum:*

Tentative steps were thereupon taken in this direction and among others, E. P. Wigner and E. Teller were informed of these considerations. They shared the opinion that no time must be lost in following up this line of development and in the discussion that followed, the opinion crystallized that an attempt ought to be made to enlist the support of the Government rather than that of private industry. Dr. Wigner, in particular, urged very strongly that the Government of the United States be advised of certain possible consequences of this as well as some other lines of work connected with uranium. With this in mind, we approached Professor Albert Einstein and Dr. Alexander Sachs, and after a number of consultations, Dr. Einstein wrote a letter to the President of the United States recommending that a person or committee be appointed to act as a permanent liaison between the Government and the physicists who are working on uranium. (August 2, 1939)

Einstein's letter and a memorandum which I was asked to write were submitted by Dr. Sachs to the President and Dr. Lyman J. Briggs was appointed as chairman of a Government committee. Dr. Sachs, Dr. Wigner, Dr. Teller and I were given an opportunity to explain to this committee why we believed that the work which is being done on uranium deserved the attention and the support of the Government. (October 21, 1939)

On this occasion a plea was made for the Government's support either financial or moral both for the work on uranium in general and for the work along the lines indicated in the paper in particular. It was stated that a lattice of uranium metal spheres embedded in graphite appeared to offer the greatest chance for immediate success; that about 100 tons of graphite and 10 to 20 tons of uranium would have to be used in a large scale experiment in order to produce a divergent chain reaction. A recommendation was made that steps be taken to prepare for the performance of such a large scale experiment and that methods of producing uranium metal from uranium oxide be explored. It was emphasized that before starting a large scale experiment the capture cross-section of carbon would be measured and that a few tons of graphite were required for this purpose. A memorandum summarizing these statements and recommendations was submitted to Dr. Briggs (October, 1939)

The value of the capture cross-section of carbon for thermal neutrons was not known at that time; only an upper limit for this magnitude of 0.01×10^{-24} cm^2 was published by Frisch, v. Halban, and Koch. To measure this value appeared, therefore, to be an urgent task, indeed. The usual methods for measuring small absorption cross-sections did not seem to be adequate for this purpose and so a method was devised which called for the study of the spatial distribution of the thermal neutron density in a large mass of graphite. The thermal neutron density in graphite obeys a diffusion equation which contains the mean free path for scattering and the ratio of the capture and scattering cross-sections. Since the thermal neutrons are produced in the graphite through the slowing down of the fast neutrons emitted from the neutron source, this diffusion equation is not homogeneous. By introducing screens which are black for thermal neutrons one can obtain, however, experimental values which obey the homogeneous diffusion equation. The first experiment of this type which was originally planned (July 5, 1939) is described in the paper.

Due to various circumstances, experimentation along this line was halted between July 1939 and March 1940. While no information on the capture cross-section of

---

* [Inserted in this form by Szilard in original—Eds.]

carbon was as yet forth-coming, an increasingly optimistic view on the chances of a chain reaction in a uranium–graphite system appeared to be justified when it was realized that the neutron absorption in carbon could be materially reduced by allowing the bulk of the graphite to heat up to high temperatures. Thus having finally reached the conclusion that we may expect a divergent chain reaction in such a system, the paper was submitted in February 1940 to the *Physical Review* for publication. After its acceptance, and after consultations with a number of my colleagues and Dr. Lyman J. Briggs, its publication was deferred at my request.

Divergent Chain Reactions in Systems
Composed of Uranium and Carbon
Submitted to The Physical Review, February 1940,
but publication withheld at the author's request.
Report A-55 of the Uranium Committee,
declassified in November 1946 as *MDDC-446* (1940)

January 31, 1956

The attached report A-55 of the Uranium Committee

is a copy of a paper which I sent to the <u>Physical Review</u>

in February, 1940. This paper was accepted for publica-

tion, but publication was deferred at my request because

of the nature of the subject.

Leo Szilard

Divergent Chain Reaction in Systems
Composed of Uranium and Carbon

A-55

SUMMARY
By L. Szilard   Copy 9

It is shown that a divergent chain reaction may be maintained in a system composed of uranium and carbon. Conditions particularly favorable for a chain reaction are obtained if instead of using a homogeneous mixture of uranium and carbon a large number of rather small spheres of uranium metal are used embedded in a mass of graphite. The small uranium spheres may form a close-packed hexagonal or cubic lattice embedded in a large sphere of graphite. The average number of fast neutrons emitted by uranium for one thermal neutron absorbed by uranium is calculated from known experimental data and is found to be about 2. In our system conditions for a chain reaction become more and more favorable as the temperature increases and it is shown that we could expect a chain reaction to be self- generating in such a system at about 900° C. even if the cross-section of carbon were as high as 0.01, its present experimental upper limit. As the intensity of the chain reaction increases with increasing temperature the system is thermally unstable. It can be controlled artificially. The time within which the control would have to respond is found to be longer than one second. As much as 100 tons of graphite and 30 tons of uranium might perhaps be required in order to reach the point of divergence at which nuclear transmutation will go on with an intensity limited only by the necessity of avoiding over heating. But in so far as the capture cross-section of carbon is likely to be below 0.01 the amount of material required will probably be smaller.

A-55

.) All numerical values of nuclear cross- sections given in this paper are in units of 10^{-24} cm.

1.) $\sigma_c(C)$ is the capture cross section of carbon for thermal neutrons .

2.) $\sigma_{sc}(C)$ is the scattering cross- section of carbon for thermal neutrons.

3.) ϵ is the ratio of the number of thermal neutrons and the number of resonance neutrons absorbed by a sinlge uranium sphere and under the following circumstances: A single uranium sphere is embedded in an infinite space filled with carbon. Neutrons are generated in the carbon and the numbers of thermal neutrons and resonance neutrons produced per cc and sec are equal and have the same value trhoughout the whole infinite mass of carbon.

4.) E_0 is the energy at which the resonance absorption line of uranium has its maximum.

5.) v is the velocity of thermal neutrons .

6.) E_f is the lower end of the resonance region of uranium below which we consider the ratio of the absorption coefficients for thermal neutrons and for resonance neutrons as constant.

7.) E_L is the upper end of the resonance region of uranium above which we neglect the resonance absorption in uranium.

8.) k^{u} is the average number of collisions which a neutron moving "diffusing" in carbon survives within the energy region

9.) k^{res} is the average number of collision which a thermal neutron diffusing in carbon survives before being captured by carbon.

10.) $\lambda(C)$ is the mean free path for scattering of thermal neutrons in carbon .

11.) ρ is the density of thermal neutrons .

12.) Q is the number of thermal neutrons produced per cc and sec. in carbon.

13.) R is the radius of a sphere of uranium or uranium oxide which is embedded in carbon.

14.) A is the range of thermal neutrons in carbon defined by

$$A = \frac{\lambda(C)}{\sqrt{3}} \sqrt{\frac{\sigma_{sc}(C)}{\sigma_c(C)}}$$

Notations 2 $H - 55$

15.) J_0^{th} is the number of thermal neutrons absorbed by a single uranium sphere embedded in carbon provided that the uranium sphere absorbs each thermal neutron which reaches its surface.

16.) B is the range of the resonance neutrons in carbon defined by

$$B = \frac{\lambda'(C)}{\sqrt{3}} \sqrt{R^{res}}$$

17.) J^{res} is the number of resonance neutrons absorbed by a single uranium sphere embedded in carbon.

18.) $\lambda'(C)$ is the mean free path for scattering for neutrons of energy E_0 in carbon.

19.) ε_0 is the value of ε for a uranium sphere which absorbs each thermal neutron which reaches its surface.

20.) J^{th} is the number of thermal neutrons absorbed by a single uranium sphere embedded in carbon.

21.) φ is defined by $J^{th} = \varphi J_0^{th}$ or $\varepsilon = \varphi \varepsilon_0$

22.) $\lambda(U)$ is the mean free path for scattering in the substance of the uranium sphere.

23.) N_u is the number of uranium atoms per c.c. in the substance of the uranium sphere.

24.) $\sigma_a(U)$ is the absorbing cross-section of uranium for thermal neutrons which includes both the cross-section for fission and radiative capture but which does not include the cross-section for scattering.

25.) $\sigma_s(U)$ is the cross-section of uranium for scattering of thermal neutrons.

26.) U is the range of thermal neutrons in the substance of the uranium sphere defined by

$$U = \frac{1}{\sqrt{3}} \sqrt{\frac{\lambda(U)}{N_u \, \sigma_a(U)}}$$

and in the particular case of pure uranium metal

$$U = \lambda(U) \sqrt{\frac{\sigma_s(U)}{3\sigma_a(U)}}$$

A^{-55}

Notations }

27.) α carbon is the fraction of the thermal neutrons absorbed by in an infinite mass of carbon which contains
a lattice of uranium spheres

28.) J^{α} is a the number of thermal neutrons absorbed per sec
by a uranium sphere within a lattice of uranium
spheres embedded in carbon.

29.) J^{res} is the number of resonance neutrons absorbed persec
by a uranium sphere within a lattice of uranium
spheres embedded in carbon .

30.) q is the fraction of thermal and slow fast neutrons
of the resonance neutrons produced in carbon which
is absorbed as a thermal neutron by the lattice of
uranium spheres if the number of thermal neutrons
and the resonance neutrons produced per cc and sec.
in the carbon are equal and have the same value
throughout the whole infinite mass of carbon.

31.) α_m is the value of α for which q becomes maximum.

32.) q_m is the maximum value of q

33.) $q_{corr.}$ is the correct value of the fraction of the resonance
neutrons produced in carbon which is absorbed as a
thermal neutron by the lattice of uranium spheres
if obtained by taking into account that the number
of thermal neutrons produced per cc and sec. in the
carbon near the uranium spheres is reduced due to the
absorption of resonance neutrons by the uranium .

34.) V Is the volume of carbon per uranium sphere.

35.) L is the distance between neighbouring uranium spheres
in a cubic or hexagonal close packed lattice .

36.) n is the number of H atoms per uranium atom in a mixture
of uranium oxide and water.

37.) τ_0 and τ_{int} is the mean lifetime of a thermal neutron in water
or a mixture of uranium oxide and water,

38.) $\sigma_c(H)$ is the capture cross section of hydrogen for thermal
neutrons.

39.) p is the fraction of the resonance neutrons generated
in a homogenous mixture of uranium oxide and water which
are captured by uranium at resonance.

Notations 4

A-55

40.) λ is the density of water in gm per cc in a homogeneous mixture of uranium oxide and water.

41.) c is the first factor in

$$\varepsilon = \left\{ \frac{A^2}{B^2} \right\} \times \left\{ \frac{1 + R/A}{1 + R/B} \, \rho \right\}$$

42.) f is the second factor in the same expression.

43.) l is the critical radius ~~of which~~ of a graphite sphere which contains a lattice of uranium spheres giving the value of the radius for which the chain reaction becomes divergent.

A-55

Divergent Chain Reaction In Systems Composed Of
Uranium And Carbon

INTRODUCTION 1

As early as 1913 H. G. Wells forecast the discovery of induced radioactivity for the year 1933 and described the subsequent advent of nuclear
transmutations on an industrial scale. It was not possible for physicists
to realize the potentialities and the limitation of nuclear physics in this
direction until after the discovery of the neutron in 1932. Owing to the
discovery of artificial radio-activity in 1933 by Joliot and Irene Curie,
and Fermi's pioneer work on neutron reactions in 1934, progress in the field
of nuclear physics was greatly accelerated. By the middle of 1934 it could
be clearly seen that within the framework of modern nuclear physics transmutation of elements on an industrial scale might be achieved by means of a
chain reaction in which neutrons form the links of the chain. If there is an
element or a mixture of elements which interacts with neutrons and from which
a neutron liberates on the average more than one neutron for one neutron which
is absorbed within the mixture we have a chain reaction and may bring about
nuclear transmutations on a large scale. If a neutron source is placed in
the center of a large sphere which is composed of such an element or mixture
the number of neutrons emerging out of the sphere will be larger than the
number of neutrons emitted by the source in the center of the sphere. If
the radius of the sphere approached a certain critical radius the number of
neutrons generated in the chain would tend to become infinite. If the radius
of the sphere is larger than the critical radius there is no stationary solution of the equation and the number of neutrons would increase exponentially
with time.

By simplifying the problem so as to be able to apply the theory of diffusion to the motion of the neutrons which are liberated in the chain reaction it
is easy to get an approximate picture of the general type of behaviour of a
chain reaction within a finite sphere of matter.

INTRODUCTION   2      A-55

Such a simplified treatment has been applied to the problem as early
as 1934 particularly with regard to the potential possibility of setting
up a chain reaction in beryllium. At that time it appeared from Bains-
bridges's value for the mass of beryllium and Aston's value for the
mass of helium that beryllium ought to be unstable and that neutrons might
be liberated from beryllium by slow neutrons in sufficient numbers to
make a chain reaction possible. This idea had to be abandoned, as far
as beryllium was concerned, when Aston's value for the mass of helium
proved to be in error.

In 1934 the transmutation of uranium by neutrons was discovered
along that of other elements by Amaldi, D' Agostino, Fermi, Rasetti, and
Segre who found that a number of radioactive elements are generated from
uranium by neutrons. An important advance was made by Irene Curie
and P. Savitch,   who found that an element which behaved chemically
apparently like radium was among those produced and later Irene Curie
and P. Savitch discovered radio active rare earths among the disintegration
products of uranium. Finally Hahn and Strassman announced in December 1938/that in reality
uranium splits into a large number of elements of medium atomic weight if
irradiated by neutrons.

As soon as this became known it must have been evident to all those
who had been thinking previously of the potential possibilities of a
chain reaction, as well as to others, that it neutrons should be emitted
from splitting uranium atoms in sufficient numbers a chain reaction might
be set up in a large mass of uranium. Independently of each other a small
number of physicists began therefore to prepare experiments with the
aim to discover whether or not neutrons are emitted from splitting uranium.

Ideas along these lines found wider circulation in America through
a semi-private meeting held in Washington in January, 1939 under the
auspices of the George Washington University and the Carnegie Institute
for Terrestrial Magnetism. At this meeting I understand, Fermi drew
attention to the potential possibility of a neutron emission from splitting

INTRODUCTION   3

uranium and some of its possible consequences. For a short time afterwards it seemed
that the absorption of uranium would over compensate any conceivable neutron emission
since it was found[1] that the cross-section of uranium for splitting by thermal neutrons
is about 2 whereas the currently accepted value for the absorption of uranium for the
thermal neutrons was 43. However the absorption of uranium for thermal neutrons was
thereupon remeasured by Fermi who discovered that its value was only about 5.

A delayed emission of neutrons from uranium under the action of neutrons bombardment was discovered in February 1939 by Roberts, Meyer, and Wang.[2]

A much stronger instaneous emission of neutrons by uranium under the action of
thermal neutrons was discovered independently and almost simultaneously by Halban,
Joliot, and Kowarski;[3] Anderson and Fermi;[4] and Szilard and Zinn[5] who reported their
observations in letters to Nature and to Physical Review dated March 8, March 16, and
March 16 (1939) respectively. An observation pointing in the same direction was also
reported by v. Droste[6] in a letter to Die Naturwissenschaften dated March 17, (1939).

The first paper to appear in print was that of Halban, Joliot, and Kowarski
which was published on Nature on March 18, (1939). A number of papers dealing with
the instantaneous neutron emission of uranium were sent to periodicals after this
date some of which presumably reporting experiments which were started independently
at an earlier date.[7-10]

The balance of neutron absorption and emission by uranium was subsequently
studied by Halban, Joliot and Kowarski;[11] Anderson, Fermi and Szilard;[12] and Halban,
Joliot, Kowarski, and Perrin.[13] This work demonstrates beyond doubt that, on the
average, more than one fast neutron is emitted for one thermal neutron absorbed by
uranium. It shows therefore, that in certain circumstances a chain reaction might be
maintained in a mixture of uranium and an element which slows down to thermal energies
the fast neutrons emitted by the splitting uranium. If it is possible to avoid that
too large a fraction of the fast neutrons emitted from the splitting uranium be
captured (without causing fission) at resonance by uranium and if it is possible
at the same time to avoid that too large a

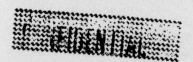

A-55

INTRODUCTION 4

fraction of the thermal neutrons be captured by the element which is used

for the purpose of slowing down, then we would have a chain reaction which

may diverge for a sufficiently large mass of uranium.

This does not necessarily mean that a chain reaction can be maintained

in a mixture of uranium oxide and water. In a homogeneous mixture of uranium

oxide and water a considerable fraction of the neutrons is absorbed by

uranium at resonance and it remains doubtful whether in such a mixture the

number of neutrons produced may exceed the number of neutrons absorbed. [9]

Fermi has investigated the question whether more favorable conditions can

be obtained in mixtures of uranium oxide and water by keeping the uranium

oxide and water in separate layers and found that a slight improvement can

thus be obtained. But even so, for the present the question whether a chain

reaction can be obtained in a system composed of uranium and water is being

left in abeyance.

QUOTATIONS   A - 5

0) H. G. Wells, The World Set Free, (1913)

1) Anderson, Booth, Dunning, Fermi, Glasoe and Slack, Phys. Rev. 55, 511,
   (1939)

2) Roberts, Meyer and Wang, Phys. Rev. 55, 510 (1939)

3) Halban, Joliot and Kowarski, Nature 143, 479, (1939)

4) Anderson, Fermi and Hanstein, Phys. Rev. 55, 797 (1939)

5) Szilard and Zinn, Phys. Rev. 55, 799, (1939)

6) v. Droste, Naturwissenschaften, 27, 198, (1939)

7) Haeny and Rosenberg, C. R. 208, 898, (1939)

8) Michiels, Perry, and G. P. Thomson, Nature, 143, 760, (1939)

9) v. Droste and Reddemann, Naturwissenschaften, 27, 371, (1939)

10) Rotblatt, Nature, 143, 852, (1939)

11) Halban, Joliot and Kowarski, Nature 143, 680, (1939)

12) Anderson, Fermi, and Szilard, Phys. Rev. 56, 284, (1939)

13) Halban, Joliot, Kowarski and Perrin, Journ. de Phys. 10, pp. 428-429, (1939)

14) F. Perrin, C. R., 208, 1394, (1939)

15) Adler and Halban, Nature 143, 793, (1939)

- 1 -   A-55

The purpose of the present paper is to show that a chain reaction can be achieved by using an element like carbon for slowing down the neutrons in certain particular systems composed of uranium and carbon. The theory which is given in the present paper can be applied to elements other than carbon but it does not give any useful information for systems composed of uranium and hydrogen.

Though one might think that carbon should be much less efficient for slowing down neutrons than hydrogen from several points of view it would be preferable to use carbon in the form of graphite rather than hydrogen in the form of water. The capture cross-section of carbon for thermal neutrons $\sigma_c(C)$ is small. An upper limit of $\sigma_c(C) > 0.01 \times 10^{-24}$ cm^2 has been reported by Frisch, Halban and Koch, but this upper limit is not sufficiently low to allow us at present to conclude that a chain reaction could be maintained in homogeneous mixtures of uranium and carbon. For neutrons it takes about 6.5 collisions with carbon atoms to reduce their energy by a factor of e. Thus a neutron which is being slowed down by carbon stays for a long time within the resonance absorption region of uranium. Consequently, very low uranium concentrations would have to be used in order to avoid that a large fraction of the fast neutrons emitted by uranium is absorbed at resonance by uranium. At such very low uranium concentrations, on the other hand, the fraction of the thermal neutrons which is absorbed by carbon might perhaps be too large to permit a chain reaction.

It will be shown, however, in the present paper that if instead of using a homogeneous mixture of uranium and carbon a large number of spheres of uranium which may form, for instance, a close-packed hexagonal or cubic lattice are imbedded in carbon, the ratio of the number of thermal neutrons and the number of resonance neutrons absorbed by the uranium can be so much increased that a chain reaction will become possible. It will be seen that this ratio strongly depends on the radius of the uranium spheres and that a rather small radius must be chosen in order to obtain most favorable conditions.

-2-

Single spheres of uranium.

We wish to first calculate ε the ratio of the number of thermal neutrons, and the number of resonance neutrons which are absorbed by a single uranium sphere, which is embedded in an infinite space, filled with carbon, in the special case in which the number of resonance neutrons and thermal neutrons produced per c.c. and second are equal and have the same value throughout the whole infinite mass of graphite. In order to obtain a conservative estimate for the value of ε we shall assume that all neutrons which have an energy between $E = 0.2\ E_0$ and $E = 2E_0$ where E_0 is the energy at which the resonance absorption of uranium has its maximum, are absorbed by uranium at resonance if they reach the surface of the uranium sphere by diffusion. That this is indeed a conservative assumption can be seen by considering an absorption line which obeys the Breit-Wigner formula and has its maximum at E_0. For such an absorption line the absorption falls off with $1/v$ in the thermal region. It reaches a minimum at $0.2\ E_0$ then it rises up to E_0 and falls again so rapidly that if E_0 is not too close to the thermal region the absorption becomes negligible for $2\ E_0$. If E_0 is higher than five volts and if the temperature of the thermal neutrons does not exceed $1/10$ of a volt then the absorbing cross-section beyond $2\ E_0$ is less than $1/10$ of the absorbing cross-section for the thermal neutrons.

A neutron which is slowed down by elastic collisions with carbon atoms and which enters the resonance at $E = E_2$ will survive on the average $k^{res.}$ collisions within the resonance region between E_2 and E_1 and we have

(1) Under the assumption which we have made above *i.e.*

$$k^{res} = 6.5\ ln\ E_2/E_1$$
$$E_2 = 2E_0 \; ; \; E_1 = 0.2\ E_0$$
$$k^{res} = 6.5\ ln\ 10 = 15$$

This may be compared with the average number of elastic collisions k^{th} which a thermal neutron will survive in carbon before being captured by a carbon atom. Since the capture cross-section of carbon is small compared with $\sigma_{sc}(C)$ the scattering cross-section of carbon for thermal neutrons the probability $g_1(h)$ that a thermal neutron will survive h collisions in carbon will be given by

$$g_1(h) = e^{-h\frac{\sigma_c(C)}{\sigma_{sc}(C)}}$$

-3-   A-55

Accordingly, k^{1R} the average number of collisions which a thermal neu on will make with carbon atoms before being captured is given by

$$k^{1R} = \int_0^\infty h\, g_1(h)\, dh$$

so that

$$k^{1R} = \frac{\sigma_{sc}}{\sigma_c}$$

2 i

A-55

-4-

Let us now first determine the number of thermal neutrons which
are absorbed by a single uranium sphere of radius R embedded in an
infinite space filled with carbon if Q thermal neutrons are produced
per cc and sec. in the carbon. If R is large compared to $\lambda(C)$, the mean
free path for elasic scattering of thermal neutrons in carbon, the density ρ of the thermal neutrons in the carbon can be calculated by treating
the problem as a diffusion phenomenon. We th a find for ρ as a function
of the distance r from the center of the s here

3 $$D(C)\frac{d^2(r\rho)}{dr^2} - S(C)\,r\rho + Q(r)\,r = 0$$

$$D(C = \frac{v\,\lambda(C)}{3} \quad , \quad S(C) = \frac{v\;\sigma_c(C)}{\lambda(C)\,\overline{\sigma_{sc}(C)}}$$

If the same number of thermal neutrons are produced everywhere in the
carbon per cc and sec. we have

4 $$\frac{dQ}{dr} = 0$$

For a sphere which absorbes each thermal neutron which reaches it
surface i.e. for a " black" sphere we have $\rho(R)=0$ and find for from 3 and 4 $r > R$

5 $$\rho(r) = \frac{Q}{S(C)}\left(1 - \frac{R}{r}\,e^{-(r-R)/A}\right)$$

6 $$A = \sqrt{\frac{D}{S}} = \frac{\lambda(C)}{\sqrt{3}}\sqrt{\frac{\sigma_{sc}(C)}{\sigma_c(C)}} = \frac{\lambda(C)}{\sqrt{3}}\sqrt{R^{lR}}$$

y^u the number of thermal neutrons which s absorbed by a single
uranium sphere per sec. is given by

7 $$y^u = D(C)\,4\pi R^2\,\rho'(R)$$

and for a black uranium sphere we find from No. 5 $$y^u = y_0^u$$

8 $$y_0^u = 4\pi\,Q\,R A^2\left(1 + R/A\right) \quad ; \quad A = \frac{\lambda(C)}{\sqrt{3}}\sqrt{R^2}$$

where A has the dimension of a length and will be called
the range of thermal neutrons in carbon.

- 5 -   A-55

Quite similarly since we assume that the uranium sphere is "black" for resonance neutrons we can write for y^{res} the number of resonance neutrons absorbed by the sphere per second with good approximation

(9)
$$y^{res} = 4\pi Q R B^2 (1 + {}^R/_B)$$

where

(10)
$$B = \frac{\lambda^{res}}{\sqrt{3}} \sqrt{k^{res}}$$

is the mean free path of resonance neutrons for scattering and B has the dimension of a length and will be called the range of the resonance neutrons in carbon.

Expression No. 9 is identical with the expression No. 9 which holds for a sphere which is "black" for thermal neutrons. Only the values for λ the mean free path for scattering and k the average number of collisions which a neutron survives within the category called thermal or resonance are different for these two categories of neutrons. It would be strictly true that J the number of neutrons belonging to a category which will reach the sphere by diffusion per second is determined in the same way for different categories by λ the mean free path and $g(h)$, the function giving the probability of surviving h collisions with carbon atoms. In reality the function $g(h)$ is different for thermal and for resonance neutrons and expression No. 9 holds in so far as we may assume that J is determined with sufficient accuracy by λ and $k = \int_0^\infty h\, g(h)\, dh$ the first moment of $g(h)$.

From No. 6 and No. 9 we find as the value of ε for a sphere which is "black" for thermal neutrons $\varepsilon = \varepsilon_0$

(11)
$$\varepsilon_0 = \frac{A^2}{B^2} \frac{1 + R/A}{1 + R/B}$$

Assuming, for example, $\sigma_{sc} = 4.8$; $\sigma_c(C) > 0.005$ and $\frac{\lambda^*(C)}{\lambda(C)} = 1.18$ we find for graphite of density 1.7:

$$A \cong 43.5\,cm \quad B = 6.5\,cm$$

and so obtain for small values of R

$$\varepsilon_0(0) = \frac{A^2}{B^2} \gtrsim 45$$

and for large values of R

$$\varepsilon_0(\infty) = \frac{A}{B} \cong 6.7$$

- 6 -   A-55

Large values of R correspond to plane layers of uranium and a comparison
of these two values for ε_o illustrates how very much superior small
spheres of uranium are to plane layers.

A real sphere of uranium having a radius ~~below 8 centimeters~~ finite is not
"black" for thermal neutrons and the number of thermal neutrons absorbed
by the sphere is smaller than J_o^{th}. We write

$$(12) \quad J^{u} = J_o^{u} \varphi$$

and accordingly we have

$$(13) \quad \varepsilon = \varepsilon_o \varphi \quad \text{and} \quad (14) \quad \varepsilon = \frac{A^2}{B^2} \frac{1 + R/A}{1 + R/B}$$

In order to calculate φ we take into account that inside the
uranium sphere the thermal neutron density ρ obeys the equation

$$(15) \quad D(u) \frac{d^2(r\rho)}{dr^2} - S(u)\, r\rho = 0$$

$$D(u) = \frac{v\,\lambda(u)}{3} \qquad S(u) = v\, N_u\, \sigma_a(u)$$

having as its solution

$$(16) \quad \rho(r) = \frac{C}{r}\left(e^{r/u} - e^{-r/u} \right) ; \quad r < R$$

$$('17) \quad \text{where } u = \sqrt{\frac{D}{S}} \quad = \sqrt{\frac{\lambda(u)}{3\,N_u\,\sigma_a(u)}}$$

and for pure uranium metal we have

$$(18) \quad u = \lambda(u) \sqrt{\frac{\sigma_c(u)}{3\,\sigma_a(u)}}$$

From equations $3, 4$ and 16 we find that J^{u} the number of thermal
neutrons diffusing into the sphere per second is given by

$$J^{u} = J_o\, \varphi$$

where

$$(19)\, \varphi = \frac{\dfrac{\lambda_a(u)}{u}\left\{ \dfrac{e^{R/u} + e^{-R/u}}{e^{R/u} - e^{-R/u}} - \dfrac{u}{R} \right\}}{\dfrac{\lambda_a(C)}{R}\left(1 + R/A\right) + \dfrac{\lambda_u(u)}{u}\left\{ \dfrac{e^{R/u} + e^{-R/u}}{e^{R/u} - e^{-R/u}} - \dfrac{u}{R} \right\}}$$

- 7 -   A-55

For uranium in its pure state we have from No. 14, 18, and 19

$$(20a) \quad \varepsilon = \left\{ \frac{A^2}{B} \right\} \times \left\{ \frac{1}{1 + R/B} \times \frac{1}{RG\left(\sqrt{\frac{3\sigma_a(u)}{\sigma_s(u)}} - \lambda(u)\right) + \frac{1}{1 + R/A}} \right\}$$

where G stands for

$$G = \frac{e^{-R/U} + e^{-R/U}}{e^{R/U} - e^{-R/U}} \quad ; \quad R/U = \frac{R}{\lambda(U)} \sqrt{\frac{3\sigma_a(u)}{3\sigma_s(u)}}$$

For $R/U > 2$ we can write $G \cong 1$, the difference being about **3.5%** for $R/U = 2$.

The first factor in expression No. 20 a increases proportionately with the reciprocal value of the capture cross-section of carbon. The second factor is practically independent of the carbon cross-section, since we have $R/A << 1$. Its value is determined by the density of graphite and uranium and the nuclear values $\sigma_a(u)$, $\frac{\sigma_a(u)}{\sigma_s(u)}$. The value of R may be so chosen as to make this factor a maximum.

All expressions for ε were so far obtained from diffusion equations involving the assumptions

$$R >> \lambda(C) \quad ; \quad R >> \lambda(u) \quad ; \quad U >> \lambda(U)$$

For small values of R the problem can no longer be treated as a diffusion phenomenon and we shall, therefore, refrain from using expression No. 20 or 20 a for values of R of less than 5 cm.

8   LATTICE OF URANIUM SPHERES

If we have now an infinitely large number of uranium spheres forming a lattice embedded in an infinite mass of carbon and want to calculate the ratio of the number of thermal neutrons and resonance neutrons absorbed by the uranium spheres, we shall again assume for the time being that everywhere in the carbon the same number Q of neutrons enter the resonance region and the thermal region per c.c and second.

In the absence of uranium the thermal neutron density in the graphite is given by $\rho_0 = \dfrac{Q}{S(C)}$. If a lattice of uranium spheres is embedded in the carbon, the average neutron density $\bar{\rho}$ in the carbon is reduced by some factor α .

$$\bar{\rho} = \alpha \rho_0$$

Since the number of neutrons captured per second by carbon is proportionate to the average neutron density, and since in the absence of uranium all the neutrons produced are captured by carbon, the fraction of the neutrons which is captured by carbon in the presence of the uranium lattice is given by α . Correspondingly, the fraction of the neutrons which are absorbed by the uranium lattice is given by $1-\alpha$.

In order to determine the number of thermal neutrons J^{th} absorbed per second by one uranium sphere within the lattice, we may consider the following: a single uranium sphere, which is embedded in carbon, does not appreciably affect the thermal neutron density at distances which are large compared to R. Equation No. 5 shows that even for a "black" uranium sphere at a distance of 2 R from the center of the sphere ρ has already reached the value of $\tfrac{1}{2}\rho_0$. For this reason, the uranium spheres within the lattice affect each other with respect to their thermal neutron absorption only in so far as the presence of these spheres in the carbon determines the average neutron density, and we have

$$J^{th} = \alpha \, J^{th}$$

(20)

Further, since the distance L between neighboring uranium spheres within the lattice will be large compared to B, the range of the resonance neutrons in carbon, we have for J^{res} , the number of resonance neutrons absorbed by a uranium sphere within the lattice

(21)
$$J^{res} = J^{res}$$

A-55

From this it follows that q the fraction of all the neutrons which are absorbed by the uranium spheres in the thermal region alone is given by

(22)
$$q = \frac{J^{th}}{J^{th} + J^{res}} (1-\alpha)$$

or

(23)
$$q = \frac{\varepsilon \alpha}{1 + \varepsilon \alpha} (1-\alpha)$$

This expression has its maximum value for $\alpha = \alpha_m$

(24)
$$\alpha_m = \frac{-1 + \sqrt{1+\varepsilon}}{\varepsilon}$$

and for the maximum value for q we have $q = q_m$

(25)
$$\alpha_m = \frac{1-q_m}{2}$$

(26)
$$q_m = 1 - 2\frac{-1+\sqrt{1+\varepsilon}}{\varepsilon}$$

or

(27)
$$\varepsilon = \frac{4q_m}{(1-q_m)^2}$$

CORRECTED VALUE

By calculating the value of q from ε we have neglected the effect of the absorption of resonance neutrons (by the uranium spheres embedded in the lattice) on the production on thermal neutrons in the carbon in the neighborhood of the uranium spheres. The absorption of resonance neutrons reduces in reality the value of Q near the spheres below the average value of Q and accordingly the correct number of thermal neutrons absorbed by the uranium spheres per sphere will be smaller than J^{th}. In order to find the value for the difference $(J^{th}_{cor} - J^{th})$ we may proceed in the following way:

If the uranium spheres in the lattice, ~~would not absorb~~, stopped absorbing resonance neutrons then a neutron which reaches a given sphere at least once while its energy is in the resonance region would have some probability γ of reaching the sphere at least once after it had been slowed down to the thermal region. On the other hand, the probability that the sphere which is

10 - - - A-55

.... Not "black" for thermal neutrons absorbs a thermal neutron which reaches
the sphere at least once is given by φ . (φ was originally defined
as the ratio of the number of thermal neutrons which a sphere of uranium
absorbs and the number of thermal neutrons which an equally "black" sphere of
uranium would absorb under the same previously specified conditions. It is
easy to see though that φ can also be defined as the ratio of
the number of thermal neutrons which are absorbed by the uranium sphere and
the number of thermal neutrons which reach the surface of the uranium sphere
at least once during their life time. That these two definitions of
are identical simply follows from the fact that a "black" uranium sphere absorbs all the thermal neutrons which reach its surface during their lifetime.)
It follows that we have

$$J^{th}_{corr} = J^{th} - \nu \varphi J^{res} \quad ; \quad 0 < \nu < 1$$

correspondingly we have

$$q_{corr} = \frac{J^{th} - \nu \varphi J^{res}}{J^{res} + J^{th} - \nu \varphi J^{res}} (1-\alpha)$$

or

$$q_{corr} = \frac{\varepsilon \alpha - \nu \varphi}{1 + \varepsilon \alpha - \nu \varphi} (1-\alpha)$$

and by introducing the value of q_m from equation No 25 and 27 we find

$$(28) \qquad q_{corr} = q_m \frac{1 - \frac{\nu \varphi (1-q_m)}{2 q_m}}{1 - \frac{\nu \varphi (1-q_m)}{1+q_m}}$$

If we neglect terms which contain powers higher than the second of
$\frac{\nu \varphi (1-q_m)}{1+q_m}$ we obtain

$$(29) \qquad q_{corr} = q_m \left\{ 1 - \frac{\nu \varphi (1-q_m)^2}{2 q_m (1+q_m)} \left(1 + \nu \varphi \frac{1-q_m}{1+q_m} \right) \right\}$$

Taking into consideration that we have $\nu < 1$ and that for the uranium spheres
which we shall consider φ has a value of about $\varphi \cong 0.5$ we assume
$\nu \varphi \leq 0.5$.   and find then for $q_m > 0.5$ from No. 28 :

$$(30) \qquad q_{corr} > q_m \times 0.9$$

With increasing q_m q_{corr}/q_m approaches 1; for $\varphi \leq 0.667$; $q_m \gtrsim 0.68$.
we have $q_{corr}/q_m > 0.97$. For $\varphi \leq 0.645$; $q_m \gtrsim 0.73$
we have $q_{corr}/q_m > 0.98$

11 ────

A-55

Spacing of the lattice.

Formula No. 26 giving the value of q_m was derived under the assumption that there is a uniform production of resonance neutrons throughout the whole mass of carbon into which the lattice of uranium spheres is embedded. We have to verify that this assumption is correct. For this reason we have to compare the distance $\sqrt{\overline{r^2}}$ to which a fast neutron "diffuses" away from a uranium sphere from which it is emitted before it is slowed down to thermal energies with the distance L between two neighboring uranium spheres in a close-packed hexagonal or cubic lattice. If the value of L which corresponds to the maximum value of q were large compared to $\sqrt{\overline{r^2}}$ then obviously equation No. 26 giving the value of q_m could not be used.

In graphite of density 1.7 we have $\sqrt{\overline{r^2}}$ about 50 cm. and we shall see that the values of L which correspond to the maximum value of q are smaller for all values of $\sigma_c(C)$ which we are going to discuss in this paper.

In order to estimate L as well as for other reasons we shall calculate the volume V of carbon per uranium sphere in the lattice.

In the lattice of uranium spheres from the QV neutrons which are slowed down per second within the volume V to resonance energies, the carbon absorbs

$\alpha\, Q V$ neutrons and the uranium spheres in the lattice absorbs $(1-\alpha)\,QV$ neutrons. Accordingly we have $J^{th} + J^{res} = (1-\alpha)\,QV$

and from this we find

$$31 \qquad V = 4\pi\, \ell\, \frac{\alpha}{q}\, A^2 R\,(1 + R/A)$$

and if q has its maximum value q_m we have $\dfrac{\alpha}{q} = \dfrac{1-q_m}{2\,q_m}$

and it is

$$32 \qquad V = 4\pi\, \ell\, \frac{1-q_m}{2\,q_m}\, A^2 R\,(1 + R/A)$$

This gives for the ratio of the volumes of carbon and uranium

$$33 \qquad \frac{V}{\frac{4}{3}R^3} = 3\,\ell\, \frac{1-q_m}{2\,q_m}\, \frac{A^2}{R^2}\,(1 + R/A)$$

and for L the distance between neighboring uranium spheres in a hexagonal or cubic close-packed lattice we have

$$34 \qquad L^3 = \left(V + \frac{4}{3}R^3\right)\sqrt{2}$$

12 -**-   A-55

$$(35) \quad L = A B R^3 \sqrt{1 + 3 \varphi \frac{1 - \varphi_m}{2 \varphi_m} \frac{A^2}{R^2} \left(1 + \frac{R}{A}\right)}$$

For large values of ε we can write

$$\frac{1 - \varphi_m}{2 \varphi_m} \simeq \frac{1}{\sqrt{\varepsilon}} - \frac{1}{\sqrt{\varphi \varepsilon_0}},$$

so that we would have for $\varepsilon \gg 1$

$$(36) \quad \gamma \simeq 4 \pi A B R \sqrt{\varphi \left(1 + \frac{R}{A}\right)\left(1 + \frac{R}{B}\right)}$$

From No. 35 we find for R = 8 cm., $\sigma_c(C) = 0.0033$,   A = 75.5, $\varphi_m = 0.68$

$\varphi = 0.65$,: ~~by this~~.

$$L = 51 \, cm$$

-13-   A-55

BALANCE OF EMISSION AND ABSORPTION OF NEUTRONS BY URANIUM

A direct comparison of the number of fast neutrons emitted by uranium with the number of thermal neutrons absorbed by uranium may be obtained from different types of experiments carried out by Halban, Joliot and Kowarski,[11] Anderson, Fermi and Szilard,[12] and Halban, Joliot, Kowarski and Perrin.[13] These experiments show that on the average more than one fast neutron is emitted by uranium for one thermal neutron absorbed by uranium.

Halban, Joliot and Kowarski[8] reported that 3.5 ± 0.7 fast neutrons are emitted per fission. Anderson, Fermi and Szilard[9] reported that about $\mu = 1.5$ fast neutrons are emitted on the average for one thermal neutron absorbed by uranium.

Halban, Joliot, Kowarski and Perrin[13] carried out an experiment of the following type: a sphere of radius r_0 is filled with a homogeneous mixture of uranium and water which contains n atoms of hydrogen per atom of uranium. This sphere is immersed in a large water tank and a photo neutron source is placed in the center of the sphere. The density ρ of the thermal neutrons is measured along a radius and the integrals of

$$I_{int} = \int_0^{r_0} r^2 \rho \, dr \quad ; \quad I_{ext} = \int_0^\infty r^2 \rho \, dr$$

are determined, giving a measure of the number of slow neutrons which are present within the mixture inside the sphere and in the water outside the sphere. In another experiment the integral

$$I_0 = \int_0^\infty r^2 \rho \, dr$$

is determined for the same neutron source in pure water.

From the values of these three integrals which these authors determined for n = 1, 2 and 3 and r_0 = 25 cm. they concluded that on the average eight neutrons are emitted from the uranium within the sphere for one photo neutron which is slowed down to thermal energies and causes a fission process in uranium. This was interpreted to show that a considerable number of secondary and tertiary neutrons were generated in this experiment.

Since the value of the fission cross-section is not well known and rather difficult to determine, it seems necessary to avoid expressing results in terms of this cross-section and to use the magnitude μ rather than the number of neutrons emitted per fission. An accurate knowledge of the value of μ is of greatest importance from the point

14 —38—

A-55

of view of estimating the chances and finding the optimum conditions for a chain reaction. In the following we shall therefore indicate in what way an accurate value for ρ can be obtained, from an experiment of the type performed by Halban, Joliot, Kowarski and Perrin:

F_{int}, the average number of neutrons which reach the thermal region inside the sphere within the uranium water mixture for one neutron emitted by the source in the center of the sphere, is given by

$$F_{int} = \frac{I_{int}/I_o}{\overline{\tau}_{int}/\tau_o}$$

where

3)
$$\frac{\tau_{int}}{\tau_o} = \frac{n\,\sigma_c(H)}{s\{\sigma_a(U) + n\,\sigma_c(H)\}}$$

is the ratio of the life-time of the thermal neutrons in the mixture inside the sphere and in pure water.

Of the neutrons which are slowed down to the resonance region inside the sphere a fraction p is absorbed at resonance and a fraction $(1 - p)$ reaches the thermal region. F_{int}^{*} the number of the neutrons reaching the resonance region per second inside the sphere is therefore given by

$$(1-p)\,F_{int}^{*} = F_{int}$$

F_{ext}^{*} the number of neutrons reaching the resonance region outside the sphere in the water is the same as F_{ext} the number of neutrons reaching the thermal region outside the sphere in the water and we have, therefore

$$F_{ext}^{*} = F_{ext} = \frac{I_{ext}}{I_o}$$

P, the total number of fast neutrons emitted by the uranium in the mixture for one neutron emitted by the neutron source in the center of the sphere is equal to the total number of the resonance neutrons produced minus 1, the neutron emitted by the source in the center. Therefore, we have

$$P = F_{int}^{*} + F_{ext}^{*} - 1 = \frac{F_{int}}{1-p} + F_{ext} - 1$$

The total number of fast neutrons produced by the uranium divided by the number of thermal neutrons produced within the sphere in the mixture is

$$\frac{P}{F_{int}} = \frac{1}{1-p} + \frac{F_{ext} - 1}{F_{int}}$$

15 ~~--12--~~ $A\text{-}55$

And finally since the ~~uranium~~ absorbs of the n utrons which reach the thermal region within the sphere the fraction

$$\frac{\sigma_a\, u}{\sigma_a\, u\; -\, n\,\sigma_c(H)}$$

we have

$$\mu = \frac{P / F_{int}}{1\, \dfrac{\sigma_a(u)}{\sigma_a(u) + n\,\sigma_c(H)}}$$

or

$$\mu = \frac{\sigma_a(u) + n\,\sigma_c(H)}{\sigma_a(u)} \left\{ \frac{1}{1-p} + \frac{\tau_{int}}{\tau_0}\; \frac{I_{int}/I_0 - 1}{I_{int}/I_0} \right\}$$

Introducing the value of $\dfrac{\tau_{int}}{\tau_0}$ from No.   we obtain

$$3\partial\quad \mu = \frac{1}{1-p} + \frac{n\,\sigma_c(H)}{\sigma_a(u)} \left\{ \frac{1}{1-p} + \frac{1}{3}\, \frac{I_{int}/I_0 - 1}{I_{int}/I_0} \right\}$$

From this equation we see that we can determine μ by determining the value of p.

A considerable fraction of the fast neutrons emitted by uranium may of course escape from within the sphere but the value of this fraction need not be determined for our purpose. On the other hand, the formula given for μ holds with good approximation only if the sphere is sufficiently large to permit us to neglect transition phenomena in the equilibrium between resonance neutrons and thermal neutrons near the surface of the sphere. Since the uranium inside the sphere absorbs resonance neutrons the density of the neutrons which have an energy below the upper end of the resonance region of uranium and above the thermal region is smaller⅄ This has the consequence that neutrons of this category will diffuse from the water across the sphere into the inside of the sphere. If the radius of the sphere were not sufficiently large so that this phenomenon could not be neglected the expression $F_{int}/_{1-p}$ would then give a too large value for the number of resonance neutrons produced inside the sphere. Correspondingly the expression No.   would give a too large value for μ . But with the proviso that a sufficiently large sphere is used we shall obtain an accurate value for μ if only the value of p is measured accurately.

⅄near the surface inside the sphere ~~than near the surface outside the sphere.~~

CONFIDENTIAL

A-55

The value of p has so far not been measured for hydrogen concentrations as low as n = 3 but we can give a lower limit for p for this concentration as we are able to extrapolate by means of theoretical considerations from values of p which have been measured for higher hydrogen concentrations such as n = 30.

In order to do this we shall make use of formulae relating to homogeneous mixtures of hydrogen and uranium uranium which will be published shortly by G. Placzek. A general treatment of the process of slowing down of neutrons in water has been given by E. Fermi. In 1936 In a uranium oxide – water mixture p, the fraction of the neutrons which is absorbed by uranium between the energies E_2 and E_1 is given by

39) $$\ln(1-p) = -\int_{E_1}^{E_2} \frac{f(E)}{E}\,dE \qquad ; \; E_1 > 1\,Volt$$

where

40) $$f(E) = \frac{1}{1 + \frac{nH}{g(E)}}$$

$g(E)$ is the capture cross- section of uranium for radiative capture and H is the scattering cross- section of hydrogen for the resonance neutrons of uranium. $(H \cong 17 \times 10^{-24} cm^2)$ If the hydrogen concentration is high enough so that the energy region in which there is an appreciable absorption by uranium is small compared with the resonance energy E_0 one may write

41) $$\ln(1-p) = -\frac{1}{E_0}\int_{-\infty}^{\infty} f(E)\,dE$$

for a single absorption line of the form

42) $$g(E) = \frac{g_0}{1 + \left(\frac{E-E_0}{\Gamma}\right)^2} \qquad ; \; g_0 = g(E_0)$$

giving

43) $$\ln(1-p) = -\frac{\pi \Gamma}{E_0} \frac{1}{\sqrt{\frac{nH}{g_0}\left(1 + \frac{nH}{g_0}\right)}}$$

For $n \lesssim 30$ we have $\frac{nH}{g_0} \ll 1$ and No 44 gives

44) $$\ln(1-p) = -\frac{\pi \Gamma}{E_0}\sqrt{\frac{g_0}{nH}}$$

From this we see that for such hydrogen concentrations for which No 44 holds we have for two different hydrogen concentrations n and n_1

45) $$\ln(1-p) = \sqrt{\frac{n_1}{n}}\,\ln(1-p_1)$$

---

* Private communication by G. Placzek June, 1939

** Private communication independently by G. Placzek and E. Fermi June, 1939

17 - 88 -   A-55

This relationship No. 45 does no longer hold for hydrogen concentrations which are as low as $n = 3$. But we may write in the case of uranium

(46)
$$\ln(1-p) = \int_{0.2E_0}^{1.8E_0} \frac{f(E)}{E} \, dE$$

and it is then easy to see that for an absorption line of the form given by No. 42 which is symmetrical in $(E - E_0)$ or even more so for a similar line which deviates from symmetry in the sense that the absorption is larger for smaller energies (as it is the case for a line which obeys the Breit-Wigner formula) we have

(47)
$$\ln(1-p) < \sqrt{\frac{n'}{n}} \ln(1-p_1) \quad \text{if} \quad n < n_1$$

The value of p has been measured by Halban, Joliot, Kowarski and Perrin for $n = 30$ and was found to be $p = 0.2 \pm 0.02$; using equation No. 47 we find from this

$$p_3 > 0.5 \pm 0.04 \quad \text{for } n = 3$$

According to No. 38 μ increases with increasing values of p and in the circumstances we should obtain a conservative value for μ by using the value of $p = 0.5$.

Using this value and the values
$$\frac{I_{int}}{I_0} = 0.72 \quad ; \quad \frac{I_{ext}}{I_0} = 0.45$$

reported by Halban, Joliot, Kowarski and Perrin for a sphere of 25 cm. radius filled with a uranium oxide--water mixture for which they had $n = 3$ and $s = 0.42$ gm./c.c. we find from No. 38

(48)
$$\mu = 2 + 0.2 \frac{3\sigma_c(C)}{\sigma_a(U)}$$

We see that the value obtained for μ if calculated from this equation is scarcely affected by the wide limits of error of the present experimental values of $\sigma_a(U)$. By attributing the value of $\frac{1}{4}$ or 1/8 to $\dfrac{3\sigma_c(H)}{\sigma_a(U)}$ we obtain $\mu = 2.05$ or $\mu = 2.02$ respectively.

In these circumstances we shall use for the present as a presumably conservative value and as the best value at the present available:

$$\mu = 2$$

18 - ~~20~~ -

A-55

Conditions For A Chain Reaction

If q denotes the fraction of fast neutrons emitted by uranium which are slowed down to the thermal region and are absorbed as thermal neutrons by uranium and if μ denotes the number of fast neutrons produced on the average by uranium for one thermal neutron absorbed by uranium then obviously

(49)
$$\mu q > 1$$

is the condition for the possibility of a chain reaction. If this condition is fulfilled then a divergent chain reaction can be maintained in a sufficiently large system from which only a small fraction of the neutrons emitted by the uranium within ~~the system~~ can escape across the boundary of the system without being absorbed within ~~the system~~.

Accordingly, the condition for the possibility of a chain reaction in a system composed of a lattice of uranium spheres embedded in carbon

(50)
$$\mu q_{corr} > 1$$

and using equation No. 30 we find
$$\mu q_m \times 0.9 > 1 \quad or \quad q_m > 1.11/\mu$$

From which we find by using equation No. 27 as a sufficient condition for the possibility of a chain reaction

(51)
$$\frac{4.44 \mu}{(\mu - 1.1)^2} < \varepsilon$$

Using the value of μ = 2 as a presumably conservative value we have as a sufficient condition

$$11.3 < \varepsilon$$

19 - ** -   A-55

In order to see now whether a chain reaction is possible we have to calculate from our formulae the numerical value of ξ . We shall do that in the fol
lowing under the assumption that the energy liberated in the chain reaction will
maintain the carbon at a temperature of about 900°. and in order to be on the con
ervative side we shall assume that the temperature of the uranium spheres in
which most of the energy is liberated is , in spite of efficient cooling, about
the same.

Since we have at room temperature $\sigma_c(c) > 0.01$ we shall have at 900 C
a capture cross-section of carbon half of this value.   The scattering cross-
section of uranium for thermal neutrons we take to be $\sigma_s(u) \cong 9$. Finally,
at room temperature we take $\dfrac{\sigma_a(u)}{\sigma_d(u)} = \dfrac{1}{2}$ and correspondingly we take at
900 C. $\dfrac{\sigma_a(u)}{\sigma_d(u)} = \dfrac{1}{4}$. For a density of graphite of 1.7 and a density of
uranium of 15 we then obtain from No. 14 $for\ R = 8cm$

$$\xi = 14$$

This being larger than the value required by No. 51 we conclude that in
the circumstances we can expect a divergent chain reaction to take place in
the system which we have investigated.

In reality the capture cross-section of carbon *is perhaps* ~~may of course be~~ much small
er than the upper limit which has so far been established and consequently
there is hope that conditions will be much more favorable for a chain reaction
than would seem from the values so far quoted.

The amount of carbon and uranium required to reach the point of divergence at which nuclear transmutation will proceed at a rate limited only by the
necessity of avoiding over-heating is essentially determined by the value of

$$(\mu q - 1)$$

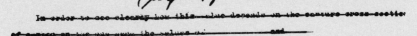

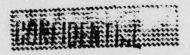

= 20 =   A-55

In the following we shall calculate how the value of this expression depends on the value of the carbon capture cross-section at room temperature.

We shall take the density of graphite to be 1.7; the density of uranium metal to be 15 and choose R = 8cm.

We then obtain for a capture cross-section of carbon at room temperature of 0.005 the following set of values: $A = 64cm$. $\frac{A^2}{B^2} = 90$; $\mathcal{Y} = 0.666$

at 900 C°

$\mathcal{E} = 27$; $q_m = 0.681$; $q_{corr} = 0.66$ and for $\mu = 2$

$$(\mu q - 1) = 0.32$$

If the capture cross-section of carbon at room temperature were .003 then at 900 C. our equations would give the following set of values:

$A = 75.5cm$; $\frac{A^2}{B^2} = 135$; $\mathcal{Y} = 0.645$; $\mathcal{E} = 39$; $q_m = 0.72$

$q_{corr} = 0.71$ and for $\mu = 2$

$$(\mu q - 1) = 0.42$$

- 21 -

A-55

FINAL DISCUSSIONS

For a large sphere of graphite which contains a large number of small spheres of uranium 1, the critical value for the radius of the graphite sphere for which the chain reaction becomes divergent may be calculated for various distributions of uranium within the graphite sphere. The optimum distribution of uranium is not uniform within the graphite sphere and will either decrease or increase with r according to whether we want to have a minimum amount of uranium or a minimum value for l. The treatment of this question may as well be postponed until the value of the carbon capture cross-section is known. It will then be possible to find the optimum distribution of uranium as a function of the distance from the center of the graphite sphere and give a value for l. In the meantime, a very rough approximation may give an idea of the order of magnitudes which are involved. In graphite of 1.7 density the average distance $\sqrt{\bar{r^2}}$ to which a fast neutron emitted by uranium diffuses away from its point of origin until it becomes a thermal neutron and reacts with uranium or carbon is about 50 cm. For $(\eta q - 1) \cong \frac{1}{3}$ we find for the critical radius l from

$$\ell \sim \sqrt{\frac{3\,\overline{r^2}}{\eta q - 1}}$$

$1 \sim 250$ cm. This corresponds to about 100 metric tons of graphite. If the carbon capture cross-section is lower, then l will be smaller and the amount of graphite required might perhaps be as low as 20 tons.

The amount of uranium required can be calculated from equation No. 33a

$$\frac{4\pi}{3} R^3 / \nu = \frac{1 - q_m}{6} \frac{R^2}{B^2} \frac{1}{1 + R/B}$$

It may be kept down by choosing a smaller value for R than the value corresponding to the maximum value of ε. For R = 5cm. and $q \sim 0.6$ gives $\frac{4\pi}{3} R^3 / \nu = 0.033$ corresponding to 30 tons of uranium for 100 tons of graphite. For larger values of q we find a smaller ratio of uranium to carbon.

For q   we would have as the ratio of weights   .

MEASUREMENTS   22   /1 ~ ~

In order to determine the critical dimensions and the most favorable
distribution of the small uranium spheres wihin a large graphite sphere it
is essential to have an accurate value for μ and for the capture cross-section of carbon for thermal neutrons.

In order to have an accurate value for μ it will be necessary to have
a direct measurement of p for small hydrogen concentrations preferably for
n = 3 . In the following we shall indicate in what way an accurate value for
p can be obtained:

Let us consider a box filled with a homogeneous mixture of uranium oxide
and water and let the box be so large that a neutron which has an energy of
less than 1000 volts in the center of the box be slowed down below 1 volt before it reaches the boundary of the box by diffusion. If such a box is then
irradiated with neutrons the energy distribution of the neutrons below
1000 volts in the center of the box will be the same as it would be in an
infinite space filled with the same uranium water mixture in case of a uniform generation of neutrons throughout the space. If a radio-active indicator
like rhodium or indium which has its lowest dominant resonance absorption line
below the dominant resonance absorption line of uranium is placed in the center of the box the activity induced in the rhodium indicator by the resonance
neutrons of rhodium will be ~~called~~ caused only by those neutrons which are slowed down
below the resonance region of uranium without being captured by uranium at
resonance. We shall call this activity the resonance activity of the indicator and designate it if a rhodium indicator is used with $(Rh^{*} \text{ in } U)$.
If, on the other hand, an indicator like, for instance, iodine is used which
has its dominant absorption lines above the uranium resonance the resonance
activity of this indicator $(J^{*} \text{ in } U)$ will, at least in the ideal case,
not be affected by the resonance absorption of uranium. The ratio of
$(Rh^{*} \text{ in } U)$ and $J^{*} \text{ in } U)$ would therefore give some measure of 1 - p if the
activity of the indicators as measured by an ionization chamber or electron
counter gave a real measure of the number of the neutrons which pass through
the resonance region of rhodium or iodine. This, of course, is by no means
the case. Therefore, a second experiment has to be performed in which the
resonance activities of the rhodium and iodine indicators are determined in
the center of another box which contains water. ~~or perhaps even better a mix-~~
~~ture of water and bismuth having approximately the same water density as the~~
~~mixture used in the first experiment. We then find~~

23 -∞- 23   A-55

(Preferably but not necessarily water of the same density as the water contained in the uranium water mixture used in the first experiment) We then find

$$1-p = \frac{(Rh^* \, in \, U) \; \Delta J^x \, in \, H_2 O)}{(Rh^* \, in \, H_2 O) \times (J^x \, in \, U)}$$

If there are uranium resonance lines of some importance above the lowest resonance lines of iodine then we have

$$1-p < \frac{(Rh^* \, in \, U) \times (J^x \, in \, H_2 O)}{(Rh^* \, in \, H_2 O) \times (J^* \, in \, U)}$$

giving a reliable upper limit for 1 - p and consequently a lower limit for p which used in conjunction with equation No. 38 will give a lower limit for the value of μ .

The upper limit which has been reported for the carbon cross-section by Halban, Frisch, and Koch is already so low that it would be difficult to improve upon it unless a method were used which is specifically designed to measure extremely small cature cross-sections. Such a method will be ~~described~~ described in the following:

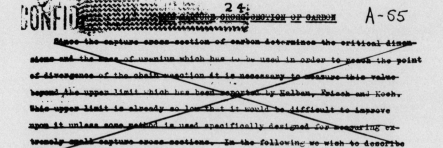

A-65

~~Since the capture cross-section of carbon determines the critical dimensions and the mass of uranium which has to be used in order to reach the point of divergence of the chain reaction it is necessary to measure this value beyond the upper limit which has been reported by Halban, Frisch and Koch. This upper limit is already so low that it would be difficult to improve upon it unless some method is used specifically designed for measuring extremely small capture cross-sections. In the following we wish to describe such a method.~~

Let us consider a sphere of carbon and a neutron source in the center of the sphere. The thermal neutron density inside the carbon will then obey equation No. only in this case Q is a function of r of which we must not assume $\frac{dQ}{dr} = 0$. Let the carbon sphere be immersed in a water tank or surrounded by paraffin wax. The thermal neutron density will then have a cer- fairly high value at the surface of the sphere and inside the sphere it will some function of r, $\rho_1(r)$ If in a second experiment the surface of the sphere in the water is covered by cadmium the thermal neutron density at the surface of the sphere is then reduced to zero and inside the sphere it will be ano— ther function of r, $\rho_2(r)$. The difference $\rho(r) = \rho_1 - \rho_2$ obeys the homogene- ous equation

$$D(C)\frac{d^2(r\rho)}{dr^2} - \delta(C)\, r\rho = 0$$

which has the solution

$$\rho(r) = C\frac{e^{r/A} - e^{-r/A}}{r}$$

We can thus find A by determining the value of ρ for two values of r, for instance, r=0 and r=r . It is

$$\frac{\rho(r)}{\rho(0)} = \frac{e^{r/A} - e^{-r/A}}{2\, r/A}$$

Or for small values of r/A

$$\frac{\rho(r)}{\rho(0)} \cong 1 + \frac{1}{6}\left(\frac{r}{A}\right)^2$$

Using a sphere of graphite of fifty to seventy centimeters of radius it should be possible to measure the range A with sufficient accuracy. There is a limit to using very large spheres which arises out of the fact that for a very large sphere most of the neutrons emitted in the center of the sphere will be slowed down to the thermal region within the carbon and the thermal neutron density near the surface of the carbon may become very low. If that

- 25 -

A-55

happens, then the difference $\rho = \int_1 - \int_2$ will become small and will therefore set a limit to the accuracy of the measurement.

- 26 -

A-55

Stabilizing the chain reaction

Soon after the discovery of an abundant neutron emission from uranium ~~and even before it was known whether a way would be found to set up a chain reaction~~ the question of stabilizing such a reaction was a subject of discussion, (4,15) but the situation as we see it appears to be rather different in practice:

If a chain reaction could be maintained in a homogeneous mixture of water and uranium or carbon and uranium it would have a certain natural stability in the sense that with rising temperature there would be a decrease in the neutron production. The reason for this is the fact that the absorption of both uranium and hydrogen obey the $1/v$ law in the thermal region and thus at higher temperatures the range of thermal neutrons in the mixture is larger. Correspondingly, at higher temperatures a larger fraction of the thermal neutron will escape across the boundary of the mixture without having reacted with the uranium in the mixture. This natural stability could even be enhanced by having bodies of strong thermal neutron absorbers inserted in the mixture. Fairly thin sheets of such absorbers as boron, for instance, are practically "black" for thermal neutrons and any strong thermal neutron absorber would stabilize equally well.

A system, on the other hand, in which uranium bodies which are almost "black" for thermal neutrons are embedded in carbon, like the system which we have considered in great detail in the present paper, has no such stability. This is due to the fact that with rising temperature the capture cross-section of the carbon decreases whereas the absorption by the uranium spheres remains almost unchanged. Accordingly, at higher temperatures, a larger fraction of the thermal neutrons is absorbed by uranium and a smaller fraction is absorbed by carbon and this leads to an increase in q_0 and thermal instability.

It is, however, quite easy artificially to stabilize the chain reaction by slowly shifting the position of absorbing bodies within the system in such a way as to reduce the average value of q whenever the intensity of the neutron radiation emanating from the chain reaction increases. One might perhaps think that the time within which such controlling action would have to take place is very short. We shall therefore now show that this is not so.

27   A-55

Of the neutrons which are ~~~~~~~~~ the chain reaction by uranium only a fraction ψ is absorbed within the system and $1-\psi$ escapes across the boundary of the system without reacting with uranium. A stationary state can be maintained as long as

$$\mu q \psi < 1$$

We write

$$q\psi = \delta \left(T_1 , x(t) \right)$$

In order to indicate that this product is a function of the temperature T and also depends on a parameter such as the position of some absorbing or scattering body near or within the system which can be shifted by some controlling mechanism and thus be made a function of time t.

In order to have a large neutron production we must maintain a chain reaction near the point

$$\mu \delta_0 = 1$$

If this product becomes larger than one, as it may well happen then there is an exponential rise in the neutron production and accordingly also in the temperature. In case of a sudden small deviation from one

$$\Gamma = \delta_0 \left(1 + \xi \right)$$

the time t_2 in which the number of neutrons doubles is given by

$$t_2 = \xi t_1$$

where t_1 is the time which a fast neutron ~~~~ emitted by a uranium atom in the system would require to produce two fast neutrons if it is slowed down and absorbed (as a thermal neutron) within the system.

For instance if we have a sudden change in δ' of $1 \permil$ as we well may have and if we have $t_1 = 4 \cdot 10^{-3}$ sec, it would take 4 seconds for the neutron production to double its value and accordingly there would be only an insignificant rise in the temperature if the control responded within 4 seconds.

It is easy to see that for a lattice of uranium spheres in carbon ξ, the mean life-time of a thermal neutron within the system is given by

$$\tau_1 = \alpha \frac{\sigma_{sc}(k)}{\sigma_c(k)} \frac{\lambda(k)}{v}$$

For $\sigma_c(k) = 0.005$ at room temperature, for instance, we have at 900 C.

$\psi_m = 0.66$ and $\alpha_m = \frac{1 - \psi_m}{2} = 0.17$ giving at 900 C.

$$\tau_1 = 4 \times 10^{-3} \text{ sec}$$

- 28 -

A-66

In calculating t_2 we did not take into consideration the fact that a fraction of the neutrons is emitted by uranium with a time delay of about ten seconds. Though this fraction is small it has a marked effect in leading to still longer times than those which we have estimated. But as for all practical purposes the time t_2 which we have found is already long enough we need not include for the present the delayed neutron emission in the treatment of the subject.

A way to stabilize the nuclear chain reaction in a lattice of uranium spheres which is "economical" from the point of view of the nuclear phenomena involved is illustrated in figure 1. The uranium sphere which is surrounded by a spherical layer of liquid bismuth (serving the purpose of cooling the uranium in a way which does not reduce the nuclear efficiency of the arrangement as expressed by φ is shown in this figure. A short rod or disc composed of an element which strongly absorbs thermal neutrons is near the center of the uranium sphere and is shielded by the uranium from thermal neutrons. This rod or disc can move within a tube or slit and its position may be controlled by the intensity of the neutron radiation emitted by the chain reaction. If the intensity of this radiation increases the rod or disc may be automatically moved away from the center of the uranium sphere and ultimately if required entirely out of the uranium sphere. It will then absorb larger and larger numbers of thermal neutrons thereby reducing the value of q and thus stabilizing the chain reaction.

A-55

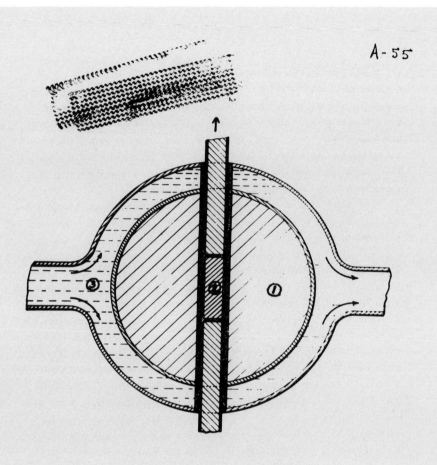

1 represents a sphere of uranium metal.   2 represents a rod, composed of a
thermal neutron absorber moving in a *vertical* bore, or a circular disc having a horizontal axis moving in a vertical slit in the uranium sphere 1. 3 is a spherical shell of liquid bismuth which surrounds the uranium sphere.

29   CONCLUSIONS   A-65

We conclude that we can expect a chain reaction to take place in a sufficiently large mass of graphite which contains, for instance, a close-packed
hexagonal or cubic lattice of uranium spheres. The capture cross-section of
carbon is likely to be smaller than the upper limit so far established and
consequently there is hope that moderately large masses of graphite and uranium or uranium oxide will be sufficient to reach the point of divergence
at which nuclear transmutation can be maintained at an intensity which is
limited only by the necessity of avoiding over heating.

Large quantities of radio-active elements will be produced directly from
the splitting uranium atoms and indirectly by the intense neutron emission.
The necessity of protecting human beings from deadly irradiations emanating
from the chain reaction will undoubtedly limit the scope of practical applications and perhaps will slow down the industrial development of this field
but it is difficult to imagine that practical applications should not follow
in due course of time the present turn of events in physics.

In so far as the production of power for practical purposes is concerned
the crucial question which will determine the scope of applications is now
whether the rare isotope of uranium 235 or the abundant isotope 238 is the
active agent in the thermal neutron reaction. If the rare isotope is the
active agent ten tons of uranium may become exhausted by the chain reaction
after having supplied as much power as can be obtained from about fifty thousand tons of coal. In case of the other alternative, ten tons of uranium
could supply more power than five million tons of coal without being used up.
Though N. Bohr put forward interesting arguments in support of the view that
it is the rare isotope which is split by thermal neutrons this question will
have to be decided by direct observations performed on small samples of the
separated isotopes. Only after some such observation shall we be able to
express a well balanced opinion upon the immediate future of "atomic engineering."

Nuclear Chain Reaction in a System
Composed of Uranium, Beryllium and Carbon*

In a previous paper dated February 14th, 1940, I have attempted to show that we may expect to be able to maintain a nuclear chain reaction in a system composed of uranium and carbon. The purpose of the present paper is to point out that we may perhaps obtain a considerable improvement of the efficiency of the system for the purpose of a chain reaction by introducing beryllium into the system. An appreciable fraction of the neutrons emitted from the uranium which is split by thermal neutrons appear[1] to have energies above 1.7 MeV, the binding energy of neutrons in beryllium, and hence we may expect that an appreciable fraction of the fission neutrons can produce an additional neutron by knocking out a neutron from beryllium. In the circumstances, by introducing beryllium into the system in such a way that it is exposed to the fast neutrons emitted from uranium we may obtain a significant increase from this knock-out process in μ, the total number of neutrons generated in the system per thermal neutron which is absorbed by the uranium in the system.

In the above mentioned paper particular attention was given to a system consisting of a lattice of uranium spheres embedded in a large mass of graphite. Formulae were derived for a lattice in which the distance between two uranium spheres is large compared to the radius of a single uranium sphere. Under these conditions, and within the limits of the approximation used in deriving these formulae, one finds the optimal radius for the uranium spheres by determining the value of R for which the expression

$$(20) \qquad \varepsilon = \frac{A^2}{B^2}\frac{1}{1+R/B}\ \frac{1}{\dfrac{\lambda(C)}{RG\sqrt{\dfrac{3\sigma_a(U)}{\sigma_{sc}(U)}}-\lambda(U)}+\dfrac{1}{1+R/A}}$$

becomes a maximum. Using uranium at a density of 16 gm per cc and graphite at a density of 1.7 gm per cc we take at room temperature the values involved as follows: $A = 53.5$ cm corresponding to $\sigma_c(C) = 0.0033$; $B = 6.5$ cm; $\lambda(C) = 2.44$ cm; $\sigma_a(U) = 5.5$; $\sigma_{sc}(U) = 11$ corresponding to $\lambda(U) = 2.25$ cm.

For a value of $R = 5$ cm we have $G \cong 1$ and we find from (20) $\varepsilon \cong 24$ which is a value close to the maximum. The corresponding value for the fraction of the neutrons which are absorbed as thermal neutrons by the uranium spheres in the lattice is given by

$$(26) \qquad q_m \cong 1 - 2\frac{-1+\sqrt{1+\varepsilon}}{\varepsilon} \cong 0.67$$

---

* In a letter dated September 4, 1942, Szilard wrote:

"The use of uranium spheres surrounded by one layer of beryllium metal was considered between June and November, 1940. A memorandum giving sizes of the uranium sphere and the beryllium layer is available." This seems to be the memorandum referred to. See Part V for the "Disclosure" on this subject.—Eds.

and for the ratio of the volumes of uranium and carbon we have

(33a)
$$\frac{4\pi R^3}{3} \Big/ V \cong \frac{1 - q_m}{6} \frac{R^2}{B^2} \frac{1}{1 + R/B} \cong \frac{1}{40}$$

giving a ratio of weights of uranium to carbon about 1 to 4.

Beryllium may now be introduced into such a system by surrounding each uranium sphere with a spherical shell of beryllium metal 4–5 cm thick. The density of beryllium is about 1.8 gm per cc, and the amount required would be about equal in weight to the amount of uranium and perhaps one tenth of the amount of graphite.

Thus the beryllium would be located at a site where the thermal neutron density is low, and the average thermal neutral density within the beryllium would be less than one half of the average thermal neutron density in the graphite. Moreover, the number of beryllium atoms would be about one-tenth of the number of carbon atoms, and in the circumstances a much larger thermal neutron absorption cross-section per beryllium atom can be tolerated for beryllium metal with its impurities than can be tolerated per carbon atom for graphite. Since the fraction of neutrons absorbed is given by α_m, an absorption cross-section of

$$\sigma_c(\text{Be}) = \xi\sigma_c(\text{C})$$

would lead to a loss of $(\xi/20)\alpha_m$ neutrons. Since we have

(25)
$$\alpha_m \cong \frac{1 - q_m}{2} \cong \tfrac{1}{6}$$

we would have a loss of perhaps 5 % if we had an absorption in beryllium six times as large per beryllium atom as the absorption in graphite per carbon atom, i.e., if we had $\sigma_c(\text{Be}) = 0.02 \times 10^{-24}$ cm^2.

A fast neutron emitted from a uranium atom within the sphere will go through the beryllium shell once and may pass through the shell again after one or more collisions with carbon atoms. During its passage through the beryllium shell it will suffer collisions with beryllium atoms. The energy of such a fast neutron will decrease by every collision with either beryllium or carbon. This process of slowing down will limit the total number of neutrons which may be liberated by a fission neutron moving in beryllium.

In order to get a better picture of this limitation we may assume for the sake of argument that one-half of the fission neutrons has an initial energy above the dissociation energy of beryllium, and that the cross-section for the disintegration of beryllium is one third of its total cross-section (and one-half of its elastic collision cross-section). A fission neutron would then in its first collision with a beryllium nucleus on the average knock out 0.166 neutrons. If we further assume, rather arbitrarily, that the fission neutrons withstand two elastic collisions with beryllium with undiminished capacity for the disintegration of beryllium, but that after the third elastic collision their energy is below the threshold, we find that a fission neutron moving entirely in beryllium would liberate about $0.5(\frac{1}{3} + \frac{2}{9} + \frac{4}{27}) = 0.35$ neutrons and not more.

In our arrangement collisions will take place with carbon atoms as well as beryllium atoms, and accordingly the total number of neutrons liberated from beryllium by one fission neutron would be smaller. It should be emphasized though that a value of 0.2 would already be very significant since it would raise μ, the value of the neutrons generated in the system per thermal neutron absorbed in uranium, from a value between 1.5 and 2 to a value between 1.8 and 2.4. The data available at present do not permit to estimate the increase in μ which we may expect from the introduction of beryllium into a system composed of uranium and carbon. Experiments using 75 to 150 lb of beryllium are in preparation for the purpose of clearing up this point.

May 17, 1946

In Place of a Summary*
L. Szilard

Note added May 1946

In July 1939, it became first apparent to the author that there is a serious possibility of maintaining a chain reaction in a system composed of uranium and graphite. At that time there was only very imperfect knowledge of the values of the nuclear constants involved but this did not stand in the way of making a comparison between a hetero-geneous uranium–carbon system and a homogeneous uranium–water system. The comparison showed that if a homogeneous water–uranium system can be constructed which comes very close to be chain reacting, then it should be possible to make a heterogeneous carbon–uranium system chain reacting provided that the absorption of carbon is lower than 0.01×10^{-24}, which also happened to be the experimental upper level for carbon absorption at that time.

The United States Government was advised of this situation in October 1939. No direct experimental evidence on a uranium–carbon system was available at that time, but experiments completed in collaboration with H. L. Anderson and E. Fermi earlier, in June 1939, had reliably indicated that a water–uranium system can indeed come reasonably close to be chain reacting. No further experiments on such systems were undertaken at Columbia University, or for that matter anywhere in the United States, between June 1939 and March 1940. But in January 1940, an experiment performed by Halban, Joliot, Kowarski, and Perrin, on uranium–water systems became known in this country. One of the uranium–water systems investigated by them came close to being capable of maintaining a chain reaction and one could also see that such systems could get exceedingly close to be chain reacting for the optimum concentrations. In the opinion of the author, this meant that we can expect a chain reaction in a uranium–carbon system under practically attainable conditions, if the capture cross-section of carbon had a value of, say, about one half of the experimental upper limit quoted above, i.e., 0.005.

Thus it appeared important to try to obtain a rough idea as to the optimal composition, geometrical dimensions, and other characteristics of such a chain reacting system and this was attempted in the present paper. Over six years have elapsed since it was written and naturally it is outdated. A few footnotes were added to draw attention to some of the shortcomings which have become evident in the meantime. Some parts of the paper have been condensed or left out entirely in an attempt to shorten it without adding anything to its original contents.

---

* In a revision prepared in 1946 Szilard included this and the following note—Ed.

Further Addition Made for 1946 Revision

It is easy to understand why uranium layers of finite thickness embedded in paraffin are preferable to layers of an infinitely small thickness, i.e., to a homogeneous mixture. If the thickness of a very thin layer of uranium is increased the thermal neutron absorption of the layer increases proportionally to the thickness. The absorption for resonance neutrons of uranium, however, increases more slowly than proportional to the thicknesses at which the thermal neutron absorption is still increasing proportionally to the thickness. This is due to the fact that for not too thick uranium layers the resonance absorption is mainly due to the first sharp resonance line of uranium. At larger thickness the absorption of thermal neutrons flattens out and there is a thickness which is optimal from the point of view of ratio of the thermal absorption and the resonance absorption of uranium. This optimum may be even more marked and more favorable for lumps of uranium than for flat layers of uranium. Since the range of the thermal neutrons in paraffin wax is of the same order of magnitude as the range of resonance neutrons, and since the thickness of the paraffin layers which may be sandwiched between uranium layers must not be made large compared to the range of the thermal neutrons, systems of this type may be considered as quasi-homogeneous since the velocity distribution of the neutrons will not vary very much within the system.

While the question whether a chain reaction can be maintained in such a system remained open it appeared of interest primarily from the point of view of possible practical applications to raise the question whether a chain reaction could be maintained in a system composed of uranium and carbon. Even if it were possible to maintain a chain reaction in a system in which the neutrons are slowed down by hydrogen the rate at which the chain reaction could be maintained would necessarily be limited by the fact that hydrogen containing substances decompose or evaporate at moderately elevated temperatures. If carbon can be used in the place of hydrogen for slowing down the neutrons for the purpose of the chain reaction, there would be no such limitation of the chain reaction rate, and it would be possible to have a sufficiently high temperature gradient available for dissipating the heat which would be generated.

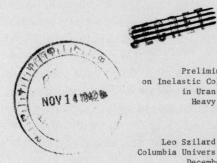

Preliminary Report
on Inelastic Collision of Neutrons
in Uranium and other
Heavy Elements.

by

Leo Szilard and W.H. Zinn.
Columbia University, New York, N.Y.
December 12, 1941

In view of the potential possibility of a chain reaction
based partly or wholly on fission caused by fast neutrons, it
appeared of interest to investigate the inelastic collision of
fast neutrons in uranium and certain other elements. In uranium
fast neutrons would be slowed down at a certain rate by inelastic
collision below the fission threshhold of uranium 238 and the main
purpose of our experiments was to determine the cross-section of
this process. This cross-section will naturally depend on the
velocity of the primary neutrons, and it would be best, although
at present not practicable, to use fission neutrons as primary
neutrons for the purpose of our experiment.

Experiments were started in February 1941, using as a
neutron source a neutron generator previously described by Zinn
and Seeley[1] in which neutrons are emitted from a heavy ice target
between 2.2 and 2.8 MEV from the D+D reaction.

---

[1] W.H. Zinn and S. Seeley, _Phys. Rev._ 52, 9 (193

-2-

In order to determine the cross-section for inelastic collisions,
as defined above, for lead, bismuth, and uranium, we proceeded
in the following way: An ionisation chamber coated with a thick
layer of uranium oxide was placed at about 26 centimeters from the
target, The neutrons emitted from the target cause the emission of
fission particles from the uranium oxide coating, and the pulses
due to these fission particles were counted serving as a measure of
the number of fissions produced by these neutrons. The decrees in
the number of these fissions which is brought about when the target
is surrounded with a lead, bismuth, or uranium sphere gives xxxly a
direct measure of the fraction of the neutrons emitted from the
target which have been slowed down below the fission threshhold of
uranium 238 while passing through a certain thickness of lead, bismuth or uranium, respectively. Since the intensity and the energy
of the neutrons which are emitted from the target vary with the angle
which the neutron velocity forms with the deuteron beam, this angle,
which is defined by the prosition of the fission chamber, was varied
from 0 to 120 degrees. The value of the integral of the fission
count over the whole sphere was extrepolated assuming cylinderical
symmetry around the deuteron beam. The lead, bismuth and uranium
spheres which were used had a hole in their centers to allow the
introduction of the target and had a radial wall thickness of bismuth, lead and uranium of 6.7 centimeters. The densities of lead,
bismuth and uranium in these spheres were as follows:

Pb. $11:15 \text{gm/cm}^3$ Bi: 9.8 gm/cm^3 U: 7.62 gm/cm^3

SECRET   -3-

For the cross section for inelastic scattering, we
obtained by the method indicated above:

Pb: 0.55;   Bi:0.64;   U:1.9 in units of 10^{-24} cm².

Similar experiments were subsequently performed by Marshall
and Szilard, using the same lead, bismuth and uranium spheres,
but using as primary neutrons neutrons from a radon-beryllium
source which have a different spectral distribution. The values
obtained for the cross-section for inelastic scattering of radon-
beryllium neutrons are the following:

Pb: 1.34; Bi: 1.85 ;   U: 2.21

We see that by changing over from D+D neutrons to radon-
beryllium neutrons, there is a conspicuous increase in the cross-
section of lead and bismuth which increase to two to three times,
whereas there is at the most a slight increase in the value for
uranium. Experiments performed by Marshall and Szilard with radon-
beryllium neutrons on another, purer, sample of uranium, give a
slightly lower value so that it might be that the value for pure
uranium and D+D neutrons should be taken as low as 1.5.

The above stated values for uranium was obtained by attributing
the observed decrease of the fission count which was caused by the
uranium sphere to inelastic scattering in uranium without taking
into account the fact that the primary neutrons may produce fission
in the uranium sphere, and that additional neutrons may be created
hereby which contribute to the fission count observed in our
experiment. In the absence of this secondary process, the decrease in
the fission count caused by the uranium sphere would be greater than

-4-

the decrease which was actually observed in our experiment. The
inelastic cross-section of uranium is therefore greater than the
value given above, and we have to add a correction term. In order to
determine the value of this correction term, we shall assume that about
2.6 neutrons are emitted per fission (both for fast neutron fission
and thermal neutron fission), xxx We shall also assume that the fission
neutrons which are generated by D + D and radon-beryllium neutrons have
the same cross-section for causing fission in uranium as fission
neutrons which are generated by thermal neutrons, xxxxxxxxxxxxxxxx
xxxxxx and that also D + D neutrons and radon-beryllium neutrons
have the same cross-section. With these assumptions we can calculate
the value of the correction term directly from experimental results
obtained by Marshall and Szilard, who observed fission caused by
fission neutrons which were generated by thermal neutrons and determined
the efficiency of this process. They reported (see Report dated
November 14, 1941) for the product of the fission cross-section of
fission neutrons and the number of neutrons emitted per fission the
value[1] of 1.3×10^{-24} cm^2. From this we find the value of $.8 \times 10^{-24}$ cm^2
for the correction term which has to be added to the above quoted
cross-section of 1.9×10^{-24} cm^2 in order to obtain the value for the
cross-section for the cross-section of uranium for inelastic scattering,
and obtain for this cross-section the value of 2.7×10^{-24} cm^2.

---

[1] Result of the first preliminary experiment.

Report No. CP-316

Copy No. ____1____

Name *Library*

PRELIMINARY REPORT ON FISSION

CAUSED BY FISSION NEUTRONS

John Marshall, Jr., and Leo Szilard, Columbia University

November 14, 1941

Abstract

Measurements of the quantity $\sigma_f \nu$ are discussed.

(σ_f = the cross section of U^{238} for fission by fission neutrons. ν = the number of neutrons emitted per thermal fission of 235.) In the experiment, fission neutrons from thermally irradiated uranium were allowed to fall upon a uranium coated ionization chamber; the preliminary result is:

$$\sigma_f \nu \cong 1.3 \times 10^{-24}$$

using $\nu \cong 2.6$, this gives $\sigma_f \cong 0.5 \times 10^{-24}$ cm^2.

November 14, 1941

PRELIMINARY REPORT ON FISSION

CAUSED BY FISSION NEUTRONS

By

John Marshall, Jr., and Leo Szilard

Columbia University

It appeared to be ~~important~~ *of interest* to determine whether
fission can be caused in uranium by fission neutrons and if
so to learn something about the cross-section of this process.

This process may play a ~~important~~ role in a chain
reaction in a system in which spheres of uranium are imbedded
in a large mass of graphite. In such a system thermal neutrons
diffuse from the graphite into the uranium sphere and lead to
the emission of fast fission neutrons. The fast neutrons thus
produced will cause a certain number of fission processes in
the same uranium sphere from which they originate and thus
produce secondary fast fission neutrons. Some of these
secondary neutrons will again cause fission in the uranium sphere
from which they originate and lead to tertiary fission neutrons,
etc. This process might increase by perhaps ~~as much as~~ a factor
of $\beta = 1.25$ the number of fast neutrons emitted by uranium per
thermal neutron absorbed by uranium. The magnitude of the
factor β depends primarily upon certain nuclear properties of
uranium and in the first approximation on the product

$\sigma_f \, (\nu - 1)$

where σ_f is the cross-section of uranium for fast fission
neutrons and ν is the number of neutrons emitted per fission
process. The magnitude of the factor β also depends on the
cross-section of uranium for inelastic collisions which slow
down fast fission neutrons below the fission threshold of U238.
This slowing-down phenomenon of uranium has been studied by
Szilard and Zinn and /forms will be/ the subject of another report.

Secondarily, the magnitude of the factor β depends
on the size and density of the uranium sphere and also on the
presence in the uranium sphere of oxygen or other elements which
slow down the neutrons by means of elastic collisions. [It
may very well be that, due to the process which is the subject
of this report, there may be a ten per cent ~~difference~~ increase/ in the
number of fast neutrons emitted by a uranium sphere per thermal
neutron absorbed by the uranium sphere if we change over from
uranium oxide at a density of about 4 to 6 gm./cc. to uranium
metal of density 18 to 20 gm./cc. [The experiment which we
performed has the purpose of measuring the quantity

$$\sigma_f \nu$$

The principle of the experiment is illustrated in the enclosed
diagram which shows the experimental arrangement. In this
diagram Be is a beryllium block which serves as a source of
photo neutrons and which is placed in the axis of a cylindrical
paraffin block. Ra represents about two grams of radium which
are placed in the center of the beryllium block. A cadmium
diaphragm A leaves a circular opening/free through which thermal
neutrons can emerge from the paraffin and can enter (in the

3.

absence of the cadmium screen B. ~~These thermal neutrons can enter~~ *at H)*
the cylindrical box U which contains uranium in the amount of
about 25 grams per square cm. A spherical ionization chamber D
which is coated with a thick layer of uranium and has a
uranium-coated surface of about 800 cm. square is used to
record fission which takes place in the uranium coating of
this chamber. This fission chamber is protected from the action
of thermal neutrons by thick ~~layers~~ *walls* of boron carbide, and/E,
one at the walls
is a boron carbide screen which can be removed if it is desired
to admit thermal neutrons to the fission chamber.

The basic experiment which we performed is the
following: With the boron screen E and the cadmium screen B *(placed) (at H*
~~in position~~, the fission chamber registers a background of about
.5 fissions per minute. Thermal neutrons which come from the
circular opening *(C)* in the center of the cadmium diaphragm A are
prevented by the cadmium screen B *at H* from reaching the uranium
during this "control" experiment. If the cadmium screen is now
brought from the position H into the position K, thermal neutrons
are admitted to the uranium, will cause fission in the uranium
and will lead to the emission of fast fission neutrons. A
considerable fraction of these neutrons passes through the *uranium*
filled box U
~~whole thickness of the uranium layer~~ and through the boron screen
E and will cause fission in the ~~fission~~ *(ionisation)* chamber D. By changing
the position of the cadmium screen from position H to position K,
we obtained a fission count which was more than double of the
background count, the difference corresponding to about .7 counts
per minute. A preliminary estimate of the quantities involved

4.

leads to the conclusion that the observed effect corresponds to
a value of about

$$\sigma_f v \cong 1.3 \times 10^{-24} \, cm^2$$

Taking for v the value of 2.6 reported by Zinn and Szilard,[*]
we obtain

$$\sigma_f = 0.5 \times 10^{-24} \, ; \quad \sigma_f (v-1) = 0.8 \times 10^{-24}$$

It has to be emphasied that this is a preliminary result, and
that ~~the experiments will be repeated, and that~~ an attempt will
be made to determine the value more accurately than it was
possible to do in the first rough ⌐experiment. In the following we
describe two methods which were used in interpreting the observed
effect. The following designations will be used:

F_f is the fission count which we obtain from the fast

   neutrons when we remove the cadmium screen B at N

F_{th} is the fission count which we obtain in the chamber

   (in the absence of the uranium, the cadmium screen B at

   or B at and the boron screen E) due to the action of

   thermal neutrons which emerge from the window in the

   center of cadmium diaphragm A, and which can be cut

   off by placing the cadmium screen in the position B.

N is the number of thermal neutrons which emerge from the

   circular window in the center of the cadmium diaphragm

   A.

a is the fraction of these thermal neutrons which are absorbed

   by the uranium in the box U.

$1 - \varepsilon$ is the fraction of the fast neutrons which are emitted

   by uranium and which are prevented from reaching the

   ionization chamber by scattering either by uranium in the wall ~~or by~~

   ~~Boit~~ in the boron screen E.

5.

per unit solid angle

$Nb/2\pi$ is the number of thermal neutrons emitted /from

the paraffin wall in the circular opening,~~in the~~ G

~~center of the cadmium diaphragm A~~ in the forward

direction\ *towards the chamber D)* ~~per unit solid angle~~.

U is the number of uranium atoms which are exposed to the

thermal neutrons.

ρ is the thermal neutron density *in* ~~of~~ the uranium.

r is the distance of a volume element of uranium in the Box U

from a surface element of the uranium layer in the

spherical ionization chamber.

σ_{th} *(fiss.)* ~~(fission)~~ is the fission cross-section of uranium for

thermal neutrons.

σ_{th} is the total absorption cross-section of uranium for thermal

neutrons.

ρ_0 is a thermal neutron density ~~XXXXX~~ *at* /the location of the

uranium surface of the spherical ionization chamber in the

absence of the uranium,~~box~~ U, the cadmium screens B and C

and the boron screen E.

The main quantities which we measure are F_{th} and F_f.

One of the methods employed makes use of the fact that we may

write:

$$F_{th} = \frac{Nb}{2\pi r_{th}^2} \sigma_{th}(\text{fission})\, n^{th}$$

$$F_f = \frac{Nb}{4\pi r_f^2} \alpha\, \varepsilon\, \frac{\sigma_{th}(\text{fission})}{\sigma_{th}} (\sigma\rho V)\, n^f$$

or

$$(1)\qquad \sigma\rho V = \frac{F_f}{F_{th}}\,\sigma_{th}\,\frac{2b}{\alpha\varepsilon}\,\frac{n^{th}}{n^f}\,\frac{\overline{r_f^2}}{r_{th}^2}$$

6.

In these formulae $N_{th}(U)$ is the number of uranium atoms which
is effective in producing an impulse in the chamber if excited
by thermal neutrons, and $N_f(U)$ is the number of uranium atoms
which is effective in producing an impulse in the chamber if
excited by fast fission neutrons. We have still assumed that
these two numbers are actually equal, i.e., that the range of the fission
particles is about the same when the fission is due to thermal
neutrons as it is when the fission is due to fast fission neutrons.

~~From these two equations, we obtain:~~

Our experiments gave for F_{th} a value of 43 counts per
minute (cadmium difference), and for F_f we obtained from 25
fifteen-minute readings with the cadmium screen at the position
H and K, each, a value of

$$Ff = 0.77 \text{ count/min}$$

above a background count of .6 per minute.

α, the fraction of thermal neutrons which are absorbed
by the uranium in the box, was estimated to be about 0.5
The uranium box contains about 27.7 gms./sq.cm. of uranium
carbide or about 25.2 gms./sq.cm. of uranium. Neglecting the If we
scattering of the thermal neutrons, then we should expect that of the thermal neutrons
passing through the uranium box parallel to the axis of the
arrangement, a fraction of .31 would be absorbed, and a fraction
of .687 would be transmitted by the uranium box. By measuring
the thermal neutron density with a vanadium indicator in front
and right behind the uranium box, we find that the thermal
neutron density actually in the uranium box drops by a factor of 8. ~~in our arrangement.~~
Large Part of this drop is due to the scattering of thermal neutrons

7.

and part of it is due to oblique passage

in the uranium box, and a very rough estimate leads us to believe that actually about .5 of the neutrons entering the uranium box are absorbed.

For b we take the value of

b = 2.53

which corresponds to ~~distribution of~~ an angular distribution of

$$\cos \varphi + \sqrt{3} \times \cos^2 \varphi$$

for the ~~spatial~~ angular distribution of thermal neutrons leaving the paraffin window G inside the cadmium diaphragm A.

For ε we took the value of .75 by estimating the effect of the scattering in the boron carbide screen E containing 3 gms./sq.cm. of boron carbide and the effect of scattering of the uranium carbide in the uranium box.

For $\dfrac{r_f^2}{r_{fh}^2}$ we took the value of $\left(\dfrac{13}{14}\right)^2 \approx 0.87$

For δ_{fh} we took the value of 5.9 x 10^{-24} sq.cm.

With these values we obtained from (1)

$$\sigma_f V = 1.24$$

The second method employed makes use of the fact that we may write

(2) $$\sigma_f V = 4\pi \bar{\rho}_0 \left(\overline{\dfrac{r_f^2}{\rho}}\right) \dfrac{1}{u} \dfrac{1}{\varepsilon} \dfrac{F^{fh}}{F_{fh}} \dfrac{n^{fh}}{n_t}$$

We estimated ρ inside the uranium, and ρ_0, the thermal neutron density in the space occupied by the ionization chamber D by observing the activity of a vanadium foil in front

8.

of the window G on both sides of the uranium box and *also* in the
absence of the uranium box, the cadmium screens at H or K
and the boron screen E with a vanadium foil taking the place
of the ionization chamber D. Assuming an exponential *fall*
of the thermal neutron density inside the uranium box, we
then find for $4\pi \bar{\rho}_0 \left(\frac{r^2}{\rho} \right) = \left(4\pi r_f^2 \right) \frac{\rho_0}{\rho} = 250$
For the number of uranium atoms in front of the window G
we take the number ~~of~~ U $= 4.63 \times 10^{24}$ corresponding to 1830 gms. of
uranium. For ε we again take the value of 0.75
With these values we obtain from (2)

$(2)_{\kappa)}$ $\eta \nu = 1.28$

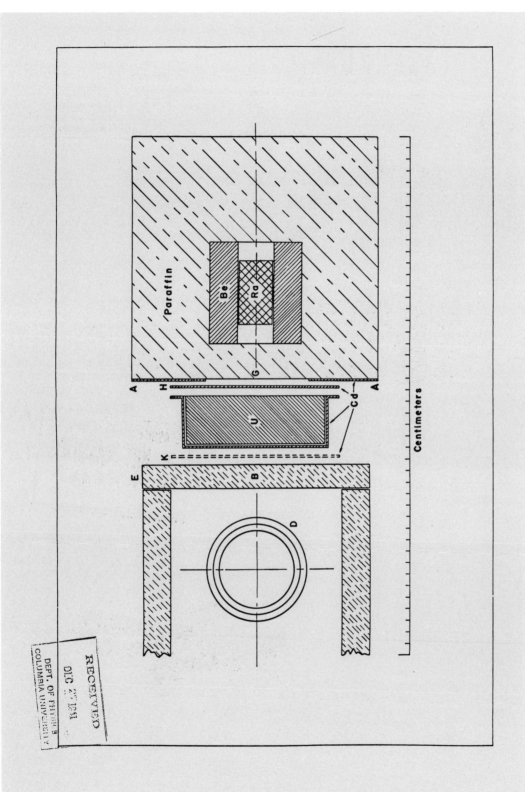

SECRET

Report No. CP-317

Copy No. _____ 1

Name *Lehman*

PRELIMINARY REPORT ON THE CAPTURE
OF NEUTRONS BY URANIUM IN THE ENERGY REGION
OF PHOTO NEUTRONS FROM RADIUM–BERYLLIUM SOURCES

J. Marshall and L. Szilard

December 5, 1941

Abstract

The cross section of uranium for the formation
of 239 has been estimated, using photo neutrons from a
radium-beryllium source. The value, after correcting
for the contribution of fission, was found to be about
6×10^{-25} cm^2. As a check, the iodine cross section
for photo neutrons was estimated and found to be about
ten times the value reported by Griffyths.

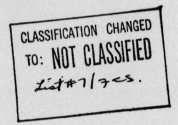
CLASSIFICATION CHANGED
TO: **NOT CLASSIFIED**
List #7/7 <s.

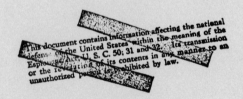

SECRET

PRELIMINARY REPORT ON THE CAPTURE
OF NEUTRONS BY URANIUM IN THE ENERGY REGION
OF PHOTO NEUTRONS FROM RADIUM-BERYLLIUM SOURCES

J. Marshall and L. Szilard

December 5, 1941

The capture of neutrons by uranium which leads to
the formation of uranium 239 plays an important role from
the point of view of the chain reaction in a system in which
a lattice of uranium spheres is embedded in graphite. If
the cross-section for this process were known as a function
of the energy, then this knowledge could be utilized in
determining the most favorable arrangement in this system.
While one may assume that this cross-section falls off inversely proportionally to the square root of the en gy
of the neutron in the region between a few hundred volts
and a few ten thousands of volts, the absolute value of the
cross-section is not known for any value of the energy.
We have for this reason attempted to determine this cross-
section for photo neutrons from a radium-beryllium source.
A preliminary rough measurement gives a cross-section of
about $\underline{5} \times 10^{-25}$ cm^2.

As a source of neutrons we used a beryllium block
8 cm. high having a diameter of 8 cm. and an inside bore
of 3 cm. diameter. A small box containing a uranium salt,
from which the natural β-active components were removed
before irradiation, was placed inside the bore of the beryllium block and two grams of radium were placed next to it
inside the bore. While this uranium sample was being irradiated, another identical uranium sample was placed in a
flat container below a Geiger counter and the growth of
its activity was observed. After irradiation (48 minutes)
the non-irradiated sample was replaced by the irradiated
sample and the activity of the latter was followed. The
difference in initial activity between the two samples
(taken at the time when the irradiation was stopped) amounted
to 0.93 expressed in units of the activity which corresponds
to one hour's growth of activity of the two samples (which
is due to the growth of uranium X in the sample). This
value has to be multiplied by 4/3 in order to correct for
infinite time of irradiation. We may write for the capture
cross-section σ of uranium leading to the formation of
uranium 239

$$(1) \quad \sigma = 0.93 \times \frac{4}{3} \frac{3}{Na} \frac{1}{852} \times \frac{1}{2.08 \times 10^{17}}$$

SECRET -2-

In this formula N is the number of neutrons emitted
by the photo source, α is the geometrical factor so defined
that αN gives the number of neutrons going per cm^2 and second
through the uranium sample. β is a factor which correlates
the penetrating power of the β rays of uranium 239 to the
β rays of uranium X; if these two β rays were equally penetrating, the factor β would have the value 1. In reality
the β rays of uranium 239 are softer, and the value of β
is roughly estimated to be about 3 for the case of our
particular geometry. This value is a rough estimate which
will be checked by later measurements. The above given
formula would hold if the total initial activity of 0.93
growth units would be due to the 24-minute period of uranium 239. In reality a fraction of about .18 of this activity is due to fission, and a corresponding correction
will be applied to the cross-section given by equation (1).

In order to determine the fraction of the initial
activity which is due to fission, and in order to determine
the value of the factor α , the following experiments have
been carried out.

(a) A box filled with 3 grams of the above-mentioned
uranium salt was irradiated by a fast neutron source (one
gram of radium mixed with beryllium) at 7 cm. distance for
48 minutes, and the initial activity was observed by switching over from a non-irradiated sample to the irradiated
sample at the end of the irradiation. This initial activity
was observed to be 1.1 growth units on a 3 gram sample placed
below the counter in the same geometry as used before.

(b) The number of fissions produced in an ionization
chamber coated with a thick uranium oxide layer was compared
for the photo neutron source and the radium-alpha-beryllium
source. We obtained 5 fissions for the photo neutron source
at a distance of 11 cm. between the center of the source
and the center of the spherical ionization chamber. We
obtained 124 fissions per minute for the radium-beryllium
mixture at a distance of 1F cm. from the center of the
spherical ionization chamber.

(c) We compared the activity obtained in iodine with
photo neutrons for two positions, one being the same position in which the uranium was irradiated by photo neutrons
in the first mentioned experiment, and the other being
10 cm. away from the center of the photo neutron source.
We found the ratio of the activities to be 13.6.

--3--

From the accessory experiments a, b and c we may conclude that about 0.16 of the initial activity is due to fission, and 0.93 - 0.16 = 0.77 is due to the 24-minute period.

From the accessory experiment c we may conclude that the factor α has the value of $\alpha = 13.6 \times \frac{1}{4\pi 10^2}$. The number of neutrons emitted by the photo source per second (n) was estimated some time ago by Anderson and Fermi who compared this photo neutron source with a fast neutron source (radium mixed with beryllium) to be about 3.8×10^6. This value is now being re-measured by B. T. Feld under the supervision of L. Fermi. With the above value and the application of the above-mentioned correction of the initial activity observed after photo neutron irradiation for fission activity, we find a cross-section for the formation of uranium 239 of about

$$\sigma = 6.15 \times 10^{-25} \text{ cm}^2$$

This value appeared to be exceptionally high if compared with the capture cross-sections of elements other than uranium which were found by Halban and Griffiths, on the other hand, for photo neutrons from radiothorium-deuterium and radium-beryllium, respectively. We have, therefore, also tried to determine the capture cross-section for iodine for which Griffiths has reported a value of 6.9×10^{-26} cm^2. Again we find a very high value, about ten times as large as the value quoted by Griffiths, and in this particular case we have compared the activity of a thin sheet of iodine irradiated by photo neutrons in a fairly well defined geometry and compared this activity with the activity of uranium X and radium L standards. It should be noted that our measurement of the uranium cross-section is not finished and, therefore, more accurate determinations are being made at present.

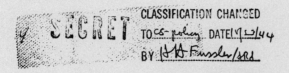

MEMORANDUM ON THE CRITICAL CONDITION FOR A FAST
NEUTRON CHAIN REACTION INSIDE A SPHERICAL SHELL
OF URANIUM METAL

Bernard T. Feld and Leo Szilard
Columbia University

December 26, 1941

ABSTRACT

The critical size for the maintenance of a fast neutron
chain reaction in a sphere of uranium 235 or other element having
high cross section for slow neutron fission is calculated when the
material is surrounded by a shell of uranium metal. The first
approximation to the appropriate diffusion equation is used, in
which we assume that there are two categories of neutrons: "slow",
or neutrons below the fission threshold of U_{238}, and "fast" neutrons.
We consider the processes of fission, inelastic scattering and
elastic scattering.

It is shown that the addition of the uranium metal shell
reduces the mass of reacting material by a factor of about 12.
The amount of core material would be 1.5 times greater if there
were no fast neutron fission in the uranium 238 shell.

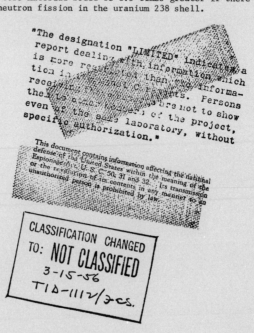

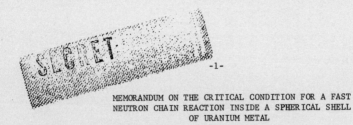

-1-

MEMORANDUM ON THE CRITICAL CONDITION FOR A FAST
NEUTRON CHAIN REACTION INSIDE A SPHERICAL SHELL
OF URANIUM METAL

Bernard T. Feld and Leo Szilard
Columbia University

December 26, 1941

If a sphere of uranium 235 or some other element which has
a high cross section for fast neutron fission is surrounded with a
spherical shell of uranium metal, we may have particularly favorable
conditions for maintaining a chain reaction with a minimum amount of
235 or one of its substitutes. Such a uranium shell acts in two ways:
it reflects neutrons back into the uranium 235 core, and it creates
additional neutrons due to fission, which is caused in uranium 238 by
fission neutrons. Fission caused in uranium 238 by fission neutrons
was observed by Marshall and Szilard, and according to the preliminary
result which they reported, the product of the cross section of this
process and the number of neutrons generated in a fission process
caused in uranium 235 by thermal neutrons is about 1.3×10^{-24} cm^2
(See Report No. CP-316 by John Marshall, Jr. and Leo Szilard, dated
November 14, 1941). This value is used in the rough estimate which
is given in the present calculations.

In the present calculations we shall distinguish between two
kinds of neutrons: "fast" neutrons which have energies above the
fission threshold of uranium 238, and, accordingly, may cause fission
both in the core and the shell, and "slow" neutrons which have energies
below the fission threshold of uranium 238, (about 1 Mev) and which,
accordingly, can cause fission in the core only. Both "fast" and
"slow" neutrons are assumed to have the same mean free path. We shall
assume that neutrons generated by the fission process in the core and
in the shell are all "fast". The average velocity of the fast neutrons
we take to be twice the average velocity of the slow neutrons. We
shall further assume that there is no appreciable slowing down of fast
neutrons in the core, but that fast neutrons are slowed down in the
shell below the fission threshold of uranium 238 with a cross section
of 2.5×10^{-24} cm^2 in accordance with the experimental value of
2.7×10^{-24} cm^2 reported by Szilard and Zinn. (See Report No. CP-285
by Leo Szilard and W.H. Zinn, dated December 12, 1941). We shall
further assume that in accordance with the above quoted results of
Marshall and Szilard, the fission cross section of uranium 238 for
fission neutrons is 0.5×10^{-24} cm^2, and further we shall somewhat
arbitrarily assume that <u>three</u> neutrons are emitted per fission of

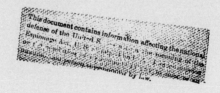

-2-

uranium 238. Similarly, we shall assume that three neutrons are
emitted per fission in the core. For the cross section for spherical
symmetrical scattering, we take the value of 4 x 10^{-24} cm^2 both in the
core and in the shell. As a fission cross section in the core, we
take the value of 2 x 10^{-24} cm^2, (this corresponds to a previously
unreported rough value obtained by Szilard and Zinn for the fission
cross section for photo-neutrons of radium-beryllium amounting to
1.3 x 10^{-26} cm^2 for natural uranium which contains one part of uranium
235 in about 140 parts of natural uranium).

Our treatment neglects the capture of slow and fast neutrons
in the shell which leads to the formation of uranium 239. It may be
that this introduces an appreciable error, since Marshall and Szilard
report for photo-neutrons of radium-beryllium a cross section of about
0.5 x 10^{-24} cm^2 for the formation of uranium 239 (See Report No. CP-317,
by John Marshall, Jr. and Leo Szilard, dated December 5, 1941). We
shall use below the following notation:

x = the density of "fast" neutrons

y = the density of "slow" neutrons

σ_f^* = fission cross section in the core

σ_{sc}^* = scattering cross section in the core

λ_* = mean free path for scattering in the core

ν^* = number of neutrons emitted per fission in the core

ν_x = velocity of the "fast" neutrons

ν_y = velocity of the "slow" neutrons

σ_{in} = cross section for inelastic collision in uranium
which slows down a fission neutron below the fission
threshold of uranium 238

σ_f = fission cross section for fission neutrons in uranium 238

ν_{238} = number neutrons emitted per fission by uranium 238

λ = mean free path for spherical symmetrical scattering
in the uranium shell

σ_{sc} = scattering cross section in the shell

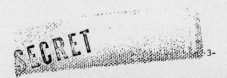

-3-

With the assumptions explained above, we may write if we use the first approximation to the appropriate diffusion equation:

(1) Inside the core describing the creation of fission neutrons, both by fast and slow neutrons:

$$\frac{d^2}{dr^2}(rx) + a_{xx}^2 (rx) + a_{xy}^2 (ry) = 0$$

$$a_{xx}^2 = \frac{\nu^*-1}{\lambda_*^2} \frac{3\sigma_f^*}{\sigma_f^* + \sigma_{sc}^*} \quad \Big\| \quad a_{xy}^2 = \frac{\nu_y}{\nu_x} \frac{\nu^*}{\lambda_*^2} \frac{3\sigma_f^*}{\sigma_f^* + \sigma_{sc}}*$$

(2) Inside the core describing the absorption of the slow neutrons (which lead through fission processes to the generation of fast neutrons described under (1):

$$\frac{d^2}{dr^2}(ry) - a_y^2 (ry) = 0 \quad : \quad a_y^2 = \frac{1}{\lambda_*^2} \frac{3\sigma_f^*}{\sigma_f^* + \sigma_{sc}}*$$

(3) In the uranium shell describing the disappearance of the fast neutrons due to combined effect of slowing down by uranium 238 and creation of new fast neutrons due to fission caused in uranium 238:

$$\frac{d^2}{dr^2}(rx) - b_x^2 (rx) = 0$$

$$b_x^2 = \frac{3}{\lambda^2} \frac{\sigma_{in} - (\nu -1)\sigma_f}{\sigma_{sc}}$$

(4) In the uranium shell describing the creation of slow neutrons generated from the fast neutrons by inelastic collisions in the uranium shell:

$$\frac{d^2}{dr^2}(ry) + b_y^2 (rx) = 0$$

$$b_y^2 = \frac{3}{\lambda^2} \frac{\nu_x}{\nu_y} \frac{\sigma_{in}}{\sigma_{sc}}$$

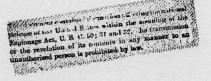

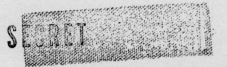

-4-

The radius R of the core must have a certain critical value in order to permit the neutron densities which obey equations 1 to 4 to remain finite, $r = 0$, and to vanish, $r = \infty$. This value of R can be considered as the critical radius at which the chain reaction could maintain itself as a stationary process. The solution of the equation must satisfy certain boundary conditions for $r = R$, inasmuch as the densities both of the slow and of the fast neutrons, and also the products of the mean free paths and the first derivatives of these quantities, have to be equal on both sides of the boundary. An equation which determines the value of R/λ has in this way been obtained by B.T. Feld (see the Appendix for details), and we may write for $\lambda_* = \lambda$

$$\frac{a_{yx}^2}{a_{xx}^2 + a_y^2} \cdot \frac{b_y^2}{b_x^2} \left[1 - \frac{a_{xx}}{a_y} \tanh \lambda\, a_y\, R/\lambda \cot \lambda\, a_{xx} R/\lambda \right] =$$

$$1 + \frac{a_{xx}}{b_x} \cot \lambda\, a_{xx}\, R/\lambda$$

We shall assume that the mean free path to be 5 cm which corresponds to a density of 20 gm cm^3, and a spherical scattering cross section of $\sigma_{sc} = \sigma_{sc}{}^* = 4 \times 10^{-24}$ cm^2. We shall further assume the following   values to be valid:

$$\frac{\nu_x}{\nu_y} = 2; \sigma_f^* = 2 \times 10^{-24} \text{ cm}^2 \; ; \; \sigma_f = 0.5 \times 10^{-24} \text{ cm}^2$$

$$\nu_* = \nu = 3; \; \sigma_{in} = 2.5 \times 10^{-24} \text{ cm}^2$$

$$\lambda_* = \lambda = 5 \text{ cms.}$$

With these values, we obtain:

$$\lambda\, a_{xx} = \sqrt{2} \; ; \; \lambda a_{xy} = \sqrt{\frac{3}{2}} \; ; \; \lambda a_y = 1$$

$$\lambda b_y = \sqrt{3.75}; \; \lambda b_x = \sqrt{1.125}$$

$$\left[\lambda b_x = \sqrt{1.875} \quad \text{(would be value if no fission would occur in shell)} \right]$$

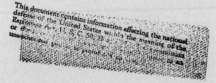

-5-

In this way we find from the equation number 5 for the radius of the core the value of $R/\lambda = 0.96$ or $R = 4.80$ cm. corresponding to a critical mass of the core of about 9.26 kg.

As a comparison, it may also be mentioned that the critical radius of the core in the absence of the shell would have a value of about 11 cm, corresponding to a critical mass about 112 kg or about 12 times larger than with a uranium shell.

If we had no creation of neutrons in the uranium shell due to fission by fission neutrons, we would obtain from the equation $R/\lambda = 1.1$ or $R = 5.5$. We see, therefore, that fission by fission neutrons leads to a reduction of the critical mass of the core by a factor of about 1.5.

APPENDIX   (By B. Feld)

The equations we must solve are:

(1) <u>inside</u>:

$$(rx)'' + a_{xx}^2 (rx) + a_{yx}^2 (ry) = 0$$

$$(ry)'' - a_y^2 (ry) = 0$$

(2) <u>outside</u>:

$$(rx)'' - b_x^2 (rx) = 0$$

$$(ry)'' + b_y^2 (rx) = 0$$

The boundary conditions are:

(a) for $r = 0$

x_i and y_i must be finite

(b) for $r = R$

$$x_i = x_o \qquad x_i' = x_o'$$

$$y_i = y_o - \qquad y_i' = y_o'$$

(c) $r \longrightarrow \infty$

$$x_o = y_o = 0$$

-6-

We have taken as solutions:

$$y_i = \frac{A_y}{r}(e^{a_y r} - e^{-a_y r})$$

$$x_i = \frac{A_x}{r}\sin(a_{xx} r) - \frac{a_{yx}^2}{a_{xx}^2 + a_y^2} \frac{A_y}{r}(e^{a_y r} - e^{-a_y r})$$

$$x_o = \frac{B_x}{r}e^{b_x r} + \frac{B_x'}{r}e^{-b_x r}$$

$$y_o = -\frac{b_y^2}{b_x^2}\left(\frac{B_x}{r}e^{b_x r} + \frac{B_x'}{r}e^{-b_x r}\right) + \frac{B_y}{r}$$

It will be noted that we have here five integration constants in four equations. (The boundary conditions ar r = 0 have already been applied, as has also the boundary condition $y_o = 0$ for r $\longrightarrow \infty$). Thus, we may solve for all of the constants in terms of any one. However, since $x_o = 0$ at r $\longrightarrow \infty$, we must not have any increasing exponential in the solution for x_o. In other words, the steady state solution of our problem requires that $B_x = 0$. This restriction will determine the critical value of R.

Solving for B_x in terms of B_x'

$$B_x = \frac{\left\{\frac{a_{yx}^2}{a_{xx}^2 + a_y^2}\frac{b_y^2}{b_x}\left[1 - \frac{a_{xx}}{a_y}\tanh(a_y R)\cot(a_{xx}R)\right] - b_x - a_{xx}\cot(a_{xx}R)\right\}}{\left\{\frac{a_{yx}^2}{a_{xx}^2 + a_y^2}\frac{b_y^2}{b_x}\left[1 - \frac{a_{xx}}{a_y}\tanh(a_y R)\cot(a_{xx}R)\right] - b_x + a_{xx}\cot(a_{xx}R)\right\}}e^{-2b_x R} B_x'$$

for $B_x = 0$

$$\frac{a_{yx}^2}{a_{xx}^2 + a_y^2}\frac{b_y^2}{b_x^2}\left[1 - \frac{a_{xx}}{a_y}\tanh(a_y R)\cot(a_{xx}R)\right] - 1 - \frac{a_{xx}}{b_x}\cot(a_{xx}R) = 0$$

We have taken for the values of the constants:

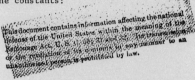

-7-

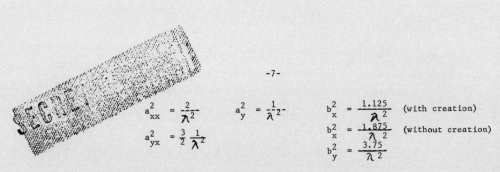

$$a^2_{xx} = \frac{2}{\lambda^2} \qquad a^2_y = \frac{1}{\lambda^2} \qquad b^2_x = \frac{1.125}{\lambda^2} \quad \text{(with creation)}$$

$$a^2_{yx} = \frac{3}{2}\frac{1}{\lambda^2} \qquad\qquad\qquad b^2_x = \frac{1.875}{\lambda^2} \quad \text{(without creation)}$$

$$b^2_y = \frac{3.75}{\lambda^2}$$

Solving:

R = 4.80 cms (with creation)

R = 5.55 cms (without creation)

R = 5.55 cms (if $b_x = b_y = 0$, i.e. - no inelastic scattering)

REPORT NO $CP-$ **189**

COPY NO. ___**2**___

NAME ~~ARGONNE LIBRARY~~

APPROPRIATE BOUNDARY CONDITIONS FOR DIFFUSION EQUATION AT

INTERFACE BETWEEN TWO MEDIA

L. Szilard, A. M. Weinberg, E. P. Wigner, and R. F. Christy

July 10, 1942

Ibser and Wheeler (Report C-88) gave a rigorous treatment of the boundary conditions at the interface between two homogeneous multiplying media. This treatment, while it is capable of giving accurate results, is rather laborious to apply. For this reason, it appears worthwhile to describe here a method of adjusting boundary conditions which was used, before Ibser and Wheeler's work became available, for the calculation of heavy water seeds. While this method does not have the accuracy of Ibser and Wheeler's method, it is much more rapidly applied and has been used recently even though the more exact formulae were available. It is more exact than the treatment underlying the Report C-11 by Breit and Fermi, which is essentially identical to the one used earlier by V. Halban and Kowarski.

The number of neutrons passing from the first into the second medium per cm^2 and sec. is $1/4 \sum n_i v_i$, where n_i is the density of neutrons of velocity v_i in the first medium. If $1/4 \sum n_i v_i$ were not equal to $1/4 \sum n_i^* v_i$ (where n_i^* is the density of neutrons of velocity of v_i in the second medium), there would be a flow of neutrons changing the total density of neutrons in both media by an amount which can be carried away by diffusion only if the gradient of n satisfied the equation

$$\sum \frac{\lambda_i v_i}{3} \; grad \; n_i = \frac{1}{4} \sum n_i v_i - \frac{1}{4} \sum n_i^* v_i \qquad (1)$$

i.e., if grad n/n were of the order $1/\lambda$, (In (1) λ_i is the mean free path of neutrons with velocity v_i). A gradient of this order of magnitude is incompatible with the ideas of diffusion theory and we write, therefore, as the first boundary condition that

$$\sum n_i v_i = \int n(E) \, v \, dE = \int N(E) \, \lambda \, dE \qquad (2)$$

be continuous. i.e. that it has the same value on both sides of the interface. In (2) n(E) is the density of neutrons per unit energy range, $N(E)$ is the number of collisions the neutrons of unit energy range suffer per cm^3 and sec. Both integrals in (2) are really an integral plus another term, referring to the thermal neutrons.

The second boundary condition expresses the fact that the flow to the boundary equals the flow away from the boundary. This gives that

$$\frac{1}{3} \sum \frac{\lambda_i v_i \; grad \; n}{1 - \cos \vartheta} = \frac{1}{3} grad \int \frac{\lambda v \; n(E)}{1 - \cos \vartheta} \, dE = \frac{1}{3} grad \int \frac{\lambda^2 \; N(E)}{1 - \cos \vartheta} \, dE \qquad (3)$$

-2-

be continuous, as above. Again, the integrals in (3) contain, in addition to the integral, an extra term corresponding to the thermal neutrons. The divisor 1 - cosϑ enters because the diffusion constant is $\lambda v/3(1-\cos\vartheta)$ (rather than $\lambda v/3$) where cos ϑ is the average cosine of the angle between the velocities of the neutron before and after collision.

We must assume now that the energy spectrum in each medium is the same at the boundary as it is in the inside of this medium. This is an arbitrary assumption which is, strictly speaking, incorrect and causes an inaccuracy of our conditions. It is this inaccuracy that has been eliminated by the above mentioned treatment by Ibser and Wheeler. If we make this assumption, n(E) becomes nν(E) where ν(E) is the ratio of neutrons in unit energy range at E and n the total density of neutrons. Similarly N(E) = q c(E) where c(E) is the number of collisions which a neutron suffers while its energy is in a unit interval at E; and q is the number of neutrons created per cm^3 and sec. Our two conditions now become that

$$ n \int \nu(E)\ v\ dE\ =\ q\ \int c(E)\ \lambda\ dE \qquad (4) $$

as well as

$$ \frac{1}{3} \int \frac{\lambda v\ \nu(E)}{1-\cos\vartheta}\ dE\ \mathrm{grad}\ n = \frac{1}{3} \int \frac{\lambda^2\ c(E)}{1-\cos\vartheta}\ dE\ \mathrm{grad}\ q \qquad (5) $$

are continuous at the interface.

Again, (4) and (5) contain two terms. The first corresponds to fast neutrons and in this c(E) = 1/(ξ E) where ξ is the slowing down power of the material (.158 for carbon). The second term refers to thermal neutrons. In this $\int c(E)\ dE$ is the ratio of total scattering cross section and total absorption cross section per unit cell, multiplied by the thermal disadvantage factor (about 1.4 in the usual arrangement). (This means, of course, that the $\int c(E)\ dE$ of this term is the number of collisions a neutron suffers while it is thermal.)

The second of these equations is capable of a very simple interpretation. The factor of grad q in (5) is evidently the mean square of the total distance traveled by a neutron from creation to absorption, i.e. the square of the migration length l. We can use a picture in which one "free path" is this migration length. Then q is the number of "collisions" per cm^3 and sec. and (5) becomes

$$ \frac{1}{3}\ l^2\ \mathrm{grad}\ q \qquad (6) $$

This is, however, equivalent to the usual formula for diffusion flow in which the density is expressed by the number of collisions per cm^3 and sec. If n is the density, q the number of collisions, 1 the mean free path, nv/l = q and 1/3 lv grad n = 1/3 1^2 grad q. One sees that the equality of the expression (5) for both sides of the interface is only another expression of the equality of the expressions for the flow in our picture.

-3-

2. As pointed out before, the preceding considerations, just as those of Ibser and Wheeler apply only at the interface of two homogeneous media. There is always a considerable ambiguity when they must be applied to the interface between two lattices. In fact, it is evident that the composition of the lattices alone does not suffice to determine the boundary conditions since the diffusion constant of the lattice depends quite considerably on the arrangement of the lumps, etc. within the damper. The following, therefore, involves a further approximation (inherent also in the work of Ibser and Wheeler).

We can express, first, $n(E)$ in terms of q in case of a lattice arrangement

$$V_q = n(E) \sum N_i V_i \sigma_{ai} v \xi_i E = n(E) \overline{N\sigma_s \xi} V_v E \qquad (7)$$

Herein, V is the total volume of the cell, V_i the volume of region i of the cell which contains N_i atoms per cm^3 with the scattering cross section σ_{si} and the slowing down power $\xi_i / \xi_i = 1$ (for H, .725 for D, .158 for C, etc.). The relative densities of the neutrons in the different parts of the lattice introduce a correction into (7) but we will disregard this because the density of resonance neutrons is nearly constant over the cell. Similarly, we assume that q is independent of energy. The density of thermal neutrons is given by the equation

$$qV = \sum n_i N_i V_i \sigma_{ai} v_t \qquad (8)$$

where n_i is the density of thermal neutrons in the i'th region, σ_{ai} the absorption cross section of the atoms of the i'th region for thermal neutrons and v_t the velocity of thermal neutrons. We can define relative densities ρ_i so that

$$\sum \rho_i V_i = V \qquad (8a)$$

Then $n_i = n\rho_i$ and n is the average number of thermal neutrons per cm^3

$$n = \frac{\sum n_i V_i}{V}$$

Introducing n into (8) gives

$$q \sum \rho_i V_i = n \sum \rho_i N_i V_i \sigma_{ai} v_t = n \overline{N\sigma_a} v_t \sum V_i \rho_i \qquad (8b)$$

Evidently, when calculating $\overline{N\sigma_a}$ from (8b), it is not necessary to use the ρ normalized by (8a) because they occur linearly on both sides. We can now write for (2)

$$q \left(\int \frac{1}{\overline{N\sigma_s \xi}} \frac{dE}{E} + \frac{1}{\overline{N\sigma_a}} \right) \qquad (9)$$

-4-

q in this is an average, i.e. the number of neutrons created per second in a cell, divided by the cell volume. The averages $\overline{N\sigma_1 \xi}$ and $\overline{N\sigma_a}$ are defined by (7) and (8b); the scattering cross section σ_1 refers in the first one to fast, the absorption cross section σ_a in the second to thermal neutrons. The first boundary condition is that (8) be continuous.

For (3) it is reasonable to write if the medium contains several regions

$$\frac{1}{3} \, grad \, \Sigma \int \frac{\lambda_i N^{m_i}(E) V_i}{(1 - \cos \vartheta) V} \, dE$$

One can write for this by (7) and (8b)

$$grad \, q \, \Sigma \, \frac{V_i}{V} \left[\int \frac{1}{3N\sigma_{si}(1-\cos\vartheta_i)} \, \frac{dE}{N\sigma_a \xi E} + \frac{D_i}{N\sigma_a \, v_{t}^2} \right] \tag{10}$$

where D_i is the diffusion constant for thermal neutrons in region i. The first term of (10) corresponds to the age, the second to the square of the diffusion length. Their sum is the square of the migration length, and (10) is identical with (6).

One can use, instead of the two boundary conditions (9) and (10) the single condition that their ratio shall be continuous. If we assume a single mean free path λ_f for fast neutrons and another λ_t for thermal neutrons, this can be written as

$$\frac{grad \, q}{q} \, \frac{\ell^2/3}{\int \frac{dE}{N\sigma_a \xi E} + \frac{1}{N\sigma_a}} = \frac{grad \, q}{q} \, \frac{\ell \, 1/3}{c_f \lambda_f + c_t \lambda_t} \tag{11}$$

where cf is the number of collisions a neutron suffers while it is fast, c_t the number of collisions it suffers while it is thermal, λ_f and λ_t are the two mean free paths. One can obtain (11) directly by dividing (6) by the right side of (4) and postulating continuity for the quantity obtained in that way.

PRELIMINARY COMPARISON OF RADON-BORON AND RA+BE

NEUTRON SOURCES

J. Ashkin, S. Bernstein, B. Feld, H. Kubitschek, and L. Szilard
January 19, 1943

ABSTRACT

The intensity of a radon-boron neutron source was compared
with a radium-beryllium neutron source (1 gm. Ra + 3 gm. Be). It
was found that, reduced to equal gamma ray intensity, the neutron
intensity of the two sources has a ratio of about 4 in favor of the
radium-beryllium source, if the intensity is measured by the n-p
reaction in phosphorous, fast neutron fission or integration of the
thermal neutron density in a water tank.

It was found that the radium-beryllium source excites the
25 minute period of iodine, which is due to radiative capture,
two or three times as strongly as the radon-boron source, if the
two sources are so normalized as to give equal fast fission counts
in a uranium fission chamber. The impression is conveyed that
radium-beryllium neutrons excite, in general, radiative capture
more strongly than radon-boron sources.

CP-412

PRELIMINARY COMPARISON OF RADON-BORON AND

RA+BE NEUTRON SOURCES

J. Ashkin, S. Bernstein, B. Feld, H. Kubitschek, and L. Szilard

January 19, 1943

A radon-boron source was prepared by B. Feld and we compared the activities induced by the fast neutrons of this source in iodine and phosphorous, with the activities induced by neutrons from a Ra+Be source (1 gm of radium mixed with 3 gm. of beryllium; first Chicago pressed source). We also compared the fast neutron fission count for these two sources by means of the Columbia spherical fission chamber. The ratio of the effect for phosphorous and for fission reduced to equal gamma activity of the sources, was about 1 to 4.2 in favor of the Be source. The ratio of the iodine activity however was 11, in favor of the Be source.

From this we see that if a radon-boron source and the Ra+Be source are so normalized as to give the same fast neutron activation of the n-p reaction in phosphorous or the same fast neutron fission in ordinary uranium, then the Ra+Be neutrons activate the radiative capture period of iodine 2 to 3 times stronger than do the radon-boron neutrons. Similar phenomena were observed with these two kinds of neutron sources by Szilard at Oxford and the impression is conveyed that Ra+Be neutrons activate the radiative capture periods of heavy elements appreciably more strongly than do radon-boron neutrons, if the sources are normalized in such a way as to give with equal intensity the n-p reaction of phosphorous. Clearly, great caution is to be observed in interpreting radiative absorption cross-sections measured by the fast neutrons of such neutron sources.

The total number of neutrons emitted by these two kinds of sources were compared by slowing down the neutrons in a water tank and measuring the thermal neutron density with a manganese detector. This very rough comparison gave a factor 4 in favor of our Ra+Be source, if we normalize our radon-boron source so as to give equal gamma activity. The comparison was actually carried out by making a direct comparison in the water tank between the radon-boron and a Ra+Be source (source No. 2) and by further comparing the fast fission count of source No. 2 and our compressed Ra-Be source. Incidentally, this comparison showed that the compressed source which contains 1 gm. of radium and only 3 gm. of beryllium emits .73 times as many neutrons as source No. 2. (Source No. 2 contains 1 gm. of radium to 10 gms. of beryllium and is reported by Anderson to emit 13.2 x 10^6 neutrons per second).

From the masses of boron 10 and 11, and the maximum energy of the alpha particles of radium in equilibrium with its daughter products, Feld and Ashkin compute that the maximum energy of the

- 2 -

neutrons emitted from boron are to be about 8,000 000 volts for
neutrons which are emitted in the forward direction. This corresponds to an energy of 4,000,000 volts for neutrons emitted in a
backward direction. Since nitrogen 13 and 14 which are formed in
this reaction may be left in an excited state, the activity
energy of the boron neutrons may be lower than it would be otherwise. These average energies are probably lower than the average
energies of fission neutrons, since Szilard found at Oxford that the
p,n reaction in aluminum is not appreciably excited by boron neutrons,
whereas the n-p reaction of phosphorous is quite strongly excited.
We intend to check this point in this laboratory. On the other
hand, it appears likely, that boron neutrons contain much fewer
neutrons at 100,000 volts energy than Ra-Be neutrons since as
mentioned above, radiative capture in iodine and other elements is
less strongly excited by boron neutrons than Ra-Be neutrons.

CF-1177
(A-1659)

METALLURGICAL PROJECT

A.H. Compton, Project Director

* * *

METALLURGICAL LABORATORY

S.K. Allison, Director

* * *

<u>PHYSICS RESEARCH</u>

E. Fermi, Division Director; L. Szilard, Section Chief

* * *

<u>NEUTRON EMISSION IN FISSION OF U^{238}</u>

L. Szilard and Physics Group II:　B. Feld--Group Leader,
J. Ashkin, S. Bernstein, L. Creutz, J. Kelsner, R. Scalettar

December 29, 1943

* * *

Table of Contents

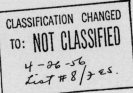

CLASSIFICATION CHANGED
TO: NOT CLASSIFIED
4-26-56
List # 8/7 ES.

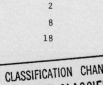

SECRET

ABSTRACT

The number of neutrons generated in a sphere of uranium by
neutrons emanating from a neutron source in the center of a uranium
sphere were determined relative to the number of neutrons emitted by
the source. This was done for a radium-boron and a radium-beryllium
neutron source. The neutrons are generated in these experiments in
the uranium sphere by fission of U^{238}. In addition to this process
an (n-2n) reaction might be involved in the case of the radium-beryllium
source.

SECRET

NEUTRON EMISSION IN FISSION OF U^{238}

L. Szilard and Physics Group II: B. Feld--Group Leader,
J. Ashkin, S. Bernstein, L. Creutz, J. Kelsner, R. Scaletter

INTRODUCTION

The purpose of the experiments reported herein was to obtain information about the number of neutrons liberated from U^{238} when fast neutrons pass through a layer of natural uranium. It would be desirable to obtain this information for fast neutrons which have the spectral distribution of neutrons emitted in thermal fission of U^{235}. The most important quantity in which we are interested is $\sigma_f (\nu - 1)$ for U^{238}. Since we have some indications that the spectrum of Ra-B neutrons comes fairly close to the spectrum of fission neutrons, we used this neutron source for most of our experiments. Earlier experiments of a similar nature were made with Ra-Be neutrons which are more abundant in high energy neutrons. Reliable knowledge of the value of $\sigma_f (\nu - 1)$ would also make it possible to improve our estimates of the contribution of fission in U^{238} to the slow neutron chain reaction.

EXPERIMENT

At the center of a paraffin cube of about 4 ft. side we had a spherical cavity, about 25 cm in diameter, at the center of which we placed a small source of fast neutrons. The indium resonance activity (indium foils surrounded by cadmium) was measured along a radius in the paraffin. A curve was then drawn by plotiing the product of the indium resonance activity and the square of the distance from the source, against distance from the source. The area underneath this curve is proportional to the number of neutrons emitted per second by the source.

-2-

-3-

A cadmium shielded uranium sphere was then placed in the cavity, with the source at the center of the sphere. We again measured the indium resonance activity along a radius and obtained the area underneath the curve. This second area is proportional to the total number of neutrons emerging per second from the uranium sphere, and ε, the ratio of the two areas, gives the number of neutrons emerging from the uranium sphere for one neutron emitted by the source in the center.

In the case where the source was surrounded by uranium, two corrections had to be applied to the measured area: a correction for the absorption of resonance neutrons by the uranium sphere; another correction for the creation of neutrons by U^{235} fission. Although the uranium sphere was surrounded by cadmium, there were still enough epicadmium neutrons penetrating into the sphere to cause an increase of about 1% in the area. However, since the resonance absorption in the uranium sphere decreased the measured area by approximately the same amount, the two corrections were considered to cancel.

A detailed description of the experiment is given in Appendix 1.

The experiment was first performed with a Ra-Be neutron source. The result, for a $5\frac{1}{2}$ cm metal sphere (there was a hole of radius 1 cm at the center of the sphere to accommodate the source) of density 18, was $\varepsilon = 1.130 \pm .015$. Since Ra-Be neutrons are known to have an appreciable high energy component, it was thought that the fission spectrum could be more closely simulated by a Ra-B neutron source. With this end in mind, we had prepared a two gram Ra-B neutron source[1]. The experiment

---

[1]A discussion of this source and its standardization is given in a separate report.

-4-

performed with this source gave $\mathcal{E}$ = 1.053 $\pm$ .015.

A comparison of the fission cross section of U^{238} for Ra-B and Ra-Be neutrons showed these to be about equal. This comparison was made in the following way: a cadmium surrounded, uranium coated ionization chamber, connected to a linear amplifier, was placed near a Ra-B source and the fast fission counting rate observed. The source was then replaced by a Ra-Be source and the counting rate again noted. The source strengths were compared by integrating the indium resonance neutron density in paraffin. From these measurements, we determined that the fission cross sections of Ra-B and Ra-Be neutrons for U^{238} are equal to within 5%.

It is of interest to note that the number of neutrons liberated from U^{238} for a given layer thickness traversed increases considerably by going from Ra-B neutrons to Ra-Be neutrons. We cannot exclude the possibility that part of this increase may have to be attributed to an (n, 2n) reaction in U^{238}. If only fission were involved, we would have to conclude that ν increases considerably by going from Ra-B neutrons to Ra-Be neutrons. This conclusion would be inescapable, inasmuch as the average fission cross sections for these two kinds of neutron energy distributions appears to be almost identical, while the value of $\sigma_f(\nu - 1)$ corresponding to the observed values of $\mathcal{E}$ is about twice as much for Ra-Be neutrons. An (n, n fission) process involving the high energy part of the Ra-Be neutron spectrum might be involved; or the number of neutrons emitted by the fission fragments might be greater for higher energies of the incident neutrons.

-5-

INTERPRETATION

      The constants primarily involved in the interpretation in the observed fast neutron multiplication are:

    (1) The average fission cross section for U^{238} , σ_f. The average fission cross section was measured for Rn-Be neutrons by Anderson and Fermi (C-83). They reported a value of .45 x 10^{-24} cm^2. Our comparison of the fast fission cross section for Ra-Be and Ra-B neutrons shows them to be about equal.

    (2) The number of neutrons per fission of U^{238}, ν. This number has not been measured before. Our experiments giving ε determine, essentially, the value of the product $\sigma_f(\nu - 1)$.

    (3) The inelastic scattering cross section of uranium, σ_i, which was reported by Szilard and Zinn to be equal to 2.7 x 10^{-24} cm^2. Recent experiments, on which we are issuing a separate report, indicate that this is a fairly good value, for both Ra-Be and Ra-B neutrons.

    Of lesser importance for the interpretation are:

    (4) The fission cross section of U^{238} averaged over the neutrons which have an energy above the fission threshold of U^{238}. This we denote by σ_f^*; we have assumed that for Ra-Be, Ra-B and fission neutrons, $\sigma_f^* = .6$ x 10^{-24} cm^2.

    (5) The cross section of uranium for spherically symmetrical elastic scattering, σ_e. The effect of elastic scattering is to increase the length of the neutrons' paths in the sphere, and thus, to increase the observed effect. Since we have, as yet, no reliable estimate of σ_e, we have made two independent sets of calculations; in the first we have

-6-

assumed $\sigma_e = 0$, and in the second, $\sigma_e = 2.0 \times 10^{-24}$ cm^2. The results of these calculations are shown in Table I[2]. In this table we have entered the calculation values of $\mathcal{E}$, for the U metal sphere used in our experiments for different values of $\mathcal{V}$. The values of the other constants used in the computation are:

$$\sigma_f = .45 \times 10^{-24}$$
$$\sigma_i = 2.7 \times 10^{-24}$$
$$\sigma_f{}^* = .6 \times 10^{-24}$$
$$\sigma_e = 0 \text{ and } 2 \times 10^{-24}$$

TABLE I

$\mathcal{V}$	$\mathcal{E}(\sigma_e = 0)$	$\mathcal{E}(\sigma_e = 2)$
1.3		1.030
1.4		1.041
(1.5)	1.036	(1.054)
1.6	1.044	1.064
[1.7]	[1.052]	1.073
1.8	1.059	1.083
1.9	1.067	1.098
2.0	1.075	1.109

For no elastic scattering, the value of $\mathcal{V}$ corresponding to the experimental value of $\mathcal{E} = 1.053 \pm .015$ is $\mathcal{V} = 1.7 \pm .2$. If the elastic scattering cross section were 2×10^{-24}, the value would be $\mathcal{V} = 1.5 \pm .15$.

---

[2] The formulae used in the computation of $\mathcal{E}$ are discussed in Appendix 2.

-7-

In the above computations it has been assumed that the constants $(\nu, \sigma_f, \sigma_i)$ for Ra-B and for fission neutrons are equal. However, since the entire fast effect is rather small, the secondary effects-- i.e., effects due to first, second, and higher generation neutrons--are quite small; so that, even if the values of σ_f, σ_i and ν were rather different for fission neutrons, the values of ε would not be much affected.

On the other hand, for reasons previously discussed, ν for Ra-Be neutrons seems to be considerably higher. Therefore in evaluating the experiment performed with Ra-Be neutrons one has to take into account that the fission caused by the first generation of fission neutrons has presumably a lower ν than the fission caused by neutrons from the source. If we assume that the other constants involved are not appreciably different we obtain for $\varepsilon = 1.13 \pm .015$ the following set of values ν (Ra-Be) = 2,8 $\pm$ 0.2 for $\sigma_e = 0$ and ν (Ra-Be) = 2.5 $\pm$ 0.2 for $\sigma_e = 2 \times 10^{-24}$.

The following table gives the values of $\sigma_f (\nu - 1)$ corresponding to the observed values of ε. It is this quantity which is primarily determined by the measured value of ε.

Source	$\sigma_f (\nu - 1)$	
	($\sigma_e = 0$)	($\sigma_e = 2 \times 10^{-24}$)
Ra-B	.32 $\pm$ .09	.23 $\pm$ .07
Ra-Be	.81 $\pm$ .09	.68 $\pm$ .09

APPENDIX I

DETAILS OF THE EXPERIMENT

J. Ashkin, S. Bernstein, B. Feld,
L. Creutz, J. Kelsner, R. Scaletter

Method

The fast neutron multiplication factor, $\mathcal{E}$, is obtained from our experiment by taking the ratio of the total number of neutrons emitted with a uranium sphere surrounding the source to the total number emitted by the source alone. The number of neutrons emitted by a source is determined by slowing the neutrons down in parrifin and finding the volume integral, in the paraffin, of neutrons passing through the indium resonance level. The total number of neutrons emitted by the source is thus given by the volume integral:

$$Q = \int q \, d \, v \qquad (1)$$

where,

q = number of neutrons passing through the indium resonance level per cm^3 per second.

Since the resonance activity of an indium foil is directly proportioned to q, the ratio $\mathcal{E}$ is determined experimentally by means of the following expression:

$$\mathcal{E} = \frac{\int A_2(r)r^2 dr}{\int A_1(r)r^2 dr} \qquad (2)$$

where,

$A_2(r)$ = Resonance activity of an indium foil with the U sphere surrounding the source.

$A_1(r)$ = Resonance activity of an indium foil with the U sphere removed.

-8-

-9-

The integrals in (2) were taken over the entire volume of
paraffin, which was large enough so that none of the neutrons escaped.

Experimental Arrangement

The experimental arrangement is shown in cross-section in
Figure 1. It consisted of a parallelopiped of paraffin 54" x 48" x 32",
made up of 8" x 12" x 12" bricks. At the center of this mass of paraffin, there was a spherical cavity of 11.8 cm radius. At the center of
the cavity was placed a uranium sphere of radius 5.5 cm and 11,200 gms.
mass. A compressed source of neutrons (Ra-α-Be or Ra-α-B)of about
2 cm diameter was placed at the center of the uranium sphere. The
uranium sphere was surrounded with a close-fitting 0.020" thick cadmium
jacket. The purpose of the cadmium jacket was to prevent any neutrons
slowed down to thermal energies in the paraffin from going back into
the sphere and causing thermal fission of U^{235}. The purpose of the
6.3 cm air gap between the outside surface of the uranium sphere and
the inside spherical surface of paraffin was to reduce the number of
epi-cadmium neutrons reflected back from the paraffin into the uranium
and there absorbed in resonance capture.

Measurements

Two complete, independent sets of measurements were taken, along
the C and D axes of Figure 1. A complete set of measurements consisted
of foil activity measurements as a function of position in the parrafin:

   (a) With the uranium sphere removed.

   (b) With the uranium sphere in place.

   Indium foils, in a close fitting .020" thick cadmium envelope,
were placed at various positions in the paraffin by means of the paraf-

-10-

fin plugs shown in Figure 1. Usually two foils were exposed simultaneously, one along each axis. More than one foil was never exposed during the same interval along the same axis. The face of the foil was perpendicular to the equator plane, with the foil center in this plane. The foils were 5.0 x 6.3 cm in size and weighed 2.340 grams to better than 1/2%. The cross section of the paraffin plugs was 9.37 cm x 7.94 cm. They were made considerably larger than the foils to reduce the effect of neutron leakage through the small clearance space between the outside surface of the paraffin plugs and the main mass of paraffin. The C and D sets of foil activities were measured, each with its own glass Geiger tube and scaling circuit. Every foil activity measurement was corrected by means of a uranium oxids standard. About 10,000 standard counts were taken immediately before and 10,000 immediately after every foil. These corrections were usually less than 1%.

Foil measurements were taken at 1/4" to 1" intervals in the paraffin to the distance at which the foil activities became about equal to the Geiger-tube background count. Three to seven foils were taken for each position to give a total number of counts of about 200,000 for the close positions to about 1,000 counts for the most distant positions.

The experiment was done first with a 1 gram Ra-Be compressed source, and then repeated for a 2 gram Ra-B compressed source.

-11-

TABLE II

Foil Activities vs Foil Position,
Ra-Be Source
(Along Radius "C" of Fig. 1)

Thickness of Parrafin, Inches	$\sqrt{\overline{r^2}}$ cm	$\overline{r^2}$ cm²	Without Uranium Sphere		With Uranium Sphere	
			$A_1(r)$ Average Saturation Indium Resonance Activity Counts per foil per min.	$A_1\overline{r^2}$ $\dfrac{\text{Counts}}{\text{foil min}}$ cm² x 10^{-3}	$A_2(r)$	$A_2\overline{r}^{-2}$
0	11.82	139.7	9515	1329	15448	2158
.25	12.44	154.7	9359	1448	14800	2290
.50	13.07	170.8	9018	1540	14363	2453
.75	13.69	187.4	8514	1596	13413	2513
1.00	14.32	205.0	7955	1631	12334	2528
1.25	14.94	223.2	7265	1622	11089	2475
1.50	15.57	242.6	6458	1567	9589	2326
1.75	16.20	262.4	5763	1512	8277	2172
2.00	16.83	283.3	5131	1454	6942	1967
2.50	18.09	327.2	3961	1296	5133	1679
3.00	19.35	374.4	3027	1133	3633	1360
3.50	20.61	424.8	2357	1001	2502	1063
4.00	21.87	478.5	1801	861.8	1781	852.2
4.50	23.14	535.3	1402	750.5	1286	688.4
5.00	24.40	595.4	1103	656.7	929.9	553.7
6.00	26.93	725.2	696.4	505.0	530.8	384.9
7.00	29.46	868.0	449.0	389.7	314.2	272.7
8.00	31.91	1023.	292.2	298.9	191.4	195.8
9.00	34.52	1192.	192.7	229.7	122.4	145.9
10.20	37.55	1410.	.....	.....	74.25	104.7
10.45	38.20	1459	108.5	158.3	......	.....
10.95	39.46	1557.	88.67	138.1	53.88	83.89
11.95	42.00	1764.	60.33	106.4	35.42	62.48
12.95	44.53	1983.	41.11	81.52	23.89	47.37
13.95	47.07	2216.	28.05	62.16	17.10	37.89
14.95	49.61	2461.	19.21	47.27	11.48	28.25
15.95	52.14	2719.	13.66	37.14		
16.95	54.67	2989.	9.178	27.43		

-12-

TABLE III

Foil Activities vs Foil Position
Ra-Be Source
(Along "D" Radius of Figure)

Thickness or Parrafin Inches	$\sqrt{r^2}$ cm	$\overline{r^2}$ cm²	Without Uranium Sphere		With Uranium Sphere	
			$A_1(r)$ Average Saturation Indium Resonance Activity, Counts per foil per min.	$A_1\overline{r^2}$ $\dfrac{\text{counts}}{\text{foil min}}\text{cm}^2$ x 10^{-3}	$A_2(r)$	$A_2\overline{r^2}$
0.00	11.82	139.7	10169	1421	16194	2262.3
0.25	12.44	154.7	9847	1523	15239	2357
0.50	13.07	170.8	9436	1612	14664	2505
0.75	13.69	187.4	8910	1670	13682	2564
1.00	14.32	205.0	8215	1684	12434	2549
1.25	14.94	223.2	7548	1685	10974	2449
1.50	15.57	242.6	6740	1635	9652	2342
1.75	16.40	262.4	6066	1592	8325	2184
2.00	16.83	283.3	5345	1514	7179	2034
2.50	18.09	327.2	4147	1357	5199	1701
3.00	19.35	374.4	3191	1195	3663	1371
3.50	20.61	424.8	2463	1046	2577	1095
4.00	21.87	478.5	1899	908.6	1863	891.4
4.50	23.14	535.3	1483	793.8	1353	724.3
5.00	24.40	595.4	1174	699.0	982.3	584.9
6.00	26.93	725.2	740.5	537.0	555.1	402.5
7.00	29.46	868.0	473.0	410.5	327.4	284.2
8.125	32.31	1044	283.9	301.6	186.3	194.5
9.125	34.84	1214	188.4	228.7	120.6	146.4
10.125	37.38	1397	128.4	179.4	76.48	106.8
10.95	39.46	1557	89.44	139.3	52.59	81.9
11.95	42.00	1764	59.46	104.9	37.79	66.6
12.95	44.53	1983	41.67	82.63	26.43	52.4
13.95	47.07	2216	28.17	62.42	15.91	35.3
14.95	49.61	2461	20.03	49.29	11.20	27.5

-13-

TABLE IV

Foil Activities vs Foil Position

Ra-B Source
(Along "C" Radius)

Thickness of Parrafin Inches	$\sqrt{\overline{r^2}}$ cm	$\overline{r^2}$ cm	Without U Sphere		With U Sphere	
			$A_1(r)$	$A_1\overline{r^2}$	$A_2(r)$	$A_2\overline{r^2}$
0.00	12.01	144.2	3571	514.9	6043	871.4
0.25	12.48	155.7	3540	551.2	5853	911.3
0.50	13.11	171.8	3447	592.2	5557	954.7
0.75	13.73	188.4	3324	626.2	5127	965.9
1.00	14.35	205.9	3162	651.1	4742	976.4
1.25	14.98	224.4	2982	669.2	4210	944.7
1.50	15.61	243.8	2782	678.3	3770	919.1
1.75	16.24	263.6	2569	677.2	3255	858.0
2.00	16.86	284.2	2356	669.6	2849	809.7
2.50	18.13	328.6	1924	632.2	2086	685.5
3.00	19.38	375.5	1549	581.6	1510	567.0
3.50	20.65	426.4	1191	507.8	1083	461.8
4.00	21.90	479.7	939.5	450.7	785.7	376.9
5.00	24.43	596.8	582.6	347.7	415.5	248.0
6.00	26.96	726.8	347.2	252.3	229.5	166.8
7.00	29.49	869.7	195.4	169.9	130.2	113.2
8.00	32.02	1026	115.4	118.4	75.20	77.16
9.00	34.55	1194	66.83	79.80	44.13	52.69

-14-

TABLE V

Foil Activity vs Foil Position
Ra-B Source
(Along "D" Radius)

Thickness of Parrafin Inches	$\sqrt{\overline{r^2}}_{cm}$	$\overline{r^2}_{cm^2}$	Without U Sphere		With U Sphere	
			$A_1(r)$	$\overline{A_1 r^2}$	$A_2(r)$	$\overline{A_2 r^2}$
0.00	12.01	144.2	3393	489.3	5763	831.0
0.25	12.48	155.7	3357	522.7	5541	862.7
0.50	13.11	171.8	3292	565.6	5300	910.5
0.75	13.73	188.4	3153	594.0	4929	928.6
1.00	14.35	205.9	2972	611.9	4503	927.2
1.25	14.98	224.4	2785	625.0	4066	912.4
1.50	15.61	243.8	2612	636.8	3631	885.2
1.75	16.24	263.6	2423	638.7	3166	834.6
2.00	16.86	284.2	2219	630.6	2750	781.6
2.50	18.13	328.6	1850	607.9	2060	676.9
3.00	19.38	375.5	1463	549.4	1503	564.4
3.50	20.65	426.4	1150	490.4	1082	461.4
4.00	21.90	479.7	897.0	430.3	784.5	376.3
5.00	24.43	596.8	531.8	317.4	417.9	249.4
6.00	26.96	726.8	320.6	233.0	224.6	163.2
7.00	29.49	869.7	187.7	163.2	124.5	108.3
8.00	32.02	1026	107.6	110.4	69.26	71.06
9.00	34.55	1194	63.73	76.09	38.61	46.10

-15-

TABLE VI
Areas Under r^2- Curves

Ra-Be Source		
	"C"	"D"
	Area under r^2 curve $\frac{counts}{foil\ min.cm^3}$	Area under "r^2" curve $\frac{counts}{foil\ min.cm^3}$
U sphere surrounding source	24810×10^3	25450×10^3
U sphere removed	21730×10^3	22780×10^3
	$\mathcal{E}$ = 1.142	$\mathcal{E}$ = 1.117
	Average $\mathcal{E}$ = 1.130	

Ra-B Source		
	"C"	"D"
	Area under r^2 curve $\frac{counts}{foil\ min.cm^3}$	Area under "r^2" curve $\frac{counts}{foil\ min.cm^3}$
U sphere surrounding source	9776×10^3	9438×10^3
U sphere removed	9384×10^3	8874×10^3
	$\mathcal{E}$ = 1.042	$\mathcal{E}$ = 1.064
	Average $\mathcal{E}$ = 1.053	

-16

Data

A summary of the data is given in Tables II, III, IV and V.

Tables II and IV include measurements along the "C" radius, Tables III

and V include measurements along the "D" radius. Column 1 gives the

thickness of paraffin between foil and source. Column 3 gives the square

of the distance of the foil from the center of the source averaged

over the surface of the foil. The values in Column 2 are the square

roots of those in Column 3. Column 4 gives the indium resonance ac-

tivity per foil as a function of distance of the foil from the source

with the sphere removed. Column 5 shows the resonance activity multi-

plied by r^2 for the sphere removed. Columns 6 and 7 apply to the case

with the U sphere surrounding the source.

Figure 2 is a plot of foil activity x r^2 vs. foil position from

which the volume integrals were determined. Figure 3 is a semi-log

plot of the "r^2" curve. The Ra-Be without-sphere curve becomes exponen-

tial after a thickness of about $2\frac{1}{2}$" of paraffin, with a relaxation dis-

tance of 9.5 cm. The Ra-B curve becomes exponential after a thickness

of about 6" of paraffin, with a relaxation distance of 6.6 cm. A dis-

cussion of the forms of these curves will be given in another report.

The areas under the r^2 curves and the values of ξ are given in

Table VI. The two independently determined values of ξ for each type

of source differ by about 2%. The average value of ξ is 1.13 for Ra-Be

neutrons, and 1.053 for Ra-B neutrons. The areas under the tail of the

curve were determined by extrapolating the exponential portions with the

relaxation distances given above. In all cases this estimated area did

not come to more than one or two percent of the total area.

-17-

We are unable to account in a definite way for the difference
of 2% between the two independently determined values of ε . The
probable error in ε , calculated from the foil activity deviations comes
to only 0.2%. The distance between two foil positions is known very
accurately, but it is not unlikely that the absolute values of r might
be in error by a considerable amount. If the values of r were off by
as much as $\frac{1}{4}$", ε would change only by 0.6%. Possibilities such as
assymmetry of the source, or the existence of blow-holes in the uranium
sphere were eliminated by alternating the taking of the with-sphere and
without-sphere data several times.

We take this opportunity to acknowledge the assistance of
Mr. C. Fiddler, by whom much of the data of these experiments were
taken.

-18-

APPENDIX 2

THE OPTICAL APPROXIMATION

J. Ashkin and B. Feld

The symbols used in this section are the same as those defined
in the body of the report, namely:

σ_f = the average U^{238} fission cross section for all the neutrons

   of the source

σ_f*= the average fission cross section for the neutrons of the

   source above the fission threshold of U^{238}

σ_i = the average cross section for inelastic scattering to below

   the fission threshold

σ_e = the average cross section for spherically symmetrical elastic

   scattering

ν = the number of neutrons per U^{238} fission

   We consider fast neutrons to have a mean free path

$$\lambda = \frac{1}{N\sigma}$$

where N = the number of atoms/cm^3

   σ =the sum of fission and inelastic scattering cross sections[3]

(for neutrons above the fission threshold).

$$\sigma = \sigma_i + \sigma_f*$$

   The"absorption" of fast neutrons is then said to take place according to an exponential law, so that the fraction of fast neutrons
penetrating through a thickness x is $e^{-x/\lambda}$.

_____

   [3]To take into account elastic scattering, we consider that this
effect is the same as a fission process, with one neutron per "fission".
We thus define an "effective fission cross section," $\sigma'_f* = \sigma_f* \sigma_e$;
and an effective number of neutrons per fission $\nu' = \dfrac{\sigma_f* \times \nu + \sigma_e}{\sigma_f}$

-19-

Let f be the fraction of fast neutrons which emerge from the uranium without undergoing an "absorption"; it may be calculated (for a sphere) in the following fashion:

Let f(r) be the f for neutrons created at a distance r from the center; then

$$f(r) = \frac{\lambda}{4r}\left[\left(1+\frac{R+r}{\lambda}\right)e^{-\left(\frac{R-r}{\lambda}\right)}-\left(1+\frac{R-r}{\lambda}\right)e^{-\frac{(R+r)}{\lambda}}-\frac{R^2-r^2}{\lambda^2}\right.$$

$$\left.\left\{-Ei\left(-\frac{R-r}{\lambda}\right)+Ei\left(-\frac{R+r}{\lambda}\right)\right\}\right]$$

where

$$-Ei(-x) = \int_x^\infty \frac{e^{-x}}{x}\,dx.$$

For a source distribution given by $\varphi(r)$, we then calculate:

$$f = \frac{\int_0^R f(r)\,\varphi(r)\,r^2\,dr}{\int_0^R \varphi(r)\,r^2\,dr}$$

Thus, if $\varphi(r) = \frac{\sinh Kr}{r}$

$$f = \frac{(1/2)Q}{e^{KR}(KR-1)+e^{-KR}(KR+1)}$$

where, for $K\lambda > 1$

$$Q = e^{KR}\left[\log(K\lambda+1)-Ei\left\{-(K\lambda+1)\frac{2R}{\lambda}\right\}+Ei\left(\frac{-2R}{\lambda}\right)\right]\left(\frac{1-KR}{K\lambda}\right)$$

$$+e^{-KR}\left[\log(K\lambda-1)+Ei\left\{(K\lambda-1)\frac{2R}{\lambda}\right\}+Ei\left(\frac{-2R}{\lambda}\right)\right]\left(\frac{1+KR}{K\lambda}\right)$$

$$+(1/2)\frac{e^{KR}}{K\lambda+1}\left[K\lambda(K\lambda-1)-2+2KR(K\lambda+1)\right]$$

$$+(1/2)\frac{e^{-KR}}{K\lambda-1}\left[K\lambda(K\lambda+1)-2+2KR(K\lambda-1)\right]$$

$$-(1/2)\frac{e^{-(K\lambda+2)\frac{R}{\lambda}}}{K\lambda+1}\left[K\lambda(K\lambda-1)-2\right]-(1/2)\frac{e^{(K\lambda-2)\frac{R}{\lambda}}}{K\lambda-1}\left[K\lambda(K\lambda+1)-2\right]$$

-20-

and for $K\lambda < 1$

$$Q = e^{KR}\left[\log(1+k\lambda) - Ei\left\{-\left(1+K\lambda\right)\frac{2R}{\lambda}\right\} + Ei\left(\frac{-2R}{\lambda}\right)\right]\left(\frac{1-KR}{K\lambda}\right)$$
$$+ e^{-KR}\left[\log(1-K\lambda) - Ei\left\{-(1-K\lambda)\frac{2R}{\lambda}\right\} + Ei\left(\frac{-2R}{\lambda}\right)\right]\left(\frac{1+KR}{K\lambda}\right)$$
$$+(1/2)\frac{e^{KR}}{K\lambda+1}\left[K\lambda(K\lambda-1) - 2 + 2KR\ (1+K\lambda)\right.$$
$$+(1/2)\frac{e^{-KR}}{1-K\lambda}\left[-K\lambda(1+K\lambda) + 2 + 2KR\ (1-K\lambda)\right]$$
$$-(1/2)\frac{e^{-(2+K\lambda)}\frac{R}{\lambda}}{1+K\lambda}\left[K\lambda(k\lambda-1)-2\right]$$
$$+(1/2)\frac{e^{-(2-K\lambda)\frac{R}{\lambda}}}{1-K\lambda}\left[-K\lambda(K\lambda+1)+2\right]$$

For the source distribution $\phi(r) = 1 + cr^2$

$$f = \frac{\lambda^3}{4R^3}\left\{\left[-1/2 + \frac{R^2}{\lambda^2} + e^{-2R/\lambda}(1/2 + R/\lambda)\right] + c\lambda^2\right.$$

$$\bullet\left[-2 + 3/2\frac{R^2}{\lambda^2} - 4/3\frac{R^3}{\lambda^3} + \frac{R^4}{\lambda^4} + e^{-2R/\lambda}\left(2 + \frac{4R}{\lambda} + \frac{5R^2}{2\lambda^2} + \frac{R^3}{\lambda^3}\right)\right\}\Bigg/1+1/5\ cR^2$$

For small R/λ, this expands to

$$f = \left\{\left(1 - 3/4\frac{R}{\lambda} + 2/5\frac{R^2}{2} - 1/6\frac{R^3}{\lambda^3} + \frac{2}{35}\frac{R^4}{\lambda^4}\right)\right.$$
$$+ \frac{cR^2}{5}\left(1-25/35\frac{R}{\lambda} +8/21\frac{R^2}{\lambda^2} - 1/6\frac{R^3}{\lambda^3} + \frac{34}{9\times63}\frac{R^4}{\lambda^4}\right)\right\}\Bigg/1 + 3/5\ cR^2$$

These formulae are useful in calculating the fast effect in a uranium-carbon lattice.

In the experiments described in this report, the primary neutron source was at the center of the sphere. For this source

$$f = e^{-R/\lambda}$$

-21-

The first generation neutrons have a source distribution

$$\varphi_1(r) = \frac{e^{-r/\lambda}}{r^2}$$

and higher generations had a source distribution more nearly approaching

$$\varphi_2(r) = \frac{1}{r^2} \ .$$

For the distributions, φ_1 and φ_2, f was calculated by performing numerical integrations.[4]

If f is known for all generations, the value of $\mathcal{E}$ may be calculated; the method is simply to determine what gets out of the sphere--both fast and slow--for each generation, and to add these up.

In the case where f is the same for the primary and all generations,

$$\mathcal{E} = \frac{1 - p}{1 - p\nu}$$

where $p = (1 - f) \dfrac{\sigma_f}{\sigma}$

If, on the other hand, the primary source leads to one f and all other generations, to another, f_1, then

$$\mathcal{E} = \frac{1-p + (p - p_1)\nu}{1 - p_1 \nu}$$

And further, if the second generation differs from the first, and to it and all higher generations there corresponds an f_2, then

$$\mathcal{E} = 1 - p + p\nu \frac{\left[1 - p_1 + (p_1 - p_2)\nu\right]}{1 - p_2 \nu}$$

This method may, of course, be carried as far as one pleases. However, for the interpretation of our experiments, we believe it sufficient to consdier that the second and all further generations had a $\dfrac{1}{r^2}$ source distribution.

---

[4]In the actual calculations, corrections were applied for the fact that the sphere had a hole in the middle to accommodate the source.

-22-

The computations for Table I were made using this formula for $\mathcal{E}$. The formulae given above assume ν to be the same for all generations. If, as is the case with the Ra-Be source, ν(Ra-Be) for the primary neutrons, differs from ν for the secondary neutrons, the last formulae becomes

$$\mathcal{E} = 1-p+ \ p \, \nu(\text{Ra-Be}) \ \frac{[1 - p_1 + (p_1 - p_2)\nu]}{1 - p_2 \nu}$$

differing only in that $p \, \nu$ becomes $p \, \nu$ (Ra-Be).

This formula was used to evaluate ν (Ra-Be) from the experimental value of $\mathcal{E}$.

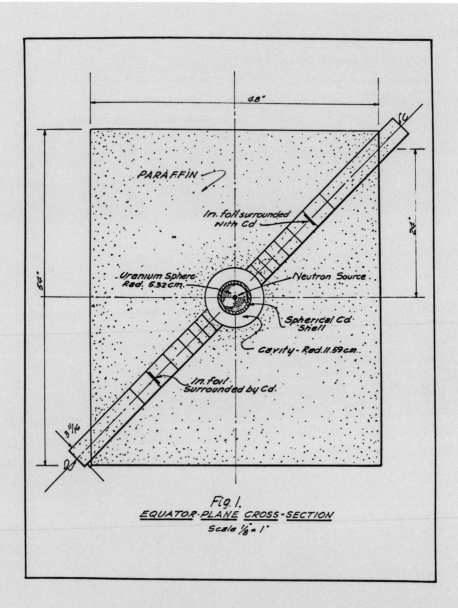

Fig. 1.
EQUATOR-PLANE CROSS-SECTION
Scale 1/8" = 1"

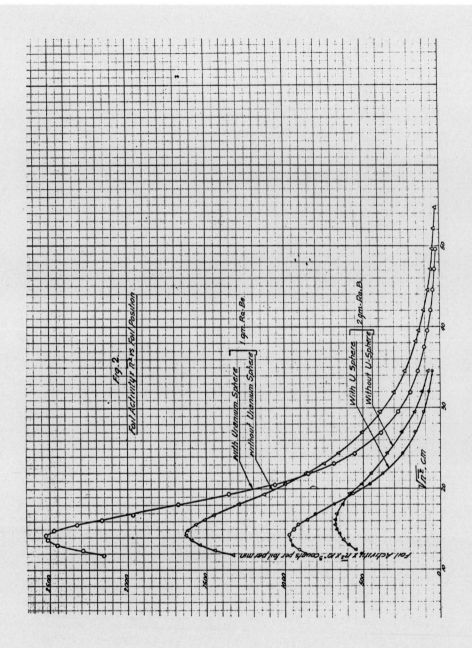

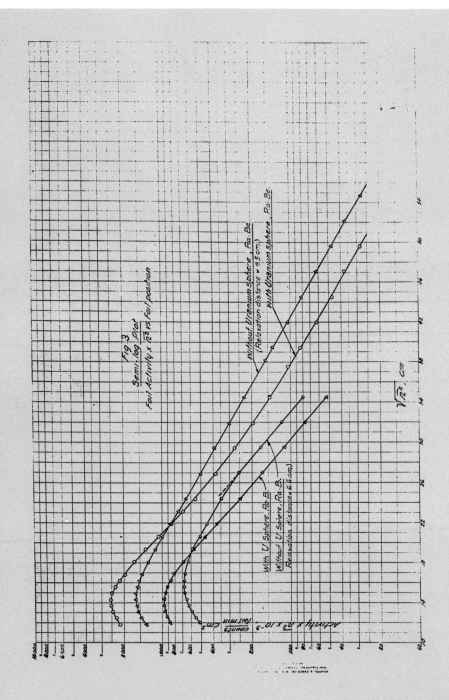

Fig 3
Semi log plot
Foil Activity x r̄² vs Foil position

MDDC - 1292

UNITED STATES ATOMIC ENERGY COMMISSION

INELASTIC SCATTERING OF FAST NEUTRONS

by

S. Bernstein
B.T. Feld
L. Szilard

University of Chicago

This document consists of 1 page.
Date of Manuscript: Unknown
Date Declassified: September 10, 1947

Technical Information Division, Oak Ridge Directed Operations

Oak Ridge, Tennessee

INELASTIC SCATTERING OF FAST NEUTRONS

By S. Bernstein, B.T. Feld and L. Szilard

When a monoenergetic beam of fast neutrons is transmitted through a "thin" scatterer, the scattered neutrons will, in general, have a line spectrum of energies less than and including the original neutron energy. From a complete knowledge of the energies and intensities of the lines in the spectrum of scattered neutrons as a function of the energy of the impinging neutrons, one obtains: (1) the energies of the low-lying energy levels of the target nucleus, and (2) the probability of decay of the compound nucleus into any one of the low-lying levels of the target nucleus. In practice, however, such a complete picture has never been achieved, since neutron spectroscopy in the high energy region is still a very inexact art.

In 1943-44 the authors investigated the trend of inelastic scattering of fast neutrons for a number of elements. As sources, we used compressed Ra-Be and RaB mixtures. It has been shown (and is to be expected) that the former source contains many more energetic neutrons than the latter. The detector was a spherical ionization chamber coated with uranium oxide and surrounded by cadmium, so as to record fissions of U^{238} due to neutrons above the threshold for this reaction (about 1 Mev). The procedure was to measure the fission rate in the chamber first with the source at a given distance from the chamber and, second, with the source at the center of a sphere of the material under investigation and at the same distance from the chamber. Corrections were applied for (1) energy loss of neutrons near the threshold due to elastic scattering, (2) the increase of solid angle subtended by the source for neutrons which suffered elastic collisions in the scattering sphere, and (3) the increased path length in the scattering sphere due to elastic scattering. To obtain a check on this correction, the experiment was performed with a least two spheres of different diameter. However, in each case we tried to keep the sphere radius small compared to the neutron mean free path.

The resulting cross sections are given in the following tabulation.

Cross section for inelastic scattering of Ra-Be or Ra-B neutrons to below the U^{238} fission threshold.

Element	Atomic number	σ (barns) Ra-Be	Ra-B
C	6	0	0
Al	13	0.4	0.4
Fe	26	0.9	0.8
Pb	82	1.2	0.9
Bi	83	1.5	1.1

---

[1]Feld, B.T., R. Scalettar, and L. Szilard, Phys. Rev. 71: 464 (A11) (1947)

MDDC - 1292

MDDC – 1536

MD... – 1536

T.. .TATES ATOMIC ENERGY COMMISSIO..

INELASTIC SCATTERING OF FE, PB, AND BI

by

L. Szilard
S. Bernstein
B. Feld
J. Ashkin

C... ...ational Laboratory

T... document consists of 11 pages
Date of Manuscript: August 27, 1947
Date Declassified: December 1, 1947

Reprinted from The Physical Review, Vol. 73, No. 11, 1307–1310, June 1, 1948
Printed in U. S. A.

Inelastic Scattering of Fast Neutrons by Fe, Pb, and Bi

L. Szilard, S. Bernstein, B. Feld, and J. Ashkin[*]
Metallurgical Laboratory, University of Chicago, Chicago, Illinois
(Received February 27, 1948)

Measurements of inelastic scattering effects of $Ra-\alpha-Be$ and $Ra-\alpha-B$ neutrons in Fe, Pb, and Bi have been made. A U^{238} fission threshold detector was used. The method consisted in measuring the fission counting rate in the detector with and without the spherical scatterer surrounding the source. From the decrease in the counting rate caused by the presence of the scatterer, values of the cross section for inelastic scattering to below the U^{238} fission threshold were calculated for several assumed values of the elastic scattering cross section.

I. INTRODUCTION

THIS report describes some experiments performed by the authors in the early part of 1943 at the Metallurgical Laboratory of the University of Chicago under the auspices of the Manhattan District. The purpose of the experiments was to measure the cross sections of Fe, Pb, and Bi for inelastic scattering of fast neutrons. As sources of fast neutrons $Ra-\alpha-B$ and $Ra-\alpha-Be$ mixtures[1] were used. For detector, the fast fission of U^{238} with a threshold of about one Mev was employed. Because fast fission of U^{238} was used as detector, the experiment dealt only with that part of the neutron source spectrum which is above the threshold energy. More exactly, the inelastic scattering cross section measured is an average for all the energies in the neutron spectrum of the cross section for scattering of these neutrons to below the threshold.

II. EXPERIMENT

The experimental procedure is patterned after that used by Szilard and Zinn.[2] Figure 1 is a schematic diagram showing the experimental arrangement. A source of fast neutrons was placed at a given distance from a spherical uranium-coated ionization chamber. The chamber was connected in the conventional manner to a linear amplifier and scaling circuit for the purpose of counting fission pulses. The chamber

* Present addresses of the authors: L. Szilard, Institute for Radiobiology, University of Chicago, Chicago, Illinois; S. Bernstein, Clinton National Laboratories, Oak Ridge, Tennessee; B. Feld, Massachusetts Institute of Technology, Cambridge, Massachusetts; J. Ashkin, University of Rochester, Rochester, New York.
¹ H. L. Anderson and B. T. Feld, Rev. Sci. Inst. 18, 186 (1947).
² Private communication.

was surrounded by Cd so that neutrons which may have been slowed down to thermal energies by collisions with the walls of the room would be prevented from causing fissions in the U^{235}. There were, possibly, some fissions in U^{235} resulting from epi-cadmium neutrons, but the number of these relative to the number of fast fissions in U^{238} must be small, since the relative amount of U^{235} in normal uranium is small, and the source emits no slow neutrons. The fission counting rate is then proportional to the fast neutron flux through the chamber. (A fast neutron in this report is defined to be one whose energy is above the U^{238} fission threshold.) The source is then surrounded by a sphere of scattering material and the measurement of the fission counting rate repeated. The ratio of counting rates, with to without scatterer, is thus determined. From this ratio the inelastic scattering cross section may be derived. Table I gives the measured values of this ratio, called F, for the various scatterers used.

III. INTERPRETATION

The ratio of fast fission counting rates, F, is equal to the ratio of fast fluxes through the fission chamber. From F it is possible to obtain the ratio, F', of the number of fast neutrons emerging from the scattering sphere to the number of fast neutrons emitted by the source. Since the presence of the sphere (and elastic scattering in the sphere) gives the emerging neutrons an angular distribution which no longer corresponds to a point source, F' and F are not precisely equal. An estimate of the upper limit of the difference between F and F' indicated that, for the small spheres and large source to

1308   SZILARD, BERNSTEIN, FELD, AND ASHKIN

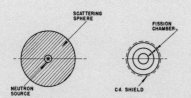

Fig. 1. Schematic diagram of experimental arrangement.

detector distances used in the experiments, the difference was always less than 1 percent. This correction has been applied in the interpretation of the data.

The simplest interpretation of the experiments assumes that the only possible process is one of inelastic scattering. For fast neutrons the capture cross section can be considered negligible in comparison. Thus, if λi is the mean free path for inelastic scattering in the sphere, the number of fast neutrons emerging is

$$F = e^{-t/\lambda i}, \tag{1}$$

where t is the thickness of the sphere (the radius minus the small amount cut out to accommodate the source).

This interpretation is satisfactory if: (1) the thickness of the sphere is small compared to the mean free path for total (elastic plus inelastic) scattering; (2) the elastic scattering cross section is small compared to the inelastic scattering cross section. The values for inelastic scattering cross section found in this approximation are given in column 6 of Table I.

Elastic scattering will cause a decrease in the number of fast neutrons emerging from the scattering sphere. For elastic scattering increases the path of the fast neutrons in the sphere and, thus, gives them a greater opportunity to suffer inelastic collisions. The larger the sphere, the more pronounced will be the effect of elastic scattering on the number of emergent fast neutrons. To calculate this effect we classify the neutrons as primary, first, second, etc., generation fast neutrons. A primary fast neutron is one which emerges from the sphere without having suffered any collisions; a first generation fast neutron has suffered one elastic but no inelastic collision, etc.

For one fast neutron starting out from the source, the number of primary fast neutrons emerging from the sphere is

$$f_0 = e^{-t/\lambda}, \tag{2}$$

where t is the thickness of the sphere, λ is now the mean free path corresponding to both elastic and inelastic scattering.

The primary neutrons act as a source of first generation fast neutrons. The number of these produced in the sphere is $(1-f_0)(\sigma_{\text{elastic}}/\sigma_{\text{total}})$. The density of first generations neutron is proportional to

$$1/r^2 e^{-r/\lambda},$$

where r is the distance from the center of the source in the sphere. The fraction of these emerging as fast neutrons from the sphere, f_1, can no longer be calculated as simply as though the source were at the center; however, it can be computed in a straightforward fashion by performing numerical integrations. The number of first generation neutrons emerging per primary fast neutron is then

$$F_1 = (1-f_0)\frac{\sigma_{\text{el}}}{\sigma_{\text{total}}} f_1 \equiv p_0 f_1. \tag{3}$$

The number of second generation neutrons produced is $p_0(1-f_1)(\sigma_{\text{el}}/\sigma_{\text{total}}) \equiv p_0 p_1$. Of these, the fraction f_2 emerges fast from the sphere. f_2 may be calculated if we know the source distribution of second generation neutrons in this sphere. For the second and all further generations, we have assumed a neutron density proportional to $1/r^2$. Using this calculated $f_2 = f_3 = \cdots = f_n$, the number of second generation neutrons emerging from the sphere, F_2, is $p_0 p_1 f_2$ and the number of third generation neutrons produced is $p_0 p_1 p_2$. The number of third generation neutrons emerging from the sphere is $p_0 p_1 p_2 f_2 \equiv F_3$.

Calculating in this way the number of fast neutrons escaping for each generation, we have for the nth generation

$$F_n = p_0 p_1 p_2{}^{n-2} f_2. \tag{4}$$

The total number of fast neutrons escaping from

INELASTIC SCATTERING OF FAST NEUTRONS 1309

TABLE I. Pairs of σ_i and σ_{el} to give measured ratios, F.

Source	Scatterer	Radius of scatterer, cm	Thickness of scatterer, cm	F	$\sigma_{el}=0$	σ_i in Barns 1	2	3	4
RaB	Bi	8.29	7.10	0.79±0.02	1.18	1.12	1.08	1.05	0.97
		12.02	10.86	0.69±0.02	1.21	1.14	1.05	0.97	0.90
	Pb	8.21	7.22	0.80±0.02	0.94	0.88	0.83	0.77	0.72
		10.72	9.55	0.71±0.02	1.12	1.04	1.00	0.88	0.80
	Fe	5.70	4.60	0.73±0.02	0.88	0.78	0.69	0.60	0.49
		8.33	7.14	0.62±0.02	0.82	0.71	0.60	0.47	0.35
Ra-Be	Bi	8.29	7.30	0.73±0.02	1.56	1.48	1.38	1.29	1.20
		12.02	11.10	0.64±0.02	1.44	1.35	1.24	1.14	1.06
	Pb	8.21	7.47	0.74±0.02	1.26	1.19	1.11	1.04	0.95
		10.72	9.79	0.64±0.02	1.42	1.34	1.23	1.12	1.00
	Fe	5.70	4.80	0.68±0.02	1.05	0.92	0.80	0.71	———
		8.33	7.38	0.57±0.02	1.00	0.83	0.72	0.62	———

the sphere is, thus,

$$F=\sum_{n=0}^{\infty} F_n = f_0 + f_1 p_0 + f_2 p_0 p_1$$
$$+ f_2 p_0 p_1 p_2 + \cdots + f_2 p_0 p_1 p_2^{n-2}, \quad (5)$$

$$F = f_0 \left\{ 1 + p_0 \left[\frac{f_1}{f_0} + \frac{p_1}{1-p_2} \frac{f_2}{f_0} \right] \right\}. \quad (6)$$

The number which have suffered an elastic collision before escaping is obtained by summing all generations except the primary.

$$\sum_{n=1}^{\infty} F_n = p_0 \left[f_1 + \frac{p_1}{1-p_2} f_2 \right]. \quad (7)$$

It is to these neutrons that the correction for angular distribution must be applied. A series of calculations has been carried out using the above formulae and different values of σ_{el}. From these, the values of σ_i, the inelastic scattering cross

section, best fitting the experimental ratio have been found for a number of different values of σ_{el}. These are given in Table I.

IV. CONCLUSIONS

If the only effects taking place in the scattering spheres were those discussed above, and if the above outlined method of calculation of these effects were adequate, then there should be one combination of inelastic and elastic scattering cross sections leading to the measured F for the two different thicknesses of scatterer used for each type of material. A glance at the table serves to show that this is not the case.

One possible reason for the failure of the simple interpretation outlined above is that there is another effect which tends to decrease the inelastic scattering cross section (as measured in this experiment) as the size of the scattering sphere is increased. This effect is in the nature of

TABLE II. Inelastic scattering cross sections (barns).

Neutron source	Detector	Lead	Bismuth	Iron	Reference
Ra-Be	$Fe^{56}(n,p)Mn^{56}$	2.13±0.24	2.32±0.20		(3)
	$Al^{27}(n,p)Mg^{27}$	1.97±0.16	1.98±0.09		(3)
	$Cu(n,n\gamma)$	0.91±0.08	0.63±0.09	0.57±0.05	(3)
	$Pb(n,n\gamma)$	0.73±0.16		0.62±0.07	(3)
Rn-Be	$Si(n,p)$	1.4	1.3		(4)
	$Al(n,p)$	1.8		1.1	(4)
	U^{238} fission	1.34	1.85		(5)
d-d	U^{238} fission	0.55	0.64		(6)

[3] D. C. Graham and G. T. Seaborg, Phys. Rev. **53**, 795 (1938).
[4] C. H. Collie and J. H. E. Griffiths, Proc. Roy. Soc. **A155**, 434 (1936).
[5] J. Marshall and L. Szilard, priv. comm.
[6] W. H. Zinn and L. Szilard, priv. comm.

1310　　　SZILARD, BERNSTEIN, FELD, AND ASHKIN

a "hardening" effect, arising out of the fact that the "cross section for inelastic scattering to below the U^{238} fission threshold" decreases with increasing neutron energy, so that the neutrons with energy close to the threshold are more rapidly removed from the fast neutron beam than those neutrons of energy far above the threshold. The sources emit a spectrum of energies going up (in the case of Ra$-\alpha-$Be) to about 12 Mev.

It should be stressed that what we call, for the purposes of this report, inelastic scattering is really only a special type of inelastic scattering wherein the original fast neutron loses enough energy to drop below the threshold of our detector. Strictly speaking, inelastic scattering is a process in which the scattering nucleus is left in any one of the (energetically) possible excited states; elastic scattering (except for "shadow" scattering) is a special case where the nucleus ends up in its ground state. Of all the possible states in which inelastic scattering can leave a nucleus, our method of measurement picks out only those scatterings in which the neutron loses enough energy to drop below the threshold. For neutrons of energy very much greater than 1 Mev, most inelastic scatterings will not be detected. Our experiment is most sensitive for neutrons close to the threshold. On the other hand, since more levels become available with increasing neutron energy, the true inelastic scattering cross section increases with energy until it reaches the value πR^2 (the cross section for formation of the compound nucleus).

Little is known about the energy spectra emitted by the sources used in this experiment beyond the fact that both emit a rather broad distribution of energies, with both the average and the maximum energy being greater for the Ra$-$Be source.[7] Because we do not know the energy distributions, nor the details of the level schemes of the nuclei involved, none of the effects mentioned above could be calculated or even estimated for our experiment. Hence the uncertainty in the results, as indicated by the values in the table. For all three elements, however, the inelastic scattering is greater for Ra$-$Be than for Ra$-$B neutrons. For the three elements investigated the inelastic scattering increases with atomic weight.

What is really needed for a determination of details of the inelastic scattering process is a series of experiments involving the scattering of monoenergetic neutrons and measurements of the energy distribution of the scattered neutrons. This has been done in one instance.[8] In most cases other observers have also used non-monochromatic neutron sources and threshold detectors. Their results are equally difficult to interpret rigorously. A compilation of values of other observers is given in Table II.[9]

[7] B. T. Feld, R. Scalettar, and L. Szilard, Phys. Rev. **71**, 464 (1947).

[8] H. F. Dunlap and R. N. Little, Phys. Rev. **60**, 693 (1941).

[9] Note added in proof: See also the recent work of H. H. Barschall, J. H. Manley, and V. F. Weisskopf, Phys. Rev. **72**, 875 (1947), and H. H. Barschall, M. E. Battat, W. C. Bright, E. R. Graves, T. Jorgensen, and J. H. Manley, Phys. Rev. **72**, 881 (1947).

[By Bernard T. Feld, R. Scalettar and L. Szilard.
Report MDDC-897 (date unknown).
Published after declassification
in *Phys. Rev.*, **71**, 464 (1947).]

A11. Use of Threshold Detectors for Fast Neutron Studies. Bernard T. Feld,[1] R. Scalettar,[2] and L. Szilard.[3] *Metallurgical Laboratory, University of Chicago, Chicago, Illinois.*—Light element (n, p) and (n, α) reactions can be used to compare, qualitatively, the energy distributions of the fast neutrons emitted by various neutron sources. Experiments are described in which a Ra-α-Be and Ra-α-B neutron source are compared by observing their relative activation of a number of light element reactions. Table I summarizes the experimental results. The thresh-

TABLE I.

Reaction	Half-life	Calculated threshold (Mev)	Calculated energy for 0.1 penetration (Mev)	Calculated energy for 0.5 penetration (Mev)	Activity $S^{32}(n, p)P^{32}$ Ra-α-Be	Activity $S^{32}(n, p)P^{32}$ Ra-α-B
$S^{32}(n, p)P^{32}$	14.3d	.93	2.9	4.0	1.000	1.000
$P^{31}(n, p)Si^{31}$	170m	1.02	2.9	3.9	.63	.51
$Al^{27}(n, p)Mg^{27}$	10.2m	1.95	3.6	4.6	.29	.058
$Si^{28}(n, p)Al^{28}$	2.4m	2.69	4.5	5.5	.95	≤.017
$P^{31}(n, \alpha)Al^{28}$	2.4m	.90	6.6	8.3	.24	<.004
$Al^{27}(n, \alpha)Na^{24}$	14.8h	2.39	7.5	9.1	.079	<.003

olds for the reactions were calculated by using the best available values of the isotopic masses. The probability for penetration of the Coulomb barrier by the product proton or alpha-particle was calculated from the Gamow formula. The relative values of the activities in the same column have no meaning, since the detectors were not intercalibrated.

[1] Now at the Massachusetts Institute of Technology, Cambridge, Massachusetts.
[2] Now at Cornell University, Ithaca, New York.
[3] Now at University of Chicago, Chicago, Illinois.

Metallurgy and Engineering

Report A-24

PRELIMINARY REPORT ON THE MELTING OF URANIUM POWDER

by

L. Szilard   August 16, 1941

Summary: Methods have been found for protecting uranium metal powder from surface oxidation, and this resulted in the production of ingots which have a density of 18 gm./cc. and melt at about 1300 degrees C.

Uranium metal has been produced in powdered form for a number of years by Metal Hydrides Inc. using a calcium hydride reduction process which had been developed by Dr. P. P. Alexander.

Dr. Alexander and others have, in the past, attempted repeatedly to fuse such uranium powder into a solid ingot. These attempts were partially successful inasmuch as occasionally a considerable fraction of the powder was molten but the results were not reproducible and a smaller or larger fraction of the powder remained unmolten. Dr. Alexander's own experiments appeared however particularly promising inasmuch as they seemed to indicate that the true melting point of uranium might be below 1300 degrees C.

It was generally thought that surface oxidation of the powder might be the sole cause preventing fusion. Such surface oxidation appeared, however, almost unavoidable since the uranium metal powder has to be leeched from a uranium calcium oxide mixture by a faintly acid solution.

Surface oxidation was also thought to be responsible for pyroforic phenomena exhibited by certain batches of uranium powder which were produced for us by Metal Hydrides.

Some batches were pyroforic from the start; other batches which were not pyroforic when received became pyroforic after a few weeks of storage. One shipment consigned to the Columbia University went up in flames in transit. Another shipment of uranium metal powder which consisted of a number of sealed cans containing the powder did not appear to be pyroforic. The first cans which we opened appeared to be all right. A few weeks later, however, when we began to open the rest of the cans, we found that the content of the can burst into flames merely upon removing the lid of the can. In the circumstances it did not seem possible to work with this metal powder and believing that surface oxidation was the cause of these phenomena it appeared imperative to devise some methods by which surface oxidation could be controlled.

Upon making inquiries I learned that it is customary to protect tungsten powder from surface oxidation by coating it with 2% of paraffine. This is done by wetting the tungsten powder with carbon tetrachloride in which some paraffine had been dissolved by allowing the carbon tetrachloride to evaporate. It seemed well worth trying to apply this method to our uranium powder. Since the presence of hydrogen would interfere with our experiments I proposed to Dr. Alexander to have uranium powder coated with paraffine in the above-described

-2-

manner, and to free it from hydrogen by pressing it into a dye and heating it
up in a vacuum furnace in order to evaporate the paraffine and sinter the powder.

A conference was arranged with Dr. Alexander and Mr. Davis for July 28,
in order to discuss this and other methods for preventing surface oxidation. At
this conference we reviewed all details involved in the production of uranium
metal powder and I thus learned that after leeching, the uranium powder was
placed into flat trays and allowed to dry in air for a few days before being
packed and shipped. This way of drying appeared to me a possible source of
trouble and I therefore proposed to Dr. Alexander and Mr. Davis that we dry the
uranium after leeching either by washing with alcohol and ether, or alternatively,
by placing the uranium powder immediately after leeching into a vacuum tank
where the water could be evaporated quickly at room temperature.

A few days later Dr. Alexander and Mr. Davis reported to me the following: They had leeched a sample of uranium metal powder and dried it in the
above-described manner with alcohol and ether and coated in the above-described
manner with paraffine. They pressed this powder in a dye at about 10 tons per
sq. in. and placed a sample in an unglazed silimanite boat and finally heated it
up in a vacuum furnace evaporating the paraffine at about 1300 degrees C. After
allowing the furnace to cool and removing the sample they found a fused uranium
ingot having a density of over 18. They further reported that the fused uranium did not appear to attack the silimanite boat at this temperature.

The determination of the hydrogen content of such ingots, which were
carried out at our request by Dr. J. C. Rodden, and Dr. Holton at the National
Bureau of Standards, showed that these ingots were free from hydrogen. It appears thus to be established that the melting point of pure uranium is at about
1300 degrees C.

In close collaboration with Dr. Alexander a number of other ingots were
prepared at Metal Hydrides under varying conditions. The purpose of these experiments is to find a method to prevent surface oxidation which is easily applied on an industrial scale. The first ingots which were successfully obtained
were made from uranium powder which was leeched in a beaker and the leeching
solution had a higher acidity than could be used in the leeching tanks which
have to be used in a large scale production process. These experiments are not
concluded but the indication is that it will be possible to use the routine procedure of leeching and that it might be unnecessary to wash with alcohol and
ether provided the freshly leeched uranium powder is at once dried at room
temperature in a vacuum tank and upon being removed from the vacuum tank is at
once paraffine-coated or otherwise protected from being attacked by oxygen or
moisture.

In the circumstances there seems to be a good chance that we shall be
able to have fusable uranium powder produced on an industrial scale possibly
without any considerable increase in cost. It also appears probably that we
shall be able to produce on an industrial scale fused uranium bodies of the
size and shape which we require without having to resort to a casting operation.
It is hoped that one crucible can be used a number of times for fusing uranium
powder into the desired shape either half spheres or cylindrical blocks. Experiments are underway to determine whether it is permissible to fuse in hydrogen. Fusion in helium is certainly permissible.

-3-

The use of an induction type furnace does not appear to be necessary and we are at present engaged in designing furnaces which appear to be more practical and which could take care of producing ingots at the rate of 700 pounds per week.

The progress made is essentially due to Dr. P. P. Alexander's unusual willingness to disclose all details of his manufacturing process. The present development has to be attributed to the close collaboration which was made possible by the confidence which Dr. Alexander showed us by giving us access to his factory and to all relevant information.

This work was greatly accelerated by the help extended to us by the National Bureau of Standards and by the speedy and accurate execution of the investigations made there under the direction of Dr. C. J. Rodden.

Professor D. P. Mitchell's kind cooperation greatly facilitated our task of starting the production of the uranium metal powder at Metal Hydrides on an intermediate scale.

MEMORANDUM

On the Cooling of the Power Plant

L. Szilard
June 15, 1942

The purpose of this memorandum is to illustrate by means of
the attached sketches certain general principles which ought to be
applied in designing a cooling system for a chain reacting uranium-
graphite power unit. These principles should universally apply to such
systems in which the cooling medium is in direct heat contact with the
uranium.

Cooling Media.--Liquid bismuth and helium were the first cooling media to be considered. The latter as well as hydrogen was first
suggested by Fermi. These two media appear to be the safest from the
point of view of chemical inertness at high temperature. In the case
of helium, direct contact between the helium and uranium metal may be
permitted. In the case of bismuth, a direct contact between liquid
bismuth and uranium oxide or uranium carbide may be permitted, but a
direct contact with uranium metal will probably lead to chemical interaction unless the surface of the metal was protected by an inert
surface layer.

Other gases which might be considered as cooling agents are
hydrogen and air. Hydrogen might be suitable for large power units
in dissipating heat of the order of magnitude of 10^6 KW, but the reactions between hydrogen and uranium metal and hydrogen and carbon
might limit the temperature range and offset such advantage as hydrogen
would otherwise have over helium. Air as a cooling medium is at

-2- SECRET

present being considered only as an expedient in the present emergency.
We could have a 30,000 KW unit cooled with air, probably in operation
within much shorter time than a helium cooled unit of the same capacity.
The only advantage of air over helium lies in the fact that the air
need not be passed through a heat exchanger but can be blown out of
the plant. The danger of poisoning by radioactive material carried
by this air current might however prevent this mode of operation.
This question is at present being studied by E. P. Wigner, and action
along this line ought to be suspended pending the outcome of his investigation.

General Requirements.--The following mode of operation would be
desirable: A power plant producing about 10^6 KW should be allowed to
run for a long period of time, perhaps 6 months or longer. Then the
chain reaction would be shut off and the cooling would be kept on
for another period of time, perhaps one month. After this period the
uranium may be dissolved in situ by circulating a solvent through the
power unit. If necessary the uranium metal could be burned to uranium
oxide in situ before it is dissolved. The solution would be pumped into
a tank, and if possible the uranium would be precipitated together with
element 94 as peroxide. The remaining solution which would contain
most of the fission products could be pumped away. The uranium oxide
could be redissolved and reprecipitated several times until it is
sufficiently inactive to permit a suitable chemical separation of
element 94 from the uranium. In order to prevent the fission products
from precipitating in quantity together with the uranium peroxide, it
may be necessary to add fairly large quantities of the stable isotopes
of the fission products to the solution each time before the uranium is

RECEIVED
JUN 17 1942
A. H. COMPTON

-3-

SECRET

precipitated as peroxide.

It should be noted, however, that the principles illustrated in the attached sketches are also consistent with a different mode of operation, i.e., the removal of the uranium from the power unit at frequent intervals for the purpose of dissolving the uranium outside the power plant.

Orders of Magnitude.--While it is desirable to have, ultimately, power units which dissipate 10^6 KW, power units dissipating up to 300,000 KW are at present considered as the immediate objective. A structure containing 40 tons of uranium metal in a graphite pile of 6 m x 6 m x 6 m which may have a cubic, cylindrical, or spherical shape, might be necessary to make the power unit operative. The cubic shape of the pile being the least favorable, will in the following be assumed in order to arrive at conservative estimates. The cooling ducts in the pile will be assumed to take up about 10 percent of the cross section of the pile, that is, about 3.6 m^2 out of 36 m^2.

The uranium metal may be present in the form of sticks about 3 mm. in diameter and 5 to 8 cm. long.

It is estimated that such a structure if operated at a friction loss of the cooling medium in the power unit of 3-4000 KW is capable of dissipating the following quantities of power:

1. At 10 atmospheres of helium, 300,000 KW; input 200° C, output 500° C.

2. At 7 atmospheres of helium, 250,000 KW; input 200° C, output 500° C.

3. At one atmosphere of helium, 100,000 KW; input 200° C, output 500° C.

4. At one atmosphere of air, 30,000 KW; input 25° C, output 150° C.

RECEIVED
JUN 17 1942
OFFICE OF
A. H. COMPTON

-4- **SECRET**

A 30,000 KW helium unit could be operated at atmospheric
pressure with a fraction loss of 3-400 KW in the power unit.

Form of Uranium Plugs.--Figure 1 shows various forms in which
the uranium metal could be used. Some of these forms are also suitable
for fused uranium oxide or fused uranium carbide.

Figures A, B and C show forms which may be called clusters, whereas Figure D shows a short cylindrical block of uranium with a number of
holes going axially through the cylinder. Such a uranium block could
be cast by using beryllium oxide sticks as cores in the casting process.

Figure 1A is a cluster which has the contour of a cube and
which consists of square bars of uranium.

Figure 1B is a cluster which has the contour of a short cylinder
of about equal diameter and length. This configuration is easily constructed by bundling together a number of thin sticks of uranium. Sticks
of about 3 mm. diameter and 5-8 cm. length, are most likely to prove
suitable.

Figure 1C shows a cluster which has the contour of a short
cylinder of about equal length and diameter. This cluster may be formed
of balls of uranium oxide or uranium carbide. More or less regularly
shaped granules can take the place of such balls. In the case of gas
cooling the arrangement shown in Figure 1C is less favorable than the
arrangement shown in Figure 1B and will be considered only if fused
uranium dioxide or fused uranium carbide cannot be obtained in some
such shapes as shown in Figures 1A, B, and C. The arrangement of Figure
1C would have no important disadvantage in the case of bismuth cooling
from the point of view of friction loss, but it would have disadvantage
from the point of view of the multiplication factor.

Double Stream Method.--Figure 3 and 3A show the cooling of the
power unit by a double stream method. The gas enters the graphite

pile on one side, for instance at the top, and the cold gas flows through
ducts designated by 2 into the graphite pile. From duct 2 the gas
passes through one or more uranium plugs designated by 4 and the hot
gas enters duct 3, which leaves the graphite pile at the opposite
side of the pile, for instance, at the bottom. The uranium plugs are
located in the graphite column 6 which, if required, can be removed
as a whole.

The modification shown in Figure 3A has a greater flexibility
than the modification shown in Figure 3. This flexibility is needed to
meet the particular requirements of a chain reacting unit. As Wigner
repeatedly pointed out, it is necessary to have a cooling system which
is adapted to deal with the specific power production three to four
times (according to whether the power unit has a spherical or cubic
shape) as large in the center of the power unit as the average value
within the power unit. The arrangement shown in Figure 3A makes it
possible to have throughout the whole power unit the same temperature
increase of the gas when it passes from duct 2 to duct 3. In order
to achieve this one has to vary the number of uranium plugs which are
connected in series, according to the position which the graphite
column 6 occupies in the pile and also according to the position of
the particular uranium plugs within the particular graphite column.

In the center of the pile the gas may flow from duct 2 to duct 3
by passing through one single uranium plug and the temperature of the
gas may rise during this passage by same value, for instance, 300° C.
In another region of the pile where the neutron density is smaller by
a factor $1/k$ than in the center we may obtain the same rise of temperature of the gas which passes from duct 2 into duct 3 by passing the
gas through n uranium plugs in series.

-6-

If the pressure difference between ducts 2 and 3 has the same
value for each graphite column 6 and if it does not vary along the
column, the velocity of the gas in the plugs will be proportional
to $\sqrt{n}$ and in order to have the same temperature of the gas at
entrance into duct 3 we have to choose

$$n \approx k^{2/3}$$

L. Szilard

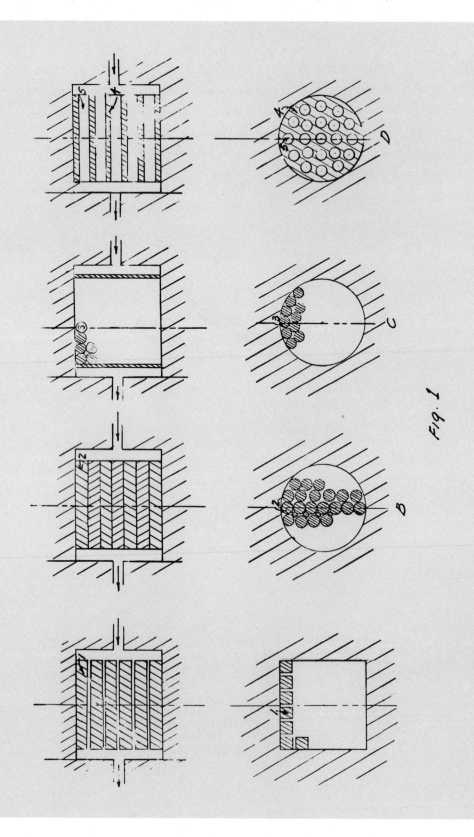

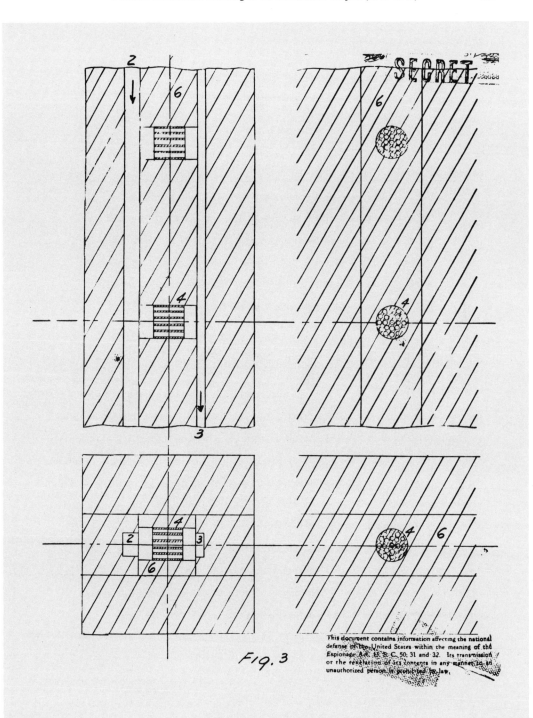

SECRET

Fig. 3

MEMORANDUM

ON THE COOLING OF THE POWER PLANT

Addition to the Memorandum Dated
June 15, 1942

L. Szilard

June 24, 1942

Correction of Typographical Error on Page 6 of Memorandum Dated June 15, 1942.

The last paragraph of the memorandum should read as follows:

"If the pressure difference between ducts 2 and 3 has the same

value for each graphite column 6 and if it does not vary along

the column, the velocity of the gas in the plugs will be pro-

portional to $\sqrt{n}$ and in order to have the same temperature of

the gas at entrance into duct 3 we have to choose

$$n \approx k^{2/3}$$ ".

Continuation of the Discussion of the Double Stream Method.

Figure 3 B shows an arrangement similar in function to the arrange-

ment shown in Figure 3 A, but differing in construction. The construction shown

in Figure 3 B makes it possible to remove the uranium from a graphite column with-

out removing an appreciable amount of graphite. In this figure we see a graphite

column having a square cross section and a length which is large compared to the

sides of the square, and which may extend if desired throughout the whole graphite

pile. This graphite column has three bores, one of them representing duct 2,

another of them representing duct 3, and the third bore, in which are placed the

uranium plugs 4. The possible constructions for the uranium plugs are shown in

Figure 1. Figure 1 B might be the most suitable if uranium metal is used.

Perforated graphite plates 9 keep the parts which constitute the uranium plug 4

in position. These perforated graphite plates are similar to the perforated

graphite plates shown in Figure 1 C. Between each two uranium plugs 4 we have a

graphite rod 5 and a bore in the center of the graphite rod 5 connects two adjacent

uranium plugs. Whereas, the example shown in Figure 3 A had groups of three

uranium plugs connected in series, and these groups connected in parallel the

example shown in Figure 3 B has all uranium plugs connected in parallel. This,

-2-

of course, will apply only in the center of the pile, and the construction

shown in Figure 3 B permits to have groups of several uranium plugs connected

in series and the groups connected in parallel as it is necessary to have in

other parts of the pile. The cooling agent enters from duct 2 through a bore

8 into the interior of the graphite rod 5, passes through a uranium plug 4 into

the interior of another graphite rod 5, and enters through a bore 7, into duct 3.

Naturally, the cross section of ducts 2 and 3 will vary along the graphite column,

the radius of duct 3 decreasing along the path of the cooling agent while the

radius of duct 2 increases correspondingly. If the graphite column is vertical,

the graphite rods 5 and the uranium plugs 4 will simply fall out of the pile

under the action of gravity if their support is removed.

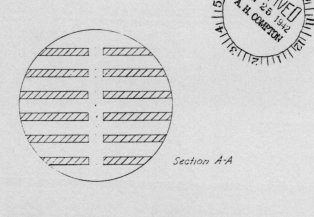

Section A-A

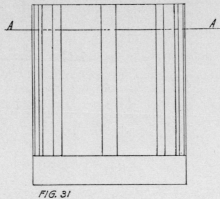

FIG. 31

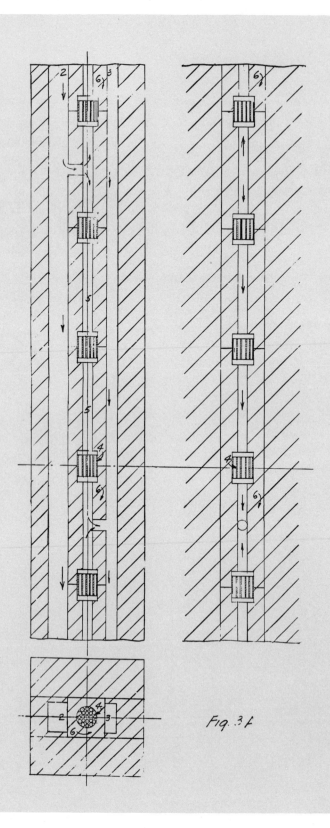

Fig. 3 f

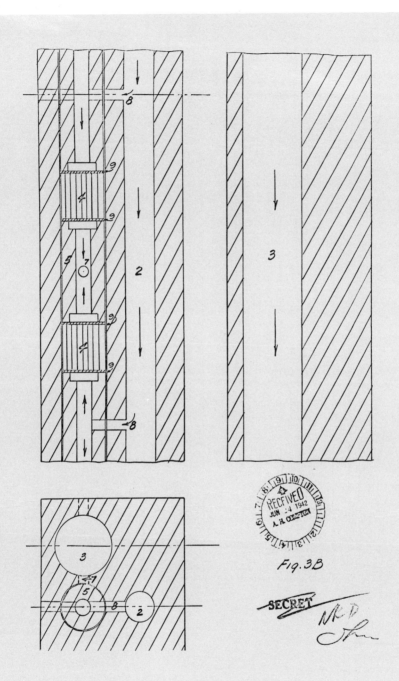

Fig. 3B

SECRET

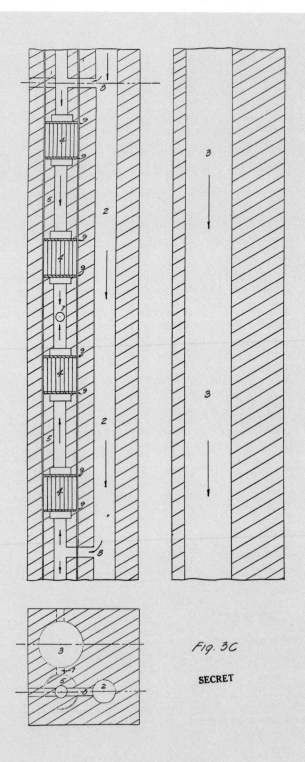

Fig. 3C

SECRET

MEMORANDUM

On The Cooling of the Power Plant

Addition to Memorandum dated

June 15 and 24, 1942

L. Szilard

June 29, 1942

Continuation of the Discussions of the Double Stream Method.

Figure 3 C shows an arrangement which is very similar to the arrangement shown in Figure 3 B. The only difference is that in the arrangement shown in Figure 3 B there are groups of uranium plugs connected in parallel and each group contains two uranium plugs connected in series.

REPORT NO. ⟨ P-302 ⟩

COPY NO. _Z A_

NAME _February_

EXAMPLES FOR PRESSURE DROP CALCULATIONS IN

PARALLEL FLOW HELIUM COOLING*

B.T. Feld & L. Szilard

June 18, 1942

With an Addition dated August 10, 1942

<u>ABSTRACT</u>

Pressure drop calculations are shown for He cooled power plants ranging from 400,000 kw to 30,000 kw.

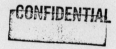

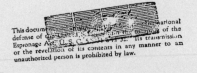

---

* For constructional details of parallel flow power unit see L. Szilard -- Memoranda dated June 15, June 25, June 29.

EXAMPLES FOR PRESSURE DROP CALCULATIONS IN

PARALLEL FLOW HELIUM COOLING

B.T. Feld & L. Szilard

June 18, 1942

With an Addition dated August 10, 1942

In this report we calculate some of the dimensions, and also the work done in circulating gas, for a number of helium cooled power plants ranging in output from 400,000 kw down to 30,000 kw. We use helium at 10 atmospheres pressure and at 1 atmosphere pressure.

I. Friction Work.

We allow the gas to have a temperature rise of $300^{\circ}C$ at the center of the machine. At the outside of the machine where the output is considerably smaller than at the center (for a cube, the thermal neutron density at the center is four times the average thermal neutron density), we keep the pressure drop the same and achieve a constant temperature rise in the gas of $300^{\circ}C$ by putting more than one plug in series; then the ratio of the heat transfer to the friction loss remains constent throughout the machine. And we need calculate it only at the center. We make our estimate a pessimistic one by taking the velocity and density throughout the plug corresponding to the highest temperature of the gas ($500^{\circ}C$). The formula for the ratio of heat transfer to friction work is:

$$Ra = 3.842 \times 10^7 \frac{k}{\mu} \left(\frac{c_p \mu}{k}\right)^{0.4} \frac{\Delta H}{v^2} \frac{T_{hot}}{T_{cold}}$$

which becomes for helium

$$Ra = 6 \times 10^7 \frac{\Delta H}{v^2} \frac{T_{hot}}{T_{cold}}$$

k = heat conductivity
μ = viscosity
c_p = specific heat at constant pressure
ΔH = Temperature difference between gas and metal = $300^{\circ}C$
v = velocity of gas at exit from the plug
T_{hot} = maximum temperature of gas = $500^{\circ}C$
T_{cold} = minimum temperature of gas = $200^{\circ}C$

If we take one-half the critical velocity*, or a velocity of 8.85×10^3 cm/sec, this ratio is equal to

$$6 \times 10^7 \times \frac{300}{v^2} \times \frac{8}{5} = 368$$

---

*The critical velocity is defined as the velocity at which the kinetic energy of the gas is equal to 1 percent of the heat transported by the gas.

-2-

or a friction loss of 0.27% of the heat transported. To this must be added 1/4% for each exit of the hot gas. Hence, if we have two plugs in series, we lose 0.5+0.27 = 0.77% in the plug. To this add the friction loss in the duct and at the hot exit from the duct. If the duct is twice the critical** length, the total friction loss for the duct and exit is 0.75%, giving about 1.5% grand total, or, for 300,000 kw, a loss of 4,500 kw. By reducing the velocity in the duct and at the duct exit by a factor of $\sqrt{3}$ we reduce the duct friction loss to 0.75/3 = 0.25, giving a total loss of about 1%, or for a 300,000 kw machine, a loss of 3,000 kw.

II. Power Output.

We will now calculate the heat transfer, using helium at 10 atmospheres pressure with a velocity equal to 1/2 the critical velocity or 8.85 x 10³ cm/sec.

1). The Uranium lump or plug consists of 0.6 cm diameter cylindrical sticks, 4.5 cm long, packed so as to have 3/4 of the cross section metal and 1/4 empty space. We cool two such plugs in series (at the center of the pile) in order to get a temperature rise of 300°C in the gas. This corresponds to a heat production of 210 calories per double stick; and, since two sticks weigh 50.9 g, to a production of 17.3 watts/g at the center of the pile, or 4.32 watts/g average. For 40 tons of metal this gives a total production of 173,000 kw.

2). We decrease the diameter of the sticks to 0.3 cm, hydraulic radius*** 0.05 cm. This gives h = 4.73 x 10⁻² calories/cm²/°C/sec, and requires us to cool single plugs of length 4 cm at the center of the pile. For this type of plug we have, at the center of the pile a production of 39 watts/g, or for 40 tons, a total production of 390,000 kw. The friction loss in this case is 0.25 + 0.27 = 0.52% per plug plus 0.75% per duct = 1.25% of the heat transport. This gives the frictional work equal to 4870 kw.

3). If we reduce the output to 300,000 kw (by reducing the velocity of the gas), the friction loss drops by the square of the velocity. However, the heat transfer drops more slowly. If we increase the pencil length in the plug slightly, so that Δ T remains 300°C, then, for a 300,000 kw machine the friction work equals 4870/1.69, or 2880 kw. For 1/2 the critical velocity, 300,000 kw output requires about 9 percent duct area. The friction loss in the ducts can be made smaller if we allow the duct area to go up to about 12 percent, corresponding to the new velocity of 8.85 x 10³ x 3/3.9 = 6.8. x 10³ cm/sec.

III. Effect of reducing pressure.

If the ducts are unchanged and we keep the same exit velocity, then, dropping the pressure by a factor of 10, reduces the output by a factor of 10.

** The critical length is that length in which the frictional loss in the tube is equal to the kinetic energy of the gas.

**The hydraulic radius is a measure of the effective radius of the empty space for heat transfer and pressure loss. It is defined as 2 x area/periphery of the space through which the gas flows.

-3-

The friction loss in the ducts is the same fraction of the transported heat as before. We may thus go from a 300,000 kw to a 30,000 kw machine. However, the heat transfer does not decrease with the pressure, but decreases more slowly ($p^{0.8}$) so that for a length of plug equal to 4.7 cm we can take pencils of diameter 0.5 cm and still have the gas temperature rise by 300°C. For this case the friction loss is 0.25+ 0.27 = 0.57% in the plug, and in the duct, for a velocity equal to the critical velocity divided by 2 x $\sqrt{3}$, the friction loss is 0.25%, giving a total friction loss of 0.77%. For a 30,000 kw machine this amounts to 230 kw. If we increase all velocities by a factor of 3, keeping $\triangle T$ constant, the friction loss becomes about 6240 kw, and the heat transferred (if the plugs are slightly changed) goes up by a factor of 3, giving an output of 90,000 kw.

Addition Dated August 10, 1942

Previously I calculated the dimensions of a 300,000 kw machine cooled by helium at 10 atmospheres. In this machine we used, at the center, a plug consisting of pencils 0.3 cm in diameter and 4 cm long. The gas was circulated with a velocity of 68 m/sec, or $\frac{1}{2.6}$ x critical velocity. The duct size on this machine is 12 percent of the total area. We now consider a 100,000 kw machine.

The duct size, if we keep the same velocity, goes from 12 percent to 4 percent, and if we take a 20 cm cell, this gives a cross sectional area of 16 cm² for each duct, and a critical length of 150 cm for this duct size. Thus, the length of our duct is about 4 critical lengths. Hence, the friction loss in the duct, including the loss upon exit, equals $\dfrac{4 + 1}{(2.6)^2}$ percent of the total heat transport--or .74 percent. In the plugs themselves the production of heat is 1/3 as much as previously considered, and so, in order to get a temperature rise of 300°C, we would put 3 plugs in series, at the center of the pile. The friction loss in the plug and on exit, is $\dfrac{.27 + .25}{1.69}$ = 0.31 percent, so that the total loss in this machine is 0.31+ 0.74 = 1.05 percent, or 1,050 kw.

If we reduce the pressure to 7 atmospheres, we would connect two plugs in series at the center of the pile and would have to increase the total duct area to 6 percent. For this size duct the critical length is 200 cm, and so we have a loss of $\dfrac{3 + 1}{6.8}$ = .59 percent in the ducts. The total loss then for this machine = 0.31 + 0.59 = 0.90 percent of the heat transported, or 900 kw.

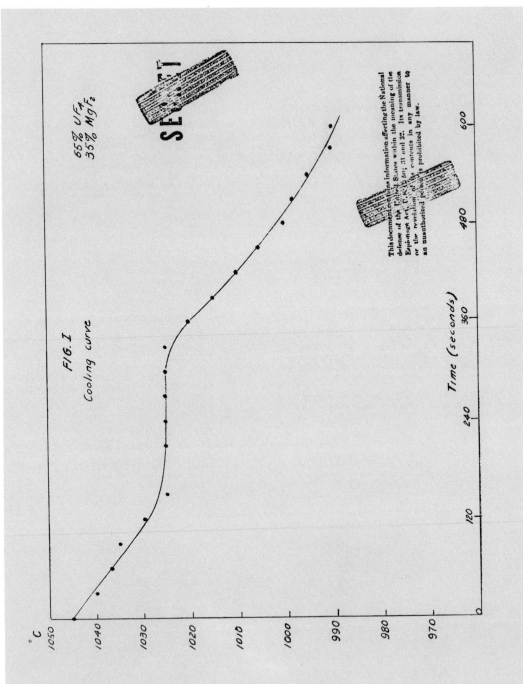

A MAGNETIC PUMP FOR LIQUID BISMUTH

B. Feld and L. Szilard

July 14, 1942

ABSTRACT

Details of the method of calculation in the design of an electromagnetic pump for liquid bismuth are given. The pump has an output capacity of 300 liters/sec at 7 atmospheres, or about 200 kw with an efficiency of about 25%. The main loss is due to induced currents in the steel shell surrounding the flowing Bi.

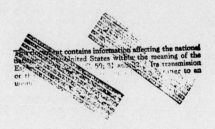

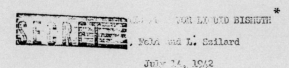

FOR LIQUID BISMUTH *

, Fehr and L. Szilard

July 14, 1942

I. Introduction.

The general idea of the pump (see Fig. A and B) is to have the bismuth flowing in the annular gap between the wall of a steel tube (1) and a magnetic iron core (2). The force on the bismuth is produced by a moving magnetic field, which in the gap is perpendicular to the direction of the flow, and which is produced by a three phase alternating current in copper coils (3) wound around the outside of the tube, and surrounded by magnetic iron (4). **The designs here calculated merely illustrate the principles involved, and bear little resemblance to a practical, working design.**

II. General Considerations.

1. Optimum Field Velocity.

The force on the bismuth obviously depends on the relative velocity between the field and the bismuth. The useful work done by the field is proportional to

$$(v_{field} - v_{Bi})\, v_{Bi}$$

The greatest losses come about through the heat production of the current flowing in the bismuth and in the steel tube. The energy dissipated by the current is proportional to

$$(v_{field} - v_{Bi})^2/\rho_{Bi} + v^2_{field}/\rho_{steel}$$

We would like the ratio of the output to the ohmic losses to be a maximum. Using this as our criterion, it is possible to calculate the optimum ratio of velocity of bismuth to velocity of field. Aside from the actual velocities, the only other quantities that enter into this consideration are the electrical resistance of the bismuth and of the steel tube, or rather, their relative values. The resistance ρ of each of these is proportional to the resistivity (micro-ohm-cm) divided by the thickness of bismuth or steel. The resistivity of bismuth is around 130, and of some steel, around 50. For our purposes, a steel tube of about 0.8 cm might be used, so that we have, if

(a) the thickness of bismuth is 2 cm, the resistance of the bismuth equals the resistance of the steel; then, the
$$\frac{velocity_{Bi}}{velocity_{field}} = 0.6 \text{ for an optimum.}$$

(b) the thickness of the bismuth is 4 cm.,
$$\frac{velocity_{Bi}}{velocity_{field}} = 0.634 \text{ for the optimum.}$$

---

*This type of pump was designed by A. Einstein and L. Szilard about 1928, and a number of models for about 2 kw were operated with Hg and liquid alloys of Na – K.

-2-

2. Maximum Wave Length of Field.

If we have a maximum field in the gap = B_{max}, then the maximum flux through the magnetic core equals

$$\frac{\lambda}{4} \cdot 2\pi R \cdot \frac{2}{\pi} B_{max} = \lambda R B_{max}$$

where λ is the field wave length, R is the radius of the iron core. However, there is a saturation value for the field in iron of about 10,000 Gauss. The saturation flux through the iron core is equal to

$$\pi R^2 B_{saturation}$$

and if we let the maximum flux equal to the saturation flux, we get

$$\lambda_{max} = \pi R B_{saturation} / B_{max}$$

and, for a B_{max} of 8,000 Gauss, $\lambda_{max} \cong 200$ cm.

III. Plan of a Machine.

1. Dimensions.

We plan to use a steel tube 50 cm in diameter, 0.8 cm thick, with a gap of 4 cm for the flow of bismuth. For this case we have shown that

$$v_{Bi}/v_{field} = 0.634 \text{ in the optimum.}$$

We take bismuth velocity of 5 m/sec, which gives a field velocity of 7.9 m/sec. and a relative velocity = v = 2.9 m/sec. The output of such a machine would equal 300 liters/sec. We have chosen B_{max} = 8,000 Gauss.

2. Pressure Produced in Bismuth.

From simple considerations we have the formula for the force on the Bi:

$$F_{max} = \frac{B_{max} I_{max}}{10}$$

which gives for the average pressure produced per unit length

$$p = \frac{1}{2} \frac{v}{\rho 1_{Bi}} B^2_{max} \times 10^{-9} \text{ dynes/cm}^3$$

where $\rho 1_{Bi}$ = resistivity in micro ohms-cms.

This gives a pressure drop of 0.071 x 10^5 dynes/cm², or 0.071 atmospheres/cm.

-3-

3. Length of Tube.

If we wish to produce a total pressure of 10 atmospheres we sould require a tube of 1.4 meter length. If we choose the wave length of the moving magnetic field to be equal to 70 cm., then the required frequency would be 11 cycles/sec.

4. Pressure drop in the bismuth due to friction.

This can be calculated from an equation of a Fanning type.

$$\frac{dp}{dx} = \frac{0.049}{(R_e)^{0.2}} \frac{\rho \, v^2_{Bi}}{\delta}$$

where δ = gap width = hydraulic radius = 4 cms. Using known constants for bismuth and for our problem, we have

$$\frac{dp}{dx} = 1.61 \times 10^3 \text{ dynes/cm}^3 = 0.00161 \text{ atmospheres/cm.}$$

This we can see is a negligible quantity.

5. Ohmic Losses.

(a) In the bismuth:

The average energy which goes into the production of heat can be written

$$w = \tfrac{1}{2} \frac{2\pi R \, v^2 \, B^2_{max}\delta}{\rho^1_{Bi}} \times 10^{-16} \text{ watts/cm}$$

In our case this gives

$$w = 1.24 \text{ kw/cm}$$

or, for 140 cm

$$w_{Bi} = 174 \text{ kw.}$$

(b) In steel:

The resistance of the steel, in our case, is twice that of the bismuth so that we can write

$$w_{steel} = \frac{w_{Bi}}{2} \times \frac{v^2_{field}}{v^2} = 4.63 \text{ kw/cm.}$$

(c) In the field-producing copper coils.

If we assume that B goes through the gap only once, and we have a gap of thickness 1, then the current required for any field is given by the formula

-4-

$$\frac{4}{10}\frac{I_{max}}{} = B_{max}\frac{1}{\mu}$$

where $l = 5$ cms, $\mu = 1$, $B_{max} = 8,000$ gauss. In our case, $I = 32,000$ amperes. Allowing 100 amp/cm^2, we would need a cross sectional area of 320 cm^2/coil, and each coil would lose

$$I^2R = 10^4 \times 320 \times 150 \times \frac{10^{-4}}{60} = 800 \text{ watts}$$

If we have 3 coils per wave length, and 2 wave lengths for our tube, the total loss is 4.8 kw. This is negligible compared to the other losses.

6. Efficiency of the Pump.

We have a pressure of 0.71×10^5 dynes/cm3. This times the area times the length of the tube times the velocity of the bismuth gives a total work output of 298×10^{10} ergs/sec., or 298 kw. The total work input is then $298 + 174 + 650 = 1122$ kw.

$$\text{Efficiency} = 298/1122 = .27 \cong 1/4$$

DETAILS OF DESIGN ON A Bi PUMP

August 27, 1942

A. Dimensions:

Tube diameter	= 25 cms	Tube thickness	= 0.5 cm
Tube length	= 150 cms	Tube resistivity	= 100 micro ohm cm
Gap width	= 4 cms		

Bi velocity	= 10 meters/sec
Output	= 300 liters/sec
B_{max}	= 5,700 gauss

$$\frac{v_{Bi}}{v_{field}} = 0.728 \quad , \quad \text{Field velocity} = 13.74 \text{ meters/sec}$$

Pressure developed	= 7 atmospheres

(1)		(2)	
λ	= 75 cms	λ = 25 cms	
frequency	= 18 cycles	frequency	= ??
B core	= 1,000 gauss	B core	= 3,600 gauss

-5-

B. Output:

Pressure	= 7	atmospheres
Pressure loss	= 0.8	atmospheres
Work output	= 212	kw
Heat loss in Bi	= 78	kw
Heat loss in steel	= 172	kw
Efficiency	$\cong$ 48 %	

C. Copper Coils:

Current per coil   = 18,140 amp.
Current density   =   300 amp/cm^2
Area per coil   =   60.5 cm^2
Heat loss per coil =   0.64 kw
Total heat loss
   in coils: (1) λ = 75 cms , 3.85 kw
              (2) λ = 25 cms , 11.55 kw

D. Iron Core:

Hysteresis loss inside and outside

(1) λ = 25 cms , 1.32 kw
(2) λ = 75 cms , 2.4 kw

Eddy current loss in the core for 1 cm laminations
   (75 laminations in the core)   = 30 kw

In actual practice, it would be entirely impossible to have a
pump, of the above design, with a gap width of over 4 cms, operate at
a frequency as high as 60 cycles. The plans for a working model using
60 cycles require a much smaller gap (~1 cm) as well as a considerably
longer wave length.

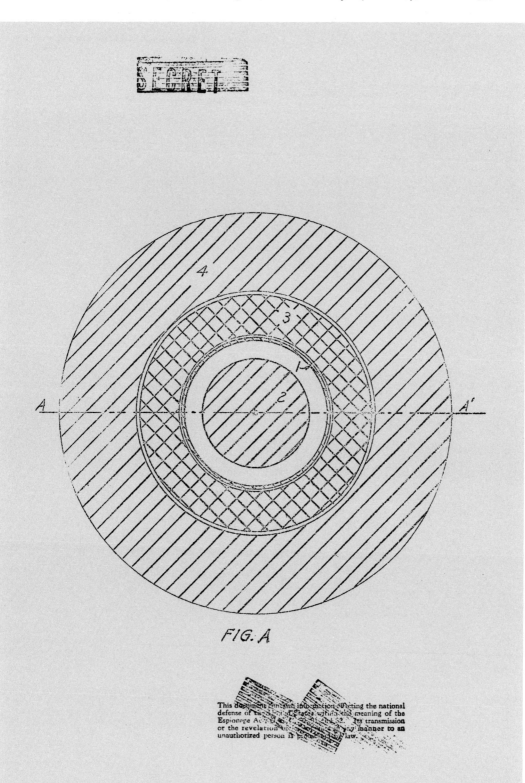

FIG. A

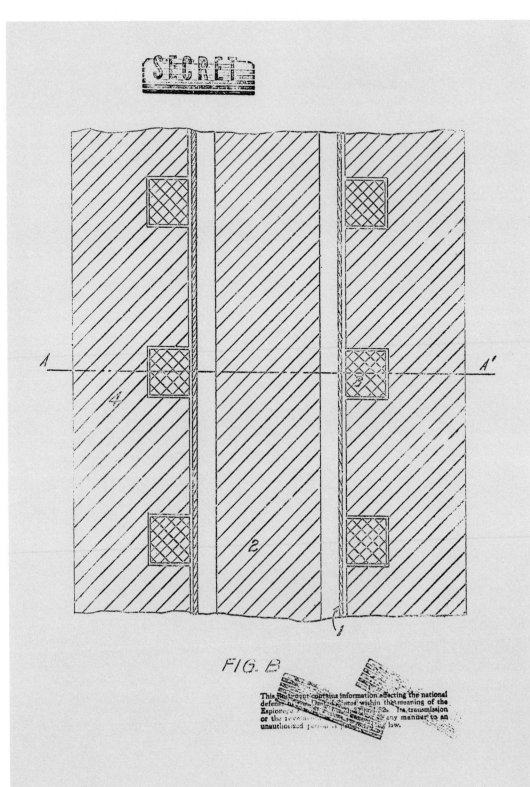

FIG. B

November 23, 1942

E. P. Wigner

L. Szilard

      Enclosed is a memorandum which you wanted Monday morning. If you
want to shorten it you could leave out the introduction on page 1, and the
"production of radioactive poisons" on the second page. Also the summary
could be left off. The rest, I am afraid, cannot be shortened without destroy-
ing the balance of the memorandum.

                              L. Szilard

SHORT MEMORANDUM ON BISMUTH COOLED POWER UNIT

L. Szilard
November 23, 1942

SUMMARY

If simple but thorough metallurgical tests on the interaction of
liquid bismuth with steel come out satisfactorily it will be possible to
build a bismuth cooled power unit that would have high operational safety.
Such a power unit can be expected to work with 1000 tons of graphite and
150 tons of uranium in the form of carbide or dioxide. More than one kilogram of 94 would be produced daily by such a power unit and it is estimated
that a total of about 200 kilograms of 94 might be produced by each such
power unit. This estimate of the total has a much smaller degree of certainty than the calculated value of the daily production of 94.

Introduction

The proposal of using bismuth as a cooling agent was part of the

original scheme of setting up a chain reaction in a uranium-graphite system.

Experiments performed since that time have greatly strengthened the view that

the neutron absorption of bismuth is sufficiently low to permit this applica-

tion. The last experiment performed on 4 tons of bismuth at Chicago gives,

according to Fermi, an about 1% loss in the multiplication factor, if the

amount of bismuth in the pile is equal to the amount of uranium. This result

bears out the contention that the commercial grade of bismuth, which was known

to be of extraordinary purity, is sufficiently pure for our purposes. The

chemical stability of a system composed of bismuth, graphite and uranium carbide,

or uranium dioxide, at high temperatures, and in the presence of strong radiation,

was one of the chief considerations in selecting this system of cooling. Com-

pared to this the good heat transfer properties of liquid bismuth are of second-

ary importance. There is reason to believe that it is safe to have liquid

bismuth in contact with steel and to pump it through steel pipes. This point

has however to be further investigated before a decision of building a bismuth

cooled power unit can be made.

-2-

The Use of a Bismuth Cooled Power Unit For:

a. production of 94.

The main purpose of operating a bismuth cooled power unit during the war is the production of about 1 ton of 94. This amount might be needed in order to win the war by means of atomic bombs, though one may hope that a smaller quantity will be sufficient. Assuming that about 1/4 of the U 235 contained in 150 tons of uranium in the power unit can be transformed into 94, 270 kg. of 94 could be obtained during the operation of the power unit. In order to produce this amount in about 200 days, it is necessary to produce about 1.3 kg./day and to dissipate about 1.3 million kw. This is quite feasible in a bismuth pile without straining matters.

b. production of radioactive poisons.

The bismuth cooled power unit could produce about 40 tons of radium equivalents in radioactive poisons outside the power unit by means of fully utilizing the neutrons which escape from within. This value holds for a time of operation which is equal to half life time of the product. Therefore some such quantity of radioactive material could be drained off every few days if a suitable product were chosen. However, a Szilard-Chalmers separation may have to be performed in order to reduce the bulk of the material which has to be transported, and this may require to put each time perhaps 15 tons of material through a chemical separation.

c. production of polonium and light sources for the armed forces.

About 10 tons of radium equivalent of polonium will be produced in 140 days of operation in the 150 tons of bismuth which will be used for cooling the power unit. Polonium mixed with a luminous compound gives a light source which is practically free from harmful rays and which can be used to serve as torches, to illuminate instruments, etc. The torches would be of very small weight and would give off light "permanently", the brilliance de-

-3-   SECRET

caying to its half value in 140 days. 250,000 such torches, each dissipating

about one watt, could be made from the quantity of polonium produced in 140

days of operation.

About four times this quantity of polonium could be produced outside

the power unit but this would involve the exposure of many hundred additional

tons of bismuth and may therefore not be expedient during the war.

In the post-war period the manufacture of polonium might become the

first important industrial application of the chain reaction.

General Features of the Power Unit.

It is proposed to maintain a chain reaction in a cylindrical graphite pile

8 m. high and 9.5 m. in diameter, weighing about 1000 tons and containing about

150 tons of uranium in the form of uranium carbide. This graphite pile would be

enclosed in hermetically sealed container filled with helium at normal pressure.

Liquid bismuth would enter at the top of the pile having a temperature of about

$300^{\circ}C$. and flow through grooves or bores in the graphite from top to bottom

under the action of gravity leaving at the bottom at about $600^{\circ}C$. The pressure

of the liquid bismuth would remain the same along the flow from the top to the

bottom of the pile and would vary according to the volume which is pumped between

1 and 2 atmosphere gauge. The velocity of the vertical bismuth flow in the pile

is estimated to be 4 m. per second. A flow of about 3.3 m^3 of liquid bismuth

per second corresponds to a heat dissipation of 1.3 million kw. and a production

of 1.3 kg. of 94 per day.

The uranium carbide is present in the form of aggregates weighing about

2 kg. Figure 30 shows one possible shape in which the uranium carbide could be

used. According to the position of these aggregates in the pile, smaller or

larger number of them are in series in a parallel flow arrangement. The principle

of such a parallel flow arrangement is illustrated in Figure 3 C taken from an

-4-

earlier memorandum. Since this figure relates to a holium cooled power unit, it does not show the dimensions which would be correct for bismuth cooling. In the case of bismuth we may have a vertical graphite column of 20 cm. x 20 cm. and the two vertical bismuth ducts have together an average cross section of 10 cm^2 near the axis of the pile. Near the axis of the pile the amount of bismuth is about equal to the amount of uranium, but closer to the periphery the relative amount of bismuth is less. Altogether about 100 tons of bismuth may be in the pile during operation.

Pumps.

The pumps lifting the liquid bismuth to about 10 m. height must have an output of about 3500 kw and assuming 50% efficiency they would have a power requirement of about 7000 kw.

Centrifugal pumps have been used for mercury under conditions of temperatures and pressures which are very similar to those required for the pumping of bismuth which will have a temperature of about $300^{\circ}C$. after passing through a heat exchanger. The General Electric Company has put at our disposal blueprints which could serve as a basis for further action. The Westinghouse Company expressed its willingness to design and build a centrifugal pump for our purpose and it was estimated that the designing work would take about three weeks and the actual building time about six months.

Another type of pump which would have certain advantages was designed by Einstein and Szilard and built and operated in 2 kw. units for mercury. This pump consists of a steel tube with an iron core and the liquid bismuth would flow through the annular gap between core and tube under the electro-dynamic action of electric windings which are outside the steel tube. This pump has the advantage of having no moving parts, of requiring no lubrication or other service, and of representing an all sealed system with a minimum of

SECRET
                                    -5-

danger of leaks. It has the disadvantage that it may have to be operated on
a low frequency and may therefore require a converter. Our estimate for the
efficiency at $12\frac{1}{4}$ cycles is about 20% for units which have an output of about
300 kw. and which would accordingly require a power input of 1500 kw. Larger
units may have a somewhat higher efficiency. The Westinghouse Company made
a preliminary investigation into these questions and expressed its willingness
to design such a pump if requested to do so. It was estimated that it would
take about two months to complete the design.

Heat Exchanger

       One may consider to have a heat exchanger at the bottom of the pile
in which the heat is transferred from the bismuth to a bismuth-lead alloy having
a melting point at 124°C. This alloy would then be pumped outside the pile and
would transfer its heat to water.

Controls

       It is proposed to control the bismuth cooled pile in a manner which
does not necessitate the moving of "control rods" through stuffing boxes.
Such an arrangement which contains no moving parts is illustrated in Figure 40.
This shows a steel tube (of about 1 cm diameter and less than 0.5 mm. wall)
going through the pile which communicates through an electrodynamic pump with
a vessel outside the pile. (The tube in the pile is cooled by a bismuth stream
indicated by arrows). This communicating system contains a low melting alloy
containing bismuth, lead, tin, and cadmium (melting point 68°C) and if the
electrodynamic pump is out of action, the steel tube in the pile is filled
with this liquid alloy as shown in the figure. The chain reaction is started
by switching on a variable transformer (Dreh transformator) which feeds the
electrodynamic pump. The liquid is then pumped out of the steel tube in the
pile into the vessel outside until the pressure difference becomes equal to
the pressure produced by the pump. This pressure is controlled by the trans-

-6-

former which in turn is controlled by the radiation intensity in such a
manner that the voltage of the transformer increases with decreasing radiation
intensity. If there is a failure in the control arrangement the pump goes out
of action and the absorbing liquid fills the steel tube inside the pile and
stops the chain reaction. Several steel tubes are connected to the same electro-
dynamic pump and two or three such systems ought to be provided for the sake of
operational safety. The power consumption at each such pump amounts to a few
kilowatts.

Start of Operation

At the start of operation the graphite pile has to be heated up to a
temperature above 300° C. Electric resistors can be used for this purpose and
a few thousand kw. input is sufficient to heat up the pile within 24 hours.
Since the bismuth pumps are not in operation during the heating up period, the
electric installation required by the bismuth pumps is capable of taking care
of the heating.

End of Operation

After operation of several months the chain reaction may be stopped
but the bismuth circulation is maintained for another month or so. After that
by admitting liquid lead in a quantity approximately equal to the quantity of
bismuth in the pile, the uranium can be cooled down before removal from the
pile to about the melting point of the bismuth-lead eutectic, 124°C.

Materials

Either of the two uranium carbides or uranium dioxide could be used
following designs which have been fairly well developed. Uranium dioxide
has to be used in a sintered form and the technique for sintering has been
worked out on a laboratory scale at Ames, Iowa. The preparing of uranium

SECRET

-7-

carbide and its casting into simple cylindrical shapes has also been worked

out on a laboratory scale at Ames, Iowa. More complicated shapes would have

certain advantages, such as the shape shown in figure 30, have so far not

been cast.

Uranium metal could possibly be used, for instance, inside of graphite

tubes, but the use of uranium metal in this form has so far not been sufficiently

studied and it is therefore uncertain whether the use of uranium in metal form

would be practicable from the point of view of operational safety.

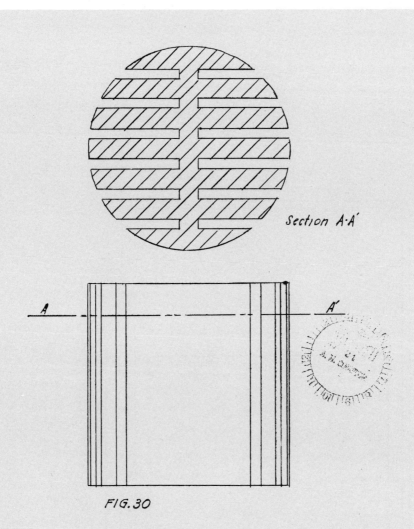

Section A·A´

FIG.30

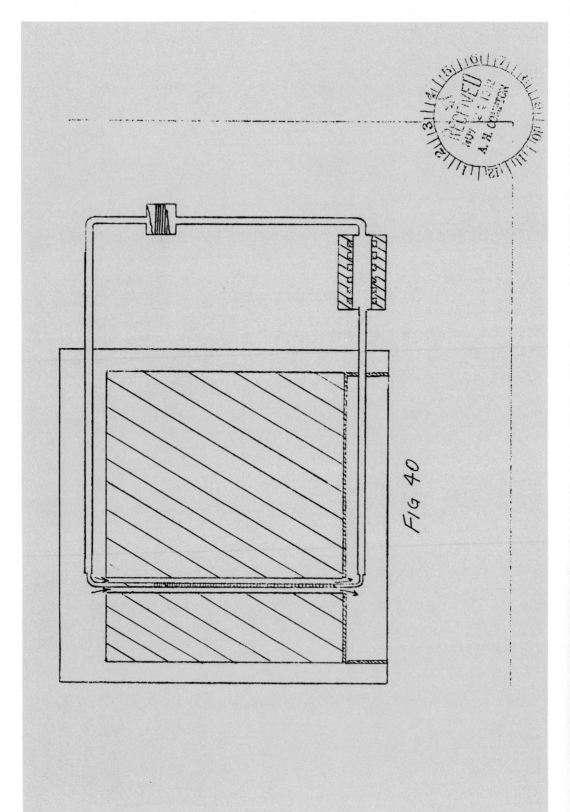

Fig 40

LIQUID METAL COOLED <u>FAST</u> NEUTRON BREEDERS   UC-LS-60

L. Szilard
March 6, 1945

This document consists of...7......
pages and ......0......figures
No..4..of..8..copies, Series...A...

This is the first of a series of memoranda outlining
a research and development program and the part which,
given favorable conditions, I might be able to play in
it. The aim of this program would be to have 10 tons
of plutonium in production within three years at a
total cost of less than $500 million and this memo
relates to what will presumably be the <u>later</u> stages of
the production process.

First type

Two different types of plutonium breeders which fall in this category are at present under discussion. To the first type belong breeders
that are based on fission neutrons which are not appreciably slowed down.
I am fairly confident that breeders of this type could be built which might
double the investment of plutonium within about a year and which would produce about one atom of plutonium in excess for one atom of plutonium that is
burned in the breeder. I suspect, however, it might prove impossible to have
units of this type <u>on a small scale</u>, i.e. containing less than about 200 kg
of plutonium which double the investment in one year. In spite of this
limitation I would wish to give considerable attention to this first type
of breeder and I believe that most of the nuclear information needed for
work on such a breeder is available or what is not available could easily
and reliably be obtained in the near future. I would like to work out a
design and hold it in readiness for use in the <u>later</u> stages of the proposed
production process.

-1-

-2-

Second type

I feel that it is also urgent to decide between the relative merits of this type of breeder and a second type of breeder in which the neutrons are slowed down into an energy region between 1, ~~and~~ 100, ~~and~~ 1000 volts. Mr. Wigner and his division show at present considerable interest in this second type of breeder.

Such breeders of the second type would indeed have the advantage of requiring a smaller quantity of plutonium per individual breeder which might double the investment within one year. But whether such breeders would in fact rapidly increase the plutonium investment cannot be determined without reliably knowing the ratio of radiative capture to fission in plutonium for energies between 1 and 1000 volts.

Various experiments can be devised for determining this ratio but it is not possible to foresee which of them will actually give a trustworthy value. I myself would therefore wish to attack this problem (if the required facilities can be put at my disposal) by observing the capture γ rays emitted from plutonium per fission in the thermal region, at the .3 volt resonance and between 1 and 1000 volts. In the thermal region and at resonance the ratio of radiative capture to fission can be measured by other comparatively simple methods and this can be used for calibrating the proposed setup so that by observing at various neutron energies simultaneously fission and the response of a counter (or a set of coincidence counters) which are sensitive to the γ rays it will be possible to obtain the pertinent ratios of radiative capture to fission.

-3-

First type:

Pending the outcome of some such experiment I personally would rather
prefer to think about breeders of the first type. ·Experiments carried out at
Site Y with fission neutrons appear to confirm the expectation that the ratio
of radiative capture to fission (α) becomes quite low for fission neutrons
so that we may indeed expect that one excess atom of plutonium can be produced for one plutonium atom burned in the breeder. It may be recalled that
it appeared a priori likely that from this point of view fast neutrons are
far superior to thermal neutrons and it was on this basis that I urged the
development of fast neutron breeders cooled by Bi-Pb eutectic or liquid Na
in the meetings held on April 26 and April 28, 1944. The estimate which I gave
for α was criticized at that time as being too optimistic; in the light of
recent experimental evidence it appears however to have been rather in error
on the conservative side. In the following I am recapitulating the views which
I put forward in those meetings as reported by Ohlinger in MUC-LAO-17 and
MUC-LAO-18. Fermi's views on the same subject are also recorded in the
same reports but are not recapitulated here. The text is as follows:

MUC-LAO-17, April 26, 1944

"Mr. Szilard was the second speaker and proposed approaching the
problem from a different viewpoint--that of assuming more optimistic values
of the constants so as to indicate other potentialities. He pointed out that
the fast reaction is preferable to the slow chain reaction for producing 49
from tubealloy and that this is probably more true if we assume more pessimistic values for ν or μ . Before discussing these values of the constants,
sketches of a possible design were distributed and described briefly. These
sketches are attached hereto.

"The sketches show two different arrangements. In sketch A, the
enriched tubealloy (enriched to where the chain reaction will go) and natural
tubealloy would be distributed in the form of rods in a cylindrical pile, in
which the enriched material would be in the center portion of the rods lying
within a circular area in the center of the pile. Part of the rods, located
within three circular areas around the center (as indicated in Fig. 1) would
be arranged so the cylindrical bundles could each be rotated about its axis.
In each of the rotating bundles, part of the rods would be natural tubealloy
and the balance of natural tubealloy with the center section enriched.

-4-   SECRET

"In the beginning, the enriched material in the three bundles would
all face the center of the pile and lie within a cylinder whose axis would
coincide with the axis of the pile and whose cylindrical surface would pass
through the three axes of the revolving bundles. By means of this arrangement, as the multiplication factor increased with the continued operation of
the pile, the enriched material could be rotated away from the center of the
pile and the natural tubealloy brought towards the center where it in turn
would be enriched. In the center of the pile would be a single tube for
introducing mercury, liquid bismuth, or some other absorbing or slowing material
for controlling the pile. The coolant for this type pile would be a bismuth-
lead alloy and would flow downward through the pile between the static and
rotating rods. The possibility of using liquid sodium in place of bismuth-
lead should also be looked into. The volumetric heat capacity of the liquid
sodium is about the same as that of the bismuth-lead alloy but its density
would be 10 times less, so that the pressure drop would be about 1/10 that
for the bismuth-lead alloy or the velocity about 3 times larger for equal
pressure drop. In the scheme just described, the following approximate conditions would obtain: (1) the bismuth-lead alloy would occupy about 1/3 of
the enriched core and would pass through the pile at a velocity of about 15
meters per sec; (2) with 1/2 cm diameter rods raised to 700° C metal temperature at the center of the central rod and with 150° C temperature increase
in the coolant, about 250,000 kw will be removed. The pumping power for the
coolant will consume about 5% of the power produced.

"In the alternative scheme B, control of the pile would be obtained
by means of a nest of tubes for the mercury or other controlling medium arranged as in Figs. 3A and 3 B and 4A and 4B. The metal rods would all be
stationary and vertical (nos. 12, 13 and 14 in Fig. 3A) and would be about
1/2 to 1 cm in diameter by about 2 meters long.

"In both designs the enriched core would be about 1/2 to 1 meter
in diameter by about the same height. The balance of the material around the
core would be ordinary tubealloy of the same rod size. The total diameter and
the height of the pile would be about 2 meters.

"The objective of such a pile must be to produce as much extra 49 as
invested. It is assumed that the production will be double the original investment. For every atom of 49 disintegrated, two atoms of 49 could be produced. Part of these will be produced in the enriched core and part in the
surrounding natural tubealloy. Some of the production in the core will tend
to leak out into the natural tubealloy and this leakage must be kept within
certain limits. Then k will increase over a period of time. As the chain
reaction goes on, the multiplication factor k will then increase so that the
controls must provide for this as well as the normal operating control of the
pile.

"In the slow chain reaction, 49 captures neutrons in radiative not
fission capture to produce a new element which we will call super plutonium
or 40-10. It is assumed there is a 50% chance that this new element will be
fissionable. If it is not fissionable, it is assumed there is 50% chance that
it will be formed only in negligible quantity in the capture of fast neutrons.
Thus, there is a 75% chance in a fast chain reaction that we may use ν and
not μ in getting the production balance (μ = 2.2 neutrons per neutron ab-

-5-

sorbed, ν_{25} = 2.2 x 1.175 = 2.6 neutrons produced per neutron absorbed). As the energy of the neutrons increases from thermal to fission energies, it is assumed there is no increase in ν . The main argument in favor of the fast chain reaction is that if a fission neutron is released in tubealloy, it causes fission in the 28 to produce 1.2 neutrons (fast effect). If all the neutrons are captured, the overall balance would be that for every atom of 49 destroyed, two atoms of 49 would be produced. One goes back into the chain reaction, the other replaces the 49 destroyed, providing a net gain in 49.

"In experiments in which a Ra-B neutron source was surrounded by 28, measurements indicated a 5.3% increase in the number of neutrons and that 63% of the neutrons remained above the fission threshold. This means that the increase in the number of neutrons for an infinite sphere would be $\frac{5.3}{1 - 0.63}$ or 19.1/2%. If the fission cross section is taken at 0.35 and the inelastic cross section at 2.7 for a ν_{28} of 2.2 to 2.6 ζ will vary from 1.18 to 1.245.

"Referring to the value above of ν_{25} of 2.6, if we were to use the more optimistic results reported by Y (that ν 49 is 20% larger than ν 25) then ν_{49} equals 3.1 neutrons produced per neutron absorbed. If we are less optimistic and assume ν 49 effective = 2.5 but use the 19 1/2% increase indicated by the experiment mentioned above, we have three neutrons produced in a mixture of 28 and 49 for one atom of 49 destroyed."

MUC-LAO-18, April 28, 1944

"The second speaker was Mr. Szilard who continued his discussion from the previous meeting. He recapped first the three possibilities as he saw them:

(1) Unseparated tubealloy→ 49 production

(2) Enriched tubealloy→ 49 production

   (a) slow chain reaction

   (b) fast chain reaction

(3) Enriched tubealloy→ no 49 production

"On the basis of Morrison's report, Mr. Szilard felt that the tubealloy should be utilized more efficiently, i.e., using the 28 and not just the 25. However, since the power production indicated in item (3) above is a long term proposition, he did not intend to discuss this phase at great length at this time.

"In item (1) above, heat is only a byproduct and not the primary object. Concerning item (2), Szilard proposed answering the question-- if an amount A of 49 were invested, how long would it be until 2A of 49 were obtained. In a fast chain reaction, if two tons of enriched ore containing 10% of 49 were used in the core surrounded by 28 at the rate of 125,000 kw, then 2A of 49 would be produced in 4 1/2 years. In order to

-6-   SEGRET

have any practical significance, this time should not be very much larger
and the readjustment of the material should be easy during the time of
operation.

"Considering first the slow chain reaction: assume a μ 49
(neutrons emitted in fission/neutrons absorbed) equal to 2.0 - 2.2.
Szilard struck out the latter figure when Fermi stated that the μ 49 is
probably lower than μ_{25}. With μ = 2, just as much 49 is being produced
as is being destroyed. In a slow chain reaction this might be improved
by usnig the fission of 28 mixed with 49 but this is not very effective
in a slow chain reaction because even from large lumps embedded for instance, in graphite,many neutrons escape, are slowed down, and do not
produce fission. Since the super-plutonium (40-10) formed might be fissionable, there is, say, a fifty-fifty chance that we can improve μ to
a μ_{eff} = 2.5.

"Considering the fast chain reaction, the situation is more
favorable. With a low concentration of 49 in the mixture with 28, experiments have shown that ε might be raised to 1.2. In addition, there is a
high energy tail producing an (n, 2n) reaction which may give a 2 1/2% increase in μ (based on observations of Turkevich).

"(The value μ has been defined as the number of neutrons pro-
duced/number absorbed in 49. Szilard uses ν_0 defined as the number of
neutrons produced/number of fissionable atoms used up.)

"In a fast chain reaction, even if 40-10 is not fissionable,
Szilard felt that it is probably true that the branching ratio for 40-10
moves in a favorable direction, or that ν_0 may be taken as ν_0 = 2.5.
He felt strongly that there is a very good chance that Pu^{240} is either
fissionable in the thermal region or at least that the branching ratio
can be counted upon to decrease by a factor of 3 as one goes from thermal
energies to, say, 1 Mev. (Fermi pointed out that the branching ratio of
49 is greater than that for 25.)

"The arguments for this belief is in part based on the rule of
thumb $(\Delta M/M)$ - $(2 \Delta Z/Z)$ (see also Morrison in Project Handbook, Chapter
IV B1.1) which gives a rough indication of the fission threshold and is
partly based on the belief that, with increasing neutron energy, the time
required for fission decreases whereas the time required for radiative
capture remains constant. Szilard therefore assumes that, in a mixture
of 238 and Plutonium, $\varepsilon \nu_0$ =1.2 x 2.5 = 3.0 neutrons emitted per thermally
fissionable atom destroyed and this would mean that there is a net gain of
one thermally fissionable atom per similar atom destroyed.

"Referring to item (3), Szilard emphasized one possibility, i.e.,
the burning of Plutonium in a slow reaction and absorbing the neutrons by
bismuth to give Polonium. Of the heat dissipated when Plutonium is destroyed to give Polonium, only about 3% would be stored in the Polonium.
However, this energy will be available for use free of γ radiation and
could be used for diving airplanes, etc.

SEGRET

-7-   SECRET

"In the discussion following, Fermi questioned the estimated value of ν_0 = 2.5 on the ground that it might be too optimistic and pointed out that there is a long range future in developing the full utilization of 28 and thorium.

"Wigner questioned the feasibility of the rotating disc arrangement described at the previous meeting on the ground of poisoning and questioned the 4 1/2 year investment return. He felt this would probably be more nearly 10 to 20 years by which time, as Mr. Morrison suggested, we may be burning water."

This ends the text taken out of the report on the meeting of April 28, 1944.

LS:s

List of Other Declassified Reports
available from Atomic Energy Commission (AEC)
or National Technical Information Service (NTIS)
not reproduced in this volume*

Physics

"Suggestions for a Search for Element 94 in Nature." Report A-45 (September 26, 1941)

"Memorandum Raising the Question Whether the Action of Explosive Chainreacting Bodies Can Be Based on an 'Expulsion' Method." Report A-56 (Columbia University) (October 21, 1941)

Metallurgy and Engineering

T. V. Moore, M. C. Leverett, C. M. Cooper, E. S. Steinbach, and L. Szilard. Engineering and Technological Divisions. Report CE-236 for month ending August 15, 1942

"Mg Reduction of UF_4: melting furnaces and casting of uranium." L. Szilard and J. Marshall, Jr. Technological Division. Report CE-271 for month ending September 15, 1942

"Internal water cooling, bismuth cooling, U238 fission reaction with fast neutrons." L. Szilard, E. Creutz, and J. Marshall, Jr. Technological Division and M.I.T. Group. Report CE-301 for month ending October 15, 1942

"Discussion of Helium Cooled Power Plant." M. C. Leverett, C. M. Cooper, T. V. Moore, E. P. Wigner, E. S. Steinbach, E. Fermi, and L. Szilard, J. A. Wheeler, S. K. Allison. Report CS-267 (September 16, 1942)

"Uranium Aggregates for Power Unit." Report CP-357 (November 23, 1942)

"Report of the committee for the examination of the Moore–Leverett design of a He-cooled plant (as given in CE-277 of September 25, 1942)." E. Fermi, S. K. Allison, C. M. Cooper, E. P. Wigner, and L. Szilard. Report CE-324 (no date).

"Memorandum on Metallurgical Problems Connected with the Power Unit which is Cooled by Liquid Metal." Report MUC-LS-1 (June 12, 1943)

"Memorandum Concerning Liquid Metal Cooling." Report MUC-LS-2 (July 30, 1943)

"Memo to S. K. Allison. Potentialities of a Beryllium Power Unit." Report MUC-LS-5 (November 15, 1943)

"Memo to A. H. Compton. Designing of production units cooled by bismuth alloy." Report MUC-LS-13 (March 28, 1944)

"New end closure for Al cans." L. A. Ohlinger, G. Young, and L. Szilard. Report N-866 (March 30, 1944)

"Proposal for use of thorium in poisoning slugs for W pile." Report MUC-LS-17 (N-962) (April 14, 1944)

"Memo to J. Chipman. Poisoning the pile with thorium metal alloy." Report MUC-LS-19 (April 19, 1944)

---

* This is not a complete list of all declassified documents which Szilard authored or co-authored, as many more were returned to his files.

"Memo to E. P. Wigner. Purifying plutonium metal." Report MUC-LS-21 (May 12, 1944)

"Memo to E. P. Wigner. Peristaltic method for purifying plutonium." Report MUC-LS-22 (May 20, 1944)

"Memo to A. H. Compton. Bismuth cooled graphite unit." Report MUC-LS-23 (May 26, 1944)

"Memo to C. M. Cooper. Swelling of slugs." Report MUC-LS-24 (June 21, 1944)

"Memo to E. P. Wigner. An extrusion process eliminating the weld on slugs." Report MUC-LS-27 (N-1346) (July 7, 1944)

"Water Moderated Pile with P-9 Core: possibility of large scale production of plutonium in a pile having a heavy water–uranium lattice at the center and a light water–uranium lattice surrounding it." Report MUC-LS-28 (N-1355) (July 7, 1944)

"Memo to S. K. Allison. Assaying 95% enriched uranium." Report MUC-LS-41 (N-1329h) (August 31, 1944)

Miscellaneous

"Memo to B. Pregel. Source of Pb from Canadian pitchblend." Report MUC-LS-4 (August 2, 1943)

"Memo to James Franck. Atomic power and Germany." Report MUC-LS-8 (December 8, 1943)

"Memorandum on the development of the atomic bomb." Report MUC-LS-61 (March 12, 1945)

"Memo to W. Bartky. Postwar plans." Report MUC-LS-63 (March 15, 1945)

"Letter to F. Oppenheimer. Petition sent to Washington." Report MUC-LS-70 (July 23, 1945)

Appendix to Part III

Reproductions from Notebooks

Szilard kept detailed records of his experiments and calculations in a set of notebooks. In the following, two typical groups of pages are reproduced. The first two pages from the notebook B 9 dated March 1st and March 2nd (1939), contain some of the observations made with W. H. Zinn which led to the discovery of neutron emission in uranium fission, as reported in the paper "Instantaneous Emission of Fast Neutrons in the Interaction of Slow Neutrons with Uranium," published in the *Physical Review*, vol. 55, p. 799, 1939. The following six pages, from B 12, represent Szilard's reworking of the data reported by Halban, Joliot, and Kowarski in the *Comptes Rendus*, vol. 208, p. 1396, 1939, on the number of neutrons emitted per fission. He arrives, by a number of alternative routes, at substantially the same result as was reported by the authors—that approximately two neutrons are emitted for each slow neutron produced fission reaction in uranium.

Experiment March 2nd

1)

"1 gram" Ra

Be

touching

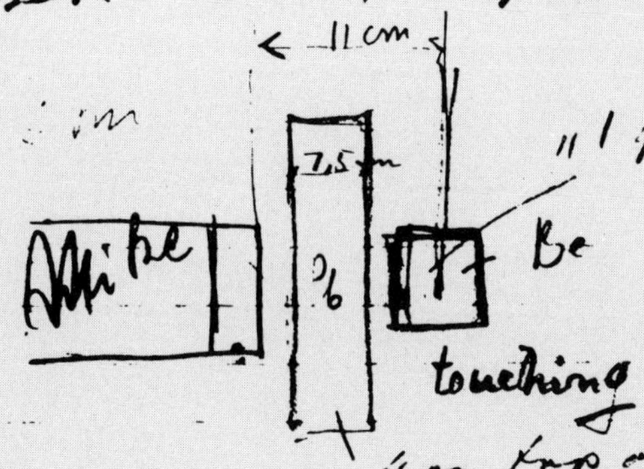

4 on top of each other

small kicks with occasional long kicks, small kicks about 2 times background in height.

52 kicks in 14 min

2.) 5:15   no Be   no Ra

1 or two kicks in 5 min

(Ra+Be in Room)

3.)   no kicks with Ra

without Be

4.)   32 kicks in 10 min

3 to 4 big kicks

Remarks to:

Joliot paper

(*Nature C. R.*
1939. 208. p. 1396)

Number of fast neutrons interacting

with $u \times 2 t = \dfrac{Q_{int}}{1-p}$ (and also absorbed free)

Number of fast neutrons produced

$Prod = \dfrac{Q_{int}}{1-p} + Q_{ext} - 1.$

R_1 Ratio (number of neutrons absorbed
by $u + n H$ / number of neutrons produced
by $u + n H$) number absorbed(

$= R_1 = \dfrac{Q_{int}}{1-p} + Q_{ext} - 1 \Big/ \dfrac{Q_{int}}{1-p}$

$$R_1 = 1 - \frac{(1-p)(1 - Q_{ext})}{Q_{int}}$$

for $p = 0.5$ $R = 1 - 0.165 = .835$
$\left(1 - \dfrac{0.55}{1.67} \cdot 0.5\right)$

R_2)

$\dfrac{Prod}{\text{Abs in thermal region}} = \dfrac{\dfrac{Q_{int}}{1-p} + Q_{ext} - 1}{Q_{int}}$
$(u + nH)$

$R_2 \uparrow$

$$R_2 = \frac{1}{1-p} - \frac{1 - Q_{ext}}{Q_{int}}$$

$$R_2 = \frac{1}{1-p} - \frac{\bar{\tau}_{int}(1 - Q_{ext})}{Q_{int} \cdot \bar{\tau}_{int}} - \cdots$$

$$R_3 = \frac{produced}{Absorbed\ thermal\ in\ U\ alone} =$$

$$= \frac{\sigma_c^u + \sigma_f^u + n\sigma_c(H)}{\sigma_c^u + \sigma_f^u} \left\{ \frac{1}{1-p} - \frac{\bar{\tau}_{int}(1 - Q_{ext})}{Q_{int}\,\bar{\tau}_{int}} \right\}$$

$$\text{where}\quad \bar{\tau}_{int} = \frac{\sigma_c(H)}{\delta\left(\sigma_c(H) + \frac{1}{n}\{\sigma_c(u) + \sigma_f(u)\}\right)}$$

where $\delta = \frac{gm}{cc}$ water in interior

$$\bar{\tau}_{int} = \frac{n\sigma_c(H)}{\delta\left(n\sigma_c(H) + \sigma_c(u) + \sigma_f(u)\right)}$$

$$R_3 = \frac{1}{1-p} + \left(\frac{n\sigma_c(H)}{\sigma_c(u) + \sigma_f(u)} \cdot \frac{1}{1-p}\right)$$

$$- \frac{1}{\sigma_c + \sigma_f} \times \frac{n\sigma_c(H)}{\delta} \times \frac{1 - Q_{ext}}{Q_{int}\,\bar{\tau}_{int}}$$

$$R_3 = \frac{1}{1-p} + \frac{n\sigma_c(H)}{\sigma_c(u) + \cdots} - \left\{\frac{n\sigma_c(H)}{1-p} - \frac{1}{\delta} \times \right\}$$

Joliot

$$\delta = 0.42 \; gm/cc$$
$$n = 3$$
$$Q_{ext} = 0.45$$
$$Q_{int} \, \sigma_{int} = 0.72$$

Joliot

$$\frac{1}{\delta} \cdot \frac{1 - Q_{ext}}{Q_{int} \, \sigma_{int}} = \frac{0.55}{0.42 \times 0.72} =$$
$$= 1.81$$

$$R_3 = \frac{1}{1-p} + \frac{3 \, \sigma_c^{\cdot} H}{\sigma_c(H) + \sigma_f(H)} \left\{ \frac{1}{1-p} - 1.81 \right\}$$

It is note worthy that & 3 becomes
0 for $p = 0.45$

for $0.45 < p < 0.6$ $\times$

$$0 < |X| < 0.2$$ if $\sigma_c(H) < 0.33 \, 10^{-24}$
$$\sigma_c(H) + \sigma_f > 4 \times 10^{-24}$$

$$R_3 \cong \frac{1}{1-p} \quad \text{in the}$$

neighbourhood of $p = 0.45$

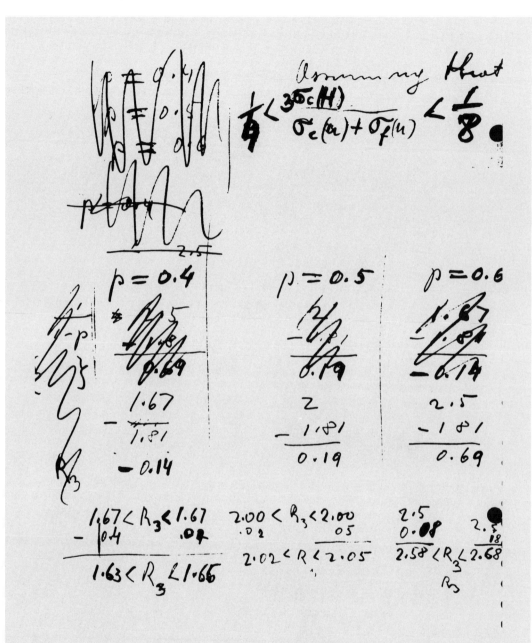

$$\frac{1}{7} < \frac{3\bar{\sigma}_c(H)}{\sigma_c(u) + \sigma_f(u)} < \frac{1}{8}$$

$p = 0.4$ $p = 0.5$ $p = 0.6$

1.67	2	2.5
$-\overline{1.81}$	$-\overline{1.81}$	$-\overline{1.81}$
-0.14	0.19	0.69

$-1.67 < R_3 < 1.67$ $2.00 < R_3 < 2.00$ 2.5
$.04$ $.04$ $.02$ 05 0.08 2.5
$\overline{1.63 < R_3 < 1.66}$ $\overline{2.02 < R < 2.05}$ $\overline{2.58 < R_3 < 2.68}$

If n is changed

if $p = 0.5$ for $n = 3$ then:

$p = 0.355$ $n = 6$

$(1-p) = 0.645$ $\dfrac{1.7 + 4}{1.}$

fraction of them

abs. ℓ_3 $\mu_3 = 1 - \dfrac{1.7}{4.+1.7} = \dfrac{4}{5.7}$

$(1-p)_3 = 0.45$ too small

$\dfrac{prod}{abs\ in\ u+6H} = 0.9$

$\underline{n = 9}$

$p = 0.29$

$(1-p) = 0.71$ $_3$ $\dfrac{4}{4+2.5} = \dfrac{4}{6.5}$

$(1-p)_3 = 0.437$

for max $(1-p)_3 \times \sqrt{\dfrac{3}{n}}$ $\Big)$ $\dfrac{\sigma_c + \sigma_p}{\sigma_c + \sigma_p + n \times 0.28}$

$\sigma(H)$

$0.28(1-p)_3 \times \dfrac{3}{n} = (\sigma_c + \sigma_p) + n \times 0.28$

$\times \frac{1}{2}(p)_3 \sqrt{3}\ n^{-3/2}$

$p = 0.6$ $n = 2.7$

for $n=9$. $p = 0.42\ 0.348$ 0.4 $= 1.08$

$(1-p) = \underset{.652}{} \times \dfrac{4}{65} = 0.365 \times 2.7 = 0.96$

Sign ...

... ...

we write ~~...~~ 2 + 2σ²ℓℓ... up to 2⁵/₂

$$R_3 = \left(1 + \frac{n\sigma_c \cdot H}{(\sigma_c + \sigma_f)}\right)\left(\frac{1}{1-p} - \text{correction} \times\right.$$

$$\left. \times \frac{1}{3}\frac{\sigma_c(H)}{\sigma_{cH} + \frac{1}{n}(\sigma_c + \sigma_f)} \frac{1 - Q e^{\lambda t}}{Q e^{\lambda t} - \text{...}} \right)$$

numerical σ_c ... 4   n = 3   p = 0.5

$$R_3 = \left(1 + \frac{0.84}{4}\right)\left(2 - \text{corr} \cdot f \times \frac{0.55}{0.42 \times 0.73} \frac{0.84}{0.84 + 4}\right)$$

$$\times \qquad \underbrace{1.81} \times \underbrace{0.174}$$

$$R_3 = (1 + 0.21)(2 - \text{corr} \cdot f \times 0.315)$$

for corr f. = 1

$$R_3 = 1.21 \times 1.685 = 2.03$$

as before

**Clipping from New York Times
on Issue of Historic First Nuclear
Reactor Patent (May 19, 1955).**

THE NEW YORK TIMES, THURSDAY, MAY 19, 1955.

PATENT IS ISSUED ON FIRST REACTOR

Reactor Inventors

Fermi-Szilard Invention Gets
Recognition—A. E. C.
Holds Ownership

Special to The New York Times.

WASHINGTON, May 18 — A historic patent, covering the first nuclear reactor, or atomic pile, has been issued by the United States Patent Office.

The inventors were Enrico Fermi, the Italian-born physicist who died last November, and Prof. Leo Szilard of the University of Chicago. The patent is owned by the Atomic Energy Commission.

Roland A. Anderson, patent counsel to the commission, classed the patent in importance with those issued to Eli Whitney for the cotton gin, to Samuel F. B. Morse for the telegraph and to Alexander Graham Bell for the telephone.

Licenses will \be issued under the patent (No. 2,708,656), but applicants will have to meet the other requirements for the civilian atomic industry prescribed by the commission on April 12. Various improvements in nuclear reactors have been made since the basic invention covered by the Fermi-Szilard patent.

The patent discloses the method by which the inventors achieved their self-sustaining chain reaction. They and their co-workers succeeded in starting the reaction—which made the atomic bomb possible—on Dec. 2, 1942, in a pile at the University of Chicago.

The patent also describes an air-cooled reactor such as X-10, which has been in operation at the Oak Ridge national laboratory since 1943 and is now used for isotope production.

Issuance of the patent establishes the priority of the Fermi-Szilard invention and protects the Government's interests.

Dr. Fermi, who came to the United States as a "self-imported exile" from Fascist Italy in 1939, joined the Columbia University faculty as Professor of Physics. In 1940, he was one of seven scientists who obtained an early patent in the atomic energy field.

Professor Szilard served on atomic energy projects at Columbia from 1940 to 1942 and later at the University of Chicago, where he has been Professor of Biophysics.

The Consolidated Edison Company of New York is the only applicant for a license to build and operate a nuclear power plant entirely with private capital.

Dr. Enrico Fermi

Prof. Leo Szilard

Part IV

Published Papers in Biology
(1949–1964)

Introduction by Aaron Novick*

The great sense of adventure so typical of everything Leo Szilard did is also seen in his work in biology. For many years he had been interested in a wide range of problems in biology and medicine, but he was preoccupied in the thirties and early forties by the developments in physics whose awesome implications for mankind he so clearly foresaw. Although the political consequences of these developments continued to occupy a large fraction of his attention until his death, he managed after the war to turn all of his scientific energies toward understanding the mysteries of biology.

In his characteristic way, he sensed that something exciting was going to happen in biology and that it would occur through the study of viruses and bacteria. He decided to set up a laboratory at the University of Chicago where President Hutchins had appointed him professor jointly in biophysics and social sciences. One spring evening in 1947 after a meeting of the Atomic Scientists of Chicago, while we walked together through the campus, Leo asked me whether I would be interested in venturing into biology with him, having decided to seek the collaboration of a younger colleague. Although he cautioned me to consider this carefully and pointed out many risks, I knew this was precisely the chance for adventure in science I had always dreamed of. He said that he would try to find some kind of position for me in the university and that meanwhile we should both take a three-week summer course in bacterial viruses which Max Delbrück had organized at Cold Spring Harbor.

To our great pleasure we found that the course gave stimulating entry to a new and exciting field in which the participants were held to very high standards by Delbrück's leadership. The clarity of the ideas being developed and the solidity of the experimental information made it possible for newcomers like us to engage in interesting and useful research following this brief course.

One of the interesting people we met at Cold Spring Harbor was Jacques Monod, who was passing through on his way from a symposium. Leo had been fascinated by Monod's observations published in his thesis that bacteria when offered the two sugars, glucose and lactose, at once would consume the glucose preferentially and utilize the lactose only after the glucose was exhausted. While discussing this with Monod, Leo suggested a scheme which he argued would oblige the bacteria to use both sugars at once, unless for some unknown reason use of one necessarily excluded use of the other. In this scheme he proposed to grow the bacteria in a continuous culture device into which glucose and lactose were to be fed continuously. He anticipated that the concentration of glucose in the growth vessel would fall as the bacteria consumed it to some point where the bacteria would begin to use the lactose. As noted later, this idea led to his discovery of the chemostat principle the following summer.

We had hardly returned to Chicago, where we had not yet set up a laboratory, when Leo suggested that we go back to Cold Spring Harbor to do some work. I took leave from my job at the Argonne Laboratory and back we went. In a few weeks we found ourselves doing experiments intended to clarify the mechanism of mating in phage which Delbrück had just discovered. But we had time only for a primitive beginning.

In January of 1948 we started our own laboratory in the basement of the former small synagogue of a Jewish orphanage whose buildings had been taken over by the University of Chicago.

Leo had been unsuccessful in securing a regular university appointment for me, therefore I was employed as a research associate on a research grant he had obtained. Subsequently I did obtain an appointment, but during all this period he was unhappy about my relatively low salary and insisted upon supplementing it out of his own pocket, although he too was astonishingly poorly paid.

One of the first things we did was to look into the mechanism of the curious phenomenon of photoreactivation—the reversal of the lethal effect of ultra-violet

* Professor of Biology, Institute of Molecular Biology, University of Oregon, Eugene, Oregon.

light by irradiation with visible light—which had just been discovered by Kelner for bacteria and Dulbecco for phage. Our conclusion, that uv produced chemical changes ("poison") which account for the lethal effect and are destroyed through the action of visible light, was a correct anticipation of present knowledge of this phenomenon.

Leo was disturbed by the fact that Delbrück and Luria hesitated to accept Lederberg's discovery of genetic recombination in bacteria. We designed what we felt should be a convincing experiment. When we completed these studies, Leo reported the results to Delbrück and to Luria in a letter in which he offered to eat his hat if this was not genetic recombination. As I recall, Luria was convinced by our findings but Delbrück remained skeptical, urging us to continue our studies. But we discovered that Lederberg had already performed an equivalent experiment whose results were contained in a table in one of his papers which typically had far more information than was seen by the usual reader.

Leo was always fascinated by paradoxes and was struck by one reported by Delbrück in his studies of genetic recombination in phage. Study of this mystery led us to the discovery of phenotypic mixing in viruses and to the realization that the genetic component of a virus is distinct from its phenotype. Hershey's subsequent brilliant demonstration of this separability and his identification of DNA as solely responsible for the genetic component was, of course, one of the dramatic discoveries of molecular biology.

During the summer of 1948, we took C. B. Van Niel's delightful course in microbiology at Pacific Grove, which again was a course set up to educate a heterogeneous group, generally including distinguished scientists as well as Stanford undergraduates. While there, Leo again thought about the continuous culture of bacteria. He had been annoyed by the fact that to get the actively growing bacteria required for most phage experiments, it was necessary to inoculate a culture two and one-half hours earlier. Ideas for experiments occurred to him so quickly and so urgently that a delay this long was intolerable to him. A continuous culture system he proposed would provide an ever-present source of growing bacteria. In thinking about the operation of such a system and perhaps recalling the idea he had proposed to Monod, he discovered the *chemostat principle*. In great excitement he described to me his realization that in a continuous culture system a stable state should be achieved if the turnover rate in the growth flask was less than the maximum bacterial growth rate. The bacteria eventually ought to grow at a rate equal to the turnover rate. In this state, the concentration of the chemical factor limiting the bacterial growth rate would be set at a constant concentration by the bacteria themselves. Therefore, he proposed to call the apparatus based on this principle the "chemostat."

On our return to Chicago at the end of the summer, we quickly verified the chemostat principle and designed a suitable apparatus. Nevertheless we were refused an N.I.H. grant on the grounds that it would not work. Later the grant was made, with apologies. Although we never used the chemostat as a source of bacteria for phage experiments, nor until much later for forcing bacteria to consume simultaneously glucose and lactose, we did use it for a series of genetic and physiological studies, some of which led to interesting discoveries.

We saw that observation of the accumulation of mutant bacteria in a population in the chemostat would give a far more accurate measure of rate of mutation than could be obtained with existing methods. Our discovery that mutation seemed to occur at a constant rate per hour independent of the growth rate of the bacteria drew considerable attention but did not provide the useful insights we wanted. We did find that a number of nontoxic compounds like caffeine increased mutation rates sharply and that some of the purine nucleosides had the novel contrary effect, they decreased rates of mutation whether spontaneous or induced by the caffeine-type mutagens. We came to realize that the mutation rate depended on the chemical environment in the cell.

Observation of the accumulation of mutants for longer periods of time led to the discovery of the phenomenon of evolution in the chemostat. Here was, in a sense, evolution in a controlled laboratory system.

The chemostat turned out to be very useful for the study of control mechanisms. By accident we observed that a tryptophan-requiring mutant excreted large quantities of a tryptophan precursor when growing at very low tryptophan concentrations. Further, we found that a sudden raising of the concentration of tryptophan immediately stopped the synthesis of this precursor. Clearly the inhibitory effect of tryptophan had to be on the activity of the enzymes in the tryptophan pathway rather than on the formation of the enzymes. Extension of these studies to an amino acid not required by the bacteria (arginine) showed again that an amino acid could block its own synthesis by interfering with the activity of the biosynthetic enzymes. We did not know that it was only the first enzyme that was affected, as Umbarger subsequently showed, but it was clear from our experiments that the inhibitory effect could not simply be a damming up of the system by mass action.

I can recall so many ideas Leo did not develop, mostly for lack of time or because of bad luck. Quite early he saw how useful it would be to be able to transfer by some kind of printing process all of the colonies on one plate to another. He built plates holding many needles—like a flat porcupine—but we made only limited use of them. Lederberg's ingenious independent development of this technique using velveteen and his applications of it showed how powerful it is and made museum pieces out of Leo's plates.

Another powerful idea which we tried to develop with not much success was to find a system where one could select a rare mutant and an equally rare reverse mutation to wild type. Seymour Benzer's discovery of the r_{II} system in phage where this can be done and his brilliant application of the technique to improve our understanding of the structure and function of the phage chromosome show how great was Leo's insight into what is important.

Our active collaboration came to an end in the summer of 1953, when I went to the Institut Pasteur for a year and Leo turned his energies toward politics and the threats facing mankind. Our friendship continued, of course, and we frequently sought each other out for the exchange of ideas.

Although he no longer had a laboratory and although he spent increasingly greater amounts of time trying to save the world, as he put it, he was still able to be very productive in biology. Among his attempts to save mankind were inventions of many ingenious birth-control devices. One astute idea was to solve the psychological problems presented by sterilization through the preservation of sperm obtained prior to the sterilization operation. He abandoned these researches when he learned that the problem had just been solved in England for bull sperm and the method was coming into commercial use.

In his typical fashion he continued to seek out the most active areas of biology and to be totally generous with his ideas. So often, as was the case in physics, great discoveries seemed to follow soon after his visits to a laboratory.

During this later period he wrote a number of theoretical papers. The phenomenon of aging had long fascinated him. At one point he considered a population of bacteria in the chemostat as a model of an aging system in that in time the fraction of cells which were mutant would increase. But this idea was discarded when we discovered that in a growing population evolution invariably occurred, i.e., a population and its accumulated mutants eventually become replaced by a new population. He remained convinced by the basic idea and prepared the theoretical papers on aging and on the effect of paternal age on sex ratios at birth.

Leo was particularly interested in the mechanism by means of which the synthesis of enzymes in cells is controlled. Here for him was not only the key for understanding the astonishing ability of mammals and many other vertebrates to recognize foreign

proteins and to make specific protein antibodies to inactivate them but also the ulti-
mate problem of differentiation of cells seen in multicellular organisms. Thus, he wrote
the two theoretical papers on control of enzyme formation and on antibody formation.

In his last work he turned to the problem which in a sense began his career when,
as a physicist, he raised the question whether the brain violated the laws of thermodynamics. Now he sought to understand how the brain could "learn." With his typical
great simplicity he applied the ideas he had developed for the control of enzyme formation and antibody synthesis to describe how a neuronal system could learn. It is too
early to tell how close to the truth he had reached, but from what has been discovered
since his death and from his incredible record I would give odds that he was right.

Reprinted from the Proceedings of the NATIONAL ACADEMY OF SCIENCES,
Vol. 35, No. 10, pp. 591–600.   October, 1949

EXPERIMENTS ON LIGHT-REACTIVATION OF ULTRA-VIOLET INACTIVATED BACTERIA

BY A. NOVICK AND LEO SZILARD

INSTITUTE OF RADIOBIOLOGY AND BIOPHYSICS, UNIVERSITY OF CHICAGO

Communicated by H. J. Muller, August 15, 1949

Many types of microorganisms are killed by ultra-violet light and the number of survivors falls off with increasing dose.   A. Kelner[1] reported recently on his discovery that, if exposure to ultra-violet light is followed by exposure to visible light, the number of survivors is very much larger.   A similar discovery was reported by R. Dulbecco[2] for bacteriophage.

592   *GENETICS: NOVICK AND SZILARD*   Proc. N. A. S.

We investigated this phenomenon of light-reactivation of ultra-violet inactivated bacteria and found some very simple regularities.

In our experiments we used a strain of coli, B/r, originally isolated by E. M. Witkin.[3] Cultures were grown in a lactate-ammonium phosphate medium under aeration to between 5×10^7 and 10^8 bacteria per cc., then transferred into saline, and incubated under aeration at 37°C. from 14 to 18 hours. The cultures were then placed in the icebox, kept at about 6°C. and used for experiments over a period of about one week.

Figure 1 shows in semilogarithmic plot the result of one series of experiments. Eight other such series were conducted with similar results.

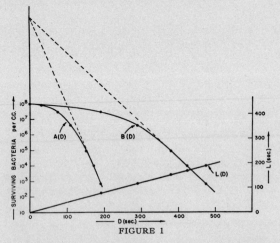

FIGURE 1

Experiment of May 11, 1949. $A(D)$, left-scale, gives in semi-log plot the number of survivors as function of ultra-violet dose, D, in seconds of exposure to 15-watt germicidal lamp at 50 cm. distance (no light reactivation). $B(D)$, left scale, gives in semi-log plot the number of survivors as function of ultra-violet dose, D, with light reactivation by 1-hour exposure to light of 1000-watt projection lamp at 8 inches distance. $L(D)$, right scale, gives in linear coordinates L as function of D satisfying equation $B(D) = A(L)$.

In these series of experiments the survivor curves, $A(D)$, were obtained in the following manner: The bacteria were exposed in saline to different ultra-violet doses, D, characterized by the number of seconds of exposure, then plated on nutrient broth agar, incubated, and the survivors determined by colony count. A 15-watt germicidal lamp was used as the source of ultra-violet radiation with the mercury line at 2536 Å. predominating in the spectrum. A 15-cm. Petri dish containing 20 cc. of the bacterial suspension was exposed at a distance of 50 cm. from the lamp and was agitated during exposure.

The survivor curves, $B(D)$, were obtained in a similar manner, except that immediately following exposure to the ultraviolet lamp the bacterial suspensions were exposed for one or two hours, at 8-inch distance, to the light of a 1000-watt projection lamp, and then plated on nutrient broth agar. The color temperature of the lamp was close to $3350°$K. The bacterial suspensions were kept at $37°$C. during exposure to the light of the projection lamp. In most of the series a 1-inch layer of a 0.05 molar solution of $CuCl_2$ was interposed between the lamp and the bacterial suspension.

As can be seen in figure 1, in semilogarithmic plot, both survivor curves A and B drop with increasing ultra-violet dose, D, slowly at first, then faster, and finally go over into a straight line.

As figure 1 shows, light-reactivation is a striking phenomenon. For an ultra-violet dose of $D = 190$ sec., for instance, the number of survivors falls in the absence of light-reactivation (curve A) from 10^8 to 700. With light reactivation, however, (curve B) the number of survivors only falls from 10^8 to 3×10^7; i.e., due to light-reactivation the titre of viable bacteria increases by a factor of about 40,000. In order to obtain the same drop in the survivors (from 10^8 to 3×10^7), in the absence of light-reactivation, a *lower* ultra-violet dose of $D = 75$ sec. would have to be employed. For every ultra-violet dose, D, which is followed by light-reactivation and thus leads to a certain number of survivors, $B(D)$, there can be found a lower ultra-violet dose, L, which in the absence of light-reactivation would lead to the same number of survivors; i.e., for which we have $B(D) = A(L)$.

Plotting L as a function of D in figure 1 we obtain for $L(D)$ a straight line going through the origin, for which we may write $L(D) = qD$, where q is a dose-independent constant smaller than 1. This means that there is a simple relationship between the survivor curves A and B which we may express by writing

$$B(D) = A(qD). \tag{1}$$

This relationship is confirmed by the nine series of experiments that were carried out. In any one of them the bacteria were light-reactivated by an exposure to the light of a projection lamp, at 8-inch distance, either of one hour or of two hours. After exposure of two hours the number of survivors obtained is, within the limits of experimental error, indistinguishable from the maximum reactivation that can be obtained with longer exposures to light. After exposure of one hour the number of survivors is lower, but the difference is not great. The q values obtained in these experiments vary slightly from series to series within the range of $0.32 < q < 0.42$. In the experiments shown in figure 1 we have $q = 0.4$.

If the light intensity used for light-reactivation is doubled by placing the bacterial suspension closer to the projection lamp, the maximum number of survivors obtainable remains the same, only it is obtained with a shorter exposure to the light.

From the relationship (1) it follows that if we extrapolate in the semi-logarithmic plot the straight line portions of the survivor curves A and B to zero ultra-violet dose, we must be led back to the *same point*. This is shown by the dotted straight lines in figure 1.

In order to interpret our results we assume for the *sake of argument* that when bacteria are irradiated with ultra-violet light a "poisonous" chemical compound, P, is produced in an amount which is *proportionate* to the ultra-violet dose, D. We assume that this "poisonous" compound is produced in two forms: a form P_z, which is not sensitive to light and which is produced in the amount x_0, and a form P_y, which can be destroyed by light and which is produced in the amount y_0, so that $D = x_0 + y_0$. We further assume that the ratio x_0/y_0 is independent of the ultra-violet dose D. And finally we assume that the number of survivors after exposure to ultra-violet irradiation, with or without subsequent light-reactivation, is determined by the amount of "poison" that is present in the bacteria *at the time* they are incubated with nutrient medium and permitted to multiply.

On the basis of these assumptions, we may now account for the relationship (1) between the survivor curves A and B by saying that if we follow up the ultra-violet irradiation of a bacterial suspension by exposure to strong light for one or two hours we destroy all the poison present in the form P_y, and leave only the poison P_z, and by further saying that the amount of the poison present in the form P_z is given by $x_0 = qD$, where q is a dose-independent constant. The amount of "poison" left in the bacteria after light-reactivation is then given by qD, and therefore the number of survivors after light-reactivation, $B(D)$, can be taken from the survivor curve obtained in the absence of light-reactivation, $A(D)$, by writing $B(D) = A(qD)$.

If we now further assume that the light-sensitive variety of the poison, P_y, which is present in an amount y, is destroyed by light at a rate which is proportionate to its amount, y, we may write $\dfrac{dy}{dt} = -\alpha y$, where α is a function of the light intensity and independent of the dose, D. From this assumption it follows that if a bacterial suspension is exposed to an ultra-violet dose, D, and immediately afterward is exposed to strong light (say, to the light of a 1000-watt projection lamp at 8-inch distance) for a period of time, t, the total amount of poison, $L = x + y$, present in the bacteria is at any time, t, given by

$$L = x_0 + (D - x_0)e^{-\alpha t}. \tag{2}$$

For an infinitely long exposure to light, we have $L_\infty = x_0$. We may then

Vol. 35, 1949   *GENETICS: NOVICK AND SZILARD*   595

write

$$\ln(L-L_\infty) = \ln(D-L_\infty) - \alpha t. \qquad (3)$$

The result of an experiment undertaken to check this relationship (3) is shown by curve I in figure 2. In this experiment we first exposed the bacterial suspension to the ultra-violet lamp for $D = 250$ sec. and then to the light of a 1000-watt projection lamp at 8-inch distance. We determined the number of survivors as a function of t, the time of exposure to the projection lamp, by plating on broth agar from aliquots taken from the bacterial suspension at different times, t, up to a time of 2 hours. From the number of survivors, we computed by means of the survivor curve, $A(L)$, the corresponding L values. As L_∞ we took the L value corresponding to the number of survivors obtained after 2 hours of exposure to the projection lamp.

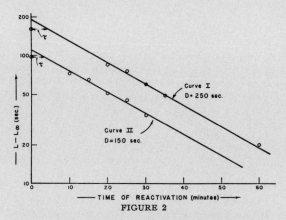

FIGURE 2

Experiment of June 23, 1949. Curve I gives in semi-log plot $L - L_\infty$ as a function of time of reactivation at 8 inches from 1000-watt projection lamp for bacteria that had received an ultra-violet dose, D, of 250 sec. Curve II is the same for bacteria that had received an ultra-violet dose of 150 sec. The interval τ represents the latent period of the process of reactivation.

As can be seen from curve I in figure 2 the points obtained fall on a straight line in accordance with equation (3). This type of relationship was confirmed in five series of experiments, giving values of α ranging from 2.19 to 2.38 per hour.

It should be noted however that the straight line I in figure 2 does not extrapolate back to the experimental point for zero exposure to light and the discrepancy corresponds to a latent period, τ, of about 3 minutes.

Line II in figure 2 differs from line I only in so far as it was obtained with an exposure to the ultra-violet lamp of $D = 150$ seconds. The fact that these two lines are parallel is in accord with the thesis that α is independent of the ultra-violet dose, D. The latent period τ manifested by line II is again about 3 minutes. Two series of experiments were performed to demonstrate in this manner the independence of α of the ultra-violet dose D.

Figure 3 shows the result of three experiments in which light-reactivation was carried out by exposing the ultra-violet irradiated bacterial suspension to the light of the projection lamp for $t = 20$, 25 and 30 min., respectively. From the number of survivors observed we computed, by means of the survivor curve, $A(D)$, the corresponding L values, and in each case we obtained for $L(D)$, as can be seen from figure 3, a straight line passing through the origin.

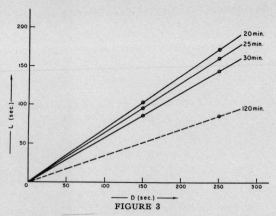

FIGURE 3

Experiment of June 23, 1949, gives L as a function of the ultra-violet dose, D, for 20, 25 and 30 minutes of light reactivation at 8 inches from 1000-watt projection lamp.

This result is to be expected provided that both α and τ are independent of the ultra-violet dose, D. Two such series of experiments were performed with identical results.

We were also interested to see in what manner the value of α depends on the intensity of the light which is used for reactivation. We varied this intensity by placing the bacterial suspension at different distances from the projection lamp. In such experiments we found that the value of α decreased with decreasing light intensity, but the decrease was less than strict proportionality with intensity would lead one to expect. The latent period τ increased in these experiments with decreasing light intensity.

We did not, however, in any of these experiments lower the light intensity very much below the value that obtains at 8inches from the pro-

VOL. 35, 1949   *GENETICS: NOVICK AND SZILARD*   597

jection lamp. We kept the intensity comparatively high in order to be able to neglect a reaction which takes place in the dark and therefore, of necessity, must also manifest itself if light-reactivation is carried out by means of very weak light. This dark reaction is easily observed if the bacterial suspension is irradiated with ultra-violet and then incubated in the dark at 37°C., for instance for 3 hours, prior to being reactivated at 8-inch distance from the projection lamp. The number of survivors after light-reactivation is then found to be much lower than is obtained when no dark incubation is interposed between ultra-violet inactivation and light-reactivation.

This concludes the report on our findings concerning the light-reactivation of ultra-violet inactivated bacteria in so far as these findings relate to the effect of light-reactivation on the number of survivors.

It is known that ultra-violet radiation not only kills bacteria but also produces mutants among the progeny of the survivors. We were interested to find out how light-reactivation affects the number of mutants which appear among these progeny. Particularly convenient for this purpose are mutants which manifest resistance to a bacteriophage, and we chose for the purpose of our investigation mutants resistant to the coli phages $T4$, $T6$ or $T1$.

M. Demerec[4] had studied the ultra-violet-induced mutations in coli to resistance to phage $T1$ and discovered that the great majority of the mutants do not appear immediately following ultra-violet irradiation, but that most of the bacteria have to go through several cell divisions before their mutation becomes phenotypically expressed.

A small minority of the mutants is phenotypically expressed prior to any cell division and A. Kelner finds that the appearance of such mutants is suppressed if the ultraviolet irradiation is followed by light-reactivation (see A. Kelner, abstracts of papers presented at the 49th General Meeting of the Society of American Bacteriologists, page 14, May 1948).

In our experiments we allowed, following irradiation, the survivors to go *through ten generations in liquid culture* and determined among the progeny the number of phage-resistant mutants per 10^8 bacteria.

If we are right in assuming that a poison is produced by the ultra-violet rays employed in our experiments, that the amount of this poison is reduced by light-reactivation, and that the amount of poison which is present in the bacteria when they are allowed to multiply determines the number of survivors, it then seems possible that this same "poison" might also determine the number of mutants—resistant to one of the phages—that appears among the progeny of the survivors. In this case we should expect that the number of mutants appearing among the progeny of the bacteria which were given a certain ultra-violet dose, D, and which were subsequently light-reactivated will be the same as the number of mutants appearing among the progeny of the bacteria which were exposed to a *lower* ultra-violet dose, qD, but which were *not* light-reactivated.

In our experiments the bacteria were exposed in saline suspension to different ultra-violet doses, D. Aliquots were then taken which were either incubated with lactate-ammonium phosphate medium at 37°C. under aeration or they were first exposed to the light of the projection lamp at 8-inch distance for one hour and were subsequently thus incubated. In either case the bacteria were allowed to go through about *ten generations* in liquid culture and were then assayed for mutants resistant to one of the coli phages *T*4, *T*6 or *T*1.

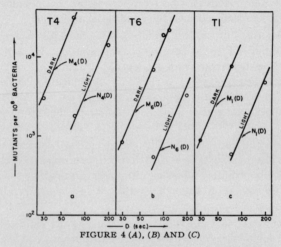

FIGURE 4 (*A*), (*B*) AND (*C*)

Experiment of July 2, 1949, gives in log-log plot the number of phage-resistant mutants per 10^8 bacteria as a function of the ultra-violet dose D both without (*DARK*) and with (*LIGHT*) light reactivation for phages *T*4, *T*6 and *T*1 in figures *a*, *b*, and *c*, respectively. The bacteria went through 10 generations in liquid culture prior to assaying for the mutants. The values shown in the graph are corrected for the spontaneous mutants which amounted to 800, 40 and 170 per 10^8 bacteria for *T*4, *T*6 and *T*1, respectively.

Figure 4 shows the result of one series of experiments in log log plot. The symbols $M_4(D)$, $M_6(D)$ and $M_1(D)$ relate to bacteria which were exposed to ultra-violet, were *not* light-reactivated, and went through ten generations in liquid culture (corresponding to an increase in bacterial titre by a factor of about 1000). Similarly $N_4(D)$, $N_6(D)$ and $N_1(D)$ relate to bacteria which were exposed to ultra-violet, were light-reactivated, and went through ten generations in liquid culture. These symbols designate the number of mutants resistant to phages $T4$, $T6$ and $T1$, respectively, which were found per 10^8 bacteria in the cultures—corrected for the spontaneous mutants that were obtained for zero ultra-violet dose—and plotted as a function of the ultra-violet dose, D.

Within the range of this experiment and within the limits of its accuracy, the lines $M(D)$ and $N(D)$ are represented in the log log plot by straight lines which are *all parallel* to each other. For each phage and for any given ultra-violet dose, D, the value of M is higher than the value of N. For every ultra-violet dose, D, which leads, when followed by light-reactivation, to a certain number of mutants, $N(D)$, there can be found a *lower* ultra-violet dose, L, which in the absence of light-reactivation leads to the same number of mutants; i.e., for which we have $N(D) = M(L)$.

Because the lines $M(D)$ and $N(D)$ are represented by *parallel* straight lines in the log log plot we may write:

$$N_4(D) = M_4(m_4 D); \qquad N_6(D) = M_6(m_6 D); \qquad N_1(D) = M_1(m_1 D) \quad (4)$$

where m_4, m_6 and m_1 are dose-independent constants. The values of m_4, m_6 and m_1 taken from the experimental series plotted in figure 4 are 0.3, 0.3 and 0.32, respectively. For the same series of experiments the value of q was also determined and was found to be 0.35. Within the limits of our experimental accuracy we have thus found: $m_4 = m_6 = m_1 = q$. Two other series of experiments which were undertaken also gave, within the limits of experimental error, identical values for m and q.

It should be noted, however, that our experiments establish the validity of relationship (4) with much less accuracy than they establish the validity of relationship (1). In general, in our experiments the determination of the number of the mutants in any one single measurement was affected by an error of $\pm 50\%$, whereas the determination of the number of survivors in any one single measurement was affected only by an error of $\pm 20\%$. And more important, due to the shape of the survivor curve $A(D)$, a large percentage error in the number of survivors leads only to a small percentage error in the corresponding L values taken from the survivor curve A, whereas an error of 50% in the number of mutants, according to the curves M and N plotted in figure 4, leads to an error in the corresponding ultra-violet dose, D, of about 25%.

We have not investigated the ultra-violet-induced mutations to phage resistance with the same thoroughness as we have the effect of ultra-violet exposure on the number of survivors. Therefore at this time we can only say that our results to date are entirely consistent with the view that the effect of light-reactivation on the appearance of mutants among the progeny of the ultra-violet-irradiated bacteria is the same as is its effect on the number of survivors and that this effect consists in the reduction of the

600   *GENETICS: NOVICK AND SZILARD*   Proc. N. A. S.

effectiveness of the ultra-violet dose by the dose-independent factor q. This makes it possible to surmise that in our experiments the killing of the bacteria and the production of the mutants might be due to the *same* chemical effect produced by the ultra-violet irradiation.

[1] Kelner, A., Proc. Natl. Acad. Sci., **35,** 73 (1949).
[2] Dulbecco, R., *Nature,* **163,** 949 (1949).
[3] Witkin, E. M., *Genetics,* **32,** 221 (1947).
[4] Demerec, M., Proc. Natl. Acad. Sci., **32,** 36 (1946); Demerec, M., and Latarjet R., *Cold Spring Harbor Symposia,* **11,** 38 (1946).

Reprinted from SCIENCE, December 15, 1950, Vol. 112, No. 2920, pages 715–716.

Description of the Chemostat

Aaron Novick and Leo Szilard

*Institute of Radiobiology and Biophysics,
University of Chicago*

We have developed a device for keeping a bacterial population growing at a reduced rate over an indefinite period of time. In this device, which we shall refer to as the Chemostat, we have a vessel (which we shall call the growth tube) containing V ml of a suspension of bacteria. A steady stream of nutrient flows from a storage tank at the rate of w ml/sec into the tube. The contents of the tube are stirred by bubbling air through it, and the bacteria are kept homogeneously dispersed throughout the tube at all times. An overflow sets the level of the liquid in the growth tube, and through that overflow the bacterial suspension leaves the tube at the same rate at which fresh nutrient enters it.

The chemical composition of the nutrient is such that it contains a high concentration of all growth factors required by the bacterium, with the exception of one, the controlling growth factor, the concentration of which is kept relatively low. The concentration of the controlling growth factor, a, in the storage tank will then determine the density, n, of the bacterial population in the growth tube in the stationary state, and it can be shown that, except for very low values of n, we have

$$n = \frac{a}{A},$$ where A is the amount of the controlling growth factor needed for the production of one bacterium.

The growth rate $\alpha = \frac{1}{n}\frac{dn}{dt}$ of a strain of bacteria is a function of the concentration, c, of the controlling growth factor in the medium, and in general we may expect the growth rate, at low concentrations c, first to increase rapidly with increasing concentration and then slowly to approach its highest attainable value, α_{max}.

The Chemostat must be so operated that the washing-out time, $\frac{w}{V}$, should be lower than the growth rate α_{max} for high concentrations of the controlling growth factor. It can be shown that in that case a stationary state will become established in which the growth rate, α, will be just equal to the washing-out rate, $\frac{w}{V}$.

What happens is that n will increase until it becomes so large that the bacteria will take up the controlling growth factor from the tube just as fast as it is necessary in order to reduce c to the point where the growth rate $\alpha(c)$ becomes equal to the washing-out rate, $\frac{w}{V}$.

Using a tryptophane-requiring strain of coli and a simple lactate medium with tryptophane added, we have used both lactate and tryptophane as the controlling growth factor. Using tryptophane, we have kept bac-

terial populations growing over long periods of time at rates up to ten times lower than normal. We are thus able to force protein synthesis to proceed very slowly while certain other biochemical processes may continue at an undiminished rate.

A study of this slow-growth phase by means of the Chemostat promises to yield information of some value on metabolism, regulatory processes, adaptations, and mutations of microorganisms. A study of the spontaneous mutations of bacteria growing in the Chemostat has been made and is being published elsewhere.

Because for most investigations a number of such Chemostats will be needed, we attempted to perfect a simple yet adequate design. Of various possible designs, we eliminated those in which changes in the barometric pressure affect the rate of flow of the nutrient from the storage tank into the growth tube. We also discarded designs that permit growth of the bacteria on the inner walls of the growth tube, or permit growth of bacteria in the Chemostat anywhere except homogeneously dispersed in the liquid nutrient in the tube. After trying out several designs, we found the one shown in Fig. 1 satisfactory.

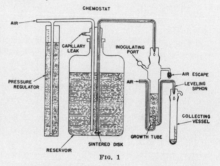

FIG. 1

A tube leading to the bottom of the storage tank is connected to a small air compressor (for example, an air pump such as is used for aerating aquaria). When the compressor is first started, the air rises rapidly in bubbles through the nutrient liquid in the storage tank and accumulates in the space above the liquid level until the pressure in the nutrient at the bottom of the tank becomes equal to the air pressure in the tube. The air space in the storage tank above the liquid level communicates through a narrow capillary with the outside air, and therefore the air will continue indefinitely to bubble through the nutrient liquid in the storage tank, but at a very slow rate (of perhaps one bubble per minute).

The pressure of the air entering the tube is regulated by a simple pressure regulator consisting of an air outlet located at the bottom of a glass cylinder filled with water

2

up to a certain level. Above this level, the air communicates freely with the outside air. By changing the water level in the pressure regulator, the air pressure can be adjusted to any value required for the operation of the Chemostat.

In this arrangement, the pressure at the bottom of the storage tank will always be greater than the pressure of the outside air by the height of the water column in the pressure regulator, and hence will be independent of the height of the level of the nutrient liquid. This is important because the level of the nutrient will gradually fall during the operation of the Chemostat.

From the storage tank the nutrient liquid is forced through a sintered glass filter into the growth tube, where it is mixed drop by drop with the bacterial suspension. The content of the growth tube is continuously stirred by aeration.

The level of the liquid in the tube is set by a siphon, and the volume of the bacterial suspension is thus maintained constant. The nutrient liquid and the bacteria suspended in it leave the tube through the syphon at the same rate at which fresh nutrient enters. The air space above the nutrient liquid in the growth tube communicates with the outside air, hence the pressure which forces the nutrient liquid through the sintered disk is at all times equal to the height of the water column in the pressure regulator.

If, after the Chemostat has been in operation for some time, the barometric pressure falls very suddenly, the pressure of the air entering into the storage tank also falls suddenly, and the nutrient liquid will rise in the air pressure tube to a certain height. If this happens, the pressure at the bottom of the storage tank will no longer exceed the outside pressure by the height of the water column in the regulator, but rather by a greater amount, and the flow of the nutrient liquid into the growth tube increases. Because of the capillary communication between the air space above the nutrient liquid and the outside air, this condition will be quickly corrected. As air flows out of the storage tank through the capillary outlet, the pressure diminishes, and the liquid which had risen into the air pressure tube in the tank is pushed out. Thus, within a short period of time, the pressure at the bottom of the storage tank is restored to its former value.

In this manner the Chemostat keeps the rate of flow of the nutrient liquid into the growth tube constant, independent of changes in barometric pressure and in the liquid level in the tank. The flow rate can be changed as desired by changing the water level in the pressure regulator.

Reprinted from the Proceedings of the NATIONAL ACADEMY OF SCIENCES,
Vol. 36, No. 12, pp. 708–719.   December, 1950

EXPERIMENTS WITH THE CHEMOSTAT ON SPONTANEOUS MUTATIONS OF BACTERIA

By AARON NOVICK AND LEO SZILARD

INSTITUTE OF RADIOBIOLOGY AND BIOPHYSICS, UNIVERSITY OF CHICAGO

Communicated by H. J. Muller, October 18, 1950

Introduction.—All bacteria require for growth the presence of certain inorganic chemical components in the nutrient, such as potassium, phosphorus, sulphur, etc., and with a few exceptions all bacteria require an energy-yielding carbon source, such as, for instance, glucose or lactate, etc.   In addition to these elements or simple compounds, certain bacteria require more complex compounds, for instance an amino acid, which they are not capable of synthesizing.   For the purposes of this presentation, any of the chemical compounds which a given strain of bacteria requires for its growth will be called a "growth factor."

In general, the growth rate of a bacterial strain may be within very wide limits independent of the concentration of a given growth factor; but since at zero concentration the growth rate is zero, there must of necessity exist, at sufficiently low concentrations of the growth factor, a region in which the growth rate falls with falling concentration of the growth factor.   It therefore should be possible to maintain a bacterial population over an indefinite period of time growing at a rate considerably lower than normal simply by maintaining the concentration of one growth factor—the controlling growth factor—at a sufficiently low value, while the concentrations of all other growth factors may at the same time be maintained at high values.

We shall describe further below a device for maintaining in this manner, over a long period of time, a bacterial population in the growth phase at a reduced growth rate and shall refer to it as the Chemostat.

If the growth rate of a bacterial population is reduced, it is not *a priori* clear whether the growth rate of the individual cells which constitute the population is uniformly reduced or whether a fraction of the total cell population has ceased to grow and is in a sort of lag phase, while the rest keeps growing at an undiminished rate.   We believe that under the conditions of our experiments, to be described below, we are dealing with the slowing of the growth rate of the individual cells rather than the cessation of growth of a fraction of the population.

By using an amino acid as the controlling growth factor we were able to force protein synthesis in the bacterial population to proceed at a rate ten times slower than at high concentrations of that amino acid.   It appears that we are dealing here with a hitherto unexplored "state" of a bacterial population—a state of reduced growth rate under the control of a suitably chosen growth factor.

Vol. 36, 1950   *GENETICS: NOVICK AND SZILARD*   709

The study of this "slow-growth-phase" in the Chemostat promises to yield information of some value on metabolism, regulatory processes, adaptations and mutations of micro-organisms; the present paper, however, is concerned only with the study of spontaneous mutations in bacteria.

There is a well-known mutant of the B strain of coli, B/1, which is resistant to the bacterial virus T_1, sensitive to the bacterial virus T_5, and which requires tryptophane as a growth factor. We used this strain and mutants derived from it in all of our experiments here reported. As a nutrient medium we used a simple synthetic lactate medium (Friedlein medium) with tryptophane added. As the controlling growth factor, we used either lactate or tryptophane.

Experiments on Growth Rates at Low Tryptophane Concentrations.—In order to determine the growth rate of B/1 as a function of the tryptophane concentration (at high lactate concentrations) we made a series of experiments in which we incubated at 37° at different initial tryptophane concentrations c, flasks inoculated with about 100 bacteria per cc. and obtained growth curves by determining (by means of colony counts) the number of viable bacteria as a function of time. Because the bacteria take up tryptophane, the tryptophane concentration c decreases during the growth of the culture and the growth rate for the initial tryptophane concentration c must therefore be taken from the early part of the growth curve.

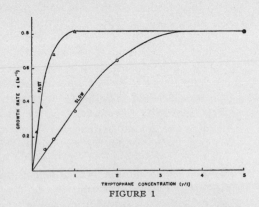

FIGURE 1

Experiment of September 18, 1950, at 37°C. The curve marked SLOW relates to strain B/1 and the curve marked FAST relates to B/1/f.

The growth rate α is defined by

$$\alpha = \frac{1}{n}\frac{dn}{dt}$$

where n is the number of bacteria per cc. The reciprocal value, $\tau = \dfrac{1}{\alpha}$, we shall designate as the "generation time." From the generation time thus defined, we obtain the time between two successive cell divisions by multiplying by ln 2.

In figure 1 the curve marked "slow" shows the growth rate α as a function of the tryptophane concentration c for 37°. At low tryptophane concentrations c, the growth rate at first rises proportionately with the concentration; with increasing concentrations, however, the growth rate approaches a limit and for concentrations above 10 γ/l. (micrograms per liter) the growth rate is no longer appreciably different from its highest attainable value. This highest value corresponds to a generation time of $\tau = 70$ min. One half of the highest value is reached at a tryptophane concentration of about $c = 1$ γ/l. This concentration corresponds to about three molecules of tryptophane per 10^{-12} cc. (The volume of one bacterium is about 10^{-12} cc.)

The proportionality of the growth rate with the concentration of tryptophane at low concentrations becomes understandable if we assume that the uptake and utilization of tryptophane by the bacterium requires that a tryptophane molecule interact with a molecule of a certain enzyme contained in the bacterium and that the uptake of tryptophane by these enzyme molecules in the bacterium becomes the rate-limiting factor for the growth of the bacterium. On the basis of this argument, we believe that down to as low concentrations of tryptophane as the proportionality of growth rate to concentration can be experimentally demonstrated, the observed growth rate of the bacterial culture represents the growth rate of the individual bacterium and that no appreciable fraction of the population goes into lag.

The Theory of the Chemostat.—In the Chemostat, we have a vessel (which we shall call the growth tube) containing V cc. of a suspension of bacteria. A steady stream of the nutrient liquid flows from a storage tank at the rate of w cc./sec. into this growth tube. The content of the growth tube is stirred by bubbling air through it, and the bacteria are kept homogeneously dispersed throughout the growth tube at all times. An overflow sets the level of the liquid in the growth tube, and through that overflow the bacterial suspension will leave the growth tube at the same rate at which fresh nutrient enters the growth tube from the storage tank.

After a certain time of such operation, at a fixed temperature, a stationary state is reached in the growth tube. We are interested in this stationary state in the particular case in which the growth rate of the bacteria is determined by the concentration in the growth tube of a single growth factor (in our specific case tryptophane). By this we mean that the concentration of a single growth factor (tryptophane) in the growth tube is so low that a small change in it appreciably affects the growth rate of the bacteria, and at the same time the concentration of all other growth factors in the growth tube is so high that a small change in them has no appreciable effect on the growth rate of the bacteria. As we shall show, under these conditions the concentration c of the growth factor in the growth tube *in the stationary state*, for a fixed flow rate w, will be independent of the concentration a of this growth factor in the nutrient liquid in the storage tank.

In order to see this, we have to consider the following:

1. For zero flow rate of the nutrient ($w = 0$), the bacterial concentration n would rise in the growth tube according to $\frac{1}{n}\frac{dn}{dt} = \alpha(c)$, where α is the growth rate which, according to our premise, is a function of the concentration, c, of the growth factor.

2. In the absence of *growth*, the bacterial concentration in the growth tube would decrease for a given flow rate w according to the formula

$$\frac{1}{n}\frac{dn}{dt} = \frac{-w}{V}$$

where $\frac{w}{V} = \beta$ may be called the "washing-out rate" of the growth tube. and $\frac{1}{\beta}$ the washing-out time.

After a while, for any given flow rate w, a stationary state will be reached in the Chemostat at which the growth rate α will be equal to the washing-out rate β $\left(\text{and the generation time } \tau \text{ equal to the washing-out time } \frac{1}{\beta}\right)$, i.e.,

$$\alpha(c) = \beta = \frac{w}{V}; \qquad \tau = \frac{1}{\beta} = \frac{V}{w}. \tag{1}$$

Thus, in the stationary state for any fixed flow rate w, the growth rate α is fixed; since α is a function of the concentration c in the growth tube, it follows that c is also fixed and independent of the concentration a of the growth factor in the storage tank.

It may be asked what is the mechanism by which, for different values of a but the same flow rate w, the same concentration c establishes itself in the growth tube in the stationary state. Clearly what happens is this: Suppose that, for a certain concentration a_1 of the growth factor in the storage tank, a stationary state with the concentration c in the growth tube has established itself and subsequently the concentration of the growth factor in the storage tank is suddenly raised to a higher value a_2. When this change is made, the concentration c in the growth tube will at first rise and along with it will rise α, the growth rate of the bacteria, which is a function of c. The concentration of the bacteria in the growth tube will thus start to increase, and therefore the bacteria will take up the growth factor in the growth tube at an increased rate. As the increase of the bacterial concentration continues, the growth rate of the bacteria will, after a while, begin to fall and will continue to fall until a new stationary state is reached at which the bacteria again grow at the same rate at which they are washed out, i.e., for which again we have $\alpha = \dfrac{w}{V}$ When this state is reached, the concentration of the growth factor in the growth tube is again down to the same value c which it had before the concentration of the growth factor in the storage tank was raised from a_1 to a_2, while the bacterial density is now higher.

In the stationary state the tryptophane balance requires that the following equation hold:

$$a = c + n\,\frac{V}{w}\,F(c) \tag{2}$$

or

$$a = c + n\,\frac{F(c)}{\alpha(c)} \tag{3}$$

where $F(c)$ gives in grams per second the amount of the growth factor which one bacterium takes up per second.

As can be easily seen, the amount Q of the growth factor that is taken up per bacterium produced is given by

$$Q = \frac{F(c)}{\alpha(c)}$$

so that, for the stationary state, we may also write

$$a = c + nQ \qquad \text{or} \qquad n = \frac{a - c}{Q} \tag{4}$$

and for the $c \ll a$ we may write

$$n = \frac{a}{Q}. \tag{5}$$

The Use of Tryptophane as the Controlling Growth Factor.—Since in the stationary state the tryptophane concentration in the growth tube of the Chemostat is always below 10 γ/l. whenever the generation time is appreciably above 70 min., we may use the approximation given in equation (5) whenever the tryptophane concentration a in the storage tank is above 100 γ/l.

In order to determine the amount of tryptophane, Q, taken up per bacterium produced, we grew bacterial cultures in lactate medium with varied amounts of tryptophane added. We found that if the initial tryptophane concentration is kept below 10 γ/l., then the amount of tryptophane taken up per bacterium produced is not dependent on the tryptophane concentration and has a value of $Q = 2 \times 10^{-15}$ gm. At higher tryptophane concentrations, however, more tryptophane is used up per bacterium produced.

From equation (5), using the value of $Q = 2 \times 10^{-15}$ gm. we obtain $n = 5 \times 10^7$/cc. for $a = 100$ γ/l. and we obtain $n = 5 \times 10^8$/cc. for $a = 1000$ γ/l.

From this, it may be seen that, by choosing suitable values for a and w, we may vary over a wide range, independently of each other, the bacterial concentration n and the tryptophane concentration c.

When we grew B/1 in a Chemostat ($V = 20$ cc.) for ten days at 37° at a generation time of $\tau = 2$ hrs. and at a bacterial density of 5×10^8/cc., we found that a change from the original bacterial strain, B/1, had taken place. The new strain, which we shall designate as B/1/f, differs from the original strain only inasmuch as it grows, at very low tryptophane concentrations, about five times as fast as the original strain. The growth rate at higher tryptophane concentrations is not perceptibly different, nor could we detect any other difference between the two strains. The curve marked "fast" in figure 1 gives the growth rate of the B/1/f strain as a function of the tryptophane concentration at 37°.

The ability of the B/1/f strain to grow faster at very low tryptophane concentrations gives it an advantage over the B/1 strain under the conditions prevailing in the growth tube of the Chemostat; and a mutant of this sort must, in time, displace the original strain of B/1.

Because in our experiments we would want to avoid—as much as possible—population changes of this type in the Chemostat, we used in all of our experiments reported below this new strain, B/1/f.

Spontaneous Mutations in the Chemostat.—If we keep a strain of bacteria growing in the Chemostat and through spontaneous mutations another bacterial strain is generated from it, then the bacterial density n^* of the mutant strain should (for $n^* \ll n$) increase linearly with time, provided that, under the conditions prevailing in the Chemostat, the new strain has the same growth rate as the original strain, so that there is no selection either for or against the mutant. In the absence of selection we have

$$\frac{dn^*}{dt} = \frac{\lambda}{\tau} n \qquad (6)$$

where n^* is the density of the mutant population, n is the density of the population of the parent strain and λ the number of mutations produced per generation per bacterium. Equation (6) holds under the assumption that back mutations can be neglected. From (6), we obtain for $n^* \ll n$

$$\frac{n^*}{n} = \frac{\lambda}{\tau} t + \text{Const.} \qquad (7)$$

From this it may be seen that—as stated above—the relative abundance of the mutants must increase linearly with time if there is no selection for or against the mutant.

714   *GENETICS: NOVICK AND SZILARD*   PROC. N. A. S.

If the growth rate of the mutant strain is smaller than the growth rate of the parent strain ($\alpha^* < \alpha$) so that there is selection against the mutant in the growth tube of the Chemostat, then the density n^* of the mutant population should—after an initial rise—remain constant at the level given by

$$\frac{n^*}{n} = \frac{\alpha}{\alpha - \alpha^*} \lambda. \tag{8}$$

Experiments on Spontaneous Mutations in the Chemostat.—Of the various mutations occurring in a growing bacterial population, mutants resistant

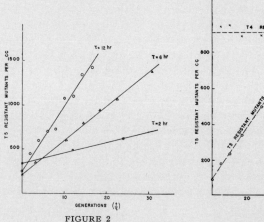

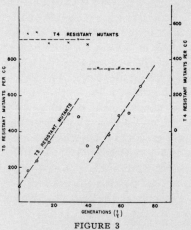

FIGURE 2	FIGURE 3
Experiments of May 3, 8, and 28, 1950, at 37°C. giving for strain B/1/f for three different values of the generation time the concentration of the mutants resistant to T_5, for a population density of 5×10^8 bacteria per cc.	Experiment of July 19, 1950, at 37°C. giving for strain B/1/f the concentration of mutants resistant to T_5 (left-hand scale) and mutants resistant to T_4 (right-hand scale) for a population density of 2.5×10^8 bacteria per cc. In this experiment oxygen containing 0.25% CO_2 was used for aeration.

to a bacterial virus are perhaps the most easily scored with considerable accuracy. In our experiments we mostly worked with mutants of our coli strain which were resistant to the bacterial viruses T_5 or T_6.

When we grow the strain B/1/f in the Chemostat with a high concentration of tryptophane but a low concentration of lactate in the nutrient in the storage tank, so that lactate rather than tryptophane is the controlling growth factor, we find—after a short initial period—that the bacterial densities of the mutants resistant to T_5 or T_6 each remain at a constant level. These levels appear to correspond to a selection factor

$\dfrac{\alpha - \alpha^*}{\alpha}$ of a few per cent.

We are inclined tentatively to assume that the behavior of these two mutants exemplifies the general rule that the vast majority of all the different mutational steps leading away from the wild type yield mutants which—under conditions of starvation for the carbon source—grow slower than the parent type.

On the other hand, if we grow our tryptophane-requiring strain in the Chemostat with a high concentration of lactate but a low concentration of tryptophane in the nutrient in the storage tank (so that tryptophane rather than lactate is the controlling growth factor) and if we run the Chemostat at a generation time well above 70 min. (the generation time at high tryptophane concentrations)—then there is no reason to expect mutants *in general* to grow appreciably slower than the parent strain, particularly if the growth of the parent strain is kept very slow by keeping the tryptophane concentration in the growth tube very low. In this case one would rather expect a mutation to affect the growth rate only if it affects the uptake or utilization of tryptophane by the bacterium or if the mutant is a very slow grower. Accordingly, we should, in general, expect the mutant population to increase linearly with time in the Chemostat when tryptophane is used as the controlling growth factor.

Figure 2 gives for 37° the experimental values for the bacterial density for the mutant population resistant to T$_5$ in the growth tube of the Chemostat as a function of the number of generations through which the parent strain has passed in the Chemostat. $\left(\text{Number of generations } g = \dfrac{t}{\tau}.\right)$ The three curves in the figure correspond to generation times of 2 hours, 6 hours and 12 hours. The slope of the straight lines gives λ, the mutation rate per generation, as 2.5×10^{-8}; 7.5×10^{-8}; and 15×10^{-8} per bacterium. We see that the mutation rate per generation for $\tau = 6$ hours is three times as high and for $\tau = 12$ hours is six times as high as it is for $\tau = 2$ hours. Thus the mutation rate per generation is, in our experiment, not constant but increases proportionately with τ and what remains constant is the number of mutations produced per unit time per bacterium. According to the above figures, we have $\dfrac{\lambda}{\tau} = 1.25 \times 10^{-8}$ per hour per bacterium.

This result is not one that could have been foreseen. If mutants arose, for instance, as the result of some error in the process of gene duplication, then one would hardly expect the probability of a mutation occurring per cell division to be inversely proportionate to the rate of growth.

If the processes of mutation could be considered as a monomolecular reaction—as had been once suggested by Delbruck and Timofeeff-Ressovsky—then, of course, the rate of mutation per unit time should be constant. The rate k of a monomolecular reaction is given by

$$k = Ae^{-W/RT}. \tag{9}$$

The value of the constant A can be calculated from the observed reaction rate k and the heat of activation W (which can be obtained by determining the temperature coefficient of the reaction).

Using the Chemostat, we have determined the rate of mutation to resistance to T_5 at 25° (for $\tau = 6$ hrs. and $\tau = 12$ hrs.) and found it to be about one half of the mutation rate at 37°. From this value and the mutation rate of $\dfrac{\lambda}{\tau} = 1.25 \times 10^{-8}$ per hour per bacterium at 37° we compute $A \approx 10^{-3}$ per sec.

In a condensed system, such as an aqueous solution, A has been found to lie between 10^5 and 10^{14} per sec. for known monomolecular reactions. Therefore if the mutation studied by us were due to a monomolecular reaction, it would have an A value 10^8 times lower than the lowest value so far found.

The density of the mutants resistant to the bacterial virus T_6 in the Chemostat, with tryptophane as the controlling growth factor, also appears to rise linearly with time for $\tau = 2$ hours, $\tau = 6$ hours and $\tau = 12$ hours, but our results so far are not sufficiently accurate to say whether this mutation also occurs at a constant rate per unit time for different generation times τ. The temperature coefficient of the mutation rate appears to be very low, but again this conclusion must await more accurate experiments.

The result obtained for mutation to resistance to the virus T_5, showing that this mutation occurs at a constant rate per unit time up to a generation time of $\tau = 12$ hours, raises the question whether this is generally true of spontaneous bacterial mutations or whether we are dealing in our case with certain exceptional circumstances. Clearly, a number of different mutations will have to be examined, different amino acids will have to be used as the controlling growth factor and other conditions will have to be varied before one would draw the far-reaching conclusion that our observation on mutation to resistance to the virus T_5 exemplifies a general rule.

Mutants Resistant to T_4.—We find that mutants resistant to T_4 are selected against in the Chemostat when grown either with lactate or with tryptophane as the controlling growth factor, i.e., the number of mutants remains—after an initial rise—at a fixed level.

It is known that of the different mutants of the B strain of coli which are resistant to the virus T_4, the most frequent one is also resistant to the viruses T_3 and T_7 and that this mutant is a very slow grower under ordinary conditions of culture. It is conceivable that this might explain why the mutants resistant to T_4 are selected against in the Chemostat even when the bacterial population grows under tryptophane control and at a much reduced rate.

Manifestation of "Evolution" in the Chemostat.—If a bacterial strain is grown over a long period of time in the Chemostat, from time to time a mutant might arise which grows faster, under the conditions prevailing in the Chemostat, than the parent strain. If this happens, practically the entire bacterial population in the Chemostat will change over from the parent strain to the new strain. We have discussed one change-over of this sort, i.e., the change-over from the strain B/1 to the strain B/1/f. There is no reason to believe, however, that no further change-over may take place when we start out with B/1/f as the parent strain and continue to grow it in the Chemostat over a long period of time.

We have seen that the mutants resistant to T_5 accumulate in the Chemostat and that their number rises linearly with the number of generations, giving a straight line, the slope of which is given by λ. If now at a certain time the population changes over in the Chemostat from the parent strain to a faster-growing strain, the accumulated mutants resistant to the bacterial virus T_5 which were derived from the parent strain should disappear from the Chemostat along with the parent strain. This should lead to a fall in the number of mutants resistant to the bacterial virus T_5 during a change-over from the parent strain to the faster-growing strain. After the change-over to the new strain, the concentration of the mutants resistant to T_5 may be expected again to increase linearly with the number of generations, giving a straight line which has the same slope as before the change-over, because the new strain which displaces the parent strain may be expected to mutate to resistance to T_5 at an unchanged rate λ.

Thus, we may in general expect, when a change-over in the population takes place, the concentration of the mutants resistant to T_5 to shift from one straight line which lies higher to another, which lies lower. The magnitude of this shift may be somewhat different from experiment to experiment, depending on when mutants resistant to T_5 happen to make their first appearance in the population of the new strain.

At the outset, the bacteria belonging to the new strain will be few in number but their number will increase exponentially with the number of generations until—at the time of the change-over—the bacteria belonging to the new strain become an appreciable fraction of the total population. If the mutation rate to resistance to T_5 is of the order of magnitude of 10^{-8}, then it is unlikely that such a mutant should appear in the population of the new strain until its population has reached perhaps 10^7. However, because an element of chance is involved, occasionally a mutant resistant to T_5 may appear earlier and, if that happens, the "shift" associated with the change-over will be smaller and in principle it might even be negative.

If a bacterial population remains growing in the Chemostat for a sufficiently long time, a number of such change-overs might take place. Each such step in the evolution of the bacterial strain in the Chemostat may be expected to manifest itself in a shift in the ascending straight line curve of the T_5 resistant mutants.

As we have seen, the mutants resistant to T_4 remain—apart from an initial rise—at a constant level in the Chemostat. However, when the bacterial population in the Chemostat changes over from a parent strain to a new strain, the T_4 resistant mutants might change over from one level to another, because the selection against the two strains might be different.

Figure 3 shows, for mutants resistant to T_5 and for mutants resistant to T_4, the number of mutants as a function of the number of generations $\frac{t}{\tau}$ in a Chemostat which was run for 300 hours at $\tau = 4$ hours with tryptophane as the controlling growth factor.

It may be seen that these two curves show a population change-over of the type just described. The curve for the T_5 resistant mutants shows a shift, P, of $P = 32$ generations.

A number of shifts of this type were observed in different experiments. We verified that these "shifts" represent population change-overs by showing in one case that (under the conditions prevailing in the chemostat) bacteria taken from the Chemostat before the change-over in fact grow slower than bacteria taken from the Chemostat after the change-over.

In order to show this, we took from the Chemostat before the change-over a bacterium resistant to T_5 and after the change-over a bacterium sensitive to T_5 and inoculated a *second* Chemostat (operated under identical conditions) with a 50-50 mixture of these two strains. We then found that the relative abundance of the resistant strain rapidly diminished. In the corresponding control experiment we took a sensitive bacterium from the Chemostat before the population change-over and a resistant one after the population change-over and again found that the strain prevalent before the change-over (this time the sensitive one) was the slower grower.

In the later stages of the change-over the concentration x of the original strain falls off exponentially with the number of generations, $g = \frac{t}{\tau}$, so that we may write $x = Ce^{-g/\gamma}$. In our experiment we obtained for γ a value of $\gamma = 3.25$.

It should be noted that the value of γ can be read also directly (though not accurately) from the curve, which gives the concentration n^* of the resistant mutants as the function of g, the number of generations. During the change-over the concentration c of the tryptophane in the growth tube goes over from an initial value c_1 to a final, lower value c_2 and it can be shown that for the midpoint of the change-over at which $c = \dfrac{c_1 + c_2}{2}$ we have

$$\gamma = \frac{P/4}{1 - \dfrac{1}{\lambda n}\dfrac{dn^*}{dg}} + \frac{1}{2} \tag{10}$$

where P is the magnitude of the shift expressed in the number of generations by which the ascending straight line of the resistant mutants is shifted in the change-over. This formula holds only if τ is large so that the rate of growth of the bacteria in the Chemostat is proportionate to the tryptophane concentration c. Because the exact position on the curve of the midpoint of the change-over on the curve n^* is not known, this formula can give only a rough indication for the value of γ.

In our case, the estimate based on it gave for γ a value of $\gamma = 2.4$ in place of the directly observed value of $\gamma = 3.25$. Within the limits of the accuracy of our curve for n^* these two values are consistent with each other.

Population change-overs manifesting themselves in a shift in the ascending straight line of the T_5 resistant mutants occurred in every experiment carried at $\tau = 4$ hrs. beyond the 50th generation. In an experiment carried to the 450th generation at a bacterial density of 2.5×10^8/cc., a number of such shifts occurred, the last one at about the 350th generation. (In the course of this experiment the mutants resistant to T_4 rose twice from a low level to a high peak, the first of which reached 4.6×10^4 and the second 4.5×10^6 mutants per cc. This phenomenon is now being investigated.)

It may be said that our strain, if grown in the Chemostat at low tryptophane concentration for a long period of time, undergoes a number of mutational steps, each one leading to a strain more "fit" than the previous one, and that each step in this process of evolution becomes manifest through the shifts appearing in the curve of the mutants resistant to T_5.

Reprinted from SCIENCE, January 12, 1951, Vol. 113, No. 2924, pages 34–35.

Virus Strains of Identical Phenotype but Different Genotype

Aaron Novick and Leo Szilard

Institute of Radiobiology and Biophysics, University of Chicago

Delbruck and Bailey (*1*) noticed an anomaly in the lysate of bacteria which was obtained by mixedly infecting the B strain of coli with the bacterial viruses T2 and T4. Subsequently, Luria (*2*) found this anomaly to be even more pronounced when he repeated Delbruck's experiment—using, however, virus T2 that had been exposed to ultraviolet irradiation.

When we undertook experiments in an attempt to understand this anomaly, we were led to the following result: If we infect a culture of the B strain of coli mixedly with the bacterial viruses T2 and T4 and incubate to permit lysis of the bacteria, there are present in the lysate 3 easily distinguishable types of bacterial viruses. Two of these, as expected, behave like the original parent strains T2 and T4, i.e., one of them behaves like T2 inasmuch as it is unable to attack the mutant strain B/2 (which is resistant to T2) but is able to grow in the mutant strain B/4 (which is sensitive to T2); the other behaves like T4, being unable to attack B/4 (which is resistant to T4) but is able to grow in B/2 (which is sensitive to T4). The third type of virus present is phenotypically like T4 inasmuch as it is capable of multiplying in the strain B/2 (which is sensitive to T4), but it is genotypically like T2 inasmuch as, after one passage in the strain B/2, it is no longer capable of growing in it but is capable of growing in the strain B/4 (which is sensitive to T2).

The presence of this third type of virus, which may be called "latent T2," can be demonstrated in the following manner: We add to a culture of the B strain of coli viruses T2 and T4 in ratios corresponding to 10 T2 and 10 T4 virus particles per bacterium, incubate to permit lysis of the bacteria, and then filter the lysate.

If we plate a sample of this lysate on agar that is inoculated with the strain B/4 (which is sensitive to T2 but resistant to T4), those virus particles contained in the lysate which have the phenotype T2 will show up as plaques on these plates. T4 virus particles will not give plaques on this plate because B/4 is resistant to T4. The number of plaques is thus a measure of the number of T2 particles in the lysate.

Using a sample of the lysate, we determine in this manner the number of plaques obtained on an agar plate inoculated with the strain B/4. When we repeat this experiment—with the difference that before plating on the B/4 plate we add to the sample of our lysate a certain quantity of the strain B/2, allow 5 min for absorption, dilute with broth, and incubate for 1 hr to permit lysis of the bacteria—then we obtain a ten to twenty-five times larger number of plaques on the B/4 plate.

This phenomenon appears to show that there is present in our lysate a virus (the "latent T2") which is capable of multiplying in B/2 and subsequently forming plaques on B/4. In order to account for our observation, the concentration of the "latent T2" in the lysate would have to be about 10% of the concentration of T2. We were not able to obtain, after one passage in B/2, any appreciable further growth in B/2 of our hypothetical "latent T2." Before drawing the conclusion that the presence of a "latent T2" is in fact responsible for our phenomenon, it is necessary to exclude alternative explanations.

As an alternative explanation of our observation, it appeared a priori conceivable that our lysate contains aggregates of virus particles formed by a T2 and a T4 particle that stick together. Such aggregates might then perhaps be capable of entering into a bacterium of the strain B/2 (by virtue of their T4 component) and, once inside, both virus particles T2 and T4 might then be able to multiply, and thus to produce the observed phenomenon. We were able to rule out this possibility, however, by performing the following experiment.

We add to a sample of our lysate a certain quantity of B/2, using an excess of B/2 so that independent infection of one bacterium by more than one virus particle can be neglected. We then allow 5 min for absorption and plate on an agar plate that has been inoculated with both B/2 and B/4. If there are present any B/2 bacteria into which has entered an aggregate of virus particles composed of T2 and T4, and in which both viruses will grow, then a certain number of clear plaques centering around such bacteria (which yield both T2 and T4) should develop on the agar plate. We were not able to find any such clear plaques, however, and found only turbid plaques (in which either the B/2 is lysed by T4 or the B/4 is lysed by T2). This rules out the alternative explanation of our phenomenon.

We ascertained that our phenomenon is produced under conditions in which we use an excess of B/2, so that independent infection of one bacterium by more than one virus particle can be neglected. We also ascertained that our phenomenon is not produced if, in place of our lysate, we use a mixture of T2 and T4.

We are thus led to conclude that the phenomenon described is due to virus particles that have the phenotype of T4, but the genotype of T2. The properties of this "latent T2" virus would seem to merit investigation.

References

1. DELBRUCK, M., and BAILEY, W. T., JR. *Cold Spring Harbor Symposia*, **11**, 33 (1946).
2. LURIA, S. Private communication (1947).

Reprinted from COLD SPRING HARBOR SYMPOSIA
ON QUANTITATIVE BIOLOGY
Volume XVI, 1951
Made in United States of America

GENETIC MECHANISMS IN BACTERIA AND BACTERIAL VIRUSES I

EXPERIMENTS ON SPONTANEOUS AND CHEMICALLY INDUCED MUTATIONS OF BACTERIA GROWING IN THE CHEMOSTAT

AARON NOVICK AND LEO SZILARD

Institute of Radiobiology and Biophysics, The University of Chicago

In an earlier paper we reported observations on the spontaneous mutations of bacteria occurring in a continuous flow device we call the Chemostat (Novick and Szilard, 1950a). In this device we maintain a bacterial population in the growth phase over an indefinite period of time by maintaining the concentration of one of the growth factors—called the controlling growth factor—at a low fixed value.

A number of different mutations will occur in a population so maintained in the Chemostat, but in the experiments reported here we are concerned only with the mutation from sensitivity to the bacterial virus T5 to resistance to this virus. In the absence of selection for or against the mutant, and if reverse mutations can be neglected, the mutant population in the Chemostat will increase linearly with time, over periods of time short enough to disregard "evolutionary" changes (Novick and Szilard, 1950a).

By plotting the number of mutants against time we obtain a straight line whose slope is proportional to the mutation rate. With tryptophane as the controlling growth factor we found for mutants resistant to T5 that the mutation rate per hour is independent of the growth rate (Novick and Szilard, 1950a) within the range of generation times, γ, between $\tau_i = 2$ hours and $\tau_i = 12$ hours.

In our present experiments we have used a modified model of the Chemostat, shown in Figure 1, instead of the previously described model (Novick and Szilard, 1950b). In the new model an electromagnetically operated valve, shown in the figure, is controlled by a clock and is opened for a short time at fixed time intervals. There are two pressure regulators with water levels at heights of h_3 and h_4 which can be adjusted at will. The sum of h_3 and h_4 must be so adjusted that when the valve is closed the nutrient liquid rises from the storage tank into a capillary to a height of h_1 which is too low to permit an overflow of the liquid into the growth tube. When the valve is opened a certain fraction of the liquid contained in the capillary is blown (by an air pressure whose magnitude is set by h_4) into the growth tube, while the rest of the liquid returns into the storage tank. The average flow rate of the nutrient into the growth tube can be regulated either by changing h_1 (the height to which the liquid rises in the capillary when the valve is closed) or by changing the time intervals at which the clock operates the valve.

The nutrient medium used in our experiments was a simple synthetic medium (Friedlein) containing lactate, ammonium chloride, phosphate buffer, and magnesium sulfate. Tryptophane was added when required. When we use phosphorus as the controlling growth factor, we employ tris (hydroxymethyl) amino methane in a $\frac{1}{60}$th molar concentration as a buffer. The B strain of *E. coli*, or mutants of this strain, was used in all of our experiments.

SPONTANEOUS MUTATIONS

In order to see whether the spontaneous mutation rate depends on the choice of the controlling growth factor, experiments were performed in which tryptophane, ammonium chloride (our nitrogen source), phosphate, or lactate (our carbon source) was used as the controlling growth factor. As may be seen from Figure 2, the spontaneous mutation rate is appreciably lower with ammonium chloride, phosphate, or lactate than with tryptophane as the controlling growth factor.

Earlier (Novick and Szilard, 1950a), with tryptophane as the controlling growth factor, the mutation rate was found to be about 1.25×10^{-8} per hour per bacterium. Our present, more accurate measurements indicate, however, that the real value is probably close to 1.5×10^{-8} per hour.

The mutation rate appears to be lowest when lactate is the controlling growth factor—about

338   *AARON NOVICK AND LEO SZILARD*

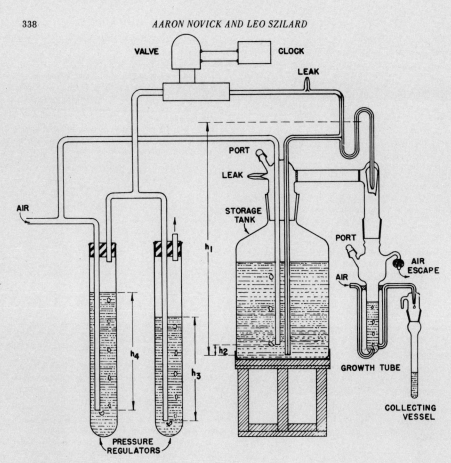

FIGURE 1.

one-third of the value obtained with tryptophane as the controlling growth factor. Because of this low value, our earlier measurements led us to believe that with lactate as the controlling growth factor the number of mutants does not increase at all with time, but stays at a constant level as the result of selection against the mutant (Novick and Szilard, 1950a). Our present findings with lactate as the controlling growth factor do not, however, indicate such selection which, if present, would lead to a curved line instead of the straight line appearing in Figure 2.

One may ask whether the high value of the spontaneous mutation rate obtained with trypto-phane as the controlling growth factor is due to the specific choice of tryptophane or whether other amino acids, when chosen as the controlling growth factor, give the same spontaneous mutation rate. In order to learn something about this point we used a mutant of our strain of *coli* which requires both arginine and tryptophane as growth factors. This strain we obtained from an arginineless mutant of B, D84, (kindly supplied to us by A. Doermann) by picking a spontaneous T1-resistant mutant which requires tryptophane. This mutant strain we grew in two Chemostats. Both of these contained arginine as well as tryptophane in the nutrient storage tank, but

SPONTANEOUS AND CHEMICALLY INDUCED MUTATIONS OF BACTERIA   339

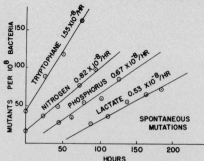

FIG. 2. Rise of spontaneously occurring T5 resistant mutants in the growth tube of the Chemostat for different controlling growth factors. For detailed data see Table 1.

one of them contained arginine in excess so that *tryptophane* was the controlling growth factor, and the other contained tryptophane in excess so that *arginine* was the controlling growth factor. As Figure 3 shows, the spontaneous mutation rate is about the same in both experiments.

If bacterial mutations arise as the result of an "error" occurring in the act of gene reproduction, one would expect the mutation rate per hour to be inversely proportional to the generation time rather than independent of it. Independence of generation time is, however, experimentally established (Novick and Szilard, 1950a) in the case of mutation to T5 resistance. In order to reconcile the error hypothesis of mutation with the experimental facts, one might say that when we slow protein synthesis by using tryptophane

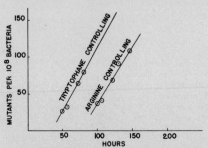

FIG. 3. Two experiments giving the rise of spontaneously occurring T5 resistant mutants with a strain requiring both tryptophane and arginine. In one experiment tryptophane was the controlling growth factor, in the other arginine. For detailed data see Table 1.

as the controlling growth factor, we might not necessarily slow the rate at which the genes reproduce; the genes might perhaps reproduce at an unchanged rate, and genes produced in excess might be discarded. Clearly, the error hypothesis of mutation demands constancy of the mutation rate per gene generation rather than bacterial generation.

Luria and Hershey (oral communications) have drawn attention to this way of possibly reconciling the observed constancy of the mutation

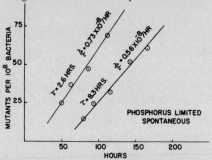

FIG. 4. Rise of spontaneously occurring T5 resistant mutants with phosphorus as the controlling growth factor for two different generation times. For detailed data see Table 1.

rate per unit time with the error hypothesis of mutation. It was further thought that if phosphorus or nitrogen were used as the controlling growth factor, the rate of gene synthesis might be slowed along with the rate of protein synthesis. Constancy of the mutation rate per gene generation demanded by the error hypothesis would then show up inasmuch as the mutation rate per hour would no longer be constant but would rather fall proportionately with increasing generation time.

In order to test this point we performed experiments with phosphorus as the controlling growth factor. In Figure 4 are represented two experiments in which the bacteria were grown with generation times of $\tau = 2.6$ hours and $\tau = 8.3$ hours. As may be seen from the figure, the mutation rate per hour is increased only slightly— if at all—by a more than three-fold increase in generation time. Thus these results do not lend any support to the error hypothesis of mutation.

With nitrogen as the controlling growth factor we again found no appreciable decrease of the mutation rate with increasing generation time.

340   *AARON NOVICK AND LEO SZILARD*

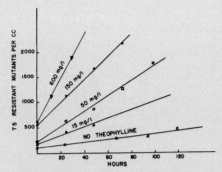

FIG. 5. Rise of T5 resistant mutants induced by different concentrations of theophylline with tryptophane as the controlling factor. For detailed data see Table 1.

The mutation rate seemed rather to increase with increasing generation time, but the accuracy of these experiments is not sufficient to say whether this increase is real.

THEOPHYLLINE-INDUCED MUTATIONS

Methylxanthines have been reported to be mutagenic for fungi (Fries and Kihlman, 1948) and also for bacteria (Bertani, Demerec and Flint, 1949). If theophylline, a dimethylxanthine, is added to the nutrient medium to bring its concentration to 150 mg/l there is no appreciable killing of the bacteria growing in the Chemostat. We find that, with tryptophane as the controlling growth factor, this concentration of theophylline raises the mutation rate from the spontaneous rate of about 1.5×10^{-8} per hour to a rate of about

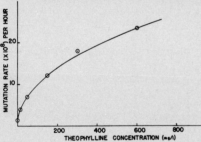

FIG. 6. Rate of theophylline-induced mutations to T5 resistance (with tryptophane as the controlling growth factor) as a function of theophylline concentration.

11×10^{-8} per hour. Measurements carried out at different theophylline concentrations are represented in Figure 5. In Figure 6 the mutation rate is plotted against the theophylline concentration.

Having found that the spontaneous mutation rate is lower when ammonium chloride, phosphorus, or lactate rather than tryptophane is used as the controlling growth factor, we performed experiments to see whether for theophylline-induced mutations there is a similar dependence of the mutation rate on the controlling growth

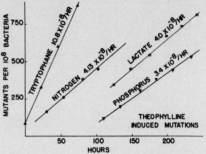

FIG. 7. Rise of theophylline-induced T5 resistant mutants in the growth tube of the Chemostat for different controlling growth factors at a theophylline concentration of 150 mg/l. For detailed data see Table 1.

factor. Figure 7 gives the results of these experiments. As the figure shows, we obtained (for the B strain) with these three controlling growth factors theophylline-induced mutation rates which are about one-third to one-half of the value obtained (for a tryptophane requiring mutant of the B strain) with tryptophane as the controlling growth factor.

Figure 8 gives for this tryptophane-requiring strain a direct comparison of the theophylline-induced mutation rates for growth with tryptophane as the controlling growth factor and for growth with nitrogen as the controlling factor. The mutation rate with nitrogen as the controlling growth factor is about one-third of the rate obtained with tryptophane as the controlling growth factor.

It may be seen from these results that the theophylline-induced mutation rate shows a similar dependence on the choice of the controlling growth factor as does the spontaneous mutation rate, being higher with tryptophane in control than with lactate, phosphorus, or nitrogen in control. This raises the question whether

TABLE 1.

	Experiment No.	Controlling growth factor	Growth factors	Strain	Bacterial density	τ hrs.	Mutagen	$\frac{\lambda}{\tau}$ (× 10⁸ hr.)
Figure 2	147	Tryptophane	1000 γ/l Tryptophane	B/t/t	4.6 × 10⁸	4.7	None	1.55
	154	Nitrogen	30 mg/l NH₄Cl	B/r	4.0 "	4.6	"	0.82
	214	Phosphorus	3 mg/l P as phosphate	B	4.5 "	5.8	"	0.67
	222	Lactate	200 mg/l Lactic acid	B	4.6 "	4.7	"	0.53
Figure 3	232	Tryptophane	{500 γ/l Tryptophane, 9200 γ/l Arginine}	D84/lt	3.0 "	2.4	"	1.76
	231	Arginine	{2000 γ/l Tryptophane, 2300 γ/l Arginine}	"	2.0 "	2.6	"	1.56
Figure 4	229	Phosphorus	3 mg/l P as phosphate	B	3.7 "	2.6	"	0.73
	228	Phosphorus	"	B	6.9 "	8.3	"	0.56
Figure 5	147	Tryptophane	1000 γ/l Tryptophane	B/t/t	4.6 "	4.7	"	1.55
	116	"	500 γ/l Tryptophane	"	2.2 "	3.3	15 mg/l Theophylline	4.0
	117	"	"	"	2.2 "	3.3	50 " "	7.0
	118	"	"	"	2.2 "	3.3	150 " "	12.0
	113	"	"	"	2.2 "	4.8	600 " "	23.8
Figure 6			Derived from Figure 5.					
Figure 7	157	Tryptophane	500 γ/l Tryptophane	B/t	2.2 "	5.5	150 mg/l Theophylline	10.8
	201	Nitrogen	30 mg/l NH₄Cl	B	2.2 "	5.5	" "	4.13
	189	Lactate	150 mg/l Lactic acid	B	2.5 "	3.5	" "	4.0
	208	Phosphorus	3 mg/l P as phosphate	B	4.3 "	6.5	" "	3.4
Figure 8	157	Tryptophane	{1 gm/l NH₄Cl, 500 γ/l Tryptophane}	B/t	2.2 "	5.5	150 mg/l Theophylline	10.8
	166	Nitrogen	{30 mg/l NH₄Cl, 10,000 γ/l Tryptophane}	B/t	3.3 "	7.2	" "	3.5
Figure 9	225	Tryptophane	500 γ/l Tryptophane	B/t/t	2.3 "	4.7	150 mg/l Caffeine	19.0
	157	"	"	B/t	2.2 "	5.5	" Theophylline	10.8
	199	"	"	B/t/t	2.5 "	6.2	" Paraxanthine	8.4
	226	"	"	B/t/t	2.3 "	4.1	" Theobromine	7.5
	184	"	"	B/t/t	2.5 "	4.8	" 8-Azaguanine	3.4
Figure 10	114	Tryptophane	500 γ/l Tryptophane	B/t/t	2.2 "	6.0	500 mg/l Adenine	5.2
	124	"	"	"	2.2 "	5.3	164 mg/l Adenine	2.95

342   *AARON NOVICK AND LEO SZILARD*

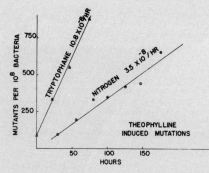

FIG. 8. Two experiments on theophylline-induced mutations to T5 resistance in a tryptophane requiring strain, B/1t, one with tryptophane, the other with nitrogen as the controlling growth factor. For detailed data see Table 1.

the processes leading to spontaneously occurring mutations have some step in common with those involved in the induction of mutations by theophylline.

PURINE DERIVATIVES OTHER THAN THEOPHYLLINE

With tryptophane as the controlling growth factor, caffeine, paraxanthine, theobromine, and 8-azaguanine in concentrations of 150 mg/l have, like theophylline, a marked effect on the mutation rate. This may be seen from the data shown in Figure 9.

The effect of adenine on the mutation rate—with tryptophane as the controlling growth factor—

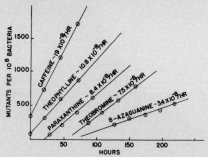

FIG. 9. Mutations to T5 resistance induced by a number of purine derivatives at concentrations of 150 mg/l with tryptophane as the controlling growth factor. For detailed data see Table 1.

was determined for two different concentrations and is shown in Figure 10.

2—6, diamino purine in a concentration of 150 mg/l did not appear to be mutagenic, at least not within the limits of our experimental error.

Of the purine derivatives which we found to be mutagenic, caffeine, theophylline, paraxanthine, and theobromine are all N-methyl xanthines. Because purines are constituents of nucleic acid and nucleic acid is a constituent of the cell

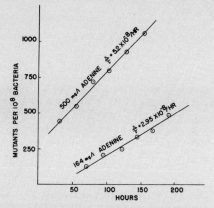

FIG. 10. Mutations to T5 resistance induced by two different concentrations of adenine with tryptophane as the controlling growth factor.

nucleus, one might think that the likeness of the mutagenic purine derivatives to the naturally occurring purines has something to do with their mutagenic action.

If this were so, one might also expect some pyrimidine derivatives to be mutagenic. We therefore tested for mutagenic action, using tryptophane as the controlling growth factor, the following pyrimidine derivatives: uracil, thymine, 6-methyl uracil, 5-amino uracil, 5-bromo uracil. None of these appeared to be mutagenic in a concentration of 150 mg/l. Experiments with other pyrimidine derivatives are in progress.

The authors wish to express their gratitude to Dr. George H. Hitchings of the Wellcome Research Laboratories, who very kindly put at their disposal many of the purine and pyrimidine derivatives used in these experiments. The authors gratefully acknowledge the support of this work by a grant from the National Institutes of Health of the United States Public Health Service.

SPONTANEOUS AND CHEMICALLY INDUCED MUTATIONS OF BACTERIA   343

REFERENCES

BERTANI, G., DEMEREC, M., and FLINT, J., 1949, Carnegie Institution of Washington Yearbook *48*: 138.
FRIES, NILS, and KIHLMAN, BENGT, 1948, Fungal mutations obtained with methyl xanthines. Nature, Lond. *162*: 573.
NOVICK, AARON, and SZILARD, LEO, 1950a, Experiments with the Chemostat on spontaneous mutations of bacteria. Proc. Nat. Acad. Sci. *36*: 708. 1950b, Description of the Chemostat. Science *112*: 715.

DISCUSSION

MARSHAK: Dr. Novick has pointed out that there is no correlation in his experiments between the mutagenicity of a compound and its presence as a constituent of nucleic acid. In this case presence or absence does not appear to be an adequate basis of comparison. Actually on the basis of tracer studies there is a very good correlation between the observations that adenine is readily incorporated into nucleic acid while the pyrimidines are not with his finding that adenine and some of its derivatives are mutagenic while the pyrimidines are not. Another parallelism may be found with the results of analysis of the composition of nucleic acids. I have found that the nuclear pentosenucleic acids of several different tissues in the same animal all have approximately the same proportions of the purines adenine and guanine, but the pyrimidine ratios vary widely.

SZYBALSKI: The idea of the "Chemostat," that is, the idea of continuous propagation of microorganisms (used often in fermentation industries) controlled by a growth limiting factor (described also in a paper of J. Monod, 1950, Inst. Pasteur *79*: 390), is obviously a very ingenious tool for studying microbial genetics.

It has however, some disadvantages since in some cases it is difficult to be certain that the use of marked, deficient strains or "one-factor starvation" has no influence on the primary effect under investigation.

It seems to me that a "Turbidostat" (if I can introduce this name) which makes use of the turbidity of the culture to control and limit growth is a more general and perhaps even simpler solution of this problem. Such a device based on the principle of photoelectric control is now being used in our laboratory for studying mutations (V. Bryson).

NOVICK: One of the characteristic features of the Chemostat is the independence of the concentration of the controlling growth factor in the growth tube in the stationary state from the concentration of this factor in the incoming nutrient. It is on account of this feature that we called the device "Chemostat." We did not recognize this feature until the summer of 1948 even though we discussed building such a device for the study of bacterial adaptations while we were working at Cold Spring Harbor in the summer of 1947. We have been working with the Chemostat since October 1948, and find it capable of a much greater variety of applications than the "Turbidostat." The possibility of varying the generation time and also the controlling growth factor gives it a greater flexibility. We do not consider "one-factor starvation" a disadvantage provided we know what the factor is and can change it at will.

(*Reprinted from Nature, Vol.* 170, *p.* 926, *November* 29, 1952)

Anti-mutagens

WHILE studying the mutagenic action of various purine derivatives on bacteria, we came across a new phenomenon : we found that certain nucleosides can act as anti-mutagens.

Following the discovery[1] that caffeine—a purine derivative—increases the mutation-rate in fungi and in bacteria, we began a quantitative study of the mutagenic action of purine derivatives. Such a study has been made possible by the use of a constant flow device, called the 'Chemostat'[2-4], which maintains a stationary bacterial population growing at a fixed rate that can be set at will. The concentration of the bacterial population maintained in the growth tube of the 'Chemostat' is determined by the input concentration of one of the required nutrients, called the controlling growth factor, and the growth-rate is fixed by the rate at which fresh nutrient flows into the growth tube.

A variety of different mutations will occur at different rates in such an otherwise stationary population, and if one plots the concentration of one particular type of mutant against time, one should obtain a straight line which rises with a slope that is determined by the mutation-rate. This holds for each type of mutant which grows at the same rate as the parent strain, that is, if there is no selection for or against the mutant. If there is selection against a mutant, the concentration of that mutant will remain stationary after an initial rise.

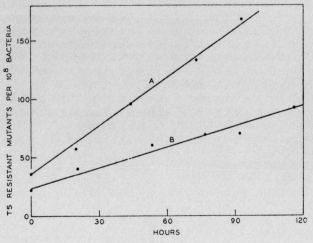

Fig. 1

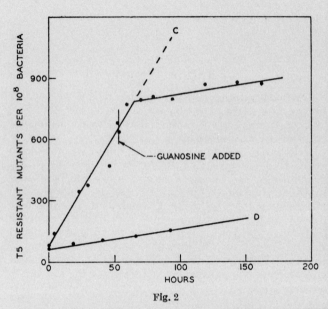

Fig. 2

In the experiments to be reported here, we used
a strain of *E. coli* (*B/lt*) requiring tryptophane, and
used tryptophane as the controlling growth-factor.
This organism is sensitive to the bacteriophage $T5$.
Mutants resistant to $T5$ present at any given time in
the population growing in the 'Chemostat' can be
scored by colony count simply by adding a small
quantity of the virus $T5$ to an aliquot at the time of
plating; in the presence of the virus, only the
resistant mutants will grow out into colonies.

As we reported earlier[3], mutation to $T5$ resistance
occurs at a constant rate independent of the rate at
which the bacteria grow, that is, independent of
the generation-time of the bacteria.   By plotting
against time the number of $T5$-resistant mutants present in the growth tube of the 'Chemostat', one obtains
curve A of Fig. 1. The slope of this straight line gives
a mutation-rate $\mu = 1\cdot4 \times 10^{-8}$/bacterium/hr.

When the nutrient medium contains theophylline
(a dimethylxanthine) in a concentration of 150
mgm./l., the number of mutants rises very quickly,
corresponding to the straight line C in Fig. 2, giving
a mutation-rate of $10\cdot7 \times 10^{-8}$/bacterium/hr. This
represents a seven-fold increase in the mutation-rate,
which we attribute to the mutagenic action of theophylline. However, if the nutrient contains, in addition
to 150 mgm./l. of theophylline, 50 mgm./l. of the
nucleoside guanosine, the number of mutants rises
much more slowly, as shown by line D in Fig. 2.
The slope of this line corresponds to a mutation-rate
of about 1×10^{-8}/bacterium/hr., indicating that

guanosine, in the concentration used, completely counteracts the mutagenic action of the theophylline.

In the experiment mentioned earlier, which is described in the upper curve in Fig. 2, the bacterial population is first grown (at a generation-time of $3 \cdot 2$ hr.) in the presence of 150 mgm./l. of theophylline with no guanosine present. After 53 hr., guanosine is added to give a concentration of 150 mgm./l. For the first 53 hr. and for a short time thereafter, the number of mutants follows the straight line C, which gives a mutation-rate of $10 \cdot 7 \times 10^{-8}$/bacterium/hr.; but afterwards the number of mutants follows another straight line which gives a mutation-rate of less than $1 \cdot 5 \times 10^{-8}$/bacterium/hr. The two straight lines intersect, not at the time when the guanosine is added, but about 12 hr. later.

In order to explain this 12-hr. delay in the fall of the mutation-rate after adding guanosine, we do not have to assume that it takes that time for the guanosine to counteract the mutagenic effect of theophylline, but may attribute the delay to the fact that mutations are not immediately expressed in the phenotype of the bacteria. When the guanosine is added, the mutations induced by theophylline may very well cease to occur; but the mutations induced prior to the addition of guanosine continue to be expressed phenotypically for a period of about 12 hr.

The results shown in Fig. 2 have to be interpreted as an actual reduction of the mutation-rate by guanosine; that is, they cannot be attributed to a selection against the bulk of $T5$-resistant mutants resulting from the presence of guanosine. It is easy to show that if such a selection were responsible for the low mutation-rate shown by line D in Fig. 2, then in the upper curve in Fig. 2 the number of mutants resistant to $T5$ should fall steeply after adding guanosine at the 53rd hour.

The concentration of guanosine needed to counteract the mutagenic effect of 150 mgm./l. of theophylline is quite low. For a concentration of about 2 mgm./l. of guanosine, the rate of mutation induced by theophylline falls to one-half.

The other normally occurring purine ribosides were examined for anti-mutagenic action at concentrations of 5 mgm./l. At this concentration adenosine and inosine are strongly anti-mutagenic against theophylline, whereas xanthosine has no such activity. In contrast to inosine itself, its components, that is, the free purine hypoxanthine and the free sugar ribose, are not anti-mutagenic even at concentrations of several hundred milligrams per litre.

A concentration of 500 mgm./l. of guanosine gives

practically complete suppression of the mutagenic action of the following purine derivatives (at concentrations of 150 mgm./l.) : theophylline, caffeine, theobromine, paraxanthine, and 8-azaguanine. But tetramethyluric acid and benzimidazole retain more than half their mutagenic effect.

One may ask what effect guanosine has on the spontaneously occurring mutations. As can be seen from line B in Fig. 1, 50 mgm./l. of guanosine gives a mutation-rate of 0.6×10^{-8}/bacterium/hr., that is, one-half to one-third as much as the spontaneous mutation-rate derived from curve A. This shows that guanosine in the concentration used reduces the mutation-rate to $T5$ resistance appreciably below the spontaneous mutation-rate.

AARON NOVICK
LEO SZILARD

Institute of Radiobiology and Biophysics,
          University of Chicago,
                Illinois.
                July 30.

[1] Fries, N., and Kihlman, K., *Nature*, **162**, 573 (1948). Demerec, M., Bertani, G., and Flint, J., *Amer. Naturalist*, **85**, 119 (1951).

[2] Novick, A., and Szilard, L., *Science*, **112**, 715 (1950).

[3] Novick, A., and Szilard, L., *Proc. U.S. Nat. Acad. Sci.*, **36**, 708 (1950).

[4] Monod, J., *Ann. Inst. Pasteur*, **79**, 390 (1950). Novick, A., and Szilard, L., *Cold Spring Harbor Symp. Quant. Biol.*, **16**, 337 (1951).

Printed in Great Britain by Fisher, Knight & Co., Ltd., St. Albans.

II. EXPERIMENTS WITH THE CHEMOSTAT
ON THE RATES OF
AMINO ACID SYNTHESIS IN BACTERIA

BY AARON NOVICK AND LEO SZILARD[1]

INFORMATION is accumulating on the pathways of the biosynthesis of amino acids in microorganisms, but it is not known what regulates the rates of the individual steps in such syntheses. As a rule, a bacterium synthesizes from ammonia and some simple carbon source all of the amino acids that are contained in its proteins, and it produces of each amino acid no more than is needed. For instance, if a strain of coli that requires tryptophane grows in an arginine-free nutrient in which the tryptophane concentration is kept above a certain minimum level, it will grow rapidly, obviously synthesizing all the arginine contained in its proteins at a correspondingly high rate. If the tryptophane concentration is maintained at a low level, the strain will grow slowly, perhaps ten times more slowly than before, and we see from our experiments (unpublished) that there is no appreciable outpouring of arginine (or ornithine or citrulline) into the medium. This leads us to believe that the rate at which arginine is synthesized is also reduced tenfold. One may ask how such a regulation of the synthetic rates is accomplished.

We may speculate that perhaps when protein synthesis is slowed, by supplying the tryptophane at a low rate, the internal "free amino acid" level of all the other amino acids rises and that somehow the increased concentration of each amino acid depresses the rate of the individual steps of synthesis leading to the formation of that amino acid.

Other experiments of ours (unpublished) indicate that a tryptophane-requiring strain of coli growing in a nutrient in which a certain very low concentration of arginine (of the order of $1 \gamma/1$) is maintained, will, with increasing concentration of arginine, synthesize a smaller fraction of the arginine contained in the bacterial proteins and will take up the rest from the medium. Again one may assume that the level of "free arginine" inside the bacterium regulates the biosynthetic steps leading to the formation of arginine, and the question arises as to how this is accomplished.

[1] Institute of Radiobiology and Biophysics, University of Chicago.

[21]

[Reprinted from *Dynamics of Growth Processes*, Princeton University Press, Princeton, N.J. (1954), pp. 21–32.]

A. NOVICK AND L. SZILARD

The normal regulation of the rate of the individual steps along a biosynthetic pathway leading to the formation of an amino acid might, however, be absent for a certain class of mutants in which a *precursor* of the amino acid, instead of being converted into the amino acid, is converted into some slightly modified compound which is poured out into the nutrient medium. The production of this precursor may then, under certain circumstances, go on at a high rate, apparently removed from the normally operating controls and limited only by the synthetic capacity of the bacterium. Such might be the case for a mutant of coli that we used in experiments described in this paper. This mutant requires tryptophane, is unable to grow on indol, and when grown slowly at very low concentrations of tryptophane pours out into the medium at a high rate a compound which is not indol and yet has a u.v. absorption spectrum closely resembling that of tryptophane. The absorption of this compound in the u.v. makes it easy to determine the rate at which it is poured out in any given experiment, and this permits us to study how the rate of this biosynthesis is affected when we vary selected factors of interest.

In these experiments, as in all the other experiments reported here, we used a special technique that utilizes a bacterial population which is maintained in the growth phase in a Chemostat (Novick and Szilard, 1950a, 1950b, 1951; Monod, 1950). In this device the bacterial suspension is contained in a growth tube which is provided with an overflow. Fresh nutrient enters the growth tube continuously at a rate of w cc/sec. The contents of the growth tube are kept stirred by aeration, and the bacterial suspension leaves the growth tube at the same rate at which fresh nutrient enters it so that the volume V of the bacterial suspension in the growth tube remains constant. The fresh nutrient entering the growth tube contains in excess all the factors needed for the growth of the bacteria except one, which is called the *controlling growth factor*. In our experiments we used tryptophane as the controlling growth factor and a strain of coli that requires tryptophane for its growth. The nutrient consisted of an ammonium lactate medium (Friedlein) with a certain input concentration a of tryptophane, which was varied from experiment to experiment within the range of $1/2$ to 1 mgm/l.

In the stationary state the tryptophane concentration c in the growth tube is very low, of the order of a microgram per liter, and at such low concentrations the growth rate α of the bacteria is an increasing function of the tryptophane concentration. Fig. 1 gives the growth rate

AMINO ACID SYNTHESIS

a as a function of the tryptophane concentration c. The growth rate a is defined as follows. In a nutrient medium in which the tryptophane concentration is maintained at a value c, the bacterial density n rises with time according to the formula $n = n_0 e^{at}$. The coefficient a in the exponent is, by definition, the growth rate. The reciprocal of this value, $1/a = \tau$, is the generation time. The doubling time of the bacteria is obtained from the generation time τ by multiplying by $\ln 2 = 0.693$.

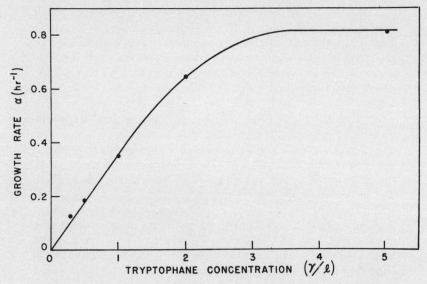

Fig. 1. Growth rate of *E. coli* B/lt as a function of tryptophane concentration.

Since in the stationary state the bacteria grow as fast as they are washed out, a tryptophane concentration c is automatically established for which we have: $a(c) = w/V$. The reciprocal of the washing out rate gives us the generation time τ of the bacteria $(\tau = V/w)$.

If the concentration c of the controlling growth factor in the growth tube is small compared to the input concentration a, then in the stationary state a change in the flow rate w will affect the generation time only and leave the bacterial density n unaffected. Such is the case if our tryptophane-less mutant strain of coli is grown with tryptophane as the controlling growth factor under the conditions of our experiments. The bacterial density in the growth tube in the stationary state is then proportional to the concentration of the controlling growth factor in

[23]

A. NOVICK AND L. SZILARD

the incoming fresh nutrient. For an input concentration of $a = 500\ \gamma/l$ we obtain a bacterial density of $n = 2 \times 10^8/cc$ as determined by colony count, and the bacterial suspension has an optical density of 0.135 at 350 mμ as determined by a Beckman spectrophotometer.

The tryptophane-requiring mutant that we used was obtained by picking from the B strain of *E. coli* a mutant resistant to the bacterial virus T1 and unable to use indol as a substitute for tryptophane. About 20% of the mutants resistant to the virus T1 fall into this well-known class. It happened, however, that we picked a particular mutant which, when grown in the Chemostat with tryptophane as the controlling growth factor, pours out into the medium a compound with a u.v. absorption very closely resembling that of tryptophane. Of ten other mutants which we subsequently isolated, none showed this phenomenon.

Why does one mutant pour out such a compound when ten other mutants belonging to the same class do not? All of them probably suffered an alteration at the same genetic locus; at least we are inclined to believe this since all of them have, in a single mutational step, acquired resistance to the virus T1 and also lost the ability to synthesize tryptophane. We might attempt to account for this difference in behavior by assuming the following. In all these mutants biosynthesis may proceed normally until a certain precursor of tryptophane containing the indol ring is reached. At that stage the precursor encounters in all these mutants some, but not necessarily the same, modified form of the normal enzyme contained in the wild type (or more precisely, in the prototroph). The modified enzyme itself can be different for the individual mutants picked. In the mutant which pours out a compound, the modified enzyme reacts with the tryptophane precursor and diverts it from the normal synthetic pathway by transforming it into a compound which is poured out by the bacterium. In the other mutants the modified enzyme may leave the precursor unchanged, and the increased concentration of the precursor might depress the rate at which it is synthesized so that its concentration inside the bacterium does not reach the threshold at which it will be poured out. This model is presented here solely for illustration and without any claim that other models might not be equally good or even better.

For simplicity, we shall refer to the compound poured out by our mutant as the "Compound." The absorption spectrum of the "Compound" in aqueous solution at pH 7 is shown in Fig. 2. It has a maximum close to 280 mμ and the ratio of the absorption at 280 and 250 mμ is about 2.2. Assuming that the molar extinction coefficient of

this compound at 280 mµ is the same as that of tryptophane, we can determine the quantity of the "Compound" in the growth tube of the
Chemostat by centrifuging off the bacteria and measuring the optical
density of the supernatant against the optical density of the original nutrient at 280 mµ. In the absence of other absorbing microbial products, the
optical density at 280 mµ may be taken as a measure of the quantity
of the "Compound" present. In each particular case, however, we have

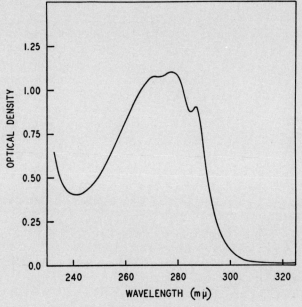

Fig. 2. Absorption spectrum of "Compound."

to check the spectrum between 250 and 280 mµ in order to see if the
presence of other u.v.-absorbing bacterial products makes it necessary
to apply a correction. A correction will be needed whenever the ratio
of the absorption at 280 and 250 mµ falls considerably below the
value of 2.2.

Fig. 3 shows the optical density of the supernatant at 280 mµ for
our bacterial strain growing in the Chemostat in the stationary state
with a tryptophane input concentration of $a = 500 \, \gamma/l$ at different
generation times τ. As the figure shows, the concentration of the
"Compound" is proportional to the generation time between $\tau = 4$ hrs

[25]

A. NOVICK AND L. SZILARD

and $\tau = 14$ hrs; i.e. between these limits the "Compound" is poured out at a constant rate corresponding to an increase in optical density of 0.141/hr. As the generation time is decreased below 3 hrs the rate of production of this "Compound" begins to fall off, and at a generation time of less than 2 hrs it is poured out at a rate of less than 0.040/hr.

By measuring the optical density of a known concentration of trypto-

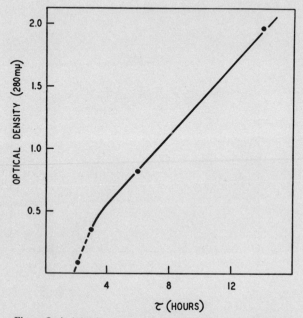

Fig. 3. Optical density of the supernatant at 280 mμ as a function of generation time.

phane at 280 mμ, we can compute on a mole-for-mole basis how much faster the "Compound" is poured out in a given experiment than tryptophane is taken up. One thus finds that at a generation of 14 hours the "Compound" is produced about 125 times faster than tryptophane is taken up. This furnishes a measure of the bacterium's capacity for tryptophane synthesis. A simple computation shows that at 37°C. tryptophane could be synthesized about four times as fast as the bacterium would need it if it grew—in the absence of tryptophane but otherwise under optimal nutritional conditions—at its maximum

AMINO ACID SYNTHESIS

growth rate, i.e. with a generation time of about 30 minutes (doubling time about 20 minutes).

It seemed of interest to find out whether our strain of bacteria continues to produce a compound even if we block protein synthesis by stopping the inflow of fresh nutrient into the growth tube of the Chemostat. In order to see this we first permitted a stationary state to establish itself at the generation time of about three hours with the bacterial population maintained at an optical density of 0.245 at 350 mμ, and then we stopped the inflow of nutrient. The small amount of tryptophane contained in the growth tube is then exhausted within a few seconds and so protein synthesis ceases almost instantaneously. As shown in Fig. 4, we found, when the inflow was stopped at zero time, that the

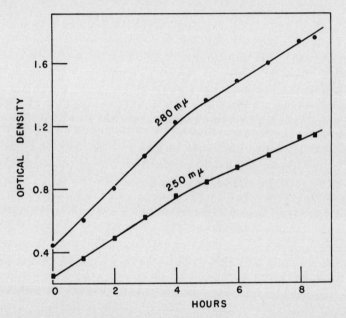

Fig. 4. "Compound" produced as a function of time after growth was stopped.

optical density at 280 mμ of the supernatant increased proportionately with time for about four hours at the rate of 0.19/hr; the rate then decreased to 0.12/hr and the "Compound" continued to pour out at this lower rate for another four hours, at which time the experiment was stopped. This means that even though a large number of steps and

[27]

A. NOVICK AND L. SZILARD

a correspondingly large number of enzymes must be involved in the synthesis of the "Compound" from lactate and ammonia, its production is continued at a high rate for more than 8 hours after protein synthesis has stopped.

In order to see how the rate of production of the "Compound" depends on the temperature when the generation time is long enough to permit us to assume that "Compound" production is going on at the maximum rate, we recorded the optical density of the supernatant in the stationary state at 280 mμ at three different temperatures. With the bacterial population maintained at an optical density of 0.272 at 350 mμ, we found the following rates at which the "Compound" is poured out:

at 25°   = 0.106/hr (τ = 13 hrs);
at 36.5° = 0.226/hr (τ ranging from 6.18 to 7.14 hrs);
at 43°   = 0.170/hr (τ ranging from 4.18 to 8.55 hrs).

It may be seen from this that the synthetic capacity for the production of the "Compound" rises from 25.5° to 36.5° by a factor of about 2 and then falls from 36.5° to 43° by a factor of about 0.75.

As Fig. 3 shows, the rate at which the "Compound" is produced rises when the generation time is increased from two hours to, say, four hours. The response of the rate of "Compound" production to a change in generation time might be instantaneous, or it might occur with a delay if, for instance, the accumulation of intermediates or adaptive phenomena are involved. To see which is the case we first maintained a stationary state in the Chemostat at a generation time of $\tau_1 = 2.3$ hrs and then suddenly decreased the flow rate w by a factor of 8/3 to give a generation time of $\tau_2 = 6.2$ hrs.

If the "Compound" is poured out at some rate A_1 at the generation time τ_1 and at a higher rate A_2 at the longer generation time τ_2, and if we then switch at zero time from the shorter generation time τ_1 to the longer generation time τ_2 by changing the flow rate w, then—assuming that the rate at which the "Compound" is poured out responds instantaneously to the change in generation time—the optical density of the supernatant at 280 mμ should be given by the formula:

$$\sigma = A_1\tau_1 + (A_2\tau_2 - A_1\tau_1)(1 - e^{-t/\tau_2})$$

If this formula holds we should obtain a straight line when we plot the optical density of the supernatant at 280 mμ against $(1 - e^{-t/\tau_2})$, and this straight line should start at zero time with $\sigma = A_1\tau_1$.

[28]

AMINO ACID SYNTHESIS

Fig. 5 shows such a plot of the results. The dotted line indicates the straight line along which the optical density would increase if the rate of production of the "Compound" remained unchanged when the generation time increased, i.e. if $A_2 = A_1$. The observed points, with the exception of the first few, lie on a straight line which rises more

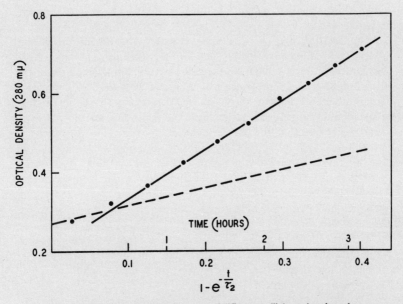

Fig. 5. Effect on the production rate of "Compound" brought about by a sudden decrease in growth rate.

steeply, corresponding to the increased rate of "Compound" production at the longer generation time. But this straight line does not start at zero time with $\sigma = A_1 \tau_1$ as does the dotted line; rather, the two straight lines intersect at a time near 30 minutes, giving the impression that the increase in rate of production of the "Compound" sets in with a lag of about a half hour.

This impression might, however, be misleading because at the shorter generation time of $\tau = 2.3$ hrs there seems to be present a bacterial product (other than the "Compound") which absorbs at 280 mμ and which makes its presence manifest by a decreased ratio of the optical densities for 280 and 250 mμ. At the shorter generation time this ratio is 1.68, whereas after switching to the longer generation

[29]

A. NOVICK AND L. SZILARD

time the ratio becomes 1.86, and under optimal conditions it may be as high as 2.2, corresponding to the absorption spectrum shown in Fig. 2. This extraneous absorption at the shorter generation time might simulate a lag when in fact there is no lag involved, and under the circumstances we can only say that if there is a lag, it is shorter than 30 minutes. When the experiment was repeated at 25°C. with a shift in generation time from $\tau_1 = 7$ hrs to $\tau_2 = 14$ hrs, no evidence for such a lag appeared.

The question may also be raised whether the production of the "Compound" responds instantaneously to a change of conditions which tends to reduce rather than increase the rate of its production. In order to see this we made an experiment in which we allowed a stationary state to establish itself in the Chemostat (at a tryptophane input concentration of 1 mg/l) at a generation time of 3.5 hrs, and then at zero time we suddenly raised the tryptophane concentration, both in the growth tube and in the incoming nutrient, to 4 mg/l. If production of the "Compound" were to stop instantaneously at zero time, the absorption of the supernatant at 280 mμ that is due to the presence of the "Compound" ought to fall off in a semilogarithmic plot as a straight line with a slope corresponding to a washing-out rate of $w/V = 1/3.5$ hrs.

The uppermost curve in Fig. 6 shows the observed points for the optical density of the supernatant at 280 mμ. Part of this optical density is due to the increased tryptophane concentration, for which we have to correct.

This correction is made as follows. When the tryptophane concentration is raised there is a corresponding increase in the absorption at 280 mμ. If tryptophane is converted into indol or some other compound containing the indol ring, there should be no change in the absorption at 280 mμ. However, the amount of free tryptophane present will be diminished by the amount of tryptophane which is bound in the protein of the bacteria. (We disregard as negligible the amount of free tryptophane that might be soaked up by the bacteria because we assume that it is not more than 1 mg/l.) From the excess of the bacterial population over the initial bacterial population we compute how much less free tryptophane is present at a given time than the amount that was added at zero time. The bacterial density is determined from the observed optical density of the bacterial suspension at 350 mμ. As Fig. 6 shows, this optical density follows in the semilogarithmic plot a straight line giving a bacterial growth rate of $a = 1/2$ hrs.

[30]

AMINO ACID SYNTHESIS

The corrected curve for the absorption at 280 mμ shown in the figure turns out to be a straight line which falls with a slope corresponding to a washing-out rate of $w/V = 1/4$ hrs. This means that the "Compound" present in the growth tube at zero time is washed out almost as fast as the growth tube is washed out by the incoming nutrient (washing-out time 4 instead of 3.5 hrs). If the two rates were exactly equal, this would mean that after zero time no "Compound" is poured out by the bacteria. Within the limits of our experimental

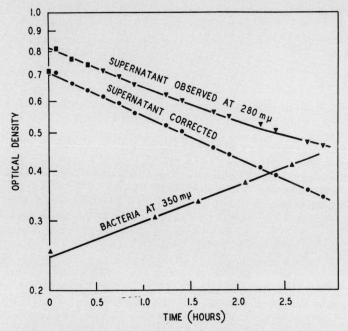

Fig. 6. Effect on the rate of production of "Compound" brought about by a sudden increase in tryptophane concentration.

accuracy we may only say that if any "Compound" is poured out after zero time the rate is less than one-third of the original rate. The corrected curve, taken at face value, shows that the decrease of the rate occurs without a lag, and within the limits of our experimental accuracy we may say that if there is a lag it is less than 10 minutes.

In all the experiments presented here, the rate at which the "Compound" is poured out is lower at generation times of about 2 hours than

[31]

A. NOVICK AND L. SZILARD

at longer generation times. But in all these experiments using the
Chemostat, bacteria growing with a generation time of 2 hours are of
necessity growing at a tryptophane concentration c higher than those
concentrations prevailing at longer generation times. For this reason
we cannot say whether the reduced rate of the production of the
"Compound" is due to a depressing effect of the higher tryptophane
concentration on this rate or whether the faster growth rate itself pre-
cludes a high rate of "Compound" production.*

The experiments described in this paper only begin to provide some
understanding of the manner in which the rates of various synthetic
processes in a bacterium are controlled. These studies have been selected,
however, to illustrate the usefulness of the Chemostat as an instrument
for the study of growing systems. This usefulness is in large part due
to the fact that the Chemostat maintains a population of constant
size, growing at a chosen rate in a constant chemical environment.

BIBLIOGRAPHY

Monod, Jacques. 1950. La technique de culture continue théorie et applica-
tions. *Annales de L'Institut Pasteur 79*, 390.
Novick, Aaron, and Szilard, Leo. 1950a. Experiments with the Chemostat
on spontaneous mutations of bacteria. *Proc. Nat. Acad. Sci. 36*, 708.
Novick, Aaron, and Szilard, Leo. 1950b. Description of the Chemostat.
Science 112, 715.
Novick, Aaron, and Szilard, Leo. 1951. Experiments on spontaneous and
chemically induced mutations of bacteria growing in the Chemostat.
Cold Spring Harbor Symp. Quant. Biol. XVI, 337.

* *Note added in proof*: We have been able to show that it is the higher concentration
of tryptophane and not the faster growth rate which suppresses the production of
"Compound." This was established by the use of mutants of our strain (kindly provided
by Dr. B. D. Davis) having another amino acid requirement in addition to tryptophane.
In one case such a mutant requiring arginine was grown in the Chemostat with arginine
as the controlling growth factor, tryptophane being supplied in slight excess. Under
these conditions the "Compound" was produced at most at one-tenth the rate observed
when the same strain was grown with tryptophane being limited and arginine being
supplied in slight excess. Similar results were obtained with another mutant requiring
histidine as well as tryptophane.

[Reprinted from THE JOURNAL OF GENERAL PHYSIOLOGY, November 20, 1955,
Vol. 39, No. 2, pp. 261–266]
Printed in U.S.A.

A DEVICE FOR GROWING BACTERIAL POPULATIONS UNDER STEADY STATE CONDITIONS*

By MAURICE S. FOX‡ AND LEO SZILARD

(*From the Institute of Radiobiology and Biophysics, University of Chicago, Chicago*)

(Received for publication, May 31, 1955)

A number of photocell-controlled devices for growing bacterial populations under steady state conditions have been described (1–3). We wish to describe here a device, similar in principle though different in operation, that was used by one of us in a series of mutation rate studies carried out in 1951–52 (4). In this device, the turbidity of an exponentially growing population of bacteria is maintained at some preset value by controlling the rate of influx of fresh nutrient solution into the culture while the total culture volume remains constant. The rate of influx of the nutrient solution into the growth tube is controlled by a photocell which responds to the turbidity of the culture, and the culture volume is held constant by means of an overflow siphon from the growth tube.

For a device of this kind to be useful in the study of bacterial populations, it is necessary that the bacteria grow solely in liquid suspension and do not grow on the walls or other surfaces of the growth tube. This consideration was of primary concern in the design of the device to be described.

The "breeder," as we call this device for the sake of brevity, includes a nutrient reservoir and an air pressure–controlled feeding system from the reservoir into the growth tube similar in design to the chemostat (5) (see Fig. 1). Feeding or addition of fresh nutrient liquid is accomplished in the following way. A pressure h_3 plus h_4 in Fig. 1 is maintained in the air space of the storage tank, this pressure head plus h_2 maintains a level of liquid, h_1, in the U-shaped capillary. At certain fixed time intervals a microswitch,[1] periodically activated by a ¼ R.P.M. synchronous motor,[2] opens a solenoid valve for a brief interval so that a pressure h_4 is allowed to act on the liquid in the capillary and force the liquid volume contained in the U-bend into the growth tube. When the solenoid valve closes, the original liquid level in the capillary is re-established. In the breeder this synchronous motor which controls feeding is switched on when the turbidity of the culture exceeds a certain set value.

* Aided by a grant from The Rockefeller Foundation.

‡ Present address: The Rockefeller Institute for Medical Research, New York.

[1] Acro snap switch SPDT, RD5L, ½ HP 125 to 250 volt, Newark Electric Co., Chicago.

[2] ¼ R.P.M., 1600 series, Hayden Manufacturing Co., Torrington, Connecticut.

262   GROWING BACTERIA IN STEADY STATE

For maintaining constant turbidity, a differential photocell circuit is used. This circuit is activated by a projection lamp,[3] whose beam passes through a 70 to 30 beam splitter. The 30 per cent fraction of the beam is reflected through a prism onto a reference photocell[4] (see Fig. 2), while the 70 per cent fraction of the beam passes through the growth tube and then onto a second photocell. By opposing the two (see Fig. 3) photocells through a variable resistance the current output across the resistance can be adjusted to zero for any given culture turbidity. An increase in turbidity registers as a positive current on an adjustable microammeter relay.[5] The relay is set to

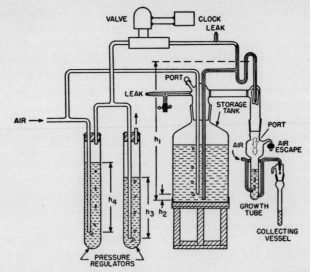

FIG. 1. The chemostat.

lock for readings greater than 7 microamperes which corresponds to a turbidity change of 5 to 10 per cent in a population of bacteria of about 2×10^8/ml. When the relay is in the locked position the synchronous motor[2] controlling the influx of fresh medium is energized.

A magnetic stirrer is used to provide adequate agitation at the bottom of the growth tube for the purpose of avoiding stagnant spaces where growth on the wall may occur.

A shaft driven by an additional synchronous motor,[6] geared down from 1

[3] Spartus 35 mm. slide projection lamp.

[4] Barrier layer selenium photocell No. 735, Photovolt Corporation, New York.

[5] Model 261-c, Assembly Products, Chagrin Falls, Ohio.

[6] 1 R.P.M. motor, model C5, Telechron, Inc., Ashland, Massachusetts.

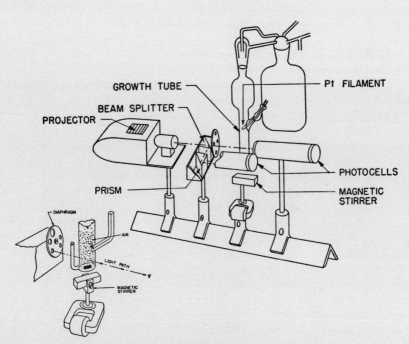

FIG. 2. The breeder.

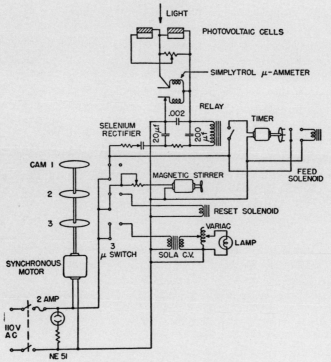

FIG. 3. Photocell control circuit.

263

264 GROWING BACTERIA IN STEADY STATE

to ¼ R.P.M. (Figs. 3 and 4), carries three cams, which activate three micro-switches[1] in succession. As a result the following sequence of events occurs:

1. The previous reading on the ammeter relay is erased and the feeding stopped if it had been going on during the previous period (microswitch on cam 1).

2. The stirring motor is turned off and a relay (reset solenoid) is activated, the vibration of which assures the restoration of the ammeter to zero (microswitch on cam 2).

3. The light source is turned on (microswitch on cam 3).

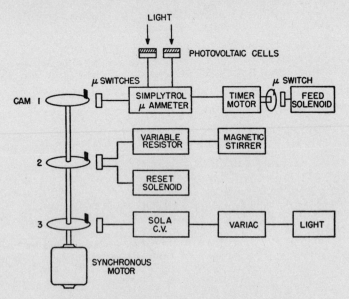

Fig. 4. Block diagram of photocell control circuit.

4. The circuit permitting response by the photocells to the light source is turned on and about 10 to 15 seconds is allowed for the culture turbidity to register on the ammeter relay (microswitch on cam 1).

5. The light is turned off (microswitch on cam 3).

6. The stirring motor is turned on (microswitch on cam 2). This sequence lasts 30 seconds and occurs once every 4 minute revolution of the cam shaft.

If the turbidity reading is above that set for the experiment the ammeter relay locks and feeding is initiated and continued for 3½ minutes. If the turbidity is below the set value no feeding is initiated.

It is thus possible to maintain a population of exponentially growing bacteria in a steady state. If n is the number of bacteria per milliliter in a growing culture, then the rate of increase of bacterial density is given by:

$$\frac{dn}{dt} = \alpha n$$

in which α is the growth rate given as one over the time required for an e-fold increase in bacterial density. If the bacterial suspension is removed and replaced by sterile nutrient in an approximately continuous manner at some given rate w/V, in which w is the rate at which fresh nutrient enters the growth tube in milliliters per hour, and V is the total volume of the culture in the growth tube in milliliters, then the rate of change of bacterial density will be given by:

$$\frac{dn}{dt} = \alpha n - n\frac{w}{V}$$

When the breeder maintains the culture at constant turbidity, $dn/dt = 0$, and $\alpha = w/V$. The growth rate, α, is thus given by the volume overflow from the growth tube per hour, divided by the liquid volume contained in the growth tube.

It is convenient to operate the breeder at a bacterial density of about $\frac{1}{10}$ of the saturation density for a given medium in order to avoid reduction of the growth rate due to metabolic limitations.

The principal difficulty encountered in the operation of the device here described is due to the tendency of the organism used (strain B of *Escherichia coli* and B/1t, a tryptophan-requiring bacteriophage T_1-resistant mutant of the same strain) to adhere to the walls of the growth tube. The first manifestation of such "wall growth" is a marked increase in the rate of overflow from the growth tube which simulates an increase of the bacterial growth rate. This can be readily understood on the basis of a dual source of bacteria in the liquid suspensions, those originating by growth in the liquid phase, and those being thrown off by the population growing on the walls of the growth tube. Soon after the first observation of such an increased growth rate, areas of bacterial growth on the walls of the growth tube become visible.

This difficulty was overcome by coating the walls of the growth tube with General Electric drifilm No. 9987, supplementing the medium with 0.05 per cent tween 80, and agitating the culture vigorously with a magnetic stirrer. The stirring element was encased in glass and coated with drifilm. The introduction of tween 80 resulted in excessive foaming of the culture under aeration. A fine platinum filament (0.003 inch) maintained just below red heat set about 1 cm. above the surface of the growing culture was found adequate to destroy the foam and to prevent its rising beyond the filament in the growth

266   GROWING BACTERIA IN STEADY STATE

tube (see Fig. 2). Control experiments demonstrated that neither the light beam used to measure turbidity, nor any of these expedients used to prevent wall growth, had any significant effect on either the growth rate of the bacteria or the mutation rates observed in the bacterial population (4).

Using the breeder, mutation studies have been carried out by one of us on the B/1t mutant of *E. coli* B in various complex and synthetic media under aerobic and anaerobic conditions. The results of these experiments are described elsewhere (4).

SUMMARY

A device for maintaining an exponentially growing population of bacteria in the steady state at maximal growth rates is described.

REFERENCES

1. Bryson, V., and Szybalski, W., *Science*, 1952, **116,** 45.
2. Anderson, P. A., *J. Gen. Physiol.*, 1953, **36,** 733.
3. Northrop, J. H., *J. Gen. Physiol.*, 1954, **38,** 105.
4. Fox, M. S., *J. Gen. Physiol.*, 1955, **39,** 267.
5. Novick, A., and Szilard, L., *Cold Spring Harbor Symp. Quant. Biol.*, 1951, **16,** 337.

Reprinted from

PROCEEDINGS
of the
NATIONAL ACADEMY OF SCIENCES

January 1959
Volume 45, Number 1

ON THE NATURE OF THE AGING PROCESS

By Leo Szilard*

ENRICO FERMI INSTITUTE OF NUCLEAR STUDIES, UNIVERSITY OF CHICAGO

Communicated by Theodore Shedlovsky, November 24, 1958

Introduction.—This paper represents an attempt to describe a hypothetical biological process that could conceivably account for the phenomenon of aging. Aging manifests itself in much the same general manner in all mammals, and we are in a position to learn enough about the aging of mammals to be able to test the validity of a theory that leads to predictions of a quantitative kind—as does the theory here presented.

We know that a gene can be responsible for the synthesis of a specific protein molecule, which in many cases has a known enzymatic activity. When we speak later of a mutant, or "incompetent," form of a gene, we mean an altered form of the gene, which cannot synthesize the specific protein molecule in its chemically active form.

Our theory assumes that the elementary step in the process of aging is an "aging hit," which "destroys" a chromosome of the somatic cell, in the sense that it renders all genes carried by that chromosome inactive. The "hit" need not destroy the chromosome in a physical sense. (See note 1 added in proof.)

We assume that the "aging hits" are random events and that the probability that a chromosome of a somatic cell suffers such a "hit" per unit time remains constant throughout life. We further assume that the rate at which chromosomes of a somatic cell suffer such "hits" is a characteristic of the species and does not vary appreciably from individual to individual.

As a result of an aging process of this nature, the number of the somatic cells of an individual organism which have "survived" up to a given age (in the sense of remaining able to fulfil their function in the organism) decreases with age. On the basis of our assumptions, spelled out below, the "surviving" fraction of the somatic cells decreases with age at an accelerating rate.

Our theory postulates that when f, the surviving fraction of the somatic cells of an individual, approaches a certain critical value, f^*, then the probability that that individual may die within a period of one year will come close to 1. On this basis, the theory establishes a relationship between the surviving fraction of the somatic cells and the age of death of the individual.

Because the young mammalian organism may be assumed to have a large functional reserve, we shall assume that the surviving fraction of the somatic cells of an individual may fall substantially before the organism loses its capacity to live, perhaps to a value somewhere between $1/3$ and $1/12$.

The precise meaning of the term "critical value," f^*, will shift as we go from the crudest form of the theory, which we shall discuss first, to a less crude form of the theory, which we shall discuss thereafter. In the crudest form of the theory, we shall assume that an adult does not die of natural causes until the surviving fraction of his somatic cells comes very close to the critical fraction f^* and that he dies at the critical age, i.e., within the year in which this surviving fraction reaches the critical fraction f^*. Thus, in its crudest form, the theory postulates that the age at death is uniquely determined by the genetic makeup of the individual.

Clearly, this cannot be strictly true, for, if it were true, identical twins would die within one year of each other. In fact, the mean difference between the ages at death of female identical twins can be estimated to be about 3.5 years. The discrepancy arises from the failure of the crude theory to take into account that in a cohort of identical individuals the number of deaths per year may be expected to rise as a continuous function with advancing age and that an appreciable number of deaths may be expected to occur at ages lower than the "critical age."

If not otherwise stated, our discussion here relates to man and, in particular, to the female of the species. In the case of man, the somatic cells of the female contain $m = 23$ pairs of homologous chromosomes. There may be in man perhaps 15,000 genes. There may be a larger number of specific DNA molecules which are inherited from generation to generation, but we designate as "genes" here only those DNA molecules which would handicap the individual if present in a mutant form. An individual who is a heterozygote for a mutant gene might not necessarily be handicapped under the conditions prevailing at present in the United States, where essentially no adult dies for lack of food or shelter and no adult has a reduced propensity to procreate because of his inability to provide food or shelter for his offspring. But such a heterozygote would have been handicapped (according to our definition of the term "gene") under conditions which were prevalent in the past—up to recent times. The present abundance of mutant forms of genes in the population may not correspond to the steady state under present conditions.

We may assume that the "genes" somehow affect differentiation and morphogenesis during the embryonic development of the individual and that a mutant form of a gene may cause, with a certain probability—appreciable even in the hetero-zygote—a developmental abnormality of the individual.

We assume that among the 15,000 genes, there are perhaps 3,000 genes which are important for the functioning of the somatic cells of the adult. We shall call these genes "vegetative genes," and a mutant form of such a gene we shall desig-nate as a "fault." Of the remainder of the genes, we shall assume that they are irrelevant for the functioning of the somatic cells of the adult organism.

We postulate that, in the course of aging, a somatic cell remains functional as long as, out of each pair of homologous vegetative genes, at least one of the two genes is competent and active and that the cell ceases to be functional when both genes are out of action. Accordingly when a chromosome suffers an aging hit, the cell will cease to be functional if the homologous chromosome has either previously suffered an aging hit or if it carries a fault.

According to the views here adopted, the main reason why some adults live shorter lives and others live longer is the difference in the number of faults they have inherited. If we assume that faults are distributed in the population at ran-dom, then we can compute the distribution of the faults, from the mean value of faults per person (which we shall designate by n). From the observed distribution of the ages at death, between seventy and ninety years of age, we shall be led to conclude that we have $n > 2$. For $n = 2$ we would obtain from the crude theory for the critical surviving fraction of the somatic cells $f^* \backsim 1/4$. For $n = 4$ we would obtain $f^* \backsim 1/12$. On this basis we shall be led to conclude that we have $n < 4$.

We shall, for the purposes of our discussion, adopt, as a reasonable value, $n = 2.5$, and then we obtain $f^* \backsim 1/6$, which would seem to be a reasonable value.

The "Surviving" Fraction of the Somatic Cells.—We shall now proceed to com-pute the "surviving" fraction of the somatic cells of a female who has inherited r faults, as a function of her age.

We designate by ξ the average number of "aging hits" that have been suffered by a chromosome of a somatic cell, and we may write

$$\xi = \frac{1}{2m} \frac{\text{age}}{\tau},\tag{1}$$

where τ is the average time interval between two subsequent aging hits suffered *in toto* by the m pairs of homologous chromosomes contained in a somatic cell. We may call this average time interval τ the "basic time interval of the aging process."

Let us now consider a female who has inherited r faults. If none of the pairs of homologous chromosomes contain more than one fault—a condition likely to be fulfilled if r is small compared to m—then we may write for the "surviving" frac-tion of her somatic cells at a given age

$$f = [1 - (1 - e^{-\xi})^2]^{m-r} \cdot e^{-r\xi}\tag{2}$$

or

$$\ln f = (m - r) \ln [1 - (1 - e^{-\xi})^2] - r\xi.\tag{3}$$

For $\xi \ll 1$ we may write, from equation (3), neglecting $m\xi^4$ and $r\xi^3$, etc.,

$$\ln \frac{1}{f} = m(\xi^2 - \xi^3) + r(\xi - \xi^2).\tag{4}$$

Writing

$$\rho = \frac{r}{2m},$$ (5)

we may write

$$\ln \frac{1}{f} = m[\xi + \rho]^2 . [1 - (\xi + \rho)],$$ (6)

provided $r \ll x \ll 2m$ (i.e., $\rho \ll \xi - 1$)

$$\ln \frac{1}{f} = m\eta^2 (1 - \eta),$$ (7)

where $\eta = \xi + \rho$. In place of equation (7) we may write, in our approximation,

$$f = [1 - (1 - e^{-\eta})^2]^m$$ (8)

We may also write inversely

$$\eta = \ln \frac{1}{1 - \sqrt{1 - f^{1/m}}},$$ (9)

or, expanding,

$$\eta = \sqrt{\frac{1}{m} \ln \frac{1}{f}} + \frac{1}{2m} \ln \frac{1}{f}$$ (10)

According to the assumption of the crude theory, f, the surviving fraction of the somatic cells, reaches the critical value f^* at the age of death, which we designate by t_r. Further, we designate by x_r the average number of aging hits suffered *in toto*, up to the age of death, by the m pairs of homologous chromosomes of the somatic cells. Thus we have, at the age of death,

$$x_r = \frac{t_r}{\tau} = 2m\xi$$ (11)

and

$$\eta = \frac{x_r + r}{2m}.$$ (12)

Accordingly, we may write at the age of death, where we have $f = f^*$, from equation (8),

$$\ln \frac{1}{f^*} = \frac{(x_r + r)^2}{4m} \left(1 - \frac{x_r + r}{2m}\right).$$ (13)

Similarly, we may write at the age of death, from equation (10),

$$x_r + r = \sqrt{4m \ln \frac{1}{f^*} + \ln \frac{1}{f^*}}$$ (14)

or

$$\frac{t_r}{\tau} + r = \sqrt{4m \ln \frac{1}{f^*} + \ln \frac{1}{f^*}}. \tag{15}$$

For the genetically perfect female, for whom we have $r = 0$, we shall designate the age at death by t_0.   We shall call t_0 the "life-span" of the species.

From equation (15) we may write for, the life-span, t_0,

$$\frac{t_r}{\tau} + r = \frac{t_0}{\tau} \tag{16}$$

or

$$t_r = t_0 - \tau r \tag{17}$$

or

$$r = \frac{t_0 - t_r}{\tau}. \tag{18}$$

As may be seen from equation (17), the addition of one fault to the genetic makeup of an individual shortens the life of that individual by $\Delta t = \tau$, so that we may write

$$\Delta t \text{ per fault} = \tau. \tag{19}$$

This expresses one of the basic results of our theory.   According to equation (19), an individual whose genetic makeup contains one fault more than another individual has a life-expectancy which is shorter by τ, the basic time interval of the aging process.   This holds true within the crude theory for individuals who have inherited a small number of faults.

Concerning the life-span, t_0, we may write, from equations (11), (13), and (16),

$$\ln \frac{1}{f^*} = \frac{1}{4m} \left(\frac{t_0}{\tau}\right)^2 \left(1 - \frac{1}{2m}\frac{t_0}{\tau}\right), \tag{20}$$

and, from equations (15) and (16), we may write

$$\frac{t_0}{\tau} = \sqrt{4m \ln \frac{1}{f^*} + \ln \frac{1}{f^*}}. \tag{21}$$

The Distribution of the Ages at Death.—The above equations hold within the framework of the crude form of the theory.   In this form of the theory, members of one cohort would die only in certain years—at the critical ages, t_r—and thus the years in which death occurs within one cohort would be separated from each other by time intervals of τ years; no deaths would occur in the intervening years.

Further, if the distribution of the faults in the population is random, then the number of deaths, P_r, occurring at each age, is given by the Poisson distribution:

$$P_r = \frac{n^r e^{-n}}{r!} \tag{22}$$

where, according to equation (18), we have $r = \dfrac{t_0 - t_r}{\tau}$ and where n stands for the average number of faults per individual.

The distribution of the ages at death in the population is actually a continuous function of the age. Even though the probability that an individual may die within a year may increase rather steeply as the surviving fraction of his somatic cells approaches the critical value f^*, genetically identical individuals do not all die at the same age. The observed mean age difference at death of identical twins may be regarded as a measure of the scattering of the ages at death, which is left out of account by the crude form of the theory and to which we shall refer as the "non-genetic scattering."

For the time being, we shall continue to leave this non-genetic scattering out of account; yet, for the sake of convenience, we shall henceforth describe the distribution of the ages at death by $P(r)$, a continuous function of r, in place of the discontinuous "Poisson" values, P_r. For $P(r)$ we may write

$$P(r) = \frac{n^r}{\Gamma_{(r+1)}} e^{-n} \tag{23}$$

where Γ represents the gamma function (which for integral values of $r + 1$ assumes the values of $r!$) and where we have

$$r = \frac{t_0 - t}{\tau}. \tag{24}$$

For the number of deaths occurring within a cohort per unit time, we may then write, according to our theory,

$$d(\text{theor.}) = - \frac{dr}{dt} \frac{n^r}{\Gamma_{(r+1)}} e^{-n}. \tag{25}$$

From equation (24) we obtain

$$- \frac{dr}{dt} = \frac{1}{\tau}. \tag{26}$$

Thus we may write, from equation (25),

$$d(\text{theor.}) = \frac{1}{\tau} \frac{n^r}{\Gamma_{(r+1)}} e^{-n} \text{ per year}, \tag{27}$$

where r is given by equation (24) and where τ is expressed in years.

The approximation used throughout this paper holds for small values of r, which correspond to high ages at death. We may say that at high ages of death the distribution of ages at death in the population is represented by a reversed Poisson distribution (27), where small values of r correspond to high ages at death.

Lower Limit for n.—We shall now proceed to compare the distribution of the ages at death, as given by our formula (27), with the actually observed distribution of the ages at death, as given by the U.S. Life Tables, based on the 1949–50 Census.

For the purposes of this comparison, we shall use Table 6 for white females, which lists the number of deaths per year, in yearly intervals, as a function of age. According to this table, the maximal number of deaths occurs between the eightieth and eighty-first year; the corresponding maximal number of deaths per year is 0.0344 per person.

The distribution of the ages at death is not symmetrical around the age at the maximum, $l^* = 80.5$ years of age; the number of deaths per year fall faster toward higher ages than toward lower ages. Thus the table lists, for the number of deaths per year, 0.0230 per person between the ages seventy and seventy-one and 0.0179 per person between the ages of ninety and ninety-one.

We may derive from this table a "normalized" distribution of the ages at death by forming R(obs.), the ratio of the number of deaths per year and the maximal number of deaths per year, 0.0344. Thus we obtain R(obs.) $= 0.667$ at 70.5 years of age; R(obs.) $= 1$ at 80.5 years of age; R(obs.) $= 0.520$ at 90.5 years of age.

We may similarly obtain from the number of deaths per year, given as a function of age by the theory, a "normalized" distribution of the ages at death, by forming R(theor.), the ratio of the number of deaths per year given by equation (27) and the maximal number of deaths per year given by $d(\text{theor.})_{\text{max.}}$:

$$d(\text{theor.})_{\text{max.}} = \frac{1}{\tau} e^{-n} \left\{ \frac{n^r}{\Gamma_{(r+1)}} \right\}_{\text{max.}} \tag{28}$$

If we designate by r^* the value of r for which this expression becomes maximal, we may write, for $n \geqq 2$,

$$r^* \approx n - 0.5. \tag{29}$$

Accordingly, we may write

$$R(\text{theor.}) = \frac{d(\text{theor.})}{d(\text{theor.})_{\text{max.}}} = \frac{n^r}{n^{(n-0.5)}} \frac{\Gamma_{(n+0.5)}}{\Gamma_{(r+1)}}. \tag{30}$$

We may now ask for what value of n would the normalized Poisson distribution, R(theor.), fit R(obs.), both above and below 80.5 years of age so that we have for a suitably chosen value of Δr, for $r = r^* + \Delta r$, R(theor.) $= 0.667$ (the value of R(obs.) at 70.5 years of age) and that we also have, for $r = r^* - \Delta r$, R(theor.) $= 0.520$ (the value of R(obs.) at 90.5 years of age).

It turns out that such a fit is possible only for a value of n which is very close to $n = 2$. For the corresponding value of Δr we obtain $\Delta r = 1.4$. For the corresponding value of τ we may write

$$\tau = \frac{10}{\Delta r} \text{ years.} \tag{31}$$

For $n = 2$ and with $\Delta r = 1.4$, we obtain $\tau = 7.15$ years.

For values of n which are substantially larger than 2, it is not possible to fit the normalized Poisson distribution R(theor.) to R(obs.) in this manner. If R(theor.) is made equal to 0.520 (the value of R(obs.) at 90.5 years of age) for $r = r^* - \Delta r$, then, for $r = r^* + \Delta r$, we have R(theor.) < 0.667 (the value of R(obs.) at 70.5 years of age).

Because there is reason to believe that, below 80.5 years of age, the crude theory gives too low values for R(theor.), we cannot exclude the possibility that we have $n > 2$. Therefore, from the fact that R(theor.) derived from the crude theory fits R(obs.) for $n = 2$ between the ages of 70.5 and 90.5, we may not conclude that we actually have $n = 2$, and we may only conclude that we have

$$n \geq 2. \tag{32}$$

Approximation of the Poisson Distribution by a Gaussian Relationship between τ and n.— R (theor.) given by equation (30) goes over into a Gaussian for $n \gg 1$. For a Gaussian, the value of R(theor.) $= 0.667$ (the value of R(obs.) at 70.5 years of age) corresponds to a distance from the maximum of 0.9 σ, where σ is the standard deviation of the Gaussian. Similarly, R(theor.) $= 0.520$ (the value of R(obs.) at 90.5 years of age) corresponds to a distance from the maximum of 1.14σ. Thus the time interval of 20 years around the maximum corresponds to 2.04σ, and hence we have

$$\sigma = \frac{20}{2.04} \text{ years} = 9.8 \text{ years} \tag{33}$$

Because the variance of a Poisson distribution is given by its mean, n, we may write (for $n \gg 1$)

$$\sigma = \tau \sqrt{n} \tag{34}$$

or

$$\tau = \frac{\sigma}{\sqrt{n}}, \tag{35}$$

and thus we obtain $\tau = \dfrac{9.8}{\sqrt{n}}$ years.

While equation (34) holds, strictly speaking, only for large values of n, the error is small even for $n = 2$.

For $n = 2$ from equation (35) we obtain $\tau = 6.82$ in place of the previously given value of $\tau = 7.15$.

For $n = 2.5$ from (35) we obtain $\tau = 6.2$ years in place of the "correct" value of $\tau = 6.3$ years, which we find by fitting the "normalized" Poisson distribution, as well as possible, to R(obs.).

Thus, for most of our purposes we may use equation (35) for values of n for $n \geq 2$.

From equation (35) we obtain

$$n\tau = \sigma\sqrt{n}, \tag{36}$$

where $n\tau$ represents the average life-shortening caused by the "load of faults," n. Since the value of σ is empirically fixed, the higher we assume n to be, the higher is the life-shortening effect which we must attribute to it. In this sense, the life-shortening effect of the mutation load increases with $\sqrt{n}$.

Correction of τ for the Non-genetic Scattering of the Ages at Death.—Because the non-genetic scattering of the ages at death has so far not been taken into account by us, the observed distribution of the ages at death may be expected to be actually somewhat broader, and, accordingly, the actual value of τ may be expected to be somewhat lower than the values given above.

The mean age difference at death between female identical twins has been reported by Franz J. Kallman to be about 2.6 years for twins dying above the age of 60. From this value we may estimate, on the basis of the Life Tables, the mean age difference at death of female identical twins who die as adults above the age of 40 to be 3.4 years.   If the distribution of the ages at death of genetically identical individuals resembled a Gaussian, then the variance of the distribution of the ages at death in the population would be equal to the sum of the variance of this Gaussian and that of our theoretical distribution of the ages at death.   By making such an assumption, for the purposes of this computation, we may then correct the values of τ, given above, as follows:

From the fact that the mean age difference at death of female identical twins may be taken to be about 3.4 years, it follows that the standard deviation of the distribution of their ages at death is about 3 years.   Using this value, we find that the nongenetic scattering increases the variance of the distribution of the ages at death by a factor of about 1.1 and, accordingly, the previously given values of τ must be reduced by 5 per cent.

Thus we may now write, for the corrected values of τ, for $n = 2$, $\tau = 6.8$ years; and for $n = 2.5$, $\tau = 6$ years.

We may also write on this basis—within the limits of the approximation—for $n > 2$,

$$\tau = \frac{9.3}{\sqrt{n}} \text{ years.} \tag{37}$$

Substituting this value of τ in equation (16), we obtain

$$\frac{t_0}{\tau} = \frac{t_r}{9.3} \sqrt{n} + r. \tag{38}$$

The Value of the Critical Surviving Fraction of the Somatic Cells f*—Upper Limit for n.—In order to compute the critical surviving fraction of the somatic cells, f^*, we shall now make use of the fact that (for white females) the maximal number of deaths per year occurs at 80.5 years of age.   Our theory demands (29) that the maximal number of deaths per year should occur for individuals for whom we have $r = n - 0.5$.   Accordingly, we may write $t_r = 80.5$ and $r = n - 0.5$.   We thus obtain, from equation (38),

$$\frac{t_0}{\tau} = \frac{80.5}{9.3} \sqrt{n} + n - 0.5 \tag{39}$$

and, from equation (20),

$$\ln \frac{1}{f^*} = \frac{1}{4m} \left(\frac{80.5}{9.3} \sqrt{n} + n - 0.5\right)^2 \left[1 - \frac{1}{2m} \left(\frac{80.5}{9.3} \sqrt{n} + n - 0.5\right)\right]. \tag{40}$$

From this equation we may now compute for a given value of n the corresponding value of f^*.   Thus we obtain for $n = 2$, $f^* \approx 1/4$; for $n = 2.5$, $f^* \approx 1/6$; for $n = 4$, $f^* \approx 1/12$.

On this basis we may then write, by assuming $f^* > 1/12$,

$$n < 4. \tag{41}$$

A value of $f^* \approx 1/6$ would seem to be rather reasonable and, therefore, we shall adopt, as a reasonable value for n, the value of

$$n \approx 2.5, \tag{42}$$

and, as a reasonable value for τ, the corresponding value of

$$\tau = 6 \text{ years.} \tag{43}$$

We shall in the remainder of the paper base all our discussions on these values of n and τ. We cannot exclude, of course, the possibility that n might be somewhat larger and that τ might be somewhat lower.

The "Physiological Age."—The general physiological age may be defined for a given population on the basis of its age-specific death rate; according to our theory, it may be defined as the age of the genetically perfect female who has the same surviving fraction of the somatic cells, f. Accordingly, we may say that two females, whose genetic makeup differs by Δ faults, differ from each other in their physiological age by $\tau\Delta$ years at sufficiently high ages, as demanded by the approximation used.

Changing the Load of Faults.—If, as a result of living under "modern" conditions, our load of faults should, in time, be doubled, then the average adult woman would live $n\tau$ years less than she does today.

For $n = 2.5$ we have $n\tau = 15$ years. Thus the physiological age of the average female at 65 would be the same as that of the average 80-year-old woman today.

If we were to assume that $n > 2.5$, then $n\tau$ would amount to more than 15 years because $n\tau$ increases, according to equation (36), with $\sqrt{n}$.

A doubling of our load of faults might conceivably occur, in time, through the exposure of the population, generation after generation, to ionizing radiation, in an intensity that doubles the mutation rate.

Such an increase in our load of faults might perhaps occur also as a result of the current practice of controlling the family size. As spelled out below this practice might conceivably eliminate one of the selection pressures which have tended to keep our load of faults low.

We may, on this occasion, also ask how much advantage the genetically perfect (faultless) female would have over the average female of today. Assuming $n = 2.5$, we may say, on the basis of considerations similar to those presented above, that the genetically perfect female would at 50 years of age have the same physiological age as the average female of 35 today. Her most probable age at death would be 92 instead of 80. If n were larger than 2.5, the advantage of the genetically perfect female would be greater.

Life-Shortening Effect of Ionizing Radiation on the Adult Offspring of the Exposed Population.—Experiments of W. L. Russell have shown that the offspring of mice which have been exposed to a dose of fast neutrons have a reduced life-expectancy. This has generally been interpreted to mean that exposure of the parents to ionizing radiation induces mutations in the germ cells of the gonads and thus "reduces the viability" of the offspring.

From the point of view of our theory, however, we have to distinguish between that reduction of the life-expectancy of the offspring which is due to an increased mortality of young animals and that reduction which is due to a decrease in the life-expectancy of the adults. All the mutations induced by ionizing radiation may contribute to the former, but only the "faults" contribute to the latter.

In the case of man, at least, it should be possible to make a fairly clean separation between these two categories of life-shortening. In the case of man the U.S. Life

Tables show that the number of deaths per year falls, from an initial high value in the first year of life, to about 40 per 100,000 per year at the age of 10. Moreover, of the few deaths occurring at 10 years of age, a substantial fraction is due to accidents. Thus we are led to believe that, in the heterozygous individual, mutant (incompetent) forms of genes may cause the death of the embryo, or of the infant below 10 years of age, while they do not cause death with an appreciable probability after the tenth year of age, unless they represent "faults." Faults increase the age-specific death rate above 10 years of age only in conjuction with aging hits, and they increase it appreciably only above 40.

If we observe the life-shortening of the adult animal in the offspring of an irradiated population, resulting from the induction of faults by ionizing radiation in one species, we may be able to predict, on the basis of our theory, the life-shortening for another species. This may be seen as follows:

We obtain from equations (19) and (21) for the relationship between the life-shortening, Δt per fault, and for the life-span of the species, t_0,

$$\frac{\Delta t \text{ per fault}}{t_0} = \frac{\tau}{t_0} = \left(\sqrt{4m \ln \frac{1}{f^*}} + \ln \frac{1}{f^*} \right)^{-1} \tag{44}$$

The right-hand side of the equation contains only the chromosome number m and the critical value f^*. Therefore, if two species of mammals have the same value of f^* and the same chromosome number m, their life-shortening per fault caused by exposure of their parents to ionizing radiation amounts to the same fraction of their life-span. We may call the ratio the "specific life-shortening" of a fault.

If the two species of mammals may be assumed to have also the same number of vegetative genes and if the sensitivity of their genes to the ionizing radiation employed may be assumed to be the same, then the number of faults produced by a given dose of radiation will be the same for the two species. Thus, according to equation (44), the radiation exposure will shorten the lives of the two species by the same fraction of the life-span.

If m, the number of their chromosome pairs, is different for the two species, then the "specific life-shortening" will be larger for the species which has the smaller chromosome number. According to (44), the specific life-shortening increases about inversely with $\sqrt{m}$. The number of chromosome pairs is 11 for the Chinese hamster and 39 for the dog. Therefore, according to (44), the specific life-shortening per fault induced may be expected for the Chinese hamster to be higher than for the dog by a factor of about 2.

The mouse has $m = 20$ pairs of homologous chromosomes and we may therefore estimate the life-shortening of man from the life-shortening of the mouse and vice versa, by postulating that the life-shortening per rep in man and in the mouse amount to about the same fraction of their life-span. (Some authors believe that man is about twice as sensitive to X-rays as the mouse and, if they are correct, then our estimated value for the life-shortening of man would be low by a factor of about two.)

W. L. Russell found that an X-ray dose of 300, as well as a dose of 600 rep, induces 25×10^{-8} gene mutations per rep per locus in the spermatogonia of mice. Assuming 15,000 loci, we may conclude that an X-ray dose of 667 rep would induce 5 mutations in the diploid offspring of exposed mice. If we assume that one-fifth of the genes are vegetative genes, then 667 rep of X-rays induce on the average one fault in the offspring. Assuming $\tau = 6$ years for man, we may thus expect in man a life-shortening of the adult offspring of $\delta^* = 3.3$ days per rep.

The number of faults induced in the offspring per rep depends on the nature of the ionizing radiation, and may be assumed higher for fast neutrons than for X-rays. Also, in the case of X-rays, the number of faults induced might conceivably depend not only on the total dose but also on the dose rate and be lower for lower dose rates.

The actual value of δ^* ought to be determined experimentally, for the different kinds of ionizing radiations which are of interest, by direct observation of the life-shortening of the adult offspring. Experimental data so far available are inadequate.

If the average mutation rate per gene per generation is 1/60,000 and if we assume for N_1, the number of vegetative genes, $N_1 = 3000$, and for N_t the total number of genes, $N_t = 15,000$ then we obtain for μ_1, the spontaneous mutation rate of the haploid set of vegetative genes, $\mu_1 = 0.05$, and for μ_t, the total spontaneous mutation rate of all genes, $\mu_t = 0.25$. We shall use these values for the purposes of our discussion below.

The Life-Shortening Effect of Ionizing Radiation on the Exposed Population.— We may expect ionizing radiation to produce gene mutations in the chromosomes of the somatic cells of an exposed individual and we shall assume that the sensitivity of the genes of the somatic cells is about the same as that of the genes in the spermatogonia and the oogonia. Because a certain fraction of these mutations, perhaps one-fifth of them, affect vegetative genes, faults are induced in the chromosomes of the somatic cells of the exposed individual. It can be shown that an exposure to ionizing radiation which induces on the average one fault per somatic cell must reduce—on this score alone—the life-expectancy of the exposed individual by about τ years. If exposure to ionizing radiation had no other life-shortening effect, the life-shortening, δ, of the individual in the exposed population would be equal to δ^*, the life-shortening of the adult offspring of the exposed population (see note 2 added in proof).

Maternal Selection Pressure against Faults?—It is conceivable that a woman who carries a particular fault in her genetic makeup ceases to be capable of bearing children $\tau = 6$ years earlier, on the average, than her counterpart who lacks that particular fault. This is what one would expect on the assumption that the termination of a woman's reproductive period is determined by her physiological age— if all other factors are equal. If this assumption is correct, then a powerful selection has operated in the past that has tended to keep the load of faults low.

In the past, infant mortality was high, the birth rate was high, and women kept on having children until the end of their childbearing period. Clearly, the maternal selection mentioned above is switched off when women have two or three children between the ages of 18 and 25 and then avoid having further children.

If such a "maternal" selection was the predominating selection of the past, then we may expect that, when this selection ceases to operate, our load of faults may double, in time. As discussed before, senescence would then set in about 15 years earlier.

However, even if we assume the worst in this respect, our load of faults still cannot increase by more than $2\mu_1 = 0.1$ per generation. This means that—at the worst—it would take 25 generations for our load of faults to double.

The effect of the "maternal" selection here discussed might be estimated as follows:

Let us single out one vegetative gene. If a woman carries, as a heterozygote, this particular gene in a mutant form, her physiological age is $\tau = 6$ years higher than that of another woman who does not carry this particular "fault" but who is otherwise identical in her genetic makeup. We now assume that the "physiological age" sets the termination of the reproductive period, and we take for the "most probable duration" of the reproductive period 30 years. Thus the fault singled out "most probably" shortens the reproductive period by one-fifth of its length. The fertility of younger women is higher than that of older women. Near the end of the reproductive period, the (average) time interval between two successive pregnancies might be by a factor $k > 1$ (of perhaps 2 or 3) longer, than the interval between two successive pregnancies, averaged over the whole reproductive period.

If we postulate that this "maternal" selection constitutes essentially the sole selection pressure against faults, then we may write for the mutation equilibrium of the fault, singled out:

$$2\frac{\mu_1}{N_1} = \frac{n}{2 \times 5 \; kN_1}, \tag{45}$$

and, for the value of $\mu_1 = 0.05$ per generation, we thus obtain $k = n$. For $n = 2.5$ we have

$$k = 2.5. \tag{46}$$

Under "natural" conditions, for young women the average time interval between two successive pregnancies might be about 1 year; toward the end of the reproductive period it might be about 5 years; and, averaged over the whole reproductive period of about 30 years, it might be about 2 years. These values correspond to $k = 2.5$. This coincidence might, of course, turn out to be fortuitous.

Refinements of the Theory—Specialized Vegetative Genes.—It appears likely that there exist genes which are not essential for the functioning of most of the somatic cells of the adult but each of which is essential for the functioning of one particular kind of specialized somatic cell. We shall call these genes "specialized vegetative" genes, and mutant forms of such genes we shall call "specialized faults."

We shall now single out, for the purposes of our discussion, specialized cells which synthesize a gene product, a particular enzyme, for instance, that serves not the needs of these cells themselves but rather those of the organism as a whole. Such specialized cells might in some cases fulfill their function in the organism by releasing the enzyme into the circulation.

We may in general assume that the normal young person has a considerable reserve of such enzymes, and we shall specifically assume, for the purposes of this particular discussion, that the maximal output of a normal young person is higher than the need of the organism by a factor of about 6.

The maximal output of enzyme by such a specialized cell may be assumed to be lower by a factor of $1/2$ in the heterozygous individual, who has inherited a mutant form of the specialized vegetative gene. Further, it may be shown that, for an individual who has inherited $n = 2.5$ faults, the surviving fraction of the somatic cells is about one-third at 54 years of age. Since, in the heterozygote, the specialized cells under discussion carry one additional fault, the surviving fraction of the specialized cells will reach one-third about 6 years earlier. Thus, at 48 years of age, the maximal output of the enzyme of the surviving specialized cells of the heterozygote will be lower, by a factor of about $1/6$, than for a normal young person.

On this basis we may then expect that around 50 years of age there may become manifest, in such heterozygotes, symptoms of disease due to the insufficiency of the output of one kind of specialized cell. The inheritance of diseases of this class may be expected to show a marked degree of dominance.

Speaking more generally, we may expect to see in certain heterozygotes, late in life, narrowly circumscribed degenerative phenomena which are caused by specialized faults they have inherited.

The Number of "Segments" per Chromosome.—Instead of assuming that a whole chromosome is "destroyed" in one aging hit, we might choose to assume that the elementary step in the process of aging consists in the random destruction of one-half of a chromosome. The formulas given above remain then unchanged, except that we have to write $2m$ in place of m. As one may see from equation (40), we then obtain, for the same value of n, a higher value for f^*. Thus for $n = 2.5$ we obtain $f^* \approx 1/3$. Apart from this, the general character of the theory remains unchanged.

However, one might ask at this point whether one could not generalize the theory, presented above, by assuming that each chromosome consists of g segments and that the elementary step in the process of aging consists in the random destruction of such segments, independently of each other. By choosing the value of g larger and larger, we might then gradually change the character of the theory and might end up with a theory which postulates that the aging process consists in a sequence of gene mutations of the chromosomes of the somatic cell.

A theory of this kind would, however, come up against difficulties, which are as follows:

As may be seen from equation (40) (where we now have to write gm in place of m), for a fixed value of f^*, n goes up roughly parallel with increasing g. A very large value of n might, however, be incompatible with the known fertility of consanguinous matings.

Further, as we increase g, we would also increase the difference of the life-expectancy of the female and the male. The male of the species has only one X chromosome, while the female has two. Let us disregard here the possibility that a substantial piece of the X chromosome might be covered, in the male, by genes contained in the Y chromosome. Let us also assume, for the sake of argument, that f^* has the same value for the male as it has for the female. On the basis of these assumptions, we may then identify the male, from the point of view of his life-expectancy, with a female who has suffered g aging hits, prior to birth. Accordingly, we may expect the adult male to live a shorter time, by $g\tau$ years, than the adult female.

Actually, according to the 1949–50 Census, the maximal number of deaths for the white male occurs between the ages of 77 and 78, i.e., three years earlier than for the white female.   This difference is three years less than what we would expect on the basis of our theory, which assumes $g = 1$ and which gives an estimate for τ of $\tau = 6$ years.   This discrepancy indicates that perhaps the value of f^* is somewhat larger for the male than for the female.

Because of the possibility that this might be the case, we conceivably have $g = 2$ ($g = 2$ would mean that the elementary process of aging consists in the "destruction" of one-half of a chromosome rather than a whole chromosome).

However, there is no reason to believe that f^* may be very much larger for the male than for the female.   Therefore, the observed small difference between the life-expectancy of the female and that of the male may rule out a modification of the theory that assumes $g \gg 1$.

Experimental Test of the Theory.—The most stringent experimental test of the validity of our theory is likely to come from experiments in which one observes a reduction in the life-expectancy of the adult offspring of, say, an irradiated mouse population.   Experiments of this sort are needed in order to determine the value of δ^*.   Experiments of this sort will also show whether among the different phenomena which generally accompany senescence, such as the graying of the hair, the loss of accommodation of the eye, etc., there are any which are determined by the general physiological age, defined on the basis of the age-specific death rate.   Arrangements for experiments along these lines are now under discussion.

I am grateful to Dr. Gertrud Weiss and to Dr. N. Conant Webb, Jr., of the Medical School, University of Colorado, for detailed criticism of some of the computations involved, and to Samuel W. Greenhouse, of the National Institute of Mental Health, for critical evaluation of some of the approximations used.   For carrying out auxiliary computations I am also grateful to Mrs. Dorothy Lathrop, Office of Mathematical Research, National Institutes of Health, and to Mrs. Norma French, Section on Theoretical Statistics and Mathematics, Biometrics Branch, National Institute of Mental Health.

* The author wrote this paper while serving as consultant to the Basic Research Program, National Institute of Mental Health, National Institutes of Health, U.S. Public Health Service, Department of Health, Education, and Welfare.

NOTATIONS

f is the fraction of the somatic cells which "survive," in the sense of remaining functional up to a given age.

f^* is the "surviving" fraction of the somatic cells at the age of death.

r is the number of inherited faults.

ρ is the number of inherited faults per chromosome.

n is the average number of faults per person in the population.

τ is the basic time interval of the aging process, defined as the average time interval between two successive aging hits suffered by the chromosomes of the somatic cell.

m is the number of pairs of homologous chromosomes of the female of the species.

x_r is the number of aging hits suffered, on the average, by the chromosomes of the somatic cells up to the age of death, by an individual who has inherited r faults.

t_r is the age, at death, of an individual who has inherited r faults.

l_0 is the life-span of the species, defined as the age at death of the genetically perfect female who did not inherit any faults.

d(theor.) is the number of deaths per year that will occur in a cohort, as given by the theory.

d(theor.)$_{max.}$ is the highest number of deaths per year that will occur in a cohort, as given by the theory.

R(theor.) is the ratio of the number of deaths per year and the maximal number of deaths per year in a cohort, as given by the theory.

R(obs.) is the ratio of the number of deaths per year and the maximal number of deaths per year in a cohort, as given by the U.S. Life Tables for white females.

σ is the standard deviation of the Gaussian that approximates the observed distribution of the number of deaths per year, between the ages of 70.5 and 90.5 .

τ/l_0 is the specific life-shortening per fault of the species.

δ is the life-shortening per rep for a population that has been exposed to ionizing radiation.

δ^* is the life-shortening per rep of the adults in the offspring of a population that has been exposed to ionizing radiation.

μ_1 is the spontaneous mutation rate of the haploid set of vegetative genes per generation.

μ_t is the total spontaneous mutation rate of all genes in the haploid set.

N_1 is the haploid number of vegetative genes of the species.

N_t is the haploid number of all genes of the species.

g is the postulated number of "segments" per chromosome.

NOTES ADDED IN PROOF

1. When we say that an aging hit "destroys" a chromosome of the somatic cell, we mean that that chromosome has been rendered inactive as far as its vegetative functions are concerned, i.e., the genes which the chromosome contains will fail to produce the corresponding gene products. The question whether the chromosome is inactivated in any other sense is left open for the present. Thus it is left open whether, if a cell containing an inactivated chromosome were to duplicate, the inactivated chromosome would or would not duplicate and whether or not it would remain inactive after such a duplication. One might, for instance, imagine that the chromosomes of the somatic cell contain DNA strands which fulfil a vegetative function in the somatic cell by producing the specific gene products but do not duplicate when the cell duplicates. Aging hits would then inactivate these vegetative "copies" rather than the genetic copies. The latter would duplicate when the cell duplicates and would then produce fresh vegetative copies. This is just one of several assumptions which one may make concerning the nature of the aging hits. For the present, we are free to choose among several such *ad hoc* assumptions.

2. In the case of exposed animals it is conceivable that their life is shortened, not only through the induction of gene mutations in the chromosomes of their somatic cells by the ionizing radiation, but perhaps also through some other effects of the ionizing radiation on their somatic cells, which may involve the chromosomes or some other components of the cell. Among such effects might be the breakage of chromosomes which may lead to the loss of a chromosome. However, the theory here presented does not cover the life-shortening effect of ionizing radiation which is due to causes other than the induction of gene mutations in the somatic cells of the chromosomes. Disregarding such other effects, the "surviving" fraction of the somatic cells of an exposed female may be computed on the basis of the faults induced in the chromosomes of her somatic cells by the ionizing radiation. For a genetically perfect female who is exposed to a dose of ionizing radiation which induces, on the average, p faults per somatic cell, we may write for the "surviving" fraction of somatic cells:

$$f = [2e^{-\xi}e^{-p/2m}(1 - e^{-\xi}) + e^{-2\xi}]^m,$$

for $p/2m \ll 1$ and $p/x \ll 1$ we may write, in analogy to (13)

$$\ln \frac{1}{f^*} = \frac{1}{4m}\left(\frac{l_p}{\tau} + p\right)^2\left[1 - \frac{1}{2m}\left(\frac{l_p}{\tau} + p\right)\right],$$

where l_p is the age of death of a genetically perfect female exposed to a dose of ionizing radiation that induced an average of p faults in the chromosomes of her somatic cells.

(*Reprinted from Nature, Vol.* 184, *pp.* 956–958, *Sept.* 26, 1959)

A THEORY OF AGEING

THE theory of ageing put forward by Szilard[1] refers explicitly to mammals. It is the purpose of the present communication to point out that this theory cannot explain ageing in *Drosophila*, since it is inconsistent with two experimental observations. This of course does not prove that it cannot explain ageing in mammals; but reasons will be given for doubting that it does so.

Szilard postulates the random occurrence of 'hits', each hit rendering ineffective the genes of a whole chromosome, or perhaps of a large segment of a chromosome. A cell becomes ineffective either when two homologous chromosomes have each suffered a hit, or when one of a pair of homologues has suffered a hit, and the other carries an inherited 'fault'. By a fault is meant a recessive gene which in homozygous condition renders the cell inviable, or incapable of performing a necessary function in the adult organism. Death occurs when some predetermined fraction of the cells initially present is in this way rendered ineffective; Szilard suggests that this fraction is of the order of 2/3 to 11/12.

It is a direct consequence of this theory that, in the author's words: "The main reason why some adults live shorter lives and others live longer is the difference in the number of faults they have inherited". This is the first consequence of the theory which is contradicted by observations on *Drosophila*. In so far as differences in adult longevity are genetically determined, by far the largest differences are those between inbred and outbred individuals[2,3]. F_1 hybrids between inbred lines live for longer than do the parental lines (sometimes for twice as long). Outbred and genetically variable wild populations have approximately the same expectation of life as do F_1 hybrids. Now inbreeding increases the proportion of loci at which individuals are homozygous. An individual which survives for an appreciable time as an adult cannot, by definition, be homozygous for a fault. Therefore inbred individuals which survive to become adults, and which do not die immediately after emergence, are not homozygous for faults at any loci, and would be expected to be heterozygous for faults at fewer loci than are members of outbred wild populations. If two inbred lines are crossed, the F_1 hybrids would be expected to carry a load of faults intermediate between the loads carried by the parental lines. Thus according to Szilard's theory, inbred lines should have a higher expectation of life than wild populations, and F_1 hybrids between inbred lines should be intermediate between their parents. Neither of these predictions is in fact true.

Further, since males have only a single X chromosome, any hit on that chromosome in a male would render the cell inviable, whereas in a female not heterozygous for a sex-linked fault both X chromosomes must be hit before a cell becomes inviable. Therefore females should live longer than males. This again is not the case in *D. subobscura*. In some strains females do live longer than males, but in other strains, both inbred and outbred, the reverse is true. This point is particularly telling since in *Drosophila* the sex chromosomes account for about one-fifth of the total chromosome material.

The other group of facts which are inconsistent with the theory concern the rate of ageing at different temperatures[4]. Female *D. subobscura* of a particular strain have an expectation of life of about 56 days at 20° C. and of 18 days at 30·5° C. The changes responsible for death at 30·5° C. are not repaired or reversed in individuals kept for a time at 20° C. Consequently the changes responsible for death at both temperatures can properly be regarded as ageing processes. If these processes were, at each temperature, those postulated by Szilard, differing only in the rate at which hits occur, it follows that individuals kept for an appreciable time at 30·5° C. should have, when returned to 20° C., an expectation of life at that temperature lower than that of individuals of the same chronological age not previously exposed to 30·5° C. In fact, exposure to 30·5° C. for periods of the order of half the expectation of life at that temperature does not alter the further expectation of life at 20° C. of males, and significantly increases that of females.

Hence, if, despite the genetic evidence to the contrary, we assume that ageing at 20° C. is due to random hits on chromosomes, then ageing at 30·5° C. cannot be explained by the same process proceeding at a higher rate. In other words, either at 20° C. or at 30·5° C. ageing must be due to a process different from that postulated by Szilard ; it is possible, and in my view likely, that such a process is not primarily responsible for ageing at either temperature.

It is perhaps unreasonable to criticize a theory intended to explain ageing in mammals by quoting observations on insects. Unfortunately the temperature experiments cannot be repeated on a homiotherm. But there is some evidence[5] in mice, as well as in *Drosophila*, that inbred individuals do not live as long as outbred ones. In addition to this purely observational point, there is one more general reason why Szilard's work has made a theory of ageing by somatic mutation less, and not more, promising than it had previously appeared to be. It is assumed that the 'target' is a whole chromosome ; a 'hit' renders ineffective all the genes carried by that chromosome. This assumption is made because, as Szilard shows, if it were assumed that the target were an individual gene, it would be necessary also to assume that each individual carried a load of faults so high as to be inconsistent with the known fertility of consanguineous marriages. There are events, particularly mitotic errors and chromosome breakages, which would deprive cells of whole chromosomes or of large segments of chromosomes, but they do not seem likely to be common enough to be the main cause of ageing. Most biologists would be happier with a theory which assumed as the unit event a hit on a gene, using the word gene here to refer to a functional unit or cistron. Perhaps the most important thing Szilard has done is to show that such a theory, at least in its simplest form, would run into difficulties.

J. MAYNARD SMITH

Department of Zoology,
University College,
London, W.C.1.

[1] Szilard, L., *Proc. U.S. Nat. Acad. Sci.*, **45**, 30 (1959).
[2] Clarke, J. M., and Maynard Smith, J., *J. Genet.*, **53**, 172 (1955).
[3] Maynard Smith, J., *J. Genet.*, **56**, 227 (1959).
[4] Maynard Smith, J., *J. Exp. Biol.*, 35, 832 (1958).
[5] Mühlbock, O., *CIBA Colloquia on Ageing*, **3**, 115 (1957).

ALL the observations quoted by Mr. Smith in his interesting communication relate to fruit flies and they fall into two classes : observations which we may expect to be able to duplicate in the case of mammals and those which we may not. Since I do not propose to discuss here whether the theory might or might not be extended to insects, I am primarily concerned with the former of the two classes.

Smith states that a genetically variable, 'wild', population of fruit flies has a substantially higher life expectancy than inbred, fairly or wholly homozygous, strains derived from it. He also states that the F_1 hybrid, obtained by crossing two different inbred strains, has a substantially higher life expectancy than the two inbred strains themselves. Smith holds that these findings are not compatible with the theory of ageing that I proposed.

It is probably true that the observations quoted

3

above could be duplicated with mammals and I am
quite prepared to accept this thesis for the sake of
argument. As I shall presently show, however, my
theory does not preclude that the homozygous inbred
strains may have a substantially smaller life expectancy than the wild type strains. Further, the theory
demands that the life expectancy of the F_1 hybrid
be appreciably higher than that of the wild type
strain, if the wild type strain carries a substantial
number of faults. In order to see this, we may
consider the following.

At present there is no evidence that a gene may be
responsible for anything except for the production
of a specific protein molecule which might be endowed
with a specific enzymatic activity. In a wild population, a given gene may be present in the form of a
variety of alleles and the corresponding enzymes may
differ in their turnover number. For the purposes
of discussion here, I shall call an allele 'weak' if
the turnover number of the corresponding enzyme
is small. If this turnover number is very small, the
allele might be a recessive lethal. A completely
homozygous strain is, of course, free of recessive
lethals, but it may contain a number of 'weak'
alleles.

Again, for the purposes of discussion here, I shall
adopt a somewhat over-simplified picture, and shall
disregard the possibility that the enzyme-levels in
the somatic cells may be determined to some extent
by the regulatory mechanisms of the cell through
enzyme induction or otherwise. On this over-simplified basis, we may then say that the somatic cells
of an inbred strain, which is homozygous for a number
of 'weak' alleles, are impoverished in the corresponding enzymes, so far as their biochemical activity is
concerned.

My theory assumes that only a small fraction of
the enzymes, less than one-fifth perhaps, is important
for the functioning of the somatic cells of the adult,
while practically all of the enzymes may be important
for differentiation and morphogenesis during the
embryonic life of the individual. Accordingly, we
may then expect that an individual of the inbred
strain (which is homozygous for a number of 'weak'
alleles) may be maldeveloped, in the sense that it may
have a much smaller reserve at birth than the wild
type individual, with respect to a number of physiological functions. Thus it is conceivable that an
individual belonging to an inbred strain may die at
an age at which f, the 'surviving' fraction of its
somatic cells, has fallen to, say, $f^* = \dfrac{1}{2 \cdot 72} \approx \dfrac{1}{e}$; whereas

4

an individual belonging to the wild-type strain may die at an age at which f, the 'surviving' fraction of its somatic cells, has fallen to about $f^* = \dfrac{1}{7 \cdot 4} \approx \dfrac{1}{e^2}$.

We may compute for this case the most probable age at death, for man, from formula (14) given on p. 33 of my paper (*loc. cit.*), which reads:

$$x_r + r = \sqrt{4m\ln\frac{1}{f^*} + \ln\frac{1}{f^*}}$$

where x_r is the number of hits at death; r is the number of the inherited faults; $m = 23$ is the number of chromosome pairs and f^* is the surviving fraction of the somatic cells at the age of death.

The most probable age at death, t_r, is given by: $t_r = 6 \times x_r$ years.

For the inbred strain we obtain t_r, the most probable age at death, by writing: $r = 0$ and $\ln\dfrac{1}{f^*} \approx 1$. We thus obtain $t_r = 63 \cdot 6$ years.

For the wild type we obtain t_r, the most probable age at death, by writing: $r = 2$ and $\ln\dfrac{1}{f^*} \approx 2$. We thus obtain $t_r = 81 \cdot 5$ years. The actual value for white females in the United States is: $t_r = 80 \cdot 5$ years.

For the F_1 hybrid we obtain t_r, the most probable age at death, by writing: $r = 0$ and $\ln\dfrac{1}{f^*} \approx 2$. We thus obtain $t_r = 93 \cdot 5$ years. This is 12 years more than the value for the wild type.

It may thus be seen that a substantially shortened life expectancy of the homozygous, inbred strain, as compared with the wild type, need not be inconsistent with the theory. However, an increased life expectancy of the F_1 hybrid as compared with the wild type strain is a necessary consequence of the theory.

This consequence of the theory could be tested by experiments on short-lived mammals, say mice. In order to render the experiment more sensitive, one may first expose to ionizing radiation a population of wild type mice over several generations and may thereby increase the number of faults in the population. Starting with such a 'wild' population, enriched in faults, one would then select two unrelated families and derive from them two inbred homozygous strains. The theory demands that the F_1 hybrid of these two inbred strains should live appreciably longer than the population from which the two families were selected. Given a suitable opportunity, I propose to arrange for experiments of this sort. A

5

negative result might well prove fatal for the theory.

I should perhaps add at this point that the observed differences in the life expectancy of the male and the female do not provide a usable criterion for the validity of the theory because $f*$, the 'surviving' fraction of the somatic cells at death, might differ appreciably for the male and the female.

Smith cites a rather peculiar effect of the temperature on the life expectancy of the male and the female in *D. subobscura*. It seems to me that any future theory of ageing that may be generally applicable to insects would be put to an unduly severe test, were one to demand that it account for this particular effect.

Because the theory of ageing that I proposed makes quantitative predictions, it is capable of being disproved by experiments and, sooner or later, such might be its fate. At present I am not aware, however, of any valid observations which contradict this theory. In these circumstances, I am not at present disposed to agree with the appraisal of the theory implied in the last paragraph of Mr. Smith's communication.

LEO SZILARD

Enrico Fermi Institute for Nuclear Studies,
University of Chicago,
Chicago, Ill.

Printed in Great Britain by Fisher Knight & Co., Ltd., St. Albans.

Reprinted from
Proceedings of the National Academy of Sciences
March 1960 Issue

THE CONTROL OF THE FORMATION OF SPECIFIC PROTEINS IN BACTERIA AND IN ANIMAL CELLS

By Leo Szilard*

THE ENRICO FERMI INSTITUTE FOR NUCLEAR STUDIES, THE UNIVERSITY OF CHICAGO

Communicated January 19, 1960

In 1953, Monod and Cohen-Bazire as well as Cohn, Cohen, and Monod showed that an amino acid may repress the *last* enzyme in the biochemical pathway which leads to the formation of that amino acid.

Four years later, H. J. Vogel found that arginine, when added to a growing bacterial culture, represses the formation of acetyl ornithinase (one of the *early* enzymes along the biochemical pathway leading to arginine), which converts acetyl-ornithine into ornithine. This observation provided rather strong circumstantial evidence in favor of the view that enzyme repression may be part of the normal regulatory mechanism of enzyme formation in bacteria.

Subsequently, it occurred to Werner Maas that inducers which enhance the formation of an enzyme when added to a growing bacterial culture may perhaps be capable of doing so *only* because there is a repressor present in the cell, and that the inducer might perhaps do no more than inhibit some enzymes which are involved in the formation of the repressor. Thus the inducer would enhance the *formation* of an enzyme only because it reduces the concentration of the repressor in the cell. (Oral communication, April, 1957.)

At that time, the best investigated case of enzyme induction was the induction of the enzyme β-galactosidase. Milton Weiner helped my understanding of the induction of this enzyme greatly by pointing out that its induction must be considered in conjunction with the biochemical pathway leading from galactose to glucose-1-P. (Oral communication, January, 1957.)

The induction of β-galactosidase in a culture of bacteria growing on succinate or lactate, as the carbon source, is inhibited by adding glucose to the growing culture. From this I was led to infer that some metabolite, intermediate in the sequence of metabolites lying between galactose and glucose-1-P, is the precursor of the repressor of β-galactosidase. Further, in line with the above quoted suggestion of Werner Maas, I was led to surmise that certain galactosides may induce β-galactosidase, by inhibiting enzymes which lie on the biochemical pathway between glucose 1-P and the intermediate metabolite that may be the precursor of the repressor of this enzyme.

I believe that some such galactosides may in fact inhibit one of these enzymes and this may explain, in part, why the rate of formation of β-galactosidase rises faster than linearly with the intracellular concentration of the inducing galactoside. However, since I was not able to explain on any similar basis the induction of the degradative enzymes by their substrate, I was led to assume that the inducer must be able to exert an effect which goes beyond inhibiting the formation of a repressor. In particular, I was led to believe that a repressor may reduce the rate of formation of the enzyme by combining with an enzyme molecule which is still attached to its enzyme-forming site, and that it may thereby somehow prevent the attached enzyme molecule from leaving its enzyme-forming site.† According to this view the repressor combines with a specific site, the controlling site, of the "attached" enzyme molecule, and an inducer may then enhance the formation of the enzyme by competing with the repressor molecule for this site.

I was further led to believe that the repressor of the enzyme β-galactosidase might be a molecule composed of two moieties. One of these (which we may call the metabolite moiety) might be a galactoside and the other (which we may call the R moiety) might perhaps be a polynucleotide. Certain galactosides would then enhance the formation of the enzyme β-galactosidase by competing for the controlling site of the attached enzyme molecule with the metabolite moiety of the repressor.

A model which was based on this mechanism of induction and repression of enzyme formation in bacteria was presented by me in a paper given at the annual meeting of the German Chemical Society in Berlin (Oct. 7, 1957) and in lectures, given in the subsequent six months, at various institutions actively interested in the problem of enzyme induction.

The model appeared to be capable of accounting not only for enzyme repression but also for some aspects of antibody formation in mammals. It was not then clear, however, whether it might provide a convincing explanation for the phenomenon of lasting immunity. Such lasting immunity manifests itself in the secondary antibody response which may be elicited in mammals such as the rabbit when it is given an injection of an antigen by which it had been immunized earlier. This secondary response can be elicited even if the second injection of the antigen follows the first injection of the antigen after a very long time interval and thus it involves a "memory" which fades away only very slowly.

In the postwar years the study of inducible enzymes received its impetus chiefly from Jacques Monod's studies of the induction of the enzyme β-galactosidase, at the Institut Pasteur in Paris. The induction of this enzyme turned out to be a very complex phenomenon, however, and until recently it could only be inferred that it involved enzyme repression.

Two years ago, when the model here discussed was first presented, it was supported only by scattered experimental facts. In particular the tenet that the R moiety might be a polynucleotide was based on rather tenuous circumstantial evidence.

In the last two years, however, very considerable progress has been made in the study of enzyme induction and enzyme repression. Thus an experiment by Arthur Pardee, Francois Jacob, and Jacques Monod has provided evidence in favor of the view that the phenomenon of enzyme repression may play a major role in the induction of β-galactosidase. Recent experiments performed by Luigi Gorini, at the Medical School of Harvard University, and by George Cohen, Francois Gros, Francois Jacob, Werner Maas, Jacques Monod, and Arthur Pardee, at the Institut Pasteur, support the view that enzyme repression may be the key to the understanding of the phenomenon of enzyme induction in general.

One of these recent experiments shows that a bacterial gene which is responsible for the repression of the enzyme β-galactosidase may exert its effect not by causing the synthesis of an enzyme, but rather by causing the synthesis of a molecule which is not a protein. This remarkable finding is consistent with our notion that this gene might cause the formation of the R moiety of the repressor of β-galactosidase and that this R moiety might be a polynucleotide.

Apparently, the results of all these newer experiments are, so far, still consistent with the model formulated two years ago, and, in this limited sense, they lend support to the model.

REPRESSION AND INDUCTION OF THE FORMATION OF ENZYMES IN BACTERIA

1. *Enzymes Leading to Glucose-1-P.*—There are numerous repressible enzymes involved in the biochemical pathways which lead from a carbon source to glucose-1-P. In general, the formation of such enzymes is inhibited by glucose. This is quite consistent with the views on enzyme repression here adopted and may be readily understood on the basis of the considerations presented by Neidhardt and Magasanik (1956).

I propose to skip the discussion of this large class of enzymes in favor of discussing two other large classes of enzymes which we may designate as "special anabolic enzymes" and "degradative enzymes."

2. *Special Anabolic Enzymes.*—The enzymes which lie along the biochemical pathways that lead to the formation of an amino acid, a purine, or a pyrimidine, we shall designate as "special anabolic enzymes." We shall single out the enzymes, E_i, involved in the biosynthesis of the amino-acid arginine as being representative for the enzymes of this class.

One of these enzymes, ornithine transcarbamylase, which converts ornithine into citrulline, has been studied by Luigi Gorini and Werner Maas. In a certain strain of coli, the formation of this enzyme may be repressed by adding arginine to a growing bacterial culture. If the intracellular concentration of arginine is lowered in a mutant which cannot convert citrulline into arginine by growing it at a slow rate in a chemostat, with arginine as a controlling growth factor, then the rate of production of the enzyme is raised by a factor of about 25.

We shall refer to enzymes whose rate of production can be thus raised as "boostable" enzymes.

The high rate of enzyme production which may be obtained in the case of such a "boostable" enzyme by lowering the intracellular concentration of a controlling metabolite may represent the "full rate" at which an enzyme-forming site is capable of synthesizing the enzyme.

There are other strains of coli in which the rate of formation of ornithine transcarbamylase cannot be boosted by lowering the intracellular concentration of arginine or any other *known* metabolite. We shall not discuss these "unboostable" strains in the present paper.

We do not assume that the molecule responsible for the repression of ornithine transcarbamylase in the "boostable" strains is arginine itself. We rather assume that the repressor REP_i of the enzymes E_i is a composite molecule which consists of two moieties. One of these, the "metabolite" moiety, is arginine; the other is a moiety which we shall designate by R_i and to which we shall refer as the R-moiety. Thus we may write for the chemical formula on the repressor REP_i

$$REP_i = (\text{arginine} - R_i)$$

As stated above, the R moiety of such a repressor molecule might possibly be a polynucleotide.

Our model for the repression of the enzymes E_i, involved in the biosynthesis of arginine, is as follows:

We assume that the polypeptide chain of the enzyme molecule is synthesized along a specific enzyme-forming site, which determines the amino-acid sequence of the polypeptide. This polypeptide folds up to form the enzyme molecule, but this molecule remains attached, perhaps through a covalent bond, to the enzyme-forming site. If there are repressor molecules present in the cell which are specific for this enzyme, then the metabolite moiety of a repressor molecule may reversibly combine with the controlling site of the attached enzyme molecule (and the R-moiety might perhaps reversibly combine with the purine and pyrimidine base of the enzyme-forming site itself). We postulate that an "attached" enzyme molecule, which is so combined with a repressor molecule, cannot leave its enzyme-forming site and accordingly the formation of the enzyme may thus be repressed.

Concerning the mechanism of the action of the repressor we may assume—*pour fixer les idées*—that there is some universal enzyme U present in the cell which can split the covalent bond that ties newly-formed enzyme molecules to their enzyme-forming site. The repressor molecule, when it is combined with an attached enzyme molecule, sets up a "steric hindrance" and may thus prevent the universal enzyme U from splitting the covalent bond.

We postulate that the enzymes E_i, lying along a biochemical pathway that leads to an amino acid, have two specific combining sites, the "catalytic" site and the "controlling" site.

In the case of an enzyme which catalyzes an early step in the biosynthetic pathway, leading to an amino acid, a purine or a pyrimidine, there need be little chemical resemblance between the substrate of the enzyme and the amino acid, the purine, or the pyrimidine, which lies at the end of the biosynthetic pathway and which may be capable of repressing the formation of the enzyme. The substrate has a specific chemical affinity to the catalytic site of the enzyme, and the end product of the biosynthetic chain, the amino acid, the purine, or the pyrimidine, has a specific affinity to the controlling site. Therefore in the case of these "early" enzymes, the combining specificity of the catalytic site and of the controlling site might be quite different.

In the case of the enzymes which lie towards the end of the biochemical pathway, however, the substrate of the enzyme is likely to be a chemical analogue of the "end product" and we may expect it therefore to have specific chemical affinity, not only for the catalytic site but also for the controlling site of the enzyme.

3. *Degradative Enzymes E_j*.—The enzymes involved in the degradation of tryptophane to β-keto-adipic acid by *Pseudomonas fluorescens* may be taken to be representative for another large class of enzymes, the "degradative enzymes." There are seven enzymes involved in the degradation of tryptophane, and the formation of each of them can be enhanced by adding the substrate of the enzyme to the growing bacterial culture.

There is a vast array of compounds which can be oxidized by bacteria such as *P. fluorescens*. R. Y. Stanier estimated their number at more than 50 and estimated that at least 200 enzymes participate in their degradation. The formation of each of these enzymes may be expected to be enhanced by adding its substrate to the growing bacterial culture, and we may therefore regard these "degradative enzymes" as "inducible" enzymes.

We postulate that the degradative enzymes E_j also have two specific combining sites, a catalytic site and a controlling site. Since in the case of the degradative enzymes the substrate of the enzyme is an inducer of the enzyme, we assume that the substrate of such a degradative enzyme has a substantial chemical affinity for the controlling site as well as the catalytic site. In the case of the degradative enzymes, E_j, we may therefore expect the two combining sites to be similar in their combining specificity for any given metabolite M.

The metabolite moiety of the repressor of a degradative enzyme might be either the substrate itself or else a metabolite further down the degradative pathway which is still a chemical analogue of the substrate.

If the metabolite moiety of the repressor is the substrate itself, and if the concentration of the repressor rises more slowly with increasing intracellular concentration of the substrate than the concentration of the substrate, then the substrate must be an inducer of the enzyme because it competes with the repressor for the attached enzyme molecule.

The concentration of the repressor would rise more slowly than the concentration of the substrate, for instance, if the limiting factor for the rate of production of the repressor were the rate of production of the R moiety of the repressor.

4. *The Coupling Enzymes C_i or C_j.*—We postulate that there may be present in the bacterial cells a class of enzymes to which we may refer as coupling enzymes, C_i and C_j, which couple a specific R moiety, R_i and R_j, to the metabolite M_i and M_j, and thus form the repressor REP_i and REP_j respectively.

Within the class of the degradative enzymes E^j, the corresponding coupling enzymes C_j might be close to being saturated with respect to the metabolite moiety of the repressor. With increasing intracellular concentration of the substrate, the concentration of the repressor would then rise more slowly than the concentration of the substrate and on this basis one may then expect the substrate to be an inducer of the enzyme.

If a compound $\bar{M}$ is a chemical analogue of the metabolite moiety M of the repressor REP, and if the cell cannot transform it into the metabolite moiety M, then such a compound may reduce the concentration of the repressor by inhibiting the coupling enzyme C and thereby enhance the formation of the enzyme E.

Equations Describing Our Model for Enzyme Repression and Induction.[‡]—In a bacterial culture, growing at a fixed rate, the total repressor concentration ρ_0, that establishes itself in the stationary state, is proportional to the rate at which the repressor molecules are formed.

We may compute ρ, the concentration of the free repressor molecules, from ρ_0, the total concentration of the repressor molecule J by writing

$$\rho = \rho_0 - z \frac{\rho/K}{1 + \rho/K} \tag{1A}$$

where K designates the equilibrium constant for the dissociation of the repressor molecule from the controlling site of the "unattached" enzyme molecules and z the concentration of the enzyme E in the cell.

In (1A) the first term represents the total concentration of the repressor in the cell and the second term represents the concentration of those repressor molecules which are combined with the controlling site of an "unattached" enzyme molecule. For the sake of keeping our formulae simple we are disregarding here the fact that the repressor may also combine with the catalytic site of the "unattached" enzyme molecules. Accordingly (1A) must be amplified when this becomes relevant to the issue considered.

We may write (1A) also in the form of (1B)

$$z = \rho_0 \left(1 - \frac{\rho}{\rho_0}\right) + K \left(\frac{\rho_0}{\rho} - 1\right) \qquad (1B)$$

We shall designate by τ the average time for which a newly-formed enzyme molecule remains tied to its enzyme-forming site, when it is *not* combined with a repressor molecule at its controlling site. We shall assume that τ is large compared to the time that it takes for the polypeptide to be formed and to fold up to form the attached enzyme molecule. In these circumstances we may say that there is practically always an enzyme molecule attached to the enzyme-forming site, and this enzyme molecule is either combined with a repressor molecule or it is not.

On the basis of our model we may then say that the rate of formation of an enzyme, when it is limited by the presence of a repressor molecule in the cell, is given by

$$\text{rate} = q/\tau \text{ per enzyme-forming site, per unit time} \qquad (2)$$

where q is the probability that the attached enzyme molecule is *not* combined at its controlling site with a repressor molecule.

We may write for this probability q, in the presence of a repressor *REP* and an inducer $\bar{M}$

$$q = \frac{1}{1 + \mu/K_{\bar{M}}{}^* + \rho/K^*} + \frac{\mu/K_{\bar{M}}{}^*}{1 + \mu/K_{\bar{M}}{}^* + \rho/K^*} \qquad (3)$$

μ and ρ are the intracellular concentrations of the inducer $\bar{M}$ and the free repressor *REP*, respectively; $K_{\bar{M}}{}^*$ and K^* are the equilibrium constants, for the reversible dissociation of the inducer molecule $\bar{M}$ and the repressor molecule *REP*, respectively, from the controlling site of the attached enzyme molecule.

In (3) the first term gives the fraction of the attached enzyme molecules which are not combined at their controlling site with either a repressor molecule *REP* or an inducer molecule $\bar{M}$. The second term represents the fraction of the attached enzyme molecules which are combined at the controlling site with an inducer molecule $\bar{M}$.

Formula (3) holds true if the average time that it takes an inducer molecule to dissociate from the controlling site of the attached enzyme molecule is short in comparison with τ.

We may write from (2) and (3) for the rate of enzyme formation per enzyme-forming site per unit time

$$\text{rate} = \frac{1}{\tau} \frac{1 + \mu/K_{\bar{M}}{}^*}{1 + \mu/K_{\bar{M}}{}^* + \rho/K^*} \qquad (3A)$$

The concentration of an enzyme in a bacterium that grows at a fixed rate is proportional to the rate at which the enzyme is formed.   Accordingly, we may write for z, the intracellular concentration of the enzyme

$$z = z^* \frac{1 + \mu/K_{\bar{M}}{}^*}{1 + \mu/K_{\bar{M}}{}^* + \rho/K^*} \tag{4}$$

where z^* is the concentration which the enzyme might attain in the cell, in the absence of any repressor.

In the absence of the inducer $\bar{M}$ we may write

$$z = z^* \frac{1}{1 + \rho/K^*}. \tag{5}$$

Equations (4) and (5) give the concentration of the enzyme in the cell independent of how many enzyme-forming sites are present in the cell which synthesize the same enzyme.   In (4) and (5), z^* represents the enzyme concentration which is obtained from these formulae when ρ, the repressor concentration, becomes zero.

It should be noted, however, that when the repressor concentration becomes very small and the concentration of the enzyme becomes correspondingly large, the repressor concentration may cease to be the limiting factor for the rate of enzyme production and something else may become rate-limiting.   For this reason the enzyme concentration in the cell might not actually reach the value z^*, if the repressor concentration goes to zero.

We may now introduce into our formulae the repression factor λ which is defined by

$$\lambda = \rho_0/K^*.$$

We may then write (5) in the form

$$z = z^* \frac{1}{1 + \lambda\rho/\rho_0} \tag{5A}$$

and this we may also write in the form

$$\rho = \frac{\rho_0}{\lambda} \frac{1 - z/z^*}{z/z^*} \tag{5B}$$

We shall throughout the rest of our discussion invariably assume that we have

$$z^* \gg \rho_0. \tag{6}$$

The Simplified Equations.—For the enzymes E for which we may write

$$K \gg z^*$$

the second term in (1A) can be neglected and we may then write

$$\rho = \rho_0.$$

In this case (4) and (5) may be written in the form of

$$z = z^* \frac{1 + \mu/K_{\bar{M}}{}^*}{1 + \mu/K_{\bar{M}}{}^* + \rho_0/K^*} \tag{7}$$

and

$$z = z^* \frac{1}{1 + \rho_0/K^*} = \frac{z^*}{\lambda + 1} \qquad (8)$$

When these simplified equations hold, then the rate of enzyme formation is independent of the concentration of the enzyme in the bacterium. Accordingly, if an inducer is added to a growing bacterial culture at a given point in time, the rate of enzyme formation will rise to a new value at the time when the inducer is added and from then on it will remain constant. This could be verified by studying the kinetics of the induction of the enzyme.

It may be seen from (7) that if the presence of the inducer $\bar{M}$ does not affect ρ_0, the concentration of the repressor molecules in the cell, then the rate of formation of the enzyme cannot rise any faster than linearly with the intracellular concentration of the inducer.

If it is found that the enzyme concentration rises faster than linearly with the intracellular concentration of the inducer, we may then expect either that the inducer inhibits an enzyme involved in the formation of the metabolite moiety of the repressor or that it inhibits the enzyme C which couples the metabolite moiety of the repressor to the R moiety of the repressor, or that it does both.

Predictions Based on the "Simplified" Equations.—On the basis of the model as described by the above given simplified equations we may expect the following:

(a) If the cells of a bacterial strain are incapable of converting $\bar{M}$ (a chemical analogue of the repressor's metabolite moiety) into M (the repressor's metabolite moiety itself), and if the concentration of M is fixed, then the chemical analogue $\bar{M}$ may enhance the formation of the enzyme, provided that it can get into the bacterial cell.

The chemical analogue $\bar{M}$ may be an inducer of the enzyme in such a bacterial strain, either because it competes with the metabolite moiety of the repressor for the controlling site of the enzyme molecule which is attached to its enzyme-forming site, or because it competes with the metabolite M for the coupling enzyme C which joins the metabolite moiety to the R moiety of the repressor, or for both of these reasons.

In the case of the bio-synthetic pathway leading to arginine the substrate of a *late* enzyme is a chemical analogue of arginine. Accordingly, we may expect such a substrate to induce the enzyme in mutant bacterial strains which cannot convert the substrate into arginine. Thus we may expect the enzyme ornithine transcarbamylase, which converts ornithine into citrulline, to be inducible by citrulline, in a mutant strain which cannot convert citrulline into arginine.

Luigi Gorini has observed that ornithine induces the enzyme ornithine transcarbamylase in such a mutant, if the intracellular concentration of arginine is kept *moderate* by growing the bacterium *at a fast rate* in a chemostat with arginine as a controlling growth factor. (Oral communications, 1959.) (If the intracellular concentration of arginine were kept low by growing the bacterium at a slow rate in the chemostat, then the enzyme level would be boosted to a high value and the inducing effect of ornithine would not be observable.) Since this mutant converts ornithine into citrulline, Gorini's observation is consistent with the views here presented.

(b) There might exist mutants in which the R moiety of the repressor is produced, but it is produced at such a low rate that its production may be the limiting factor for the production of the repressor when the intracellular concentration of the metabolite moiety of the repressor is not too low.   On the basis of (7), we may then say that in such a mutant the formation of the enzyme should be enhanced by adding the repressor's metabolite moiety to the growing bacterial culture.   Accordingly, for such a mutant the metabolite moiety of the repressor may be an inducer of the enzyme.

Thus, in such a mutant, arginine (for instance) should be an inducer of the enzyme ornithine transcarbamylase and it should be an inducer of this enzyme for one reason only, i.e. because it may compete with the repressor for the controlling site of the attached enzyme molecule.

(c) There may exist mutants which are not capable of producing the R moiety of the repressor.   In such a mutant the enzyme may be produced at the full rate and accordingly the enzyme level would be very high.   In such a mutant it should not be possible to repress the enzyme by adding the metabolite moiety of the repressor to the growing bacterial culture.   Mutants of this type may be designated as "absolute constitutive strains."

Both Luigi Gorini and Werner Maas have obtained from a strain of coli, in which the enzyme ornithine transcarbamylase is repressible by arginine, mutants in which the enzyme is always maintained at a high level and is not repressible by arginine.   (Oral communications, 1959.)   Conceivably these might be mutants in which the R moiety of the repressor is not formed, i.e., they might be absolute constitutive strains.

The Case of $z^* \gg K$. — We may postulate here, for the sake of argument, a set of constants for "wild type" bacteria which might be as follows:

$$z^* = 10^{-4}\,\text{mol/l}, \qquad \rho_0 = 10^{-6}\,\text{mol/l};$$
$$K = 10^{-5}\,\text{mol/l}; \qquad K^* = 10^{-10}\,\text{mol/l}$$

The postulated value of $z^* = 10^{-4}$ mol/l would mean that a fully boosted enzyme amounts to about 10 per cent of the cell proteins.

The value of $\rho_0 = 10^{-6}$ mol/l postulated would be consistent with the assumption that there may be one thousand different repressors present in a bacterium and the assumption that the R moiety of these repressors is a polyribonucleotide of a molecular weight of about 2,000, without having to assume a larger amount of soluble RNA in the bacterium than is usually found in bacteria.

The postulated value of $K = 10^{-5}$ mol/l means that the free energy change for the combination of the repressor with the controlling site of an unattached enzyme molecule is about $\Delta F = 7,000$ cal/mol.   This appears to be a reasonable value if we assume that only the M moiety of the repressor has a specific chemical affinity to the controlling site of the unattached enzyme molecule.

The postulated value of $K^* = 10^{-10}$ mol/l corresponds to a free energy change of $\Delta F = 14,000$ cal/mol for the combination of the repressor with the enzyme molecule that is attached to its enzyme-forming site.   This appears to be a reasonable value if we assume that the M moiety of the repressor has a specific chemical affinity to the controlling site of the enzyme molecule and that the R moiety of the repressor has a specific chemical affinity to the purine and pyrimidine bases of the enzyme-forming site itself.

For the above postulated set of constants we have

$$z^* \gg K$$

and accordingly the simplified equations do not strictly speaking hold.

Nevertheless, for this set of constants, ρ, the free repressor concentration, would be very close to ρ_0, because the enzyme is strongly repressed. We may write for the repression factor λ, for our set of constants,

$$\lambda = \rho_0/K^* = 10^4$$

and we have accordingly for the above postulated set of values

$$\frac{z^*}{K(\lambda + 1)} = 10^{-3} \ll 1.$$

In these circumstances ρ may be close to ρ_0.

Limiting ourselves, for the moment, to a set of constants where we have

$$K \gg \rho_0$$

we may now say the following:

If we have to deal with a *mutant* in which the repressor is produced at a very low rate, so that the repression factor λ is small, and if we have

$$\frac{z^*}{K(\lambda + 1)} > 1,$$

then ρ, the free repressor concentration, may be much smaller than ρ_0 and, accordingly, the enzyme level in the cell may then be rather high. This may be seen from (1A) as follows:

We may write for the second term in (1A)

$$z \frac{\rho/K}{1 + \rho/K} > \frac{z^*}{\lambda + 1} \frac{\rho/K}{1 + \rho/K} \approx \frac{z^*}{\lambda + 1} \frac{\rho}{K}$$

and thus we obtain from (1A)

$$\frac{\rho_0}{\rho} > 1 + \frac{z^*}{K(\lambda + 1)}$$

and from (5A) we obtain

$$z > \frac{1}{1 + \lambda/\{1 + z^*/[K(\lambda + 1)]\}} z^*.$$

Accordingly, strains of this type may maintain an enzyme level which may be high and which might come close to the fully boosted enzyme level z^* of an "absolute constitutive mutant."

Paraconstitutive Enzymes.—If for an enzyme we have

$$\rho_0 \gg K$$

and if the other constants fall within certain ranges, then the concentration of such an enzyme may be maintained in the cell either at a low stable value or at a high stable value—in the absence of any inducer.

An enzyme for which this holds we shall designate as "paraconstitutive."

The concentration of such a paraconstitutive enzyme may be maintained indefinitely at a low level in a proliferating cell culture. But, if once the enzyme concentration is raised to a sufficiently high level and maintained there long enough to establish a stationary state, from there on a high rate of production of the enzyme may be permanently sustained—even in the absence of an external, or internal, inducer. Further, when such a cell, which sustains a high enzyme level, divides, then the daughter cells which it generates will also sustain a high enzyme level. Thus the property of sustaining a high enzyme level is hereditary even though the inheritance is not genic.

It is not possible to say whether paraconstitutive enzymes exist in bacteria. If they did exist, it would be somewhat difficult to recognize them, because if they have arisen in the normal course of evolution then it is likely that by now the enzyme would be sustained at a high level, and the paraconstitutive mutant would thus give the appearance of a constitutive mutant.

It might well be, however, that the condition $\rho_0 \gg K$ is not fulfilled for any of the bacterial enzymes. Perhaps, in order to have this inequality hold for an enzyme, it is not sufficient for the controlling site of the enzyme to have a specific combining affinity for the metabolite moiety of the repressor, but it is also necessary for it to have a specific combining affinity for one or more of the units which constitute the R moiety of the repressor.

Paraconstitutive Enzymes and Cellular Differentiation.—Even though the inherent instability of the enzyme-forming system which is described by our equations might play no role in bacteria, it may still be of great interest. It is conceivable that it might play a major role in certain types of cellular differentiation, in higher organisms such as mammals.

There is no need to assume (and in the case of amphibia there may indeed be reason to doubt) that the early cellular differentiation involved in the formation of ectoderm, mesoderm, and endoderm, is of this type. Still, cellular differentiation in the later phases of embryonal development, where an organizer is involved, might conceivably be based on the inherent instability of the enzyme-forming system here discussed. It is not possible, however, to substantiate this at present on the basis of available experimental evidence, and it may be very difficult to substantiate it even through experiments devised for the purpose.

Experimental confirmation of the view that the model here discussed may provide the molecular basis of a certain type of cellular differentiation in mammals might, however, be obtained in the case of antibody formation in mammals or birds. An antibody need not have any enzymatic activity, but we shall assume that it is in some way related to certain degradative enzymes E_j, present in the mammalian cell, and that the rate of formation of an antibody A_j is under the control of the repressor which is specific for the related enzyme E_j.

Whether an enzyme makes its appearance in cellular differentiation, provoked by an organizer, or whether a specific antibody is formed in response to the injection of an antigen, in either case we may assume that the event is triggered by a transient reduction of the concentration of a specific repressor.

If antibodies are paraconstitutive proteins, then, once the concentration of a particular antibody is raised to a sufficiently high level in a lymphatic cell, that cell and all its daughter cells may maintain a high rate of production of that antibody. This would then represent the kind of memory which may form the basis of the so-called secondary response.

The Enzyme-Forming System of the Mammalian Cell.—We assume that the enzyme-forming system in the mammalian cell is described by the same equations which we have postulated for the bacteria. There are certain differences, however, between the mammalian cells here discussed and the bacterial cells. Under physiological conditions bacterial cells as a rule proliferate. Accordingly, in bacteria there is no need for protein turnover, nor do we have any reason to assume that the R moieties of the repressors are hydrolyzed in bacteria at an appreciable rate. In these circumstances, in bacterial cells ρ_0, the total concentration of a particular repressor in the cell, and z^*, the concentration attained by an enzyme which is produced at the full rate, are both determined by the rate of two processes, i.e., the rate at which these entities are formed and the rate at which these entities are diluted through the growth of the bacteria.

In the case of mammalian cells which are not in a state of rapid proliferation, we may in general assume that ρ_0, the concentration of the repressor, and z^*, the concentration of an enzyme which is produced at the full rate, are also determined by the rate of two processes, i.e. the rate at which these entities are produced and the rate at which they are hydrolyzed.

The rate of hydrolysis of the proteins is presumably determined by the level of activity of the proteolytic enzymes in the cell. The rate of hydrolysis of the repressors is presumably determined by the level of activity of some hydrolytic enzyme H which may universally hydrolyze the R moieties of all of the repressors.

We shall assume that the R moiety of the various specific repressors is hydrolyzed in the cell at the same rate whether the specific repressor is free or is combined with the controlling site of an "unattached" enzyme molecule. This is not an unreasonable assumption because the repressor molecule combines with an unattached enzyme molecule in large part by virtue of the chemical affinity of the repressor's M moiety to the controlling site of the enzyme molecule. If the R moiety is indeed a polynucleotide, then the first, or the first few, bases of the polynucleotide may also combine with the controlling site of an unattached enzyme molecule, but the rest of the nucleotide will presumably remain freely exposed to the hydrolytic action of our hypothetical enzyme H.

We shall presently show that if the relevant constants fall within a certain range, quoted below, then according to equations (1) and (5) the enzyme concentration z exhibits the characteristics which we have attributed to paraconstitutive enzymes.

If we equate z given by (1B) and given by (5A), we obtain a cubic equation for ρ. For the below-quoted range of the constants this cubic equation has three positive roots which represent the self-sustaining free repressor concentrations. Accordingly, there may be three self-sustaining pairs of values for ρ and z which we shall designate (in the order of increasing values of z and decreasing values of ρ) with ρ_1, z_1, ρ_2, z_2, and ρ_3, z_3.

Of these, three self-sustaining pairs of values the first and the last pair, ρ_1, z_1 and

ρ_3, z_3 represent stable points.   The intermediate pair ρ_2, z_2 represents an unstable point.

We shall refer to z_3 as a high stable enzyme concentration and to ρ_1 as the high stable repressor concentration.   To ρ_2 we shall refer as the "critical" repressor concentration.

In the case of a paraconstitutive enzyme, if ρ, the free repressor concentration, is lowered below ρ_2, the critical free repressor concentration (for a period of time which is sufficiently long for the establishment of a stationary state), and if thereafter the free repressor concentration is permitted to find its own level, then the free repressor concentration will decrease to its low stable value ρ_3.   Correspondingly, the enzyme concentration will rise to its high stable value z_3.

For any arbitrarily chosen value of ρ which we maintain in the cell long enough to establish a stationary state, we can determine from equations (1B) and (5A) whether the free repressor concentration would thereafter fall or rise in the cell when it is allowed to find its own level.   In order to make this determination we substitute the arbitrarily chosen value ρ (to which we shall refer as the test value) into (1B) and (5A), and find from both of these expressions the corresponding value for z.

The rate of enzyme production, for a given free repressor concentration, is expressed by (5A) and if the value for z obtained from this expression is higher than the value for z obtained from (1B), then we may say that the repressor concentration will decrease when it is permitted to find its own level.   In the opposite case we may say that the repressor concentration will increase when it is permitted to find its own level.

We propose to utilize below this simple rule, in order to establish the fact that there exist two stable self-sustaining values ρ_1 and ρ_3, if the constants fall within the range quoted below.

We shall now show that if we have

$$\frac{z^*}{K(\lambda - 1)} > 3 \tag{9}$$

and

$$\rho_0/z^* < 10 \tag{10}$$

then there is a stable point at which we have for the low self-sustaining free repressor concentration ρ_3

$$0 < \rho_3 < \rho_0/\lambda.$$

In order to show this we choose for our test concentration $\rho = \rho_c/\lambda$.   Substituting this value of ρ into (5B) we obtain

$$\frac{z(\text{from 5B})}{z^*} = \frac{1}{2}$$

Substituting the same value for ρ into (1A) we obtain

$$\frac{z(\text{from 1A})}{z^*} = \frac{\rho_0}{\lambda}(1 + \lambda) + \frac{K(\lambda - 1)}{z^*}.$$

If we take (9) and (10) into account we see that the value for z (from 1A) is less than the value for z (from 5B). Therefore, ρ, when it is permitted to find its own level (after being held at $\rho = \rho_0/\lambda$ for an adequate period of time), will decrease and move to its low stable value, ρ_3.

For the corresponding high stable value of the enzyme concentration z_3 we may write

$$z^*/2 < z_3 < z^*.$$

Next, we propose to determine ρ_1, the high stable concentration of the free repressor, and ρ_2, the critical concentration of the free repressor. We obtain these self-sustaining values from (1B) and (5A) by writing

$$z \text{ (from 1B)} = z \text{ (from 5A)} \tag{11}$$

and by finding the roots of this equation.

Writing out (11) explicitly we obtain

$$\rho_0\left(1 - \frac{\rho}{\rho_0}\right) + K\left(\frac{\rho_0}{\rho} - 1\right) = z^* \frac{1}{1 + \lambda\rho/\rho_0} \tag{11A}$$

which we may also write in the form

$$\lambda = \frac{1}{\rho_0/z^*(1 - \rho/\rho_0)\rho/\rho_0 + K/z^* - K/z^*(\rho/\rho_0)} - \frac{1}{\rho/\rho_0}. \tag{11B}$$

If we have

$$\frac{\rho}{\rho_0}\frac{\rho_0}{z^*} \gg \frac{K}{z^*} \tag{12}$$

and

$$\frac{z^*}{\rho_0}\frac{1}{1 - \rho/\rho_0} \gg 1, \tag{13}$$

then we may simplify (11B) by leaving off the second term and we thus obtain

$$\lambda = \frac{1}{\rho_0/z^*(1 - \rho/\rho_0)\rho/\rho_0 + K/z^* - K/z^*(\rho/\rho_0)} \tag{14}$$

or

$$(\rho/\rho_0)^2 - (1 - K/\rho_0)\rho/\rho_0 + z^*/\rho_0\lambda - K/\rho_0 = 0 \tag{14A}$$

or

$$\frac{\rho_1}{\rho_0} = \frac{1 - K/\rho_0 + \left|\sqrt{[1 - K/\rho_0]^2 - 4[z^*/\rho_0\lambda - K/\rho_0]}\right|}{2} \tag{14B}$$

and

$$\frac{\rho_2}{\rho_0} = \frac{1 - K/\rho_0 - \left|\sqrt{(1 - K/\rho_0)^2 - 4[z^*/\rho_0\lambda - K/\rho_0]})\right|}{2}. \tag{14C}$$

The quantity under the square root in (14B) and (14C) is positive if we have

$$(1 - K/\rho_0)^2 > {}^4\!/_5 \tag{15}$$

and

$$\lambda > 5z^*/\rho_0. \tag{16}$$

We shall now, for the purposes of our discussion, assume a set of constants which satisfy the inequalities that we have assumed above and actually compute the ρ_1/ρ_c and ρ_2/ρ_0 from (14B) and (14C) in order to obtain the values for ρ_1 and ρ_2.

For the purpose of our discussion we assume the following set of constants:

$$z^* = 10^{-3} \text{ mol/l}; \ \rho_0 = 10^{-8} \text{ mol/l}; \ K = 10^{-10} \text{ mol/l};$$
$$K^* = 10^{-14} \text{ mol/l}.$$

This set of values satisfies the inequalities assumed above (which represent a sufficient condition for an enzyme to be paraconstitutive) and, in particular, we obtain for this set of constants $\lambda = \rho_0/K^* = 10^6$ and $z^*/K(\lambda - 1) \approx 10$.

The following comment may be made concerning the particular values chosen for our constants:

$x^* = 10^{-3}$ mol/l means that if, even at zero repressor concentrations, the enzyme were formed at the rate corresponding to (5A), then the amount of enzyme in the cell would just about equal the total protein content of the cell.

$K = 10^{-10}$ mol/l corresponds to a free energy change of $\Delta F = 14,000$ cal/mol for the combination of a repressor molecule with the controlling site of an "unattached" enzyme molecule. It seems likely that both the metabolite moiety and the R moiety of the repressor would need to have a substantial chemical affinity to the controlling site of the unattached enzyme molecule in order to have a free energy change of this magnitude.

For the above quoted values of the constants we obtain from (14B) for the high stable repressor concentration $\rho_1 \approx 0.9 \ \rho_c$ and from (14C) for the critical repressor concentration $\rho_2 \approx 0.11 \ \rho_0$.

This means that it would be sufficient to lower the free repressor concentration to about one tenth of ρ_0 in order to trigger the enzyme-forming system and cause the cell thereafter to maintain, indefinitely, the enzyme at a high concentration, i.e. at a concentration lying somewhere between $z^*/2$ and z^*.

The set of values which we have assumed for our constants was selected because it appears conceivable that a quite similar set of values might hold for antibodies formed in the lymphatic cells of the adult rabbit that are capable of forming antibodies which are specific for an antigen injected into the rabbit.

These cells might be characterized—in comparison to the lymphatic cells of the newborn rabbit and the non-lymphatic somatic cells of the adult rabbit—by an increased level of activity of the hypothetical hydrolytical enzyme H. On this basis, we may assume that the value of ρ_0 and of the repression factor λ that holds for the lymphatic cells of the adult rabbit which are capable of forming antibodies is perhaps ten times lower than the corresponding values in those other cells, which are not capable of forming antibodies.

This possibility is discussed in detail in the following paper, "The molecular basis of antibody formation," which attempts to explain a number of phenomena involved in antibody formation on the basis of the phenomenon of enzyme repression in bacteria.

It is conceivable that in attempting to build a theory of antibody formation on this foundation we may be building a house of cards, for in spite of the rapid progress made in the last two years with respect to enzyme repression in bacteria, many of the conclusions drawn from the experiments are still largely based on circumstantial evidence. This holds in particular for the conclusion, here adopted, that the repressor controls the rate at which the enzyme is formed by the enzyme-forming site rather than the rate of formation of the enzyme-forming site itself.

Clearly, we cannot attempt at present to say how antibodies are actually formed; at best we may be able to say how antibodies might conceivably be formed. But to be able to say even this much might be of some value.

I had the privilege of discussing the thoughts expressed in this paper with Dr. Maurice S. Fox, The Rockefeller Institute, New York, and it is a pleasure to acknowledge his help in clarifying the issues involved.

* The author started working on this paper while serving as a consultant to the Basic Research Program, National Institute of Mental Health, National Institutes of Health, U.S. Public Health Service, Department of Health, Education, and Welfare. In later phases the work was supported by a Research Grant of the National Institutes of Health.

† A similar model was proposed independently by O. Maaløe and presented at an informal seminar at the Cavendish Laboratory in Cambridge in 1958; it will be described in *Microbial Genetics* (Cambridge University Press, 1960).

‡ Notations:

REP	is the specific repressor which controls the rate of formation of an enzyme and which is composed of two moieties, an M moiety and an R moiety.
M	is a metabolite, which forms the metabolite moiety of the repressor.
R	is the R moiety of a repressor, which is specific for enzymes lying along a certain stretch of a given biochemical pathway.
K^*	is the equilibrium constant for the dissociation of a repressor molecule from the controlling site of the corresponding enzyme molecule that is attached to its enzyme-forming site.
K	is the equilibrium constant for the dissociation of a repressor molecule from the controlling site of the "unattached" enzyme molecule, present in the cell.
$K^*_{\bar{M}}$	is the equilibrium constant for the dissociation of a chemical analogue $\bar{M}$, of the metabolite M, from the controlling site of an enzyme molecule attached to its enzyme-forming site.
$K_{\bar{M}}$	is the equilibrium constant for the dissociation of the chemical analogue $\bar{M}$ from the controlling site of an unattached enzyme molecule contained in the cell.
ρ_u	is the concentration of the molecules of the repressor REP in the cell.
ρ	is the concentration of the free repressor molecules in the cell, i.e. of those repressor molecules which are not combined with an unattached enzyme molecule.
μ	is a concentration of the chemical metabolite $\bar{M}$ in the cell.
z	is the concentration of the "unattached" enzyme molecules in the cell.
z^*	is the concentration that the enzyme would attain in the cell, for $\rho = 0$, if we disregard the fact that factors, which our formulae do not take into account, may limit the rate of the formation of the enzyme for high values of z.
λ	is the repression factor defined by $\lambda = \rho_u/k^*$.

Reprinted from the Proceedings of the NATIONAL ACADEMY OF SCIENCES
Vol. 46, No. 3, pp. 293-302. March, 1960

THE MOLECULAR BASIS OF ANTIBODY FORMATION*,

By LEO SZILARD

THE ENRICO FERMI INSTITUTE FOR NUCLEAR STUDIES, THE UNIVERSITY OF CHICAGO

Communicated by Theodore Shedlovsky, January 19, 1960

In a preceding paper we have discussed the phenomenon of enzyme repression in bacteria and presented a model for a kind of cellular differentiation which might conceivably play a role in embryonic development.   In the present paper we propose to discuss the molecular basis of antibody formation in the light of this model.

The Experimental Facts.—We shall list in the following the major immunological phenomena that we may regard as well established.   Throughout this paper we shall restrict our discussion to one animal, the rabbit, and to one class of antigens, soluble proteins.

(1) When a soluble protein antigen is injected into an adult rabbit for the first time, the rabbit responds after about 5 days with the formation of antibodies which are specific for this antigen.   Any remaining antigen will be rapidly eliminated from the circulation, and for a while there will be a substantial concentration of antibody present in the circulation.   Subsequently the concentration of the antibody in the circulation will gradually decrease and after a while it may become no longer detectable.   This is the so-called primary response.

(2) If, a few weeks after we have injected into the rabbit a soluble antigen, such as a foreign serum protein, we inject the same antigen again, then three days later the rabbit may respond with a copious production of antibody and the antigen will be rapidly eliminated from the circulation.   The concentration of the antibody in the circulation may reach a high level, then fall off rather slowly and remain appreciable for a long time.   This is the secondary response.

In response to the same amount of antigen injected, a much larger amount of antibody may be produced in the secondary response than in the primary response.

The secondary response can be elicited even a very long time after the antibody has disappeared from the circulation, subsequent to the first injection of the antigen. The readiness of the rabbit to exhibit such a secondary response represents some sort of a memory which fades away only very slowly.

(3) Albert Coons has found in the rabbit that if one evokes the secondary response in the manner described above, then, after 48 hours, clusters of cells which contain antibodies specific for the antigen may be found in the lymph nodes which are involved.   The clusters found 4 days or 8 days after the injection are on the average larger than the clusters found after 48 hours.   The cells which compose these clusters are small and round and have the appearance of plasma cells.   Mitotic figures can be seen in the clusters, indicating that there is proliferation of the cells producing the antibody.   The clusters vary in size; the larger ones consist of about a hundred or perhaps a few hundred cells.

(4) For a few days after birth, the rabbit is not capable of forming antibody in response to the injection of an antigen.   If a newborn rabbit is injected with a *large* quantity of a soluble protein antigen, then later on when this rabbit becomes an adult and is capable of forming antibodies in general, it will still remain incapable of forming antibody against the antigen which was administered to it immedi-

ately after birth, even though by that time the concentration of the antigen in the circulation may have fallen to a very low level.   This is the phenomenon of enduring immune tolerance.

(5) If an adult rabbit is given an X-ray dose of about 400 r to its whole body and a few days later an antigen is injected into the rabbit, then the rabbit will not form antibodies against this antigen.   After a while the effect of the X-ray exposure will wear off and the rabbit is then again capable of forming antibodies.

If after the X-ray exposure, at the time when the rabbit is not capable of responding with the formation of antibody, a *large* quantity of an antigen, such as a foreign serum protein, is injected into the rabbit, then the rabbit will remain incapable of forming antibodies against this particular antigen, even after the effect of the X-ray exposure has worn off and the concentration of the antigen in the circulation has fallen to a level which is no longer detectable.   Thus the X-rayed adult rabbit exhibits the phenomenon of enduring immune tolerance.

(6) If the antigen is injected into the rabbit prior to the exposure to the X-ray dose, then this exposure will not prevent the rabbit from forming antibodies to the antigen.

(7) If a rabbit has been pre-immunized with an antigen, then exposure of the rabbit to an X-ray dose of 400 r a few days prior to the injection of the antigen will not block the secondary response, i.e. in such a rabbit the injection of the antigen will evoke the secondary response even though the injection is given a few days following the administration of the X-ray dose.

The Enzymes of the Lymphatic Cells.—We postulate that, in general, the somatic cells of the rabbit contain a number of diverse biochemical pathways, j, $j + 1$ $j + 2$, etc., and that a number of enzymes lie along each such pathway.   This might be represented symbolically by writing

$$M \begin{cases} M_j^1 \xrightarrow{E_i^1} M_j^2 \xrightarrow{E_i^2} M_j^3 \xrightarrow{E_i^3} \ldots \ldots M_j^k \xrightarrow{E_i^k} M_j^{k+1} \\ M_{j+1}^1 \xrightarrow{E_{j+1}^1} M_{j+1}^2 \xrightarrow{E_{j+1}^2} M_{j+1}^3 \xrightarrow{E_{j+1}^3} \ldots \ldots M_{j+1}^k \xrightarrow{E_{j+1}^k} M_{j+1}^{k+1} \\ M_{j+2}^1 \xrightarrow{E_{j+2}^1} M_{j+2}^2 \xrightarrow{E_{j+2}^2} M_{j+2}^3 \xrightarrow{E_{j+2}^3} \ldots \ldots M_{j+2}^k \xrightarrow{E_{j+2}^k} M_{j+2}^{k+1} \end{cases}$$

As this scheme indicates, the metabolites M_j; M_{j+1}; M_{j+2}; etc. of several of these stray biochemical pathways might have a common precursor M.

Let us now consider one of these enzymes, for instance E_j^k.   We assume that this enzyme is under the control of the repressor

$$REP_j^k = \{M_j^k - R_j\}$$

We further assume that there is a coupling enzyme C_j^k present in the cell which joins the R moiety of the repressor, R_j, to the M moiety of the repressor, M_j^k. Symbolically this might be indicated by writing

$$M_j^k + R_j \xrightarrow{C_j^k} \{M_j^k - R_j\}.$$

We shall assume in our discussion that the coupling enzymes C are not under the control of any of the repressors here considered.

We shall further assume that both the catalytic site and the controlling site of all these degradative enzymes, E_j, can specifically combine with M_j, the substrate of the enzyme.   (See preceding paper.)

The Genes G^* and the Antibodies A.—We shall designate by G_j the genes which determine the identity of these degradative enzymes E_j, and we postulate that for each enzyme E_j there is only one corresponding gene G_j contained in each of the two haploid sets of chromosomes. The total number of genes of this category might perhaps be of the order of 10,000 per haploid chromosomal set.

One might be tempted at this point to assume that the enzymes E are themselves the antibodies. We prefer, however, to postulate the following: the genes G have a tendency to get doubled and in the course of evolution each gene G_j may have doubled many times. We assume that when a gene G undergoes doubling, the genes G^* which are formed will in general lack a part, presumably a small part, of the gene G. We assume in particular that the genes G^* differ from the genes G in two respects, which are as follows:

(a) the genes G^* lack the tendency of the genes G to undergo doubling;
(b) the proteins which are determined by the genes G_j^* resemble closely the enzyme E_j which is determined by the corresponding gene G_j but they may lack the catalytic activity of the enzyme. *We postulate that these proteins are the antibodies A_j.*

The controlling site of an antibody A_j has the same specific combining affinity to the metabolite M_j as has the controlling site of the enzyme E_j. Also the catalytic site of an antibody A_j may have the same specific combining capacity for the metabolite M_j as the catalytic site of enzyme E_j, to which it is related, *and if this is the case then we are dealing with a divalent antibody.*

The spontaneously occurring doubling of the genes G might be balanced by spontaneously occurring deletions of the genes G^*, and an equilibrium might be maintained on this basis, in the absence of genetic deaths. In such an equilibrium there might be present a number of genes G^* corresponding to each gene G. However, deletions of the genes G and their mutations to incompetence *would* have to be balanced by the occurrence of genetic deaths, in a state of mutational equilibrium.

There is a limit to the amount of genetic deaths which we may assume to occur per generation, because no species of mammals could remain in existence if the amount of genetic deaths were too high. This consideration does not permit us, however, to set an upper limit for the number of genes G^* present per gene G, because we are assuming that with respect to the genes G^* mutational equilibrium is maintained in the absence of genetic deaths.

We may set an upper limit for the total number of genes G^* on the basis of the amount of *DNA* present in the cell, but this comes out to be very high; *if we assume that the weight of the mammalian gene is about the same as the weight of the bacterial gene, then the amount of DNA in the mammalian cell would be sufficient to account for one million genes.*

The Nature of the Antigen.—For the purpose of our discussion an antigen P_j may be represented as a molecule which is composed of a non-antigenic protein molecule P_0 to which are coupled m identical groups $\bar{M}_j$ (to which we shall refer as a hapten). Accordingly, we may represent such an antigen P_j symbolically, by writing

$$P_j = \{P_0 - (\bar{M}_j)_m\}$$

One can prepare rather simple artificial antigens of this type by diazo-coupling a hapten, some small molecule, arsanylic acid for instance, to a protein which is not antigenic in the rabbit. If m, the average number of such haptens per protein molecule, is made large enough ($m \geq 10$) we may have a good antigen which will elicit the formation of antibodies specific for the hapten.

If the hapten $\bar{M}_j$ is a chemical analogue of the metabolite M_j, then the antibody molecule A_j is capable of combining with the antigen molecule P_j by virtue of the specific chemical affinity of the catalytic site and of the controlling site of the antibody molecule A_j to the hapten $\bar{M}_j$.

In the case of a natural protein antigen, such as a foreign serum protein, we have a more complicated situation because one protein molecule might carry a certain number, m_1, of one kind of determinant group and also a certain number, m_2, of another kind of determinant group, etc. To these determinant groups we shall, for the sake of brevity, also refer as haptens.

More complex artificial antigens may be prepared by coupling a hapten to a natural protein which is itself antigenic in the rabbit.

The Rabbit Antibody-Antigen System.—If antibody which is obtained from the rabbit is mixed with the antigen for which it is specific, then within certain concentration limits a precipitate may be formed in which each antigen molecule may be combined with a number of antibody molecules, and each antibody molecule may be combined with two antigen molecules. The concentration limits within which such a precipitate is formed define the so-called equivalence zone where the supernatant over the precipitate contains only small quantities of both the antigen and the antibody. The free antigen concentration is exceedingly low in equilibrium with such a precipitate.

A precipitate may also be formed if the rabbit antibody is present in large excess. Such a precipitate may be represented as consisting of units in which a number of antibody molecules (the number depending on the size of the antigen molecule) are combined with one antigen molecule; the precipitate forms because such units stick to each other. The free antigen concentration is low in equilibrium with a precipitate of this type also.

No antigen-antibody precipitate is formed if antigen is present in great excess. Presumably in this case each divalent antibody molecule is combined with two antigen molecules. One of the antigen molecules is combined with the antibody molecule at the controlling site, and the other is combined with it at the catalytic site. Each antigen molecule is combined, however, with one antibody molecule only.

Proteins which may have a tendency to stick to gamma globulins may be co-precipitated when an antigen antibody precipitate is formed in their presence. An example for proteins which may be co-precipitated in this manner are the serum "complements"; which are "fixed" when an antigen-antibody precipitate is formed in their presence.

The Coupling Enzyme-Antigen System.—If the hapten $\bar{M}_j$ of the antigen P_j is a chemical analogue of the metabolite M_j, then the antigen can reversibly combine with the coupling enzyme C_j by virtue of the specific chemical affinity of the hapten to a part of the catalytic site of the coupling enzyme.

We postulate that the coupling enzymes resemble those antibodies of the rabbit which form a precipitate with antigen when the antibody is in excess. Accordingly, at low antigen concentrations a number of coupling enzyme molecules might be combined with one antigen molecule and such units may then stick to each other and form a precipitate. We may assume that the free antigen concentration can be very low in equilibrium with such a precipitate.

The "Sensitive" Lymphatic Cells.—We assume that the cells of the lymphatic system go through a maturation process that carries the cell from its initial form, presumably the stem cell, to its mature form, presumably the plasma cell. When a lymphatic cell reaches a certain phase in this maturation process it becomes "sensitive" in the sense that it becomes capable of responding to the exposure to a soluble antigen with the formation of a specific antibody. We assume that what renders the cell sensitive, when it reaches this phase of its maturation process, is a general lowering of the repression factors to a value of perhaps $\lambda = 10^6$. (For definition of the repression factor see preceding paper.) This might be brought about by an increase in the activity of an enzyme which can universally hydrolize the R moieties of the repressors. In the preceding paper we have postulated the existence of such an enzyme and designated this hypothetical enzyme as the "hydrolase."

The set of constants assumed in the preceding paper, for the enzyme-forming system of the mammalian cell, we assume to hold for the antibody-forming system of the "sensitive" lymphatic cells. In the immature form of the lymphatic cells and in the rest of the somatic cells the concentration of the repressors ρ_0 and the corresponding repression factors λ presumably have a much higher value.

The Primary Response.—We assume that when a soluble antigen is injected intravenously or intraperitoneally into the rabbit, it penetrates into the cytoplasm of all the lymphatic cells. We further assume that if the antigen stays in the circulation long enough, the free antigen concentration in the lymphatic cells will be the same as the free antigen concentration in the circulation. This means that the chemical potential of the antigen inside the cytoplasm of the lymphatic cell is the same as the chemical potential of the antigen in the circulation.

According to the views here adopted, the primary response comes about in the following manner: when a soluble antigen P_j is injected into the rabbit it will diffuse into the lymphatic cells and precipitate the coupling enzyme C_j. As the result of this, the rate at which the repressor REP_j is formed will be reduced, and the concentration of this repressor will begin to fall. As the concentration of the repressor falls, the rate of formation of the antibodies A_j will increase.

Antibody molecules A_j present in the cell bind molecules of the repressor REP_j because of the specific affinity of the repressor molecule to their controlling site. Accordingly, in a sensitive cell, as the concentration of the antibody molecules A_j rises and the concentration of the repressor molecules REP_j falls, at some point in time the enzyme-forming system will lock and from then on the cell will produce the antibody molecules A_j at a high rate, even in the absence of antigen.

This sustained antibody production, by the cells in which the enzyme-forming system has locked in the primary response, represents a kind of memory, and it may account for the sustained immunity manifested by the rabbit.

Possibly a cell might lock simultaneously for the production of antibodies specific for two different haptens. But once a cell has locked for the production of an antibody, which is thereafter produced at a high rate, then subsequently a new stationary state establishes itself in the cell. The cell is then no longer "sensitive," so that if it is exposed to another antigen it will not lock for the production of the corresponding antibody.

The Complexities of the Primary Response.—When the primary response is elicited in the rabbit by injecting an antigen which consists of a non-antigenic protein P_c to which are coupled m haptens $\bar{M}_j$, there will combine with an antigen molecule not only molecules of the coupling enzymes C_j but also molecules of the enzyme E_j and of the antibodies A_j.

The phenomena accompanying the primary response are even more complicated if the artificial hapten is coupled not to a protein which is non-antigenic in the rabbit but, for instance, to a foreign serum protein which *is* antigenic in the rabbit. When such an antigen diffuses into the lymphatic cells, various coupling enzymes corresponding to the various determinant groups carried by the antigen molecule, and also the corresponding enzymes E and antibodies A may combine with the antigen and form a precipitate.

On this basis it is possible to explain why a conjugated protein carrying an artificial hapten elicits more antibody directed against the artificial hapten if the protein is a foreign protein which is a good antigen in the rabbit, rather than if it is one of the rabbit's own serum proteins.

The Secondary Response.—We shall postulate that the cells of the lymphatic system contain a hypothetical enzyme S which, when present at high concentration, will inhibit cell division. We shall further postulate that this hypothetical enzyme S resembles complement, inasmuch as it can be co-precipitated if an antigen-antibody precipitate is formed in its presence.

When an antigen P_j is injected into a rabbit for the first time, a certain number of the sensitive lymphatic cells will lock and henceforth produce the antibodies A_j at a high rate. When a lymphatic cell locks and produces such proteins at a high rate, it will from then on produce most other proteins, including the hypothetical enzyme S, at a low rate. Thus, following the locking of the lymphatic cell in the primary response, the concentration of the enzyme S will fall, and it may reach a new stationary level within a few days or weeks.

If, about four weeks after the first injection of the antigen P_j, the same antigen is injected into the rabbit again, it will diffuse into the lymphatic cells and it will form a precipitate with the antibodies A_j in those lymphatic cells which have locked at the time of the first injection of the antigen. By the process of co-precipitation a certain quantity of the enzyme S will thereby be fixed, and, inasmuch as at the time of the second injection the level of this enzyme in the locked cells is low, the concentration of the enzyme may fall to the point where it can no longer inhibit cell division.

Thus, the second injection of the antigen will lead to proliferation of those cells which have locked at the time of the first injection of the antigen. This is our explanation of the secondary response.

If the views here adopted are correct, we may expect that the secondary response would be elicited by any compound carrying a hapten, which forms a precipitate with the antibody that is directed against this hapten, whether or not the compound is capable of eliciting the primary response.

The finding of compounds which are weak antigens, in the sense that they will

Vol. 46, 1960   *BIOCHEMISTRY: LEO SZILARD*   299

elicit only a weak primary response, but are good antigens, in the sense that they will elicit the full secondary response, would lend support to our theory.

A conjugated protein, obtained by coupling an artificial hapten to rabbit serum albumin, is supposed to be a very weak antigen in the rabbit, in the sense that it does not elicit in the primary response the formation of an appreciable quantity of antibody directed against the artificial hapten. On the other hand, a conjugated protein obtained by coupling an artificial hapten to a foreign serum globulin, which is antigenic in the rabbit, is supposed to be a very good antigen in the sense that it will elicit in the primary response the formation of a substantial quantity of antibody directed against the artificial hapten. (Oral communication, Herbert Anker, 1959.)

On this basis, our theory predicts that if we pre-immunize the rabbit with a conjugated foreign serum globulin and evoke the secondary response with the conjugated rabbit serum albumin, we should obtain in the secondary response a substantial quantity of antibody directed against the hapten. If, however, we pre-immunize the rabbit with the conjugated rabbit serum albumin and evoke the secondary response with the conjugated foreign serum globulin, then we should obtain in the secondary response a less substantial production of the antibody directed against the hapten.

In interpreting the results of an experiment of this type, one must keep in mind that when an antigen is injected for the second time one obtains not only a secondary response but also a "primary response," in which sensitive lymphatic cells will lock for the formation of the specific antibody. It should be possible, however, to distinguish these two responses, because the release of the antibody into the circulation from the secondary response presumably precedes its release from the "primary response."

The Decay of the Primary Response.—In the primary response the release of antibody into the circulation does not persist long at a substantial rate after the antigen has been eliminated from the circulation. It is conceivable, though by no means certain, that, as far as the release of antibody into the circulation is concerned, the phenomena which characterize the secondary response may play a part in the primary response also.

When a cell locks for the production of an antibody in the primary response the concentration of the antigen present in the circulation might be high enough to lead to the formation of an antibody-antigen precipitate in the locked cell and to a co-precipitation of the enzyme S. This might cause a proliferation of the locked cell as described above, in connection with the secondary response.

The locked cells which divide may disintegrate as fast as they are produced and release their antibody content into the circulation.

Soon after the antigen disappears from the circulation, this enforced proliferation of the locked cells may cease, but we may assume that the locked cells will keep on dividing, even though rather slowly, and releasing their protein content into the circulation. The amount of antibody produced which is specific for any given antigen may not be appreciable, but the total gamma globulin production of the adult animal may be a measure of the rate at which the locked lymphatic cells divide and disintegrate.

Newborn rabbits are not capable of forming antibodies and we may perhaps assume that in the lymphatic cells of the newborn rabbit hydrolase activity is low, and therefore the repression factors λ are high. On this basis we may understand why lymphatic cells of the newborn rabbit cannot lock for the production of an antibody when they are exposed to an antigen.

The fact that in the young rabbit, just as in many other young animals, the rate of production of gamma globulin is low supports the view here adopted that the gamma globulins are produced by the locked lymphatic cells. The lymphatic cells which have not locked might not divide at all or might divide exceedingly slowly.

Immune Tolerance Induced in the Newborn Rabbit.—If a large amount of an antigen P_j is injected into a newborn rabbit, which cannot form antibodies, the antigen will diffuse into the lymphatic cells and there will be a condition of antigen excess both with respect to the coupling enzymes C_j and the antibodies A_j.

At the time, perhaps two weeks after birth, when some of the lymphatic cells of the rabbit become "sensitive" and capable of forming antibodies, the antigen concentration in the circulation will still be high.

The presence of a high concentration of the antigen P_j will prevent the locking of these cells with respect to the production of the antibodies A_j, because the haptens $\bar{M}_j$ of the antigen P_j compete with the repressor REP_j for the controlling site of the antibodies A_j.

The concentration of the antigen in the circulation of the rabbit will slowly fall and after several months it might reach a very low value. Nevertheless, the rabbit may still exhibit at that time specific immune tolerance with respect to the antigen. We may attempt to account for this phenomenon as follows:

If an antigen molecule is combined with the controlling site of the "attached" enzyme molecule we may assume that it will set up a steric hindrance and repress the formation of the antibody molecule in much the same way as would the specific repressor molecule.

We shall designate the enzyme-forming site which is specific for the formation of an antibody A_j as the ribosome B_j^*. For the purpose of our discussion we may assume that corresponding to each gene G_j^* which determines an antibody A_j there is present in the cell one and just one ribosome B_j^*. Since we have assumed that there are a number of genes G_j^* which correspond to the gene G_j, there will also be a number of ribosomes B_j^* for each gene G_j. (The antibodies A_j which are made by the different ribosomes B_j^* may differ somewhat from each other in their specific combining capacity with respect to the metabolite M_j and the hapten $\bar{M}_j$.)

We shall assume that each antibody A_j is formed inside the corresponding ribosome B_j^*, and that the ribosome is a semi-closed structure.

We assume that water, salts, and small protein molecules, including the hypothetical enzyme U postulated in the preceding paper, may all diffuse freely in and out of the ribosome. The antibody molecule A_j which is produced in the ribosome can, however, not diffuse out of the ribosome, and its concentration will therefore rise until the osmotic pressure inside the ribosome exceeds the osmotic pressure in the cytoplasm outside of the ribosome. At that point water will begin to diffuse into the ribosome and as the hydrostatic pressure inside the ribosome increases, the ribosome may open up sufficiently to permit escape of a few antibody molecules. On the basis of such a model we may expect that even in cells which are not locked for the production of an antibody each ribosome contains the antibody, for which it is specific, at a high concentration.

The concentration of the antigen injected into the newborn rabbit will remain high in the circulation for an extended period of time, and during this time enough antigen P_j will diffuse into the ribosome $B_j{}^*$ to form an antigen-antibody precipitate. The "attached" antibody molecule A_j in the ribosome may be combined at its controlling site with an antigen molecule and may form part of the precipitate. Because the concentration of the free antigen in equilibrium with such a precipitate is low, the precipitate may still persist in the ribosome at the time when the concentration of the free antigen in the circulation has fallen so low as to be undetectable. In this manner we may account for enduring immune tolerance.

Accordingly, enduring immune tolerance requires the presence of the antigen in the lymphatic cells, but does not require the antigen to be present in the circulation at a detectable concentration.

Specific immune tolerance could not endure for long if the lymphatic cells of the rabbit which have *not* locked for the production of any antibody were to divide at a substantial rate, because new ribosomes $B_j{}^*$ are presumably formed by the genes $G_j{}^*$ when the cell divides, and antibody production in these new ribosomes would not be blocked. (Cells which have locked for the production of an antibody may be assumed to undergo divisions at an appreciable rate, but these cells are not capable of responding to the exposure of an antigen with the formation of the specific antibody, and would therefore not abolish immune tolerance.)

The Case of the X-Rayed Rabbit.—We may attribute the inability of the rabbit to respond with the formation of antibody, if the antigen is injected several days after the rabbit has been exposed to an X-ray dose of 400 r, to the inability of its lymphatic cells to lock for the formation of an antibody.

On this basis we may expect that, if a large quantity of an antigen is injected into the rabbit, several days after the rabbit has been exposed to an X-ray dose of 400 r the rabbit will exhibit enduring immune tolerance with respect to this antigen.

If an X-ray dose of 400 r is administered to the rabbit after the injection of an antigen, then by the time the X-ray exposure takes effect many of the lymphatic cells of the rabbit will have locked for the production of antibody directed against the antigen. These locked cells will proceed to form antibody, unaffected by the X-ray exposure. Accordingly, in this case the X-ray exposure will not prevent the rabbit from responding to the injection of the antigen with the formation of antibody.

Since the antibody released into the circulation in the secondary response is released by cells that have locked for the formation of the antibody at the time of the primary response, or by their descendants, there is no reason to expect that exposing the pre-immunized rabbit to an X-ray dose of about 400 r should inhibit the secondary response.

Concluding Remarks.—If the model for cellular differentiation, presented in the preceding paper, should in fact correctly describe the molecular basis of antibody formation, then we could be rather confident that the above given explanations for the major immunological phenomena are essentially correct. This holds in particular for the role we have attributed to the "locking" of sensitive lymphatic cells in the primary response and for the notion that the primary response and the second-

302   *BIOCHEMISTRY: S. COHEN*   PROC. N. A. S.

ary response are basically different phenomena. We cannot be equally confident, however, of the correctness of various assumptions which relate to details and which we introduced for the sake of the concreteness of the discussion. Among these are the assumptions which relate to the mechanisms by which the antigen lowers the repressor concentration and the assumptions which relate to the structure of the ribosomes B^*. As far as such details are concerned it might well be that our assumptions will have to be modified later on, in the light of future experimental data.

In the process of formulating the thoughts expressed in this paper I had the privilege of discussing the subject with Dr. Maurice S. Fox, The Rockefeller Institute, New York; Dr. Howard Green and Dr. Baruj Benacerraf, New York University College of Medicine; and Dr. Herbert Anker, The University of Chicago. It is a pleasure to acknowledge their helpful suggestions and criticism.

* This paper is the continuation of a paper entitled "The Control of the Formation of Specific Proteins in Bacteria and in Animal Cells," pages 277–292 of this issue of these PROCEEDINGS. The concepts, notations, and equations of the first paper carry over to this paper.

† This work was supported by a research grant of the National Institutes of Health, U. S. Public Health Service.

(Reprinted from Nature, Vol. 186, No. 4725, pp. 649–650, May 21, 1960)

Dependence of the Sex Ratio at Birth on the Age of the Father

RECENTLY I have attempted to develop a general theory of ageing applicable to mammals[1]. This theory assumes that, during the life-time of the animal, whole chromosomes, which are contained in the somatic cells, suffer a total loss of function in a single random event. It further assumes that the probability of such an event remains constant throughout life-time and that it is the same for all individuals of a given species.

From the American life-tables I have derived, on the basis of the theory, the frequency with which such random events occur in man and found that on the average one chromosome, of the haploid chromosomal set of the somatic cell, suffers such a 'hit' in about twelve years. If we assume that this holds also for the spermatogonia in man, we should then expect that the ratio of boys and girls, at birth, decreases appreciably with the age of the father. This conclusion is based on the following notions:

We may assume that any cell of the male becomes non-functional if the X chromosome for which the cell is hemizygous suffers an ageing hit, that is, the absence of the X chromosomes is lethal for the somatic cell in the male. The absence of the Y chromosome is not lethal for the somatic cell and, as a matter of fact, it need not be lethal even for the zygotes; there exist individuals whose cells contain one X chromosome and no Y chromosome and these exhibit Turner's syndrome.

Accordingly, if the X chromosome in a spermatogonium suffers an ageing hit, then the cell is eliminated; if the Y chromosome in a spermatogonium suffers an ageing hit, then the cell remains fully functional and will give rise to spermatozoa, but only to spermatozoa which contain an X chromosome.

The length of the Y chromosomes amounts to about 2 per cent of the total length of the haploid set of the autosomes[2] and on this basis we may perhaps assume that an average of 2 per cent of the Y chromosomes of the spermatogonia suffer an ageing hit in a 12-year period.

We may assume that some spermatogonia are lost due to causes other than ageing and are replaced by other spermatogonia undergoing divisions. Then if

Table 1

15–19 yr. (∼ 46,000): 106·7;		20–24 yr. (∼ 410,000): 105·7;
25–29 yr. (∼ 590,000): 105·0;		30–34 yr. (∼ 440,000): 105·0;
35–39 yr. (∼ 250,000): 105·0;		40–44 yr. (∼ 120,000): 104·1;
	45–49 yr. (∼ 45,000): 103·6	

spermatogonia which lack a functioning Y chromosome divide at the same rate as spermatogonia which carry a functioning Y chromosome, the ratio, at birth, of boys and girls would decrease by 2 per cent for a 12-years increase in the age of the father, and by 4 per cent for a 24-years increase. However, if the spermatogonia which lack a functioning Y chromosome undergo divisions at a somewhat lower rate, then the ratio of boys and girls would, initially, decrease with the father's age at the rate quoted above, but the ratio would approach a fixed value with increasing age of the father.

One may compute on the basis of birth data[3] for the United States, 1955, the dependence of the ratio of boys and girls on the age of the father. For the age-group of the father indicated and for the sample sizes—for boys alone or girls alone—given in brackets, the ratios are as given in Table 1.

These data do not take into account the age of the mother, and there is, of course, a strong correlation between the age of the father and the age of the mother. What one would want to know is the dependence of the sex ratio at birth on the age of the father, for a fixed age of the mother. However, if one tabulates the available data on this basis the sample sizes become too small.

From the data given in Table 1 it would appear that the ratio of boys and girls does fall with increasing age of the father. It will not be possible, however, to draw conclusions which are relevant from the point of view of the theory until the data based on larger samples become available.

LEO SZILARD

Enrico Fermi Institute for Nuclear Studies,
University of Chicago,
Chicago 37.

[1] Szilard, Leo, *Proc. U.S. Nat. Acad. Sci.*, **45**, 32 (1959).
[2] Tijo, J. H., and Puck, T. T., *Proc. U.S. Nat. Acad. Sci.*, **44**, 1929 (1959).
[3] "Vital Statistics of the U.S.", **1**, 213 (1955).

Printed in Great Britain by Fisher, Knight & Co., Ltd., St. Albans.

Reprinted from the Proceedings of the National Academy of Sciences
Vol. 51, No. 6, pp. 1092–1099.   June, 1964.

ON MEMORY AND RECALL*

By Leo Szilard

THE SALK INSTITUTE FOR BIOLOGICAL STUDIES, LA JOLLA, CALIFORNIA

Communicated April 27, 1964

The subject matter of this paper is a hypothetical biological process on which the capability of the central nervous system to record and to recall a sensory experience might conceivably be based.   It may be open to doubt whether one knows enough about the living cell to be able to say anything with reasonable assurance about the molecular processes that the brain employs.   Still, with luck, one might perhaps guess correctly the general nature of these processes.   To what extent we may have succeeded in doing so remains to be seen.

The Efficacy of a Synapse Bridging Two Neurons.—Our neural network models involve excitatory neurons and inhibitory neurons (of the kind which exert a post-synaptic inhibitory effect).

Let us consider an excitatory neuron which contacts through a synapse another neuron.   If such an excitatory neuron sends a volley of nerve impulses to this synapse, then a certain quantity of an excitatory "transmitter substance" is released in the vicinity of the presynaptic membrane which diffuses across a gap—the synaptic cleft—into the postsynaptic neuron and raises the level of excitation of that neuron by a certain amount.   We shall designate this excitatory transmitter substance as "acetylcholine" (in quotes).   The "acetylcholine" which diffuses into the postsynaptic neuron is destroyed, in the vicinity of the postsynaptic membrane, by an enzyme which we shall designate as "choline esterase."

The rate at which "acetylcholine" is released in the vicinity of the presynaptic membrane is a function of the frequency of the nerve impulses which reach the synapse, and we shall designate this rate as the "signal intensity."   For the sake of simplicity, we shall assume that the signal intensity is for all synapses the same function of the frequency of the nerve impulses which are fed into the synapse.

The rate at which "acetylcholine" is destroyed in the postsynaptic neuron is proportional to the product of the concentration of "acetylcholine" and the concentration of the enzyme "choline esterase" in the vicinity of the postsynaptic membrane. Therefore, if, at a given point in time, nerve impulses of a certain frequency begin to arrive at a synapse, the "acetylcholine" concentration will begin to rise and will asymptotically approach a limit in the vicinity of the postsynaptic membrane, which is proportional to the signal intensity and inversely proportional to the concentration of "choline esterase," prevailing in the vicinity of the postsynaptic membrane. The "acetylcholine" concentration which is asymptotically approached at the postsynaptic membrane constitutes the "excitatory input," which is received from the synapse by the postsynaptic neuron. On this basis we may then say that, for any given signal intensity, the excitatory input received from a given synapse by the postsynaptic neuron is inversely proportional to the "choline esterase" concentration prevailing in the vicinity of the postsynaptic membrane of that synapse.

We assume that the enzyme "choline esterase" is inactivated at the postsynaptic membrane in different synapses at different rates and that this rate of inactivation is determined by the chemical specificities of the two neurons which are bridged by the synapse. We shall assume, for the sake of simplicity, however, *that the enzyme "choline esterase" is produced at the same rate in all excitatory neurons.*

We designate as the efficacy of the synapse the excitatory input which a postsynaptic neuron receives from that synapse, *per unit of signal intensity*. On the basis of the above assumptions we may then say that the efficacy of the synapse is proportional to the rate at which "choline esterase" is inactivated at the postsynaptic membrane which in turn is determined by the chemical specificities of the two neurons which are bridged by the synapse.

The Rate of Inactivation of "Choline Esterase."—We assume that neurons which differ from each other in their response-specificity contain a different set of certain specific proteins in their cell membrane. We shall refer hereafter to these proteins as the "specific membrane proteins."

We postulate that to each specific membrane protein, there exists a complementary specific membrane protein and that a specific membrane protein molecule can combine with its complementary counterpart, just as an antibody molecule can combine with a molecule of its antigen. Accordingly, two complementary specific membrane proteins may behave as if they were, so to speak, each other's antibodies, as well as each other's antigens.

When an antibody molecule combines with an antigen molecule, it undergoes an allosteric transition, and an antibody molecule, when it is thus "dimerized," can bind complement. We assume that quite similarly a molecule of a specific membrane protein, when it combines with its complementary counterpart, undergoes an allosteric transition, and, when it is thus "dimerized," it can bind—and not only bind but also inactivate—the enzyme "choline esterase."

The gap (synaptic cleft) between the presynaptic membrane and the postsynaptic membrane is estimated to be about 200 Å wide. According to the notions here adopted, there must be, however, a number of places within the active zones of the two synaptic membranes at which this gap is narrowed down, so that the presynaptic membrane and the postsynaptic membrane are in physical contact. We assume that at such a point of contact, a molecule of a specific membrane protein, located in the postsynaptic membrane, can "dimerize" across the synaptic gap, with its complementary counterpart, located in the presynaptic membrane.

Let us now consider two neurons, A and B, which are bridged by a synapse. Neuron A is characterized by a set (*a*) of specific membrane proteins, which are present in its cell membrane, and neuron B is characterized by another set, (*b*). *We shall designate as the "overlap number" of these two neurons the number of specific membrane proteins contained within the set (a) which have their complementary counterpart contained within the set (b) [or vice versa].*

From this overlap number we may compute the efficacy of the synapse which bridges these two neurons. In order to simplify this computation we shall assume that the area of the active zone of the synaptic membrane is the same for all synapses, and also assume that the concentration of each specific membrane protein in the cell membrane is the same for any given neuron. On the basis of these simplifying assumptions, we may then say that the number of "dimers" contained within the active zone of the membrane of a synapse, which bridges neuron A and neuron B, is given either by the ratio of the "overlap number" to the total number of specific membrane proteins of neuron A, or by the ratio of the "overlap number" to the total number of specific membrane proteins of neuron B—*whichever ratio is smaller.* We shall designate *the smaller* one of these two ratios as the "overlap fraction" of the neurons A and B. *Accordingly, we may then say that the efficacy of a synapse bridging two neurons is proportional to the overlap fraction of the two neurons. This is the first fundamental postulate of our model.*

We assume that the same holds true also for the synapses of our inhibitory neurons, except that in this case the "transmitter substance" which diffuses across the synaptic gap into the postsynaptic neuron lowers, rather than raises, the level of excitation of the postsynaptic neuron.

The Transprinting of Neurons.—We divide neurons of the central nervous system into two broad classes: the "congenitally determined" neurons and the "memory" neurons. We designate neurons which attain their full chemical specificity of their cell membrane during the development of the individual (mostly during embryonal life and at the latest during the early postnatal period) as "congenitally determined neurons." If all the neurons of the central nervous system were of this sort, then the individual would not be able to learn and his behavior would be wholly governed by the inborn reflexes. According to the notions here adopted, an adult can learn, and recall what he has learned, because his central nervous system contains memory neurons and each of these can, once in his lifetime, acquire an additional set of specific membrane proteins, through a process which we designate as "transprinting."

We assume that there is a class of "congenitally determined" neurons which are capable of participating in the transprinting of a memory neuron and that if a "congenitally determined" neuron of this class fires, then those parts of its cell membrane (covering the boutons of the branch fibers of its axon), which form the active zones of the presynaptic membranes become permeable for the specific membrane proteins. Similarly, we assume that when a memory neuron fires, then those parts of the cell membrane (covering its cell body and its dendrites) which constitute the active zones of the postsynaptic membranes, become permeable for the specific membrane proteins. Accordingly, if a "congenitally determined" neuron of this class contacts a memory neuron through a synapse and if both neurons fire "simultaneously" so that for a period of time both the presynaptic and the postsynaptic membrane is permeable for the specific membrane proteins, then the specific membrane proteins of the presynaptic "congenitally determined" neuron will diffuse through the presynaptic and the postsynaptic membrane into the postsynaptic memory neuron. *We postulate that if a specific membrane protein penetrates in this fashion into a memory neuron, it induces in the memory neuron the complementary specific membrane protein— just as an antigen induces its antibody, if it penetrates into a certain lymphatic cells of the rabbit.*[1] If several such presynaptic neurons fire simultaneously with the memory neuron, *then the memory neuron will on such an occasion acquire the sets of specific membrane proteins which are complementary to the sets of all of these presynaptic neurons.*[2] *This is the process of transprinting. Its occurrence as an "all or none" process constitutes our second fundamental postulate.*

We shall refer to memory neurons before they are transprinted as transprintable neurons and thereafter we shall refer to them as transprinted neurons. Like congenitally determined neurons, transprinted neurons may also participate in the transprinting of a transprintable neuron.

If a neuron participates in the transprinting of a transprintable neuron, then we may expect this neuron and the transprinted neuron to have a large overlap fraction and, accordingly, we may expect synapses bridging these two neurons to have a high efficacy.

The Conditioned Response.—In order to illustrate how transprinting may take place, we shall use as an example the classical (Pavolovian) conditioning of the salivary reflex of the dog.[3] We shall indicate on this occasion, however, only rather sketchily what takes place during conditioning.

When "food" is introduced into the mouth of a dog, the dog responds with salivation. This is the inborn, or unconditioned, response. Let us now expose the dog to a compound stimulus which has an auditory as well as a visual component, and let us—before the compound stimulus is turned off—place food into the mouth of the dog. If, after several such "conditioning exposures," the dog is then presented for the first time with the compound stimulus, unreinforced on this occasion by the introduction of food into its mouth, the dog may be expected to salivate. This is the conditioned response.

We assume that there is a neuron F in the central nervous system, characterized by the set (f), which preferentially responds to the stimulus of "food in the mouth." *Moreover, we shall assume in particular that the signal to which the neuron F responds is the onset of this stimulus.* As shown in Figure 1, the neuron F is connected through a synapse to an effector neuron, which innervates the salivary gland. This effector neuron is characterized by the set $(\bar{f})$, where $(\bar{f})$ denotes a set of specific membrane proteins which is complementary to the set (f). Because the overlap fraction of the neuron A and the effector neuron is *one*, the synapses which bridge these two neurons have a high efficacy. Therefore, placing food in the dog's mouth may be expected to cause the dog to salivate.

In order to account for the conditioned response, we postulate the existence of a number of groups of transprintable neurons E. Each of these groups may consist of several hundred neurons E, all of which have the following in common: The neuron F contacts through a synapse each of the neurons E and in turn each neuron E contacts, through a synapse, an interneuron F$\bar{\text{I}}$ [characterized by the set $(f) + (\bar{i})$] which in turn contacts, through a synapse, the effector neuron.

Until something happens which is "significant" from the point of view of the

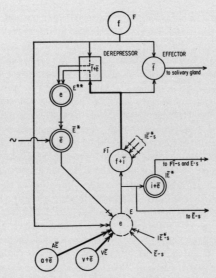

FIG. 1.—Excitatory neurons are represented by circles, and inhibitory neurons are represented by double circles. Excitatory synapses are represented by simple arrows, except if they belong to neurons which are capable of transprinting, in which case they are represented by double arrows. Inhibitory synapses are represented by arrows with a crossbar. The transprintable neuron E is represented by a dotted circle.

salivary reflex, all the transprintable neurons E are repressed, because they are inhibited by signals which are continuously being sent out by the inhibitory neuron $\bar{E}^*$ [characterized by the set $(\bar{e})$]. This inhibition is assumed to be strong enough to prevent a transprintable neuron E from firing, even if it should receive a substantial aggregate "excitatory input," because the overlap fraction of the inhibitory neuron $\bar{E}^*$ and of the transprintable neuron E is *one*. It should be noted, however, that after the neuron E is transprinted and acquires a set of specific membrane proteins which is composed of a large number of such proteins, then its overlap fraction with the inhibitory neuron $\bar{E}^*$ is reduced by a substantial factor, and the efficacy of the synapse bridging the two neurons is also reduced by the same factor. Accordingly, such a "transprinted" neuron E may be caused to fire in spite of receiving inhibitory signals from the neuron $\bar{E}^*$.

The transprintable neurons E get *derepressed* if the inhibitory neuron $\bar{E}^*$ is inhibited by signals emanating from a neural network designated as the "derepressor." This will happen if the derepressor sends out signals which are sufficiently strong to excite the inhibitory interneuron E**, which in turn will inhibit the inhibitory neuron $\bar{E}^*$. The derepressor network may receive an input signal from the neuron F and it may also receive an input signal, via the interneuron $F\bar{I}$, from neurons E. These two input signals counteract each other within the derepressor, however, and they cancel out if the intensity of both input signals is about the same. *Accordingly, the derepressor will send out strong signals only if the intensities of these two input signals differ from each other substantially.* In our second paper we shall describe a very simple "neural network" which would function in this fashion.

As will be seen later, the derepressor network may be expected to send out strong signals if food is introduced into the mouth of an unconditioned dog and to send out strong signals also when a dog, whose salivary reflex has been fully conditioned to a certain stimulus, is presented with that stimulus, without having, on this occasion, food placed into its mouth. The depressor network will not send out signals, however, if the fully conditioned dog is presented with the correct stimulus and food is placed into its mouth. Accordingly, no additional neurons E would be transprinted as the result of such "routine exposures."

It is probably generally true that a sensory experience is recorded only if there is "significance" attached to that experience. In our model of the conditioned salivary reflex there is significance attached to the stimulus "food in the mouth" for an unconditioned (or not fully conditioned) dog, while for a fully conditioned dog, there is significance attached to the conditioned stimulus, but only if that stimulus is *not* accompanied by the signal "food in the mouth."

We shall try to indicate next in what manner a conditioned salivary response may be established to a compound stimulus which has a visual and an auditory component. To this end we assume that in the central nervous system there is a neuron $A\bar{E}$ which responds preferentially to the auditory component of the compound stimulus, and another neuron $V\bar{E}$ which responds preferentially to the visual component. These two neurons are characterized by the sets $(a) + (\bar{e})$ and $(v) + (\bar{e})$, respectively. We assume that the number of different specific membrane proteins contained in the neurons $A\bar{E}$ and $V\bar{E}$, which we designate by $n(A\bar{E})$ and $n(V\bar{E})$, respectively, are large compared to the number of different specific membrane proteins contained in the transprintable neuron E, which we designate by $n(E)$. Accordingly, we have: $n(A\bar{E}) > n(E)$, and $n(V\bar{E}) > n(E)$.

We assume that, out of a group of several hundred neurons E, a certain fraction is contacted through a synapse by the neuron $A\bar{E}$, a certain fraction is contacted through a synapse by the neuron $V\bar{E}$, and a certain fraction is contacted by both the neuron $A\text{-}\bar{E}$ and the neuron $V\bar{E}$. Because the neuron $A\bar{E}$ as well as the neuron $V\bar{E}$ has an appreciable—even though small—overlap fraction with the transprintable neurons E, we may assume that if either of these two neurons fires, or if both of them fire, at a time when the transprintable neurons E are derepressed, then one or more transprintable neurons E will fire also and will on that occasion be transprinted by the neurons $A\bar{E}$ or $V\bar{E}$ or both. If at the same time the neuron F fires also, then these same neurons E will be transprinted also by the neuron F. The firing of neuron F alone would, however, not cause the neurons E to fire, even when the neurons E happen to be derepressed, because the neurons F and the transprintable neurons E have zero overlap.

If the unconditioned dog is for the first time exposed to the compound stimulus, and at the same time food is introduced into its mouth, the derepressor network will send out a strong signal and one, or several, of the transprintable neurons E will be caused to fire. These neurons E will then be transprinted, on this occasion, with the sets (f), (a), and (v). If this dog is exposed, for the second time, to the compound stimulus and at the same time food is introduced into its mouth, then the signal sent out by the derepressor network will be somewhat weaker than the first time.   This is so because the neurons E which have been transprinted at the time of the first conditioning exposure, and which will be excited at the time of the second exposure, have a large overlap fraction with the interneuron $F\bar{I}$ and will, therefore, send a signal to the derepressor network, which counteracts the signal received by this network from the neuron F.   As the conditioning process is continued and the dog is repeatedly subjected to such "conditioning" exposures, the neurons E which are transprinted with the sets $(\bar{f})$, $(\bar{a})$, and $(\bar{v})$ will increase in number.   Finally, the derepressor network will no longer send out a signal when the dog is exposed to the compound stimulus and at the same time "food" is introduced into its mouth.   Such a dog is then fully conditioned and continuing the conditioning exposures would not make the conditioning any deeper.

Let us now expose such a fully conditioned dog to the compound stimulus, unreinforced on this occasion by the introduction of food into its mouth.   The neurons E which have been transprinted during the previous conditioning exposures with the set $(\bar{f})$ as well as the sets $(\bar{a})$ and $(\bar{v})$ will now be caused to fire.   Because of the substantial overlap of the transprinted neurons E, which contain the set $(\bar{f})$, with the interneuron $F\bar{I}$, the firing of the neurons E will lead to the firing of the interneuron $F\bar{I}$ and this in turn will lead to the firing of the effector neuron. Accordingly, on the occasion of this unreinforced exposure of the dog to the compound stimulus, the dog will salivate.   This is the conditioned response.

Incidentally, on this occasion, when the interneuron $F\bar{I}$ fires, it will cause the derepressor network to send out a strong signal because this network does not on this occasion receive a signal from the neuron F.   Accordingly, on this occasion, one or more neurons E will get transprinted with the sets $(\bar{a})$ or $(\bar{v})$ or both, but none of them will be transprinted with the set $(\bar{f})$.   Therefore, if the dog is repeatedly exposed to the compound stimulus in such a fashion, i.e., without reinforcement, then the number of neurons E which are transprinted with the sets $(\bar{a})$ or $(\bar{v})$ or both, but not with the set $(\bar{f})$, will increase on each such occasion.   The overlap fraction of these transprinted neurons E with the interneurons $F\bar{I}$ is zero, and therefore the excitation of these transprinted neurons E would not contribute to the excitation of the effector neuron.   Their activation would, however, contribute to the excitation of the inhibitory neurons $I\bar{E}$*, with which the neurons E have an appreciable overlap.   This is the reason why the accumulation of neurons E which are transprinted with $(\bar{a})$ or $(\bar{v})$ or both, but not with $(\bar{f})$, will extinguish the previously established conditioned response to the compound stimulus.

Note to the First Model.—One more thing needs to be said at this point: it seems to be a fact that if we establish a conditioned salivary response in the dog to a compound stimulus, which has an auditory as well as a visual component, and if we subsequently extinguish the response, say to the visual component, we thereby automatically extinguish the conditioned response to the auditory component also.   It can be shown that, in order to account for this fact, we must assume in our model that the central nervous system contains, in addition to a number of neurons E which are characterized by the set (e), about an equal number of neurons $\bar{E}$ which are characterized by the complementary set $(\bar{e})$, and that the neurons $\bar{E}$, characterized by one of these two sets, must contact through synapses the neuron E characterized by the complementary set (and vice versa).   Presumably, this would mean that quite generally neurons characterized by complementary sets of specific membrane proteins must be present in about equal numbers in the central nervous system of the individual.

The Second Model.—We may escape this complication (if a complication it is) by assuming that *every specific membrane protein is complementary to itself.* According to this second alternative model, any set of specific membrane proteins is then identical with the complementary set of specific membrane proteins. Thus we may write (*e*) in place of (*ē*) and (*f*) in place of (*f̄*), etc. Accordingly, in our second model, *the overlap number of two sets is then defined as the number of specific membrane proteins which the two sets have in common, and when transprinting takes place, the transprinted neuron incorporates the sets of specific membrane proteins of the transprinting neurons.* Whatever functions the neural network, represented in Figure 1, would be capable of fulfilling on the basis of our first model, it would fulfill on the basis of our second model also, and the remainder of our discussion will be couched in terms of this second model, rather than the first one.

The Orderliness of the Inborn Neural Code.—According to the notions here adopted, we assume that two neurons in the central nervous system, which preferentially respond to two different sensory stimuli that "resemble" each other, must have a large overlap number. We assume that in the code of the congenitally determined neurons there is an orderly transition to smaller and smaller overlap numbers, as we go from one neuron to other neurons which differ from it more and more in their response-specificity. If it were otherwise, our model could not account for the phenomenon of the "generalization of stimuli" in the conditioned salivary reflex of the dog, first described by Pavlov.

Postscript.—If our two fundamental postulates are correct, then it ought to be possible to devise a neural network which would fully account for the phenomena exhibited by the conditioned responses of the autonomic nervous system. (The network described by Figure 1 represents a first attempt in this direction.) If one wanted to see, however, whether higher mental functions could be explained on the basis of our two fundamental postulates, then one would first have to invent adequate neural networks. Thus, if one wanted to see whether one could explain on this basis the mental functions which man is capable of performing, but the primates are not, one would perhaps have to invent the very same networks which are contained in the brain of man, but not in the brain of the primates. Clearly, this would be no mean task.

The "mental capacity" of suitable neural network models, operating on the basis of our two fundamental postulates, might be very high. For instance, the recording of information such as may be contained in a simple sentence would have to tie down only one "transprintable" neuron. Thus, if one were to expose an individual to a simple sentence every 4 seconds, 24 hours a day, and if, on each such occasion, one would tie down one transprintable neuron, then one would tie down just about 10^9 neurons over a period of 100 years. This is one tenth of the number of neurons believed to be contained in the human brain.

* This work was supported by a research grant, administered by The University of Chicago, of the General Medical Sciences Division of the National Institutes of Health.

[1] Anker, H. S., *Nature*, **188**, 938 (1960); Szilard, Leo, these PROCEEDINGS, **46**, 293–302 (1960).
[2] No neuron may, however, incorporate into its cell membrane the complementary counterpart of a specific membrane protein which its cell membrane already contains.
[3] Pavlov, I. P., *Conditioned Reflexes* (Oxford University Press, 1927).

**Memorandum to Cass Canfield by William Doering
and Leo Szilard (January 11, 1957)** *

MEMORANDUM

From: William Doering and Leo Szilard
To: Cass Canfield

A proposal to create two interdependent research institutes operating in the general area of public health, designated as:

Research Institute for Fundamental Biology and Public Health
and
Institute for Problem Studies

There appear to be a number of important problems in the general area of public health which could be solved today and yet for some reason or other progress toward their solution is conspicuously slow. In some cases the state of scientific knowledge is far enough advanced to make it possible to set up projects aimed at finding a practical solution to the problem; in other cases more fundamental research in the biological area involved is needed.

Throughout the United States at present there are scattered many spirited men among scientists who pursue fundamental research in biology with manifest success. These men engage in fundamental research primarily because it gives them pleasure to do so. Their interest is not limited in any permanent and irrevocable sense to a single specialty, and they are quite prepared to switch from one technique to another when the time comes to shift the emphasis from one set of questions to another. Many of these men are keenly interested also in the acute problems of our times, but they have rarely the opportunity to do something about them. Occasionally one or the other strikes out on his own from the realm of pure science into work on one of these problems. Jonas Salk, for instance, responded to the challenge of producing a polio vaccine by taking off several years from pure research (and he did this in the face of the prevailing notion that no dead virus can be successfully used for immunization). Usually such diversions from pure research involve a great personal sacrifice, and those who engage in them must struggle against heavy odds.

We propose that a setting be created which would offer men of this sort an opportunity for two kinds of activities:

(a) To pursue biological problems of great intrinsic interest that lie in the general areas which may ultimately have a bearing on problems of public health (with the same kind of freedom that scientists enjoy in their research at universities), and

---

* For background of this memorandum, see Part I, p. 14—Editors.

(b) To shift their attention from time to time to problems of public importance in the general area of public health and to work on projects aimed at finding practical solutions to such problems.

We believe that within such a setting the abilities of creative men could be brought to bear on problems of public importance—both recognized and unrecognized—where progress is conspicuously lacking.

Recognized Problems

Clearly it would be highly desirable to develop some biological method of birth control, adequate for the needs of the underdeveloped areas of the world. About three years ago the Conservation Foundation conducted a study in order to determine whether the problem was ripe for an attack on a programed research basis. The results are reported in *The Physiological Approach to Fertility Control*, the Conservation Foundation, 30 East 40th Street, New York City, April 1, 1953.

We participated in this study and concluded that there were a number of promising leads around which further research could be centered. Certain recommendations were made concerning the magnitude of the funds that were needed and the way in which they ought to be spent.

Subsequently the Population Council became interested in this problem and took some steps similar to those which had been recommended in the Conservation Foundation report. Currently, the Population Council is making available about $200,000 a year in the form of grants-in-aid for fundamental research in the field of mammalian reproduction. This money, added to the huge amount available at the National Institutes of Health to support pure research in this field, has led to a situation where there is probably more money available now than there are worthy takers. In spite of this, progress is slow and it is by no means certain that any of the present developments, aimed at finding a really satisfactory method for birth control, are moving in the right direction. Too few of the men active in this field have the kind of imagination and productivity that one finds among those attracted by fundamental biological problems of intrinsic interest. Too many are inclined to look upon the solution of the problem as a lifetime job.

The lag in progress in the field of birth control is not unique. It is just one example—though probably the most conspicuous one—of a general phenomenon.

Another example—taken again from the field of public health—is provided by cigarette smoking. For almost thirty years there has been reason to suspect (Carl Doering and Herbert Lombard, 1928) that cigarette smoking is harmful to health, and by now there is a strong suspicion that smoking one pack per day may shorten a man's life by at least five years. It would not take large funds to settle this question once and for all within a period of three years. There is a great public interest attached to knowing the answer and yet somehow it is not forthcoming with the desirable speed.

There is much public discussion of the topic but no concerted effort to establish the relevant facts. If smoking one pack a day shortens a man's life by five years or more, then cigarette smoking is the single most important public health issue in the United States, for even if we found a cure for cancer we could do no more than to add two and a half years to a man's life expectancy.

More than one-third of the deaths of middle-aged men in America are due to coronary disease. In Italy the contribution of this disease to the total death rate is believed to be one-twelfth and in Japan one-thirtieth. If the rate in the United States could be reduced to that of Italy, one might be able to add three years to men's life expectancy. For a long time now it has been suspected that the high coronary disease rate in America is associated with the diet. There are plenty of clues but no organized effort to find out what acceptable changes in the diet might remedy the situation. There is no reason why this question could not be settled within a period of five years, if a responsible group of men were to put their minds to it. The path from finding an answer to the relevant questions to the actual suppression of coronary disease in America may be quite long (it may take a generation's time before this can be accomplished), but the immediate, necessary steps seem clear.

The examples chosen above were taken from among the problems which are publicly recognized. It might well be that some of the unrecognized problems are of even greater importance. What are these unrecognized problems is, of course, not possible to say but an attempt is made in the Appendix [to this paper] to discuss what some might be.

It appears likely to us that progress will remain slow unless we create a setting that will provide for the scientists, whose help is needed, incentives—intellectual, moral, and material—sufficient to induce them to do something actively about these problems.

Most of these problems have in common that they require work on two levels:

(1) The knowledge, which is lacking and without which no effective action can be contemplated, must be produced. In some cases this knowledge can be obtained by a more or less direct attack on the problem—by setting up a "project." In other cases a direct study of the problem would be premature and progress must await advances in fundamental biological research in the general area.

(2) Once adequate knowledge is available, its application to the problem must be promoted. Frequently this may involve merely bringing to the attention of the agencies which are responsible for practical action the knowledge that has become available.

Work at both of these levels might begin to move very fast if the kind of scientists whom we have in mind can be induced to enter upon the stage. We say this with some degree of confidence because we have seen what can happen when an old stagnant field is invaded by "outsiders."

The Achievements of Outsiders

In the last fifteen years a revolution has taken place in the field of microbiology, which had been stagnant in the care of the classical bacteriologists for over thirty years. When suddenly biochemists, geneticists, and physical chemists—most of them young—invaded the field, things began to happen.

During the war, soon after the Radiation Laboratory was set up at M.I.T., nuclear physicists moved in and began to develop radar while the classical electrical engineers watched from the sidelines.

In the development of atomic energy, theoretical physicists invaded the field of chemical engineering and presented a process design for the Hanford plutonium plant which was accepted, while the design produced by the engineers proved to be useless.

It is conceivable that "classical" nutritionists will solve the problem of "coronary disease and fat metabolism," or that gynecologists will come up with a really satisfactory method of contraception, but it is more likely that an invasion of "outsiders" may provide the right answers.

What kind of organization would it take to provoke such invasion of outsiders?

No organization was needed to provoke the invasion of microbiology. Here were problems of great intrinsic interest which could be pursued even though the new men working in this field were scattered all over the country. There was no need to induce these men to leave their universities and gather under one roof; there were no problems of public importance that required the setting up of projects aimed at their solution.

The situation was different in the case of radar and atomic energy. In these cases it was necessary to establish closer collaboration in order to have a coherent group capable of a concerted attack. And it was possible to organize the invasion because the universities were willing to grant their men leave of absence "for the duration" and because the emergency of war provided the men with a compelling motive to respond to the call.

Our present situation is quite different. There may be at present an emergency also —but the emergency is not recognized. If we are to say what kind of a setting it would take to provoke the needed invasion, we cannot derive our answer from any precedent and have to rely solely on our imagination. In the following we shall attempt to describe on this basis an organization that we believe might be adequate to accomplish our purpose.

Organization

There shall be created two interdependent institutes having independent administrations and budgets but being otherwise closely related.

One of these—the Research Institute for Fundamental Biology and Public Health—will engage in pure laboratory research. Its regular Staff Members will have tenure and will be free to work on any biological questions of their choosing. Thus the selection of these Staff Members will automatically determine the areas in which the Research Institute will be active. In selecting the Staff Members we would, therefore, have to keep in mind that we want to see the Institute active in areas which are both of intrinsic interest from the point of view of fundamental knowledge and at the same time of relevance for unsolved public health problems. An attempt has been made to indicate in the Appendix [to this paper] what some of these areas might be.

The Research Institute will have six to eight regular Staff Members and perhaps nine to twelve Affiliate Members. In addition the Research Institute will have twelve to sixteen Research Associates on its staff, appointed for a maximum of ten years. The Affiliate Members will be distinguished scientists who serve in an advisory capacity. They will have a major role in determining the direction of the development of the Institute.

The second institute—the Institute for Problem Studies—will have no permanent scientific staff. It will have only a small staff of administrators who need not necessarily all be scientists. The choice of problems on which this Institute may work will be determined by the members of the Research Institute, both regular and affiliate.

It is assumed that the Institute for Problem Studies will take up only problems where progress is lagging and where there is a gap that it can fill. It is further assumed that in the next five years perhaps two-thirds of the attention of the Institute may be devoted to the problem of birth control and that most of the other problems tackled will also lie in the field of public health in the broad sense of the term. We may envisage two possibilities in regard to the scope of problems to be considered: The Institute for Problem Studies might:

(a) Be limited by charter to the field of public health in the broad sense of the term, or

(b) Be left free, by the charter, occasionally to take up problems that go outside the field of public health. Thus the Institute could pursue, for instance, some of the major problems of under-developed areas which lie in the area of "political thought" rather than "biological sciences." (This would involve bringing in for limited periods of time scholars whose interests are outside the field of science. See Appendix [to this paper].) But in this connection the charter shall provide that the Institute may go outside the field of public health only with the approval of two thirds of the Affiliate Members, as well as the approval of two-thirds of the Board of Trustees.

The Institute for Problem Studies will operate by bringing in from time to time, for a limited period, groups of men who may wish to collaborate with each other on a given project. If the Institute sets up a project that appeals to the imagination of an individual on the staff of the Research Institute, such an individual may, for as long as he wishes, go off the payroll of the Research Institute and transfer to that of the Institute for Problem Studies, where he may work jointly with others on a project. It is assumed that in the Institute for Problem Studies, of those who are employed on a temporary basis, about one-third to one-half will be on loan from the Research Institute, and the rest will be drawn elsewhere. The regular Staff Members of the Research Institute might spend on the average perhaps one-third of their time with the Institute for Problem Studies.

With respect to the problems assigned to it, the Institute for Problem Studies will assume the responsibility of creating or otherwise procuring and assembling the knowledge that is required in order to make effective action possible. The Institute will assume responsibility also for bringing this knowledge to the attention of those agencies which are in a position to take effective action. In this sense the Institute will promote activities that become possible through the newly derived knowledge. The Institute may have no clinical facilities of its own but it may have to assume responsibility for clinical tests in co-operation with medical schools and it may have to assume responsibility for field tests in co-operation with such agencies as the Population Council, Planned Parenthood, U.S. Public Health Service, the World Health Organization, or individual national governments.

Dissemination of information to the public, general education, propaganda, or the influencing of legislation will remain outside the scope of the Institute's activities.

The combination of these two Institutes will, we believe, attract the type of scientist whose help we need to enlist. Neither industrial laboratories nor research institutes of the classical type, nor universities, offer scientists comparable opportunities.

In the industrial laboratory, where the emphasis is heavily on commercial applications, the scientist rarely has the opportunity of satisfying his curiosity about fundamental aspects of nature, and he is almost invariably isolated, having few or no stimulating colleagues. He rarely survives.

In a pure-research institute a staff member is free to follow his inclination and move in his work anywhere his scientific curiosity may lead him, but this great freedom—which he possesses in theory—is to some extent thwarted—in practice—by his being under moral pressure to produce. He is not burdened by any teaching obligation, and he is free to spend all his time laying golden eggs—except that it is difficult to lay golden eggs if you are expected to do so. He may be permitted to follow up his basic discoveries to the point where they can be applied to the solution of a problem of public importance, but he is not encouraged to do so, for setting up projects within a pure-research institute would exert a disruptive influence.

In a university a scientist earns his "right to exist" through teaching; therefore the moral pressure to produce is much less strong. Thus it is easier there for a scientist to leave an area of research which is safe (where he can produce results by just turning a crank) and to venture into uncharted seas, where he runs the risk that he may obtain no publishable results for many years to come, but where he also has a chance of making a really great discovery. His situation is quite satisfactory if he likes to engage in formal teaching. However his teaching load, even if moderate, leaves him insufficient time to devote attention and energy to the solution of the great acute problems and to develop his basic discoveries to the point where he can see their fruits.

In contrast, the Research Institute for Fundamental Biology and Public Health will offer a man the advantage of being able to combine his interest in fundamental knowledge with his desire to see knowledge that has become available applied to the solution of problems which are of public importance. In his research he will be as free as in a university. But he will not be required to teach, and he may, if he so desires, earn his "right to exist" by joining a "project" which appeals to his imagination by going off the payroll of the Research Institute and on to the payroll of the Institute for Problem Studies.

The attraction of the Institutes for scientists—of the kind needed—would, in our opinion, be further enhanced should some of the projects of the Institute for Problem Studies extend into the area of "political thought." If this were the case, the Institute would presumably bring in, from time to time, for short periods, men whose main field of intellectual activities lies outside the realm of science. Our reason for believing that this would enhance the attraction of the Institute is based on the impression that, even in an institution which is as strong in science as the California Institute of Technology, spirited men among the permanent staff suffer from the one-sidedness of their contacts. They are deprived of the company of historians, economists, and other scholars, and they are aware of this deprivation.

Affiliate Members

The Affiliate Members will not be employees of the Research Institute and need not spend very much time at the Institute even though they shall play a major role in guiding its work. They will, together with the regular Staff Members whom they shall outnumber, control the appointment of new members. They will also, together with regular Staff Members, determine the projects which the Institute for Problem Studies may undertake. In order to enable them to fulfill these roles, the Affiliate Members may be expected to spend at least one week each year at the Research Institute, and it is hoped that many of them will spend more time than this. The Research Institute shall make secretarial facilities available to them during their stay, and they may, if they wish, make use of their visit to prepare manuscripts for publication away from the disrupting intrusions of their daily duties at their home base. It is proposed that Affiliate Members be paid their expenses and a fee of $3,000 a year.

Regular Staff Members

The regular Staff Members of the Research Institute will enjoy tenure like a professor at a university. Each will have a budget from $35,000 to $50,000, from which his own salary and those of one or two research associates may be paid. It is assumed that each regular Staff Member will, in addition, have outside grants-in-aid averaging perhaps $40,000. It is proposed that each regular Staff Member have at his disposal laboratory space between 2,500 and 5,000 square feet.

Research Associates

Research Associates will be chosen by the individual regular Staff Members and may serve in this capacity no longer than ten years. A change of status from Research Associate to regular Staff Member shall not be impossible but shall require the concurring vote of three-fourths of the Affiliate Members. This precaution is proposed in order to avoid the danger of inbreeding.

Staffing of the Research Institute

Following the selection of the first seven Affiliate Members and their acceptance, we may begin to discuss the selection of regular Staff Members. In order to do this intelligently, it will be necessary first to reach a consensus among the Affiliate Members concerning the areas in which they think the Research Institute ought to be active. Upon reaching such a consensus we may then begin to think of individual candidates for regular Staff Membership.

Our aim should be to recruit individuals with a strong overlap of interest, men who are likely to work harmoniously with each other under one roof. If appointments were offered to such men one by one, the most desirable candidates might be less likely to accept than if an offer were made to them as a group. What we must seek is a group of men who look upon fundamental research primarily as a source of pleasure and upon the application of knowledge to the acute problems of our times primarily as an opportunity for adventure; those others who are spurred to activity mainly by their sense of "duty" are not likely to be the most imaginative and productive ones.

The decision to offer initially an appointment to some such group will rest with the Affiliate Members of the Research Institute.

Procedure Concerning the Appointment of Members

The procedure that may be adopted concerning the election of new regular Staff Members and Affiliate Members (beyond the appointment of the two initial groups) may well determine whether in the long run the Research Institute may remain productive, or whether it will decay. It is proposed to discuss this point with the Affiliate Members prior to the drafting of the charter of the Research Institute.

Financial Need of the Institutes

The Research Institute will need an endowment of $10,000,000. An endowment of $5,000,000, plus perhaps a fund of $6,000,000 to be spent in ten years, would also be satisfactory. In this latter case the income from the endowment would not be spent for the first ten years but rather be added to the capital in order to build up the endowment. An endowment is necessary if the Research Institute is to grant tenure, which it must.

The Institute for Problem Studies will require $5,000,000, spendable in ten years.

It is assumed that the cost of building a laboratory may amount to $3,000,000, half of which may come from the federal government out of a Congressional appropriation of $90,000,000 (to be spent over the next three years). Under the provisions of this appropriation, half the costs of facilities devoted to medical research or public health may come from the federal government.

Housing

The character of the housing problem will depend on the location of the two Institutes which in turn shall be determined by the preference of the regular Staff Members of the Research Institute.

We believe that it is essential for the success of the Institutes that the scientific staff live within walking distance of the laboratory. If they do, many of them will return to the laboratory after dinner, perhaps two or three times a week, as the need arises, and they will have many more informal contacts with each other (proved so valuable to creative individuals).

The Boards of Trustees

It is proposed that one-half of the Trustees of each of the Institutes be drawn from the Affiliate Members of the Research Institute, and that the Affiliate Members take turns in serving on the two Boards of Trustees. Who the other members of the two boards might be is not for us to say, but we do wish to submit the following point of view:

For the Institute to be successful it is desirable that the non-scientist members of the Boards of Trustees and the Affiliate Members be congenial. In looking over the boards of trustees of other foundations, it would appear as if their members had been selected primarily on the basis of "respectability." For a member of a board of trustees to be respected is necessary but in our particular case he also needs to be imaginative and courageous. This holds particularly for the Trustees of the Institute for Problem Studies; the activities of the Research Institute will not be controversial in any way, but those of the Institute for Problem Studies probably will be.

Perhaps in thinking of prospective non-scientist Trustees it might be a good guiding principle to aim at bringing together a group of men who will take real pleasure in exchanging ideas with each other and with the Affiliate Members.

January 11, 1957

Appendix to Memorandum of Doering and Szilard*
by L. Szilard

Introduction

The areas in which the Research Institute will work will be determined by the predilection of its regular Staff Members. Affiliate Members of the Institute can exercise an influence in this regard only by their role in the selection of the regular Staff Members.

---

* This part of the memorandum was labeled "Confidential" because it mentioned specific persons by name. Paragraphs mentioning names were deleted by us—Editors.

Therefore, prior to the appointment of any regular Staff Members, the Affiliate Members ought to arrive at a projection and decide in their own minds in what areas of intrinsic scientific interest they would like to see the Research Institute active. In order to make it easier to arrive at a consensus in this regard, I have prepared this document which I hope may serve as a basis of discussion.

The Institute for Problem Studies will have no permanent scientific staff. It will be left to the Members of the Research Institute, both regular and Affiliate, to determine what problems this Institute shall take up. This Appendix is written in part with the purpose of serving as a basis of discussion in this regard also.

I. Mammalian Reproduction

The attraction of the problem of mammalian reproduction to younger scientists is mainly due to the recognition of its overwhelming importance for population control in the underdeveloped areas of the world. A good solution would be a drug that might be administered once a month to women in the form of a pill. Even better, perhaps, would be a drug that could be mixed in with certain staple foods, such as for instance rice, and made accessible to large families who live in poverty. Such an "infertility brand" of staple food might perhaps be sold—with a government subsidy—at a price below that of the "commercial brand" of the staple. This type of drug administration would demand the use of a drug that is without any detrimental physiological effect for both women and men.

It is conceivable that such a drug might be found. It might for instance be possible to find some chemical analogue of progesterone that is physiologically inactive everywhere in the body, with the single exception of the endometrium of the uterus, where it might compete with progesterone and thus prevent progesterone from preparing the endometrium to "receive" the ova.

At present clinical experiments are being conducted with certain progesterone analogues which suppress ovulation, presumably through their effect on the pituitary. Whether drugs of this kind might have some long-range effects which rule out their general administration, we cannot say at present.

I believe there would be little point in my presenting here my own evaluation of the relative merits of the different biological approaches to the birth control problem, because as soon as a really imaginative group of people enter this field they may produce a host of new ideas, many of them wholly unpredictable today.

The Institute for Problem Studies if it takes up this problem—as I assume it will—need not remain idle in this field while waiting for fundamental research to open up new avenues of attack. There are a number of "non-revolutionary" methods that need to be explored in the meantime. It is even conceivable that some minor inventions and improvements applied to such a "conventional" method of birth control as foam tablets might make this method practicable in some of the areas where there is an urgent need of birth control.

Developing methods for birth control to the point where other agencies can take over will involve subjecting such methods to clinical tests and field tests. The Institute for Problem Studies will not have clinical facilities of its own, but it will work in this field in close collaboration with medical schools both in America and abroad.

The testing of fundamentally new biological methods of birth control is difficult in America where a married couple, if it does not want to have a child, wants to use a method of birth control that is close to 100% effective. Such effectiveness cannot be guaranteed in the case of a brand new biological method. Here the best we may do is first to establish through toxicity tests in animals that the drug is harmless and then arrange for the clinical testing in countries where couples may be satisfied with slowing the rate at which the children come. We therefore assume that the Institute for Problem Studies will have to maintain close collaboration in this field with medical schools abroad.

This holds in particular for clinical experiments which are aimed at finding a satisfactory drug that will produce an early abortion. There should be no difficulty in arranging for clinical testing of such drugs in Sweden, or in Japan, where surgical abortion is at present the method of choice for population control. The drugs available at present are of limited usefulness, however, because if the therapeutic dose fails to cause abortion, surgical abortion becomes mandatory inasmuch as these drugs cause malformations in the surviving embryo....

At some point it will have to be decided by the Members of the Research Institute whether the Institute for Problem Studies shall concern itself with those problems of birth control that are peculiar to developed countries where there is no real emergency. In a sense these problems are peculiarly tempting because in thinking about them we think about our own future. For this reason these problems are even more controversial than those which relate to underdeveloped countries. What might these problems be?

Looking into the future, one is tempted to predict in developed countries that the method chosen for population control may one day be the cutting of the vas deferens. This method was not satisfactory in the past because it is irreversible, and one would not ever want to preclude any man, permanently, from the possibility of begetting children. Today, when sperm can be preserved indefinitely at liquid-air temperature, we may predict that sperm banks will be established at some time in the future, and that such banks will carry on deposit the sperm of men who have had, say, 3 children and who subsequently chose to have the vas deferens cut. If they should decide later on, that they wanted to have more children, they could always draw on their deposit.

From such a practice there is only one step to a new social custom which, if it does develop, might mark the beginning of a new chapter in human evolution.

Artificial insemination, which is frequently practiced today because often a couple cannot have children on account of the sterility of the husband, is practiced in the most primitive fashion—as a "blind date." The choice of the sperm which is used is that of the doctor, and the woman has no control over what type of man the father of her child shall be.

When sperm banks come to be established, sperm may be sold on the basis of a detailed catalogue and women who wish to conceive through artificial insemination may pick the type of man whose child they wish to bear. Clearly this could be the beginning of a rapid eugenic development of mankind. The direction of the development would not be controlled by any preconceived notion of the authorities (who may *think* they know what kind of men they want to breed), but rather by the preferences of the women who would follow their "instincts."

As far as we know, there might be no inherent reason—apart from deeply ingrained custom—why a woman should look for the *same qualities* in a man whom she may want for a husband, as in a man whom she may want for the father of her child.

The study of the behavior of sperm, which may open up vistas toward practical applications of great importance, might be regarded as a branch of microbiology. (Some of the methods which might be used in such a study, as well as the kind of men who might use them, will be described in Section II under the label "Protein Synthesis.") What are these vistas?

The sperm is haploid (i.e., it carries one set of chromosomes), and it might be possible to separate in a sperm population the sperm carrying a Y-chromosome from the sperm carrying no Y-chromosome. This would mean that we should be able to change the sex ratio in the offspring, a possibility that could provide us with a novel means of achieving a really satisfactory solution to the population control problem in the United States.

It is likely that in the United States, with increasing economic security and prosperity, we shall have a steady and rapid increase in the population—even when all families achieve control over the family's size. People who are wealthy and secure want to have more children—so it would seem—than is optimal from the point of view of the country as a whole. To discourage them, by punitive taxation, is hardly reconcilable with one of the as yet unformulated basic human freedoms—the freedom to breed. The community could try to correct this tendency to overbreed by making life so attractive for bachelors that the marriage rate shall fall. This, however, would be somewhat unfair to the women! For, as Lichtenberger has said, "All women should get married, but no man."

One "natural" solution to this problem is to change the sex ratio. Many families may desire to have two sons and one daughter, and presumably an over-all ratio of 1.5 to 1 would solve the problem of an excessive proliferation of the population in the United States. The excess men would remain bachelors. This excess would raise the value of girls, and the law of supply and demand would lead to a stable, "optimal," ratio without any interference with individual freedom. Quite on the contrary, we would have created a new basic human freedom hitherto undreamed of—the freedom of choosing the sex of one's children.

Clearly these observations about the law of supply and demand must be taken with a grain of salt. The ratio of boys and girls might be stabilized at a *high* level where so few girls would be born that the population would decline fast. In that case presumably a law would have to be passed limiting the supply of sperm that is enriched in

Y-chromosomes. Frankly, it is difficult to know what one would do about the boot-legging of such sperm, and one is somewhat reminded of the story of Daedalus and Icarus. Still, we have no choice, I believe, but to brave "the brave new world."

I am told that experiments on rabbits carried out at the University of California in Berkeley have demonstrated the possibility of shifting the sex ratio, but I have not as yet seen the relevant paper. The economic importance of changing the sex ratio in cattle would be very great, for one could increase the ratio of cows to bulls.

II. Protein Synthesis

An area of fundamental biological research which is of great intrinsic interest, and has therefore strong attraction for the kind of scientists whom we would want to see active in the Institute, might be somewhat sloppily labeled "Protein Synthesis." The men active in this field, whom we have in mind, are those who participated in the revolution that occurred in the field of microbiology in the last fifteen years. . . .

A few others who started out in pure microbiology are now shifting their interests to various aspects of the problem of protein synthesis. In particular they want to find out in what manner antibodies are formed and how "delayed sensitivity" is produced. Usually the field of antibody formation is labeled "immunology" and the study of delayed sensitivity is labeled "allergy." Of these two related areas the layman is more familiar with immunology, i.e., the formation of circulating antibodies in response to infection, rather than with the appearance of delayed sensitivity which is responsible for the fact that a skin transplanted from one individual to another will be rejected—after an initial period of healing—unless the transplant is made from one identical twin to another.

New methods which have become available should make it possible to make a *concerted* attack on the problem of antibody formation and the production of delayed sensitivity. In either case a specific chemical substance is presumably produced in response to the injection of an "antigen." The question how the animal accomplishes this feat is of great intrinsic interest and might furnish a clue to the basic problem of protein synthesis. . . .

A few words may be said on the possible practical applications of this kind of basic research:

(1) Advances in the field of immunology or allergy might open up new avenues for looking for a cure for cancer. The reason for saying this is as follows:

Malignant tumors are very frequently found at autopsies in persons who died of some other cause and in whom the growth of the malignant tumor was clearly arrested. This natural resistance to cancer might well be due to a natural immunological response, and the overt disease "cancer" might well be due to the breakdown of this immunological defense.

Since the problem of cancer is a difficult one, one man's guess may be as good as another's concerning the most promising approach to it. Still, if I had to bet on where the solution is going to come from, I should bet on the immunological approach rather than on the approach (heavily emphasized at present) which looks towards a chemotherapy of cancer.

Because of the great attention that the search for a cure for cancer is already receiving, I am inclined to think that the Institute for Problem Studies should concentrate on other fields. However, if some unexpected advance, made in the field of "immunology and allergy" in the Research Institute should open up a new avenue, one would not want to discourage a Staff Member of the Research Institute from pursuing a lead opened up by his discoveries.

(2) The field of "immunology and allergy" is presently developing fast. It has been recently shown, for instance, that 10 mg of bovine serum albumin injected into a newborn rabbit will impair the ability of that rabbit to form antibodies against bovine serum albumin. It is thus possible to make the rabbit "tolerant" against this antigen.

Whether a newborn baby is young enough to be made tolerant in this manner is now being investigated. In the somewhat unlikely case that newborn babies are "young enough" there would arise a number of interesting possibilities. I shall mention one of these, as an example. One type of mental defect is due to a disease of the newborn—erythroblastosis foetalis. This disease occurs because Rh-positive red cells of the infant evoke the production of antibodies in the mother, who is Rh-negative. If it should turn out that in man, as in rabbit, it is possible to impair the ability of the individual to form antibodies against an antigen by injecting that antibody within a few days after that individual is born, then it might be possible to get rid of erythroblastosis foetalis altogether. One would merely have to inject the newborn girls—if they are Rh-negative—with Rh-positive blood and thereby establish tolerance.

There is even a remote possibility that this field may conceivably have some bearing on the problem of birth control. The foetus may be regarded as a homotransplant. It does not sensitize the mother and the transplant is not rejected, perhaps because the sensitizing agent of the foetus can not get through the placenta. If the mother is however sensitized against the foetus—by receiving a skin transplant from the "father"—and if the substances thus evoked in the mother are able to pass the placenta, then it is conceivable that an early abortion may occur. This, admittedly, is a remote possibility, but arrangements are now being made with Dr. Howard Green for testing this possibility in rabbits. We mention this here mainly in order to emphasize the diverse connections that might *unexpectedly* appear between advances in "immunology" and some of the major unsolved problems in public health.

(3) The new techniques of studying virus growth in animal cultures have opened up the possibility of developing new powerful methods in the search for a chemotherapy for certain virus diseases. Some work along these lines might be carried out at the Research Institute but it is not likely that the Institute for Problem Studies needs to be concerned with this problem. Because of the obvious commercial interest of drugs capable of curing virus diseases, the pharmaceutical industry is likely to take over as soon as the practical significance of the advances becomes apparent. . . .

III. Fat Metabolism and Coronary Disease

Our knowledge of fat metabolism in mammals is very limited. Fat metabolism is not without intrinsic interest and therefore a few men on the staff of the Research Institute might well work in this field. The Institute for Problem Studies might also concern itself with this problem—because of its great practical importance to coronary disease in the United States. If the Institute takes up this problem, the staff of the Research Institute would almost certainly wish to guide the explorations of the Institute for Problem Studies in this field.

There has been much public discussion of this topic for a number of years but no concerted effort devoted to a really promising approach. Rather there was a strong emphasis on trying to establish the predictive value of a high blood cholesterol level, for individual middle-aged men. This is a rather unproductive approach—at least from the point of view of the major public health problem involved. Concerning this major problem hopes have been raised through the following observations:

(1) The coronary death rate in Norway fell by perhaps 30% within 18 months after the Germans occupied the country, during the last war, and consumption of animal fats dropped to about half.

(2) The blood cholesterol level and the level of beta-lipoproteins appear to be low in populations where fat consumption is low, and so is the coronary attack rate.

(3) Bronte-Stewart and his co-workers report that the blood cholesterol level can be lowered—temporarily perhaps, but conceivably permanently—by adding unsaturated vegetable oils to the diet.

On the basis of these observations—which needless to say do not permit the drawing of any conclusion—some people hope that by cutting down the fat consumption to perhaps 10% of the caloric intake (or to about 20 gm of fat a day), a middle-aged man can appreciably reduce his chances of suffering a coronary attack. Others hope that he can accomplish this even though he maintains his total fat intake at perhaps 40% of the caloric intake (or 80 gm of fat per day), provided only that this additional fat intake is in the form of unsaturated vegetable oils.

If the Institute for Problem Studies were to concern itself with this problem, and if it succeeded in bringing about a concerted effort to settle it, we could have the answer to all the relevant questions probably within five years. This would involve:

(a) Animal experiments to determine the effect of various diets on both blood levels and athero-sclerosis;

(b) Clinical experiments on groups of a few hundred patients with a coronary history who would volunteer to go on a prescribed, tightly controlled diet and whose blood levels would be closely followed, as well as their survival rate;

(c) Observations on large groups of volunteers (of the order of 20,000 men or more) who agree to modify their diet, if so requested, or to remain on their old diet for 18 months, if asked to do so. Blood tests on samples of these volunteers would indicate how much "cheating" goes on, and the survival rate would be followed. Significant results could be available at the end of a five-year period.

IV. The Effect of Cigarette Smoking on Longevity

I have reason to believe that some of the large life insurance companies have some difficulty in investigating this topic themselves and that they would be glad to co-operate with some such organization as the Institute for Problem Studies. With their co-operation this particular issue could be settled within three years.

It is conceivable that we could, with the co-operation of a large insurance company, settle within a five-year period some more refined questions of great practical importance, such as:

(a) Can a middle-aged man who stops smoking cigarettes thereby decrease his age-specific death rate, and if so, how fast would such an effect take hold?

(b) Is there any relevant difference between the effect of smoking cigarettes with and without filter tips?

The Institute for Problem Studies need not enter this field of vital statistics, if other suitable organizations are willing to carry the ball. This Institute could, nevertheless, take an interest perhaps in the relevant problems of experimental psychology. Should it turn out that cigarette smoking is as dangerous as it now would appear, the Institute might possibly take a hand in finding methods for making adolescents psychologically immune to the pressures inducing them to take up cigarette smoking.

Assuming that the parents want their children to be immune, and the children themselves are willing to be made immune, are there any psychological tools that will accomplish this? Hypnosis, for instance, does not seem to be very effective when used to break the habit of smoking—but would it be effective if used to condition children against taking up smoking? These are questions which to some extent can be decided by experiment, and such experiments might interest some of the staff of the Research Institute itself.

V. Advances in Neurology—Pharmacology of the Nervous System—Mental Health and Sleep

Some recent advances in neurology seem to attract a number of imaginative younger men and therefore it might well be that some work in the Research Institute might be carried out in this field. Advances in the field are likely to find practical applications also.

One of the advantages which the Institute offers to its staff is an opportunity to pursue really imaginative and unconventional goals. Such a pursuit is rarely encouraged in the existing type of research institutes. Among several possibilities in this regard we wish to mention one which might soon come within the reach of available methods: the elimination of sleep.

Because during evolution the best use for the hours of darkness consisted in spending the time asleep, we have developed biological mechanisms forcing us to sleep about eight hours a day. There is no evidence that sleep fulfills an inherent biological function other than doing obeisance to a built-in mechanism that compels us to sleep.

Perhaps no greater enrichment of human life could be brought about, at a single stroke, than getting rid of the tyranny of sleep (at least for extended periods of time, when other forms of activity could be substituted to great advantage). . . .

VI. The Problem of Aging

The general phenomenon of aging is of great intrinsic interest and it is therefore possible that some of the men in the Research Institute might want to work on the aging of mammals. I do not think it likely, however, that the Institute for Problem Studies will set up a "project" on aging, at least not until there are some really good leads. No such leads have so far been detected—to my knowledge.

In the last few years there has been an enormous amount of talk on the subject, and there is pressure on the U.S. Public Health Service to go into this field in a big way. But at the same time there is a conspicuous dearth of ideas.

Problems Outside the Field of Public
Health—in the Institute for Problem Studies

The advances of science occasionally create problems which may require for their solution advances in the realm of political thought. Advances in the field of public health have led to a rapid fall in infant mortality and thus overpopulation in Southeast Asia, and this is probably the most important single cause blocking the path of these areas toward progress. It is conceivable that if the overpopulation problem were solved by the development of an adequate method of birth control, these areas would be on their way toward a rapid, orderly industrial development.

Infant mortality is now also beginning to fall in a number of British colonies and former British colonies in Africa, but clearly in these areas overpopulation is only one of several major obstacles that must be overcome. How to achieve political stability in African colonies which are about to receive self-government is another of these problems. And the dearth of men who possess the qualifications needed for the administration of such areas poses a related problem calling for a solution.

The development of underdeveloped areas is likely to be of concern to the Members of the Research Institute, and they may want the Institute for Problem Studies to deal not only with the issue of birth control, but to give attention to all major aspects of the problem. This may cause the Institute for Problem Studies to venture beyond the field of public health and into the area of "political thought." With this possibility in mind it is suggested in the Memorandum that, if the Institute is not limited by charter to the area of public health, then it may, with the approval of two-thirds of the Affiliate Members and two-thirds of the Board of Trustees, attempt to take up certain problems outside of it.

In the following I shall discuss in some detail—as an example—a problem in the area of "political thought" that appears to be of major importance for certain underdeveloped areas. I shall try to examine why no real progress has been made toward the solution of this problem and to describe what procedures the Institute for Problem Studies could adopt if it were to take up this problem.

The underdeveloped areas which have recently achieved self-government, and those which may achieve self-government in the near future, can develop successfully only under a governmental system that will assure them a certain measure of political stability. Clearly it is desirable that these areas be administered with the assent of those governed and be—in this sense at least—under a democratic government. There is grave doubt, however, whether the parliamentary form of democracy is a suitable form of democracy for many of these areas. Yet, in spite of this doubt no other forms of democracy that might be more suitable have ever been devised and adequately examined, let alone tested.

If such forms of democracy were devised, there might be an opportunity to try some of them out in the next two generations in Africa, where one colony after another may be expected to be given self-government by England.

There would be no need for the Institute for Problem Studies to concern itself with this problem were it not for a certain lack of leadership on the part of our universities in the general field of "political thought." Why social scientists in our universities have not tackled this problem, and why none of the foundations has set up any project adequate to fill this gap, is not clear. Perhaps the explanation lies in the fact that in the social sciences too much emphasis is being placed today upon "research." Problems that do not require "research" but require thought and reflection are therefore frequently neglected.

Unfortunately, the areas where insight can be gained through research in the field of social studies are, by and large, the unimportant ones, and thus their current predilection for "research" leads many social scientists to work on unimportant problems. Many of them are more interested in methods than in results, and clearly we cannot turn to this type of social scientist if we wish to gain insight into certain important problems. In the circumstances we may have to make a new start and to begin pretty much where Plato has left off. Who could do this work?

In America, the study of the humanities has tended to wither away, and it might therefore be impossible to find enough men who are led by inclination to think about the kind of problem here described. Accordingly it may be necessary, in order to assemble perhaps 10 to 15 scholars who may wish to work on the problem, to go beyond the confines of the United States and look for additional men abroad, particularly in England where there exists more of a tradition of "political thought."

If the men who are needed can in fact be found, the Institute for Problem Studies might bring them in on a full-time basis for a limited period of time, perhaps three months, in the hope that they might come up with convincing answers. The same group of men, perhaps with slight variation in membership, might be gathered repeatedly, for a period of three months each, over a period of perhaps two years. It is not likely that there will be any universal answer; the form of democracy which appears best suited for governing a given underdeveloped area, where the problem has arisen or is about to arise, might not be suitable for another such area. For this reason it seems probable that in the more advanced phases of this study it will be essential to obtain the participation of men who know from personal experience the political and administrative difficulties that have arisen in areas emerging from the colonial status.

If the Institute for Problem Studies is in a position to offer "solutions" to this problem, it would presumably want to promote the "field test" by acquainting the British Colonial Office with its work. Perhaps the British Colonial Office—having been forced in the meantime to recognize that there is a need—might be willing to try out new political systems in the colonial areas which are about to be given self-government.

The following is another example of an inquiry that the Institute for Problem Studies might effectively conduct. It relates to the problem described above and is, in fact, part of it.

One of the main problems in many underdeveloped areas is the scarcity of qualified men on whom the elected officials may lean for the administration of governmental operations. To build up a civil service utilizing native talent will take a long time. Accordingly, it would greatly help if, in the meantime, the governments of under-developed countries could obtain on loan for limited periods qualified men to fill the gap. A beginning has been made in this regard by the Technical Assistance Administration of the UN. Could one go further and create an adequate international civil service, say under the auspices of the United Nations, that could loan its members to governments of countries which ask for such help?

Clearly it is impossible to answer this question without first determining by an inquiry—

(a) Is it possible to persuade men who have the required qualifications to enter such civil service as a lifetime career, and

(b) What would it take to make such civil service attractive to the kind of men in the United States and other politically developed countries who have the desired qualities?

The United States has a large number of men who are attracted by public service. There have been, for example, many able young lawyers who have entered government service in the past twenty-five years and who after a number of years of distinguished service found that serving the government is not a satisfying lifetime career in the United States. They then reluctantly left the government to go into private practice. But those who are of a certain bent of mind find private practice devoid of the kind of satisfaction to be derived from public service. Thus it would seem that there is a wealth of human material upon which an international civil service could draw in this country if it offered a satisfactory lifetime career. The same might hold for other politically developed countries.

There are serious difficulties, however. In such an international civil service a man would be abroad most of the time and could not count on staying longer than perhaps five years in any one country. Wives do not like to be shifted around in this manner, and the schooling of the children presents a major problem. Would the idea of an international civil service founder on such difficulties? Or could such a service draw, to some extent, on a dedicated minority?

A discussion of these topics which I recently had with Sir Robert Jackson (former Associate Director of UNRRA and the man in charge of the huge Volta dam project on the Gold Coast) leads me to think that there is an increasing general awareness of the problem here discussed. It is therefore conceivable that if the Institute for Problem Studies were able to make practical proposals to the governmental agencies which are in a position to take appropriate action, such proposals would not fall on deaf ears.

The End

Part V

Patents, Patent Applications, and Disclosures
(1923–1959)

Introduction by Julius Tabin*

In reviewing the patent applications that were filed by Dr. Leo Szilard one is struck by the fact that they document an important part of the technical achievements of this great man. I hope that the following few remarks will help in understanding how such patent applications came to be filed and provide at least some realization of the extent of their scientific importance.

To understand Leo Szilard's interest in patents one must know the man himself. Szilard had an inquisitive, probing mind and devoted his life to thinking and to spurring people into action. He was a man of many facets, somewhat abrasive in manner, unconventional, independent in spirit and action, and concerned with all that was happening in science and politics and its effect on mankind.

Szilard's independent spirit was well suited to his role as an inventor. His inquisitive, brilliant mind was honed by training in both engineering and physics. He was no sooner exposed to a problem than he was thinking of the solution. Szilard was once asked to state the qualities of a scientist. He replied that "the creative scientist has much in common with the artist and the poet. Logical thinking and an analytical ability are necessary attributes of a scientist but they are far from being sufficient for creative work. Those insights in science which have led to a breakthrough operate on the level of the subconscious. Science would run dry if all scientists were crank turners and if none of them were dreamers."[1] Szilard in these words described his own capabilities and the reasons why he was truly a creative scientist and inventor.

Szilard was a man of great creativity, generally many years ahead of others. As ideas occurred, he would promptly write them down, then fold and seal the paper and post it to himself by mail.[2] By this means he documented the dates when these ideas occurred to him. These ideas ran the gamut from simple solutions to practical everyday problems to complex theories which advanced the frontiers of science. Only a very small proportion of these ideas, those which appeared to have significant commercial potential, were later expanded and written in the form of patent applications.

Szilard's earliest patent applications were filed between 1923 and 1931.[3] At that time he was working in Berlin and dividing his time between the University of Berlin and the Kaiser Wilhelm Institute. While at the University of Berlin he was closely associated with Einstein and certain of the applications filed during this period listed Einstein as a joint inventor.

During this period Szilard directed his energies to working on problems that he believed might have commercial application, such as mercury vapor lamps, liquid metal pumping systems, refrigeration systems, the electron microscope, machines for accelerating charged particles, etc. He interested several German companies, including Siemens and the German General Electric Company (AEG), in certain of his ideas and apparently supplemented his income by acting as a consultant to these companies.

Several of Szilard's early German patents related to noiseless household refrigerators which operated without moving mechanical elements or linkages and to liquid metal pumping systems for such refrigerators. (It may be interesting to note that Einstein collaborated with Szilard on certain of these developments.) As it turned out, such refrigerators were never commercially utilized because of the rapid advances made in mechanical refrigerators which eliminated their objectionable noise, the dangers from leakage of the poisonous refrigerant, and erratic operation. The liquid metal pumping systems developed by Szilard are of particular interest.[4] For many

---

* Patent Attorney, Chicago, Illinois.
[1] Excerpt from taped interview 1963.
[2] List of known disclosures presently in Szilard's files is included later in this part.
[3] List of all of Szilard's known patent applications and issued patents is included later in this part.
[4] See German Patent No. 476,812 and British Patent No. 303,065, both reproduced in this part. The British patent, filed with Albert Einstein, is the same as the German Patent No. 554,959, "Vorrichtung zur Bewegung von flüssigem Metall, insbesondere zur Verdichtung von Gasen und Dämpfen in Kältemaschinen" (filed December 28, 1927, issued June 30, 1932).

years there did not appear to be any other practical use for such pumping systems but with the advent of atomic energy their need became evident (first to Szilard) and much effort has since been expended in their further development.

In 1928 Szilard began to direct his thoughts to atomic physics. Disintegration of the atom required higher energies than were available up to that time. He therefore considered ways to accelerate particles to high speeds, and this led to his development of the basic principles underlying the linear accelerator and the cyclotron. He documented these concepts in the form of patent applications which were filed in the German Patent Office. The application relating to the linear accelerator was filed on December 17, 1928 (Application No. S 89 028 VI/40c)[5] and the application relating to the cyclotron was filed a few weeks later on January 5, 1929 (Application No. S 89 288 VIIIa/21g).[6] His cyclotron application (which appears to predate the work of Lawrence) contains a description of the basic principles of the cyclotron including a discussion of the stability of the orbit brought about by having the magnetic field decrease in strength with increasing radius. Unfortunately, it appears that the existence of these applications (which were never issued into patents) was not known by other scientists working in this field and apparently did not influence the later development of the cyclotron and linear accelerator.

In 1933 Szilard moved to London and his interest remained focused in the area of atomic physics and its potentialities. In 1934 and 1935 he collaborated with T. A. Chalmers in experiments on induced radioactivity which were carried out in the Physics Department of the Medical College associated with St. Bartholomew's Hospital. In 1935 he obtained a fellowship at the Clarendon Laboratory at Oxford and continued with his work in the area of induced radioactivity.

During this period he considered the possible consequences of filing basic patents in the atomic field. From his prior experience in the area of patents, he recognized that one could obtain a patent without actually testing or proving the validity or operability of the system being patented. He saw the possibility of filing patent applications in the area of atomic physics which might dominate future developments in this area. He further felt that such patents might be exploited so as to provide a source of income disproportionate to the technical contribution involved. After careful consideration he arrived at a unique solution to his problem. He decided that patents in the area of atomic physics should be turned over to a nonprofit research corporation and the income therefrom should be used for further research.

Szilard's thoughts on this subject are contained in letters to Drs. Fermi and Segrè in 1936.[7]

Several applications which he filed in Great Britain during this period are worth special mention.

In the beginning of 1934 Szilard again considered ways of accelerating particles and on 21 February 1934 he filed a patent application in the British Patent Office (Application No. 5730)[8] entitled "Asynchronous and Synchronous Transformers for Particles." This application describes the principle of operating an accelerator in which the frequency of the accelerating voltage increases with time. It explains the principle of phase stability bringing about the variation of frequency. Again it is unfortunate that this application did not issue into a patent or come to the attention of those working with cyclotrons, because some ten years later these same principles were apparently independently rediscovered and utilized in completely revolutionizing the design of the cyclotron.

Another application that was entitled "The Transmutation of Chemical Elements" was originally filed on March 12, 1934 (British Application No.

[5] Reproduced in this part with brief explanation by Szilard.
[6] Reproduced in this part with a brief explanation by Szilard.
[7] Pertinent portions of this correspondence are reproduced in this part.
[8] Reproduced in this part.

7840).[9] This application, which issued into Patent No. 440,023,[10] originally included three inventions: the generation of radioactive elements by means of neutrons, the concept of a nuclear chain reaction, and the chemical separation of radioactive elements from nonradioactive isotopes. This separation method was later experimentally demonstrated (in collaboration with T. A. Chalmers) in the case of iodine, by separating iodine from ethyl iodide, and was reported in *Nature*, Vol. 134, September 22, 1934. This method has become generally known as the Szilard–Chalmers method or reaction.

Subsequently, because of his conviction that if a nuclear chain reaction could be made to work it might be used as an instrument of war to set up violent explosions, Szilard separated that part of the application which related to the nuclear chain reaction and incorporated it in modified form into a later filed application, No. 19157, which he assigned to the British Admiralty in order to prevent its publication.

It is worth noting that the following prophetic passage was included in Patent Application 7840 as filed in March 1934:

"(a) Pure neutron chains, in which the links of the chain are formed by neutrons of the mass number 1, alone. Such chains are only possible in the presence of a metastable element. A metastable element is an element the mass of which (packing fraction) is sufficiently high to allow its disintegration into its parts under liberation of energy. Elements like uranium and thorium are examples of such metastable elements; these two elements reveal their metastable nature by emitting alpha particles. Other elements may be metastable without revealing their nature in this way."[11]

The aforementioned application, No. 19157, was originally filed on June 28, 1934. This application, later issued as Patent No. 630,726,[12] was particularly directed to the idea of the nuclear chain reaction in which more than one neutron is emitted per neutron absorbed. It was assigned to the British Navy in 1936 (for reasons set forth below). After World War II it was reassigned to Szilard, and the Patent Specification was finally published in September 1949.

The genesis of this application covering the chain reaction is described by Szilard in material that he wrote for later use. It is quoted here[13] because of the insight it gives into the creative mind of Leo Szilard.

"In the fall of 1933, I found myself in London. I kept myself busy trying to find positions for German colleagues who lost their university positions with the advent of the Nazi regime. One morning I read in the newspapers about the annual meeting of the British Association where Lord Rutherford was reported to have said that whoever talks about the liberation of atomic energy on an industrial scale is talking moonshine. Pronouncements of experts to the effect that something cannot be done have always irritated me. That day as I was walking down Southampton Row and was

[9] We reproduce later in this part the original application which was subsequently considerably modified and abridged. The passage quoted appears on page 615.
[10] Reproduced later in this part.
[11] A related United States patent application Serial No. 10,500 was filed on March 11, 1935. Once again, in order to prevent the publication of information that might have important military consequences, he deleted that portion which related to the nuclear chain reaction including the quoted passage. The expurgated application later issued as U.S. Patent No. 2,161,985.
[12] Reproduced later in this part.
[13] From interview taped in 1963. [To be published in forthcoming Volume II—Eds.]

stopped for a traffic light, I was pondering whether Lord Rutherford might not prove to be wrong. As the light changed to green and I crossed the street, it suddenly occurred to me that if we could find an element which is split by neutrons and which would emit two neutrons when it absorbed one neutron, such an element, if assembled in sufficiently large mass, could sustain a nuclear chain reaction, liberate energy on an industrial scale, and construct atomic bombs. The thought that this might be possible became an obsession with me. It led me to go into nuclear physics, a field in which I had not worked before, and the thought stayed with me even though my first hunches in this regard turned out to be wrong.

I had one candidate for an element which might be unstable in the sense of emitting neutrons when it disintegrates, and that was beryllium. As it turned out later, beryllium cannot sustain a chain reaction and it is, in fact, stable. What was wrong was that the published mass of helium was wrong. This was later discovered by Bethe and it was a very important discovery for all of us because we did not know where to begin to do nuclear physics if there could be an element which should disintegrate, but doesn't.

In the Spring of 1934 I applied for a patent which described the laws governing such a chain reaction. It was the first time, I think, that the concept of critical mass was developed and that a chain reaction was seriously discussed. Knowing what this would mean—and I knew it because I had read H. G. Wells—I did not want this patent to become public. The only way to keep it from the public was to assign it to the Government. So I assigned this patent to the British Admiralty."[14]

Szilard came to America in January 1938. He was still intrigued with the possibility of a chain reaction and continued to look for elements which might be useful for this purpose. In January 1939 he learned of the discovery of the fission of uranium. Because of his prior thinking with respect to a chain reaction, he immediately thought of the possibility that neutrons might be emitted in the fission process and that this process might sustain a chain reaction. All his energies were now directed toward this single goal.

Szilard moved swiftly to prepare a new patent application covering uranium as a neutron source for the production of radioactive elements and as a means of producing power by using a sufficient amount of uranium to provide a chain reaction. This patent application was filed in the United States Patent Office on March 20, 1939, Serial No. 263,017.[15] Szilard was able to file this application so quickly because it incorporated much of the work contained in his prior British patent application on a chain reaction. This United States application was subsequently abandoned in 1941. The available records are not entirely clear as to the circumstances for this abandonment. There are indications from correspondence with his patent attorneys that he intended to file an updated patent application which incorporated the substance of this patent application before it was abandoned. It is not clear why this updated application was not filed.

Szilard played a central role in the historical development of atomic energy in the United States. He persuaded scientific colleagues to withhold the publication of their papers on nuclear fission in order to keep information from the German scientists and was the "ghost writer" of the Einstein letter to Roosevelt which initiated the atomic bomb project. Between early 1939 and November 1940, Szilard was not directly

[14] Letter from Szilard to C. S. Wright dated February 26, 1936. Reproduced later in this part.
[15] Reproduced in this part.

affiliated with any University or other organization. When he was not attending scientific meetings or visiting scientific friends such as Fermi, Wigner, Teller, etc., he could be found working at the King's Crown Hotel in New York. During this period he was granted guest privileges in the Physics Department of Columbia University where he carried out certain fission experiments in collaboration with Walter Zinn who had built equipment which was suitable for this purpose. In February 1940 Szilard wrote an important paper covering a chain reaction using a uranium-carbon system. This paper was originally intended for publication in the *Physical Review* but was withheld from publication at his request.[16] In November 1940 Szilard went on the payroll of Columbia University which had just received a contract from the Government to explore the possibility of setting up a chain reaction. From this point until the end of the war, Szilard was associated with the government-sponsored atomic bomb program, later referred to as the Manhattan Project.

In February 1942 Szilard left Columbia University to join a small group that was then being assembled at the University of Chicago to direct the effort toward the construction of a controlled chain reaction. The Chicago branch of the Manhattan Project was known as the Metallurgical Laboratory. Immediately following the actual demonstration of a controlled chain reaction in Chicago on December 2, 1942, Szilard was informed that the Project desired to file patent applications on inventions relating to the chain reaction. After lengthy discussion and negotiation concerning the terms and conditions of his continuation on the project, Szilard finally assigned to the U.S. Government the patent rights to his inventions on the chain reaction made prior to his employment by the Government on November 1, 1940. He received the sum of $15,417.60, an amount equal to the expenses that he incurred in connection with the development of these inventions, the arrangements relating to their transfer to the Government, and to compensate him for the time during which he worked on these inventions without a salary or other financial consideration.

Szilard remained at the Metallurgical Laboratory until the end of World War II. His many contributions to the Project are attested to by the numerous patent applications that were filed by the Government in his name. Of these applications the most notable is the basic application covering a graphite-moderated nuclear reactor, which listed Fermi and Szilard as co-inventors, and issued as Patent No. 2,708,656 on May 17, 1955.[17]

After the war Szilard turned his attention to biophysical problems. He became a Professor of Biophysics at the University of Chicago, and, although he was seldom in residence, he remained on the University staff until his retirement in 1963. Szilard then accepted a position as a resident fellow at the Salk Institute for Biological Studies in La Jolla, California. He moved to California and in the last few months of his life plunged into work with renewed energy.

Despite the frustrations from his past efforts, Szilard's optimism in the value of patents remained unabated. He filed at least two patent applications in the United States Patent Office in the postwar period. Application Serial No. 264,263 was filed on December 29, 1951, for a "Process for Producing Microbial Metabolites," and Application Serial No. 320,816 was filed (jointly with A. Novick) on November 15, 1952, for "Caffeine-Containing Products and Method of Their Preparation." Both of these applications were later abandoned.

I should like to conclude with my own feeling of regret for the fact that Leo Szilard was never adequately compensated for his many contributions to the world. This is the purpose for which the patent system was designed. However, as seen in the case of Szilard, the system is clearly inadequate in the case of a visionary who is many years ahead of his time.

---

[16] Report A-55 reproduced in Part III.
[17] Patent included later in this part.

DEUTSCHES REICH

**AUSGEGEBEN AM
3. JUNI 1929**

REICHSPATENTAMT

PATENTSCHRIFT

№ 476812
KLASSE **31c** GRUPPE 26
S 72993 VI/31c
Tag der Bekanntmachung über die Erteilung des Patents: 8. Mai 1929

Dr. Leo Szilard in Berlin-Dahlem

Verfahren zum Gießen von Metallen in Formen unter Anwendung elektrischer Ströme

Patentiert im Deutschen Reiche vom 20. Januar 1926 ab

Beim Gießen von geschmolzenen Metallen in Formen erfolgt die Beförderung des flüssigen Metalls in die Form unter dem Einfluß der Erdschwere, oder es wird, wie z. B. beim Spritz-
5 gußverfahren, ein erhöhter Druck verwendet. Beim Spritzgußverfahren wird das flüssige Metall in den Fällen, in welchen es sich um leicht schmelzende Legierungen handelt, zumeist mit Hilfe einer mechanischen Vorrichtung, etwa
10 eines sich in einem Zylinder bewegenden Kolbens, in die Form hineingepreßt. Dieses mechanische Preßverfahren führt beim Spritzen von höher schmelzenden Metallen und Legierungen zu großen Schwierigkeiten, so daß
15 man in diesen Fällen das Spritzen zumeist mit Hilfe von Druckluft bewerkstelligt. Jedoch sind auch bei Anwendung von Druckluft bekanntlich beträchtliche Schwierigkeiten technischer Art zu überwinden.
20 Die vorliegende Erfindung gestattet es nun, unter Vermeidung von mechanischen Vorrichtungen mit beweglichen Teilen oder der Anwendung von Druckluft, das flüssige Metall auf elektrischem Wege in die Gußform zu bringen und da-
25 bei im Bedarfsfalle jeden praktisch in Betracht kommenden Druck zu erzielen. Sie ermöglicht es auch, noch bei sehr hohen Temperaturen und unter veränderlichem Druck zu spritzen. Die Bewegung des flüssigen Metalls und der ge-
30 gebenenfalls gewünschte hohe Druck wird erfindungsgemäß dadurch erzeugt, daß man elektrischen Strom durch das flüssige Metall schickt unter gleichzeitiger Einwirkung eines Magnetfeldes auf das stromdurchflossene Metall.

Abb. 1 zeigt ein Ausführungsbeispiel der Er- 35
findung im Schema gezeichnet, welches zur Erläuterung des Grundgedankens der Erfindung dienen soll. 1 ist eine Flüssigkeitssäule des geschmolzenen Metalls, in welche zwei Elektroden 2 und 3 hineinragen, zwischen denen ein Strom 40 quer durch die Metallsäule hindurch aufrechterhalten wird. Senkrecht zu den Stromlinien im Metall wird durch die Magnetpole 4 und 5 ein Magnetfeld aufrechterhalten, welches auch zu der Flüssigkeitssäule senkrecht steht. Hier- 45 durch entsteht nun eine Kraftwirkung auf das flüssige Metall, und zwar steht die Kraft jeweils senkrecht sowohl zu den Stromlinien wie auch zu den magnetischen Kraftlinien. Ist die Polarität der Elektroden und des Magnetfeldes 50 so, wie dies in Abb. 1 gezeichnet ist, so entsteht in der Flüssigkeitssäule ein Druck nach unten.
Abb. 2 zeigt ein anderes Ausführungsbeispiel, im Schema gezeichnet. 6 ist der Eisenkern eines Transformators. 7 ist die Primärwick- 55 lung des Transformators, welche an eine Wechselstromquelle angeschlossen wird. Die Sekundärwicklung wird ·durch den Leiter 8 gebildet, und sein Stromkreis wird durch die Säule flüssigen Metalls zwischen den beiden 60 Backen 9 und 10 geschlossen. Das Magnetfeld wird durch die Pole 11 und 12 eines Elektromagneten erzeugt, dessen Kraftfluß senkrecht zu den Backen 9 und 10 gerichtet ist, in Abb. 2a also senkrecht auf der Zeichenebene zu denken 65 ist, und welcher durch dieselbe Wechselspannungsquelle gespeist werden kann wie die Primärwicklung 7 des Transformators. Man

2   **476 812**

wird dafür sorgen, daß Sekundärstrom und Magnetfeld möglichst in Phase sind, damit sie beide gleichzeitig ihre Richtung wechseln, so daß die Kraft, die auf das geschmolzene Metall
5 wirkt, stets dieselbe Richtung hat und ein möglichst hoher mittlerer Druck erzielt wird.

Wie Abb. 2 zeigt, kann die sekundäre Wicklung vollkommen innerhalb eines aus einem Isolator oder schlechtem Leiter bestehenden
10 Gefäßes verlaufen und so angeordnet sein, daß sie trotzdem einen geschlossenen Eisenkern umschließt. Auf diese Weise wird eine Durchführung von Elektroden durch die Gefäßwand vermieden, was besonders dann von Vorteil
15 ist, wenn das Gefäß unter Vakuum steht oder gasdicht geschlossen sein soll.

Abb. 3 zeigt ein Ausführungsbeispiel, bei welchem keine besonderen Elektroden verwendet werden, sondern bei welchem das geschmolzene
20 Metall sich innerhalb einer Röhre befindet, deren Wand ein elektrischer Leiter ist, so daß der Strom durch die Gefäßwand hindurch in das geschmolzene Metall eintreten kann. Abb.3b zeigt einen Querschnitt. 13 und 14 sind die
25 Pole des Elektromagneten, 15 ist der Querschnitt des Rohrs, in welchem sich das flüssige Metall befindet. 16 und 17 sind die stromführenden Zuleitungen zu der Rohrwandung, welche angelötet oder angeschweißt sind. In
30 Abb. 3a sieht man dasselbe in der Ansicht von vorn. Das Rohr 15 ist abgebrochen, so daß oben und unten der Polschuh 13 sichtbar wird. Im Fall des Ausführungsbeispiels nach Abb. 3 fließt ein Teil des Stroms durch die Rohrwand
35 selbst von 18 zu 19, und nur ein Teil des Stroms durchflutet also das geschmolzene Metall. Um eine möglichst hohe Stromdichte im Metall zu erzielen, wird man die Rohrwand aus einem Material mit möglichst hohem elektrischem
40 Widerstand herstellen.

Bei den Ausführungsbeispielen 1 bis 3 liegen die Verhältnisse so, daß, homogenes Magnetfeld und gleichförmige Verteilung des elektrischen Stroms vorausgesetzt, das Linienintegral
45 gral der wirkenden Kraft entlang jeder innerhalb des flüssigen Metalls verlaufenden geschlossenen Linie gleich Null ist. In diesem bemerkenswerten Fall hat das Kraftfeld ein Potential, und es entsteht überhaupt keine Strömung im
50 flüssigen Metall, falls man den Ausfluß durch die Ausflußöffnung verhindert. Der erzielte Druckunterschied zwischen zwei Punkten im flüssigen Metall ist dann gleich dem Linienintegral der Kraft zwischen diesen beiden
55 Punkten, genommen entlang einer beliebigen, im flüssigen Metall verlaufenden Linie. Da die Kraftwirkung auf das geschmolzene Metall der elektrischen Stromdichte und dem wirkenden Magnetfeld proportional ist, ist die Berechnung
60 des erzielten Druckunterschiedes in diesem Falle sehr einfach. Zum Beispiel ist im Fall des

Ausführungsbeispiels nach Abb. 2 bei gegebener Stromdichte und gegebener magnetischer Feldstärke der erzielte Druckunterschied einfach der Länge der Flüssigkeitsstrecke zwischen den 65 beiden Backen 9 und 10 proportional.

Bei anderen Ausführungsbeispielen, die zum Teil aus praktischen Gründen den Ausführungsbeispielen 1 bis 3 vorgezogen werden können, ist das Linienintegral der Kraft entlang ge- 70 schlossener Linien, welche innerhalb des flüssigen Metalls verlaufen, von Null verschieden. Bei diesen Ausführungsbeispielen entsteht auch bei abgesperrter Ausflußöffnung eine Bewegung im geschmolzenen Metall, so daß man zur Be- 75 stimmung des Druckunterschiedes zwischen zwei Punkten auch bei abgesperrter Ausflußöffnung nicht rein statisch vorgehen kann, sondern auch die Strömungsverhältnisse und im besonderen die Reibungskräfte mit berück- 80 sichtigen muß.

Ein solches Ausführungsbeispiel ist in Abb. 4 dargestellt. Das geschmolzene Metall wird aus einem in der Abbildung nicht gezeichneten Behälter durch ein verzweigtes Rohr 20, welches 85 man in Abb. 4b im Schnitte sieht, zur Gußform geführt. Der Zufluß aus dem Behälter kann etwa vom unteren Ende her erfolgen, während der Zufluß zur Gußform dann auf der oberen Seite liegt. In diesem verzweigten Rohr 90 wird dann mit Hilfe des Eisenkerns 21 eines Transformators ein Strom induziert. In der Gegend des Verzweigungspunktes 22 ist das Rohr abgeflacht, um hier eine möglichst hohe Stromdichte zu erzielen. Hier wirkt ein durch 95 einen Elektromagneten 23 erzeugtes Magnetfeld auf das stromdurchflossene Metall, so daß beim Gießen oder Spritzen eine Strömung bzw. ein Druck im Metall nach oben entsteht.

Wird das Rohr etwa aus Stahl oder einem 100 anderen magnetischen Material hergestellt, so verläuft ein Teil der Kraftlinien in der Rohrwand selbst, ohne das flüssige Metall zu durchfluten. Es ist dann zweckmäßig, den Gesamtfluß hoch zu wählen, damit in der Rohrwand 105 eine Sättigung und entsprechend eine geringe Permeabilität erzielt wird. Auch wird man möglichst ein Material mit geringer magnetischer Sättigung wählen.

Abb. 5 zeigt ein Ausführungsbeispiel, bei 110 welchem der Strom in einem ringförmigen Rohr 24 transformatorisch erzeugt wird. An der abgeflachten Stelle zwischen 26 und 27 wirkt ein Elektromagnet 25 auf das stromdurchflossene Metall. Die Zuführung des flüssigen Metalls 115 erfolgt an der Stelle 26, während die Abführung zur Gußform bei 27 erfolgt. Das geschmolzene Metall wird zum Teil durch den transformatorisch erzeugten Strom warm gehalten; außerdem können aber Teile der Rohrleitungen, im 120 besonderen die Ausflußöffnung, durch besondere Vorrichtungen elektrisch erhitzt werden.

476 812   **3**

Dies ist durch die Drahtwindungen 28 und 29 angedeutet.

Bei den angeführten Ausführungsbeispielen 4 und 5 kommt zwar die Flüssigkeitsmasse auch dann in Bewegung, wenn ein Ausfluß an der Mündung verhindert wird, weil ja das Integral der wirkenden Kraft entlang einer im flüssigen Metall verlaufenden, geschlossenen Linie nicht verschwindet; jedoch ist diese Flüssigkeitsbewegung keineswegs die Ursache des auftretenden Drucks. Es könnten aber auch Ausführungsbeispiele angegeben werden, bei welchen letzteres der Fall ist und bei denen der Druck etwa dadurch zustande kommt, daß das flüssige Metall unter der Wirkung des Magnetfeldes innerhalb eines feststehenden Gehäuses in Rotationsbewegung gerät, wobei dann ein Druck infolge der Zentrifugalkraft auftritt. Angesichts der großen Reibung, welche zwischen dem flüssigen Metall und dem feststehenden Gehäuse dann entstehen würde, erscheinen Ausführungsbeispiele von diesem Typus wenig aussichtsreich.

Bei Verwendung von Wechselstrom muß dafür gesorgt werden, daß der Phasenunterschied zwischen dem Sekundärstrom im flüssigen Metall und dem Magnetfeld, welches auf das Metall wirkt, nicht in der Nähe von 90° liegt. Der Idealfall liegt vor, wenn Sekundärstrom und Magnetfeld gleichzeitig ihr Vorzeichen ändern. Dann ändert nämlich die auf das Metall wirkende Kraft überhaupt nicht ihr Vorzeichen, und der erzielte mittlere Druck im Metall hat den größtmöglichen Wert, während er im Falle einer Phasenverschiebung von 90° den Wert Null hätte. Man erreicht zumeist schon genügend günstige Phasenbeziehungen dadurch, daß man die primäre Wicklung des Transformators und die Wicklung des Magneten in Serien schaltet. Das Magnetfeld ist in Phase mit dem Primärstrom des Transformators und damit auch angenähert in Phase mit dem Sekundärstrom des Transformators, vorausgesetzt, daß der cos. φ des Transformators nicht allzusehr von 1 abweicht. Falls Drehstrom zur Verfügung steht, kann man die Wicklung des Transformators und des Magneten an verschieden verkettete oder an eine verkettete und eine Phasenspannung anlegen und dadurch die gewünschten Phasenbezeichnungen erreichen.

Bei den bisher angeführten Ausführungsbeispielen 1 bis 5 waren stets zwei getrennte Stromkreise vorhanden. Der Strom, der durch das geschmolzene Metall geschickt wurde, war verschieden von dem Strom, der jenes Magnetfeld erzeugte, dessen Einwirkung auf das stromdurchflossene Metall den Druck hervorbrachte. Es werden nun Ausführungsbeispiele angegeben, bei welchen der Strom, der durch das geschmolzene Metall geschickt wird, bereits ein solches Magnetfeld erzeugt, daß unter dessen Ein-

wirkung im stromdurchflossenen Metall der gewünschte Druck entsteht.

Ein einfaches Modell von dieser Art zeigt Abb. 6. Ein zylindrisches Rohr 30 ist mit dem geschmolzenen Metall angefüllt. Durch das darin befindliche geschmolzene Metall wird in der Richtung der Zylinderachse Gleichstrom oder Wechselstrom hindurchgeschickt. Dies kann mit Hilfe der Stromzuleitungen 31 und 32 geschehen, wobei dann der Strom durch das Rohr rechts und links verschließenden Wände hindurch in das geschmolzene Metall eintritt. Wird der Zylinderinhalt in der Achsenrichtung durch den elektrischen Strom homogen durchströmt, so entsteht ein Druckgefälle von der Peripherie nach der Rohrachse hin. Dieses Druckgefälle bewegt dann das geschmolzene Metall durch die Ausflußöffnung hindurch zur Gußform. Die Nachlieferung des Gießmetalls wird durch das Rohr 34 aus einem größeren Behälter, etwa unter der Wirkung des Drucks der Atmosphäre, erfolgen.

Man kann diesen gegen die Mittellinie des Rohrs hin entstehenden Druck auffassen als die Wirkung des durch den Strom selbst erzeugten Magnetfeldes auf das stromdurchflossene Metall und kann ihn einfach aus der Stromstärke und dem Zylinderdurchmesser bekannten Gesetzen der Elektrodynamik entsprechend berechnen.

Abb. 7 zeigt ein zweites Ausführungsbeispiel dieser Art. 35 ist ein kurzes, flaches Rohr. Durch das darin befindliche geschmolzene Metall wird mit Hilfe der Stromzuleitungen 36 und 37 ein elektrischer Strom aufrechterhalten. Das Druckgefälle, welches dann entsteht, treibt das geschmolzene Metall von den peripherischen Teilen 38 und 39 nach der Mittellinie hin, so daß das Metall durch die Ausflußöffnung 40 zur Gußform getrieben wird. Die Anordnung der Ausflußöffnungen in der Mittellinie ergibt sich aus der Tatsache, daß bei der symmetrischen Anordnung nach Abb. 7 der höchste Druck in der Mittellinie entsteht.

Abb. 8 zeigt ein Ausführungsbeispiel, welches dem in Abb. 7 dargestellten ähnlich ist. Durch die nicht symmetrische Anordnung eines Eisenkörpers 41 ist jedoch das Magnetfeld verändert, so daß sich, gleiche Stromstärke und Rohrdimensionen vorausgesetzt, ein wesentlich höherer Druck ergeben kann als beim vorigen Ausführungsbeispiel. Bei dem Modell nach Abb. 8 tritt der höchste Druck nicht in der Mittellinie auf. Darum ist auch die Zuführung 42 zur Gußform Abb. 7 gegenüber entsprechend verschoben angebracht.

Wird Wechselstrom verwendet, so wird man keinen massiven Eisenkörper benutzen, sondern einen Eisenkörper aus Lamellen. Die Lamellen werden so angeordnet, daß die magnetischen Kraftlinien möglichst nicht durch die volle

4   **476 812**

Fläche der Bleche geschlossen werden. Im Fall eines Ausführungsbeispiels nach Abb. 8 würden z. B. die Lamellen gerade so liegen, daß sich die Bleche in Abb. 8a verdecken. Die
5 Bleche werden voneinander am besten mit Glimmer oder Asbestpapier isoliert.

Bei sehr hoher Temperatur läßt sich keine sehr hohe magnetische Sättigung mehr bei Eisen erzielen. Ersetzt man jedoch das Eisen durch
10 Kobalt oder Kobaltlegierungen, deren Umwandlungspunkt wesentlich höher liegt als der Umwandlungspunkt des Eisens, so kann man auch beim Gießen von Messing in der hier geschilderten Weise eine wesentliche Verstärkung
15 des Magnetfeldes erzielen.

Abb. 9 zeigt ein weiteres Ausführungsbeispiel. Abb. 9a ist ein Schnitt durch den Apparat, in welchem der Druck im geschmolzenen Metall erzeugt wird. Der Eintritt des geschmolzenen
20 Metalls erfolgt durch das Rohr 43 und der Ausfluß zur Gußform durch das Rohr 44. Dazwischen fließt das Metall in einem Spalt in einer eingefügten Scheibe 45. Die Scheibe 45 besteht aus einem ferromagnetischen Material mit
25 hohem magnetischen Sättigungswert. Abb. 9b zeigt einen Schnitt durch die Scheibe 45, aus welchem die Form des genannten Spaltes ersichtlich ist. Der elektrische Strom tritt durch die ebene Gefäßwand 48 bzw. 49 in das ge-
30 schmolzene Metall ein. Die Stromzuführung zu der Gefäßwand erfolgt durch die Kupferleiter 46 bzw. 47. Um zu vermeiden, daß zwischen den beiden als Elektroden wirkenden Gefäßwänden ein großer Teil der Stromlinien
35 sich durch die Scheibe 45 schließt, ist die letztere von den genannten Gefäßwänden 48 und 49 durch eine zwischengelegte Schicht, etwa Glimmer oder Asbestpapier, isoliert; dies ist durch die Zwischenräume 52 bzw. 53 ange-
40 deutet. Um zu vermeiden, daß ein wesentlicher Teil der Stromlinien zwischen den beiden Kupferzuführungen 46 und 47 sich durch die Gefäßwandung selbst schließt, sind jene Teile der Gefäßwand, welche als Elektroden im geschmol-
45 zenen Metall wirksam sein sollen, durch tiefe Spalte 50 und 51 getrennt. Indem man diese Spalte mit einer geeigneten feuerfesten Substanz vollfüllt, verhütet man das Eindringen der Schmelze in dieselben.
50 Wird nun zwischen den beiden Kupferleitern 46 und 47 eine Spannung aufrechterhalten, so fließt der Strom im wesentlichen nur durch das geschmolzene Metall im Spalte der Scheibe 45. Die durch den Strom erzeugten magnetischen
55 Kraftlinien, welche in der Scheibe 45 verlaufen, schließen sich durch den Spalt und erzeugen dort ein starkes Magnetfeld. Infolge der Einwirkung dieses Magnetfeldes auf das stromdurchflossene Metall entsteht ein Druckgefälle ent-
60 lang des Spaltes, welches die Schmelze zur Mündung des Rohrs 44 treibt.

Man wird am besten die beiden Kupferzuführungen 46 und 47 zu einer aus einer einzigen Windung bestehenden Sekundärwicklung eines Transformators ergänzen, welche man gegen 65 den Eisenkern und die Primärwicklung des Transformators mit Chamotte oder Asbest gegen Wärmeübergang isoliert.

Man kann auch dadurch ein verstärktes Magnetfeld und entsprechend eine erhöhte Druck- 70 wirkung erzielen, daß man die stromführenden Kupferleiter in der unmittelbaren Nachbarschaft der stromdurchflossenen Schmelze einige Male hin und her führt, etwa so, daß eine Spule entsteht, in deren Mitte sich die stromdurch- 75 flossene Schmelze befindet, wobei dann die Spule und das geschmolzene Metall von demselben Strom durchflossen werden.

Bei den Ausführungsbeispielen 6 bis 9 läßt sich durch Kommutieren des Stroms keine Um- 80 kehrung des Drucks erzielen, da sich alsdann auch die Richtung des Magnetfeldes von selbst umkehrt. Doch' kann man durch Veränderung der Stromstärke den Druck variieren und so das Spritzen unter veränderlichem Druck vor- 85 nehmen.

Falls zwei Stromkreise vorhanden sind, wie etwa in den Ausführungsbeispielen 1 bis 5, dann kann man dadurch, daß man nur das Magnetfeld oder nur den Strom im geschmolzenen 90 Metall umkehrt, das Ausfließen der Schmelze ebenso verhindern wie mit einem mechanischen Absperrventil. In diesen Fällen kann auch beim Fehlen jeglicher mechanischen Absperrventile das an die Gußform anstoßende Mundstück 95 tiefer liegen als der Flüssigkeitsstand der Schmelze im Vorratsbehälter. Auch kann man, wenn man die Form evakuiert, das Ansaugen der Schmelze verhindern.

100

PATENTANSPRÜCHE:

1. Verfahren zum Gießen von Metallen in Formen unter Anwendung elektrischer Ströme, dadurch gekennzeichnet, daß unter 105 gleichzeitiger Einwirkung elektrischer Ströme durch Erzeugung von Magnetfeldern auf das stromdurchflossene Metall ein Druck ausgeübt wird, welcher das Metall zur Gußform treibt. 110

2. Verfahren nach Anspruch 1, dadurch gekennzeichnet, daß zur Verstärkung der Wirkung des magnetischen Feldes in der Nähe des stromdurchflossenen Metalls ferromagnetische Körper angeordnet werden, 115 zweckmäßig innerhalb der das geschmolzene Metall umgebenden Gefäßwände.

3. Verfahren nach Anspruch 2, dadurch gekennzeichnet, daß als ferromagnetische Masse Kobalt oder Kobaltlegierungen verwendet 120 werden.

4. Verfahren nach Anspruch 1 bis 3, da-

476812   5

durch gekennzeichnet, daß als Elektrode eine das geschmolzene Metall umgebende Gefäßwandung verwendet wird, welche aus einem elektrisch leitenden Material hergestellt ist.

5. Verfahren nach Anspruch 4, dadurch gekennzeichnet, daß jener Teil der Gefäßwandung, an welchem der Strom in das geschmolzene Metall eintreten soll, von dem übrigen Teil der Wandung durch einen 10 Spalt getrennt ist.

6. Verfahren nach Anspruch 1 bis 3, dadurch gekennzeichnet, daß die Schmelze selbst einen geschlossenen Sekundärkreis bildet, in welchem ein Strom transforma- 15 torisch induziert wird.

Hierzu 3 Blatt Zeichnungen

BERLIN. GEDRUCKT IN DER REICHSDRUCKEREI

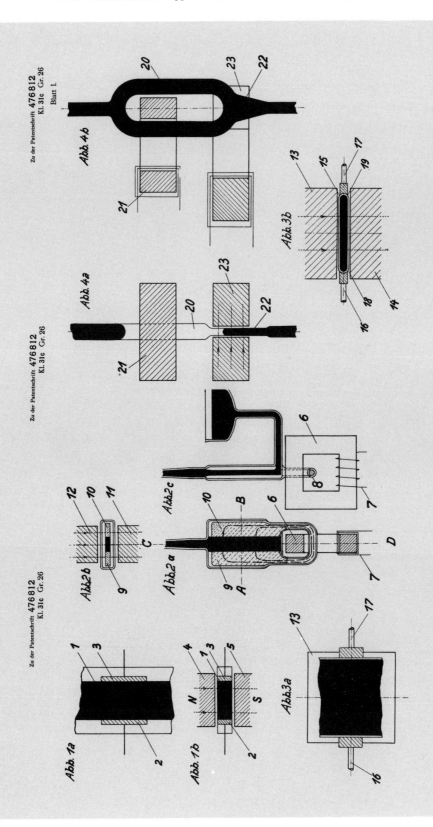

Zu der Patentschrift **476812**
Kl. 31c Gr. 26
Blatt I.

Zu der Patentschrift **476812**
Kl. 31c Gr. 26

Zu der Patentschrift **476812**
Kl. 31c Gr. 26

Abb. 4b

Abb. 4a

Abb. 3b

Abb. 2c

Abb. 2b

Abb. 2a

Abb. 1a

Abb. 1b

Abb. 3a

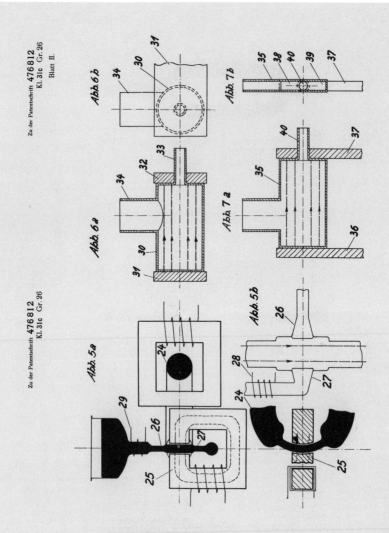

Zu der Patentschrift 476812
Kl. 31c Gr. 26
Blatt II.

Zu der Patentschrift 476812
Kl. 31c Gr. 26

Abb. 6b

Abb. 7b

Abb. 6a

Abb. 7a

Abb. 5a

Abb. 5b

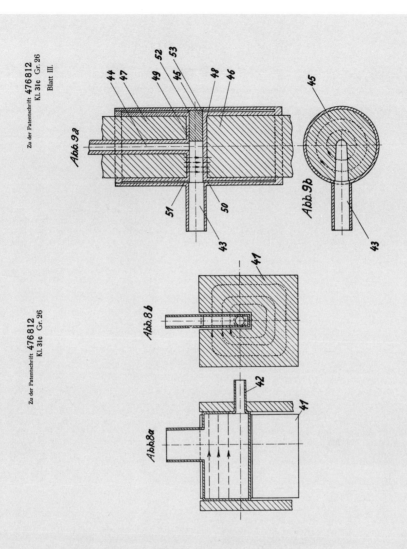

Zu der Patentschrift **476812**
Kl. 31c Gr. 26
Blatt III.

Abb. 9a

44
47
49
52
53
45
48
46

51
50
43

Abb. 9b

45
43

Zu der Patentschrift **476812**
Kl. 31c Gr. 26

Abb. 8b

41

Abb. 8a

42
41

PATENT SPECIFICATION

Convention Date (Germany): Dec. 27, 1927.

303,065

Application Date (in United Kingdom): Dec. 24, 1928.   No. 38,091 / 28.

Complete Accepted: May 26, 1930.

COMPLETE SPECIFICATION.

Electrodynamic Movement of Fluid Metals particularly for Refrigerating Machines.

We, Prof. ALBERT EINSTEIN, of Swiss Nationality, of 5, Haberlanderstrasse, Berlin, Germany, and Dr. LEO SZILARD, of Hungarian Nationality, of 95, Prinzregentenstrasse, Berlin-Wilmersdorf, Germany, do hereby declare the nature of this invention and in what manner the same is to be performed, to be particularly described and ascertained in and by the following statement:—

This invention relates to apparatus in which fluid metal moves forward under the influence of a magnetic field on the liquid through which electric current is passing, more particularly in which fluid metal is pumped from a chamber that is under low pressure into a chamber under higher pressure. Such apparatus can be employed for pouring molten metal into a mould or it may be used in refrigerating machines for feeding mercury or other liquid metals to the device. If the electric current is not passed into the liquid through electrodes but is induced in it in such manner that the stream lines of the electric current are wholly in the liquid and form closed lines therein, difficulties are avoided that are inherent in the transmission resistance between electrode and liquid, but generally a field of ponderomotive force is created in the liquid which is not free from eddies. If this field of force in the fluid metal is not free from eddies, there is a great loss of energy owing to useless agitation of the liquid.

The invention relates to apparatus in which the field of ponderomotive force is free from eddies within the liquid; according to the invention the field of ponderomotive force, which acts upon the liquid, results from a magnetic field of which the lines of force cut or cross an annular chamber through which the liquid stream is caused to flow, the magnetic field being produced by at least two or more coils energized by electric current.

Figure 1 of the accompanying drawings illustrates diagrammatically a form of apparatus according to the invention, by way of example.

An iron core 2 is inserted into an iron cylinder 1. Mercury flows in the cylin-

[Price 1/-]

drical annular chamber between the iron core and the tube 1, under the influence of the magnetic fields produced by the windings 4, 5, 6 and 7, in the direction of the longitudinal axis of the cylinder, and, if the polarity be suitably chosen, from top to bottom. The windings 4 to 7 surround the tube 1. The currents which flow in the adjacent windings are about 90° out of phase relatively to each other, while on the other hand the windings 4 and 6—likewise the windings 5 and 7— may be connected up in series. 8 and 9 are sheet iron plates shown in laminated form on the section line A—B. When the polarity is correctly chosen, the magnetic field in the mercury in the cylindrical annular chamber is moved from top to bottom; the rate of change of the magnetic field is obtained by multiplying the frequency by the identity distance apart of the windings. In the mercury an electric current is induced which circulates around the iron core 2. Such a line of force is shown in this figure of the drawings. The ponderomotive force that influences the mercury is at all points parallel to the axis of the cylinder, and the ponderomotive field is practically free from eddies.

A 90° displacement of phase between the currents in adjacent windings is produced in known manner as has been proposed for the production of the artificial phase for asynchronous motors.

Figure 2 illustrates diagrammatically a refrigerating machine according to the invention in which 10 is a device for causing mercury to move by electro-dynamic means. The mercury is forced into the tube 11 and fed to the mercury jet pump 12. The vapour of a cooling agent (for example, methyl alcohol or a suitable hydrocarbon) is drawn off through the pipe 13, compressed in a vertically extending pipe, and forced into the vapour separating chamber 15. The mercury passes out of this chamber, through the downwardly directed pipe 16, and back into the device 10, while the vapour of the cooling agent flows through the pipe 19 into the air-cooled condenser 17, where

2   303,065

it is liquefied; the cooling medium then flows through a throttle 20 into the vaporiser 18.

If a three phase current is available, the 5 apparatus shown in Figure 1 is connected to the source of current in such manner that a uniformly moving magnetic field is produced—that is to say, the windings 4, 5 etc. are connected in similar manner 10 to the windings of a three phase current motor and there is produced a magnetic field moving in a straight line instead of a rotating field.

The transmission of energy from the 15 mercury to the vapour that is to be compressed may be variously effected in the refrigerating machine. For example, water may be drawn in by means of a mercury jet pump and the vapour com- 20 pressed by the water. Alternatively, by intermittently reversing the direction of motion of the magnetic field, the mercury can be caused to flow intermittently into a vessel which communicates by valves with 25 two chambers that are under different pressures, so that the mercury compresses the vapour, and forces it into the chamber under the higher pressure, while the vapour is drawn off by the mercury from 30 the chamber which is under the lower pressure.

In order that this action may be better understood one form of such an apparatus is hereinafter more fully described by way 35 of example and illustrated diagrammatically in Figure 3 of the accompanying drawings.

10 is the electro-magnetic device in which the direction of movement of the 40 fluid metal is reversed by changing the polarity of one part of the winding. In this way the fluid metal is drawn out of the cylinders 21 and 22, or is forced into them, alternately. The valves 23 and 24 45 permit the compressed vapour to pass into the pressure pipe 25, while the vapour is drawn by suction out of the suction pipe

26 through the valves 27 and 28 into the cylinders 21 and 22.

Having now particularly described and 50 ascertained the nature of our said invention and in what manner the same is to be performed, we declare that what we claim is:—

1. Apparatus for moving fluid metals, 55 intended more particularly for use with refrigerating machines, in which a magnetic field influences metal traversed by electric current, characterised in that a stream of the liquid metal is caused to 60 flow through an annular chamber which is cut or crossed by the lines of force of a magnetic field produced by at least two or more coils energised by electric current, the magnetic field inducing electric 65 currents in the annular chamber, which currents circulate around the axis of the chamber.

2. Apparatus according to claim 1, characterised in that the currents in two 70 adjacent coils have a phase difference, thus producing a magnetic field which moves in a straight line parallel to the axis of the annular chamber in the manner that the field in a poly-phase 75 motor moves in a circular path.

3. Apparatus according to claim 1 or claim 2, comprising a tube having an iron core disposed therein, an annular space being formed around the core and 80 iron sheets disposed outside the annular space.

4. Apparatus according to claim 1, claim 2, or claim 3, characterised in that in a refrigerating machine mercury is 85 moved by the apparatus.

5. Apparatus according to claim 1, claim 2 or claim 3, characterised in that fluid metal is poured into a mould by the apparatus. 90

Dated this 24th day of December, 1928.

EDWARD EVANS & Co.,
27, Chancery Lane, London, W.C. 2,
Agents for the Applicants.

Redhill: Printed for His Majesty's Stationery Office, by Love & Malcomson, Ltd.—1930.

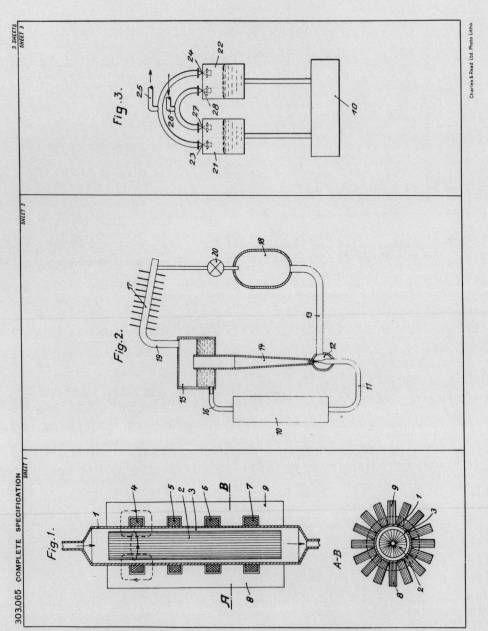

Szilard's Description of Patent Applications S.89 028 and S.89 288

Both the enclosed patent applications describe methods for the production of fast protons without the use of high voltage by means of high frequency oscillations.

The first application has been filed before Wideroe's paper on this subject had been published. It describes the method which has been used by Wideroe and subsequently by Lawrence.

The protons travel according to this method straight through a long tube and are passing through a series of grids or tubes to which a high frequency oscillation is applied.

The second patent application which has been filed a few weeks later describes the use of the magnetic field in which path the proton is coiled up. The method is based on the fact that the time of revolution of a proton in a uniform magnetic field is independent of the velocity of the proton and determined by the strength of the magnetic field alone. This method has some years later been used by Lawrence with success.

B e s c h e i n i g u n g .

Herr Dr. Leo Szilard in Berlin-Wilmersdorf,
Prinzregentenstr. 95, hat im Deutschen Reiche ein Gesuch
um ein Erfindungspatent für

" Beschleunigung von Korpuskeln "

am 17.Dezember 1928 beim Reichspatentamt hinterlegt.

Die angehefteten Stücke sind eine richtige und ge-
naue Wiedergabe der ursprünglichen Unterlagen dieser
Patentanmeldung.

Berlin, den 23.Juli 1932.

R e i c h s p a t e n t a m t .

S.89 028 VI/40 c.

Dr. Leo S z i l a r d , Wilmersdorf,Prinzregentenstr.95

"Beschleunigung von Korpuskeln."

   Die Erfindung betrifft ein Verfahren
bezw. einen Apparat, der es gestattet, Korpuskularstrahlen von hoher Energie zu erzeugen ohne dass es
notwendig wäre,so hohe elektrische Potentialen zu verwenden, wie dies notwendig wäre,wenn man durch einfaches Beschleunigen in einem elektrischen Feld den Kor-
puskularstrahl auf hohe Energie bringen wollte.

   Es ist bekannt, dass die von den radio-
aktiven Körpern emittierten α Teilchen Atome vieler
Substanzen zertrümmern können und man ist deshalb seit
langem bestrebt durch Beschleunigung positiv geladener
Korpuskeln, wie z.B. Heliumionen oder anderer positiver
Jonen, diesen Jonen eine ebenso hohe kinetische Energie

-- 2 --

zu verleihen, wie es die Jonen radioaktiven Ursprungs
haben.Dabei war man in der Ansicht befangen,dass man z.B.
das Heliumion ein sehr hohes elektrisches Potential durchlaufen lassen muss, etwa vier Millionen Volt um seine kinetische Energie derjenigen eines α Teilchens gleichwerden
zu lassen.In der Tat ist diese Zahl richtig, wenn man annimmt, dass die kinetische Energie die dem geladenen
Partikel in einem elektrischen Feld verliehen wird,gleich
dem Produkt aus seiner Ladung und dem durchlaufenden Potential ist.Diese Annahme ist jedoch nicht zutreffend,wenn
man zeitlich veränderliche Felder verwendet und man kann
erfindungsgemäss geladene Partikel eine sehr hohe kinetische Energie verleihen ohne dass man im Sinne der oben
erwähnten Annahme entsprechend hohe elektrische Spannungen verwendet.Damit fallen dann die grossen Schwierigkeiten hinweg, die sich der Anwendung hoher Spannungen ent-
gegenstellen und bisher alle Bemühungen künstlichα Teilchen, oder andere ähnliche energiereiche Partikel,durch
Beschleunigung von Jonen im elektrischen Felde herzustel-
len gescheitert sind.

Erfindungsgemäss werden zeitlich rasch
veränderliche elektrische Felder in verschiedenen Teilen
des Raumes aufrecht erhalten und ein Korpukularstrahl wird
durch diese elektrischen Felder hindurchgeschickt und zwar
so,dass es aus einem Raumteil,indem es beschleunigt wurde,
heraustritt bevor oder kurz nachdem sich die Feldrichtung
in diesem Raumteil umkehrt und in den nächsten Raumteil

-- 3 --

wieder ein Feld vorfindet, das ihn beschleunigt und diesen

zweiten Raumteil wiederum verlässt, bevor bezw. kurz nachdem

sich die Feldrichtung in diesem zweiten Raumteil umkehrt

u.s.w.

Fig. 1 zeigt ein Ausführungsbeispiel

der Erfindung im Schema gezeichnet. 8 ist eine Kanalstrahl-

röhre aus dem Kanalstrahl von Helium,Quecksilber oder irgend

eine andere Substanz, durch die Bohrung 11 der Kathode 12

in das Rohr 9 eintreten.Dieses Rohr ist durch die Gitter

1 bis 6 in fünf Abschnitte geteilt und es wird an diese

Gitter eine Wechselspannung sehr hoher Frequenz (1 bis 100

Millionen Schwingungen in der Sekunde und darüber) ange-

legt und zwar liegt der eine Pol an den Gittern 1,3,5 und

der andere Pol an den Gittern 2,4,6, so, dass die Gitter

1,3,5 untereinander in Phase schwingen und die Potential

stets 180 Grad vom Potential der Gitter 2,4,6 abweicht.

Wie man in der Figur sieht, sind die Abstände der Gitter

von links nach rechts monoton wachsend, d.h. klein dort,

wo die Geschwindigkeit der Kanalstrahlen noch klein und

gross dort, wo die Geschwindigkeit der Kanalstrahlen

schon gross ist.

Die Wirkungsweise der Anordnung ist wie

folgt: Betrachten wir ein positives Jon,das im Kanalstrahl

fliegt und in einem solchen Augenblick durch das Gitter 1

hindurchfliegt, in welchem das Potential das Gitter 1 gegen

das Gitter 2 gerade Null und im Wachsen begriffen ist.Das

Jon wird dann im Raume zwischen den Gitter 1 und 2 be-

-- 4 --

schleunigt und passiert das Gitter 2 (in dem Idealfall,
den wir jetzt betrachten) gerade in dem Augenblick als
der Potentialunterschied zwischen 1 und 2 wieder Null geworden ist.Dann ist auch der Potentialunterschied zwischen 2
und 3 Null und das Potential von 2 gegen 3 im Wachsen begriffen, sodass das Jon zwischen 2 und 3 wieder beschleunigt
wird. u.s.w.

        Falls wie im vorliegenden Falle vorgese-
hen,die Frequenz für alle Gitter dieselbe ist,muss der Abstand zwischen 2 und 3 grösser sein , als der Abstand zwischen 1 und 2, u.s.w., damit das Jon jeweils im richtigen
Augenblick in den nächsten Abschnitt eintritt. Bei unserer Anordnung lässt sich das elektrische Feld zusammen-
setzen aus einem von links nach rechts beschleunigt bewegten und aus einem von rechts nach links verzögert bewegten elektrischen Feld (hierzu käme noch evtl. entstehendes
elektrisches Feld als Korrektion, welches uns jedoch nicht
weiter interessiert.) Die Einrichtung wird nun zweckmässig
so getroffen,dass die Geschwindigkeit des beschleunigten
$J_o$ns an jedem Punkt mit der dort herrschenden Geschwindigkeit des von links nach rechts bewegten Feldes übereinstimmt.
Die $G_e$schwindigkeit des elektrischen Feldes berechnet sich
dabei als Produkt aus der Frequenz der an das Gitter geleg-
ten elektrischen Schwingung und dem Abstand der beiden in
Phase schwingenden Gitter (also z.B. 1 und 3).

-- 5 --

14 ist ein Lenardfenster, durch welches
die Korpuskeln aus der Röhre 9 austreten können und auf
die zu bombardierende Substanz 15 im Freien oder in einem
anderen abgeschlossenem Raume fallen können.7 ist die
Schwingungsquelle z.B. eine Elektronenröhre.Man kann
auch so vorgehen, dass man eine Reihe von Gittern etwa die
ersten 50 Gitter, welche das Jon zuerst passiert, und infolgedessen auch mit einer verhältnismässig geringen Geschwindigkeit passiert, mit einer anderen Röhre betreibt als
die nächsten 50 Gitter, die das Jon schon mit höherer
Geschwindigkeit passiert.Dabei wählt man die Schwingungszahl der zweiten Röhre höher als die Schwingungszahl der
ersten Röhre, sodass die Gitterabstände der letzten Gitter noch verhältnismässig niedrig ausfallen.

Arbeitet man nur mit einer Frequenz,
z.B. 10^8 Schwingung pro Sekunde und beträgt die Anfangs-
geschwindigkeit der Kanalstrahlen 10^8 cm pro Sekunde,wo
würden die Gitterabstände (z.B. der Abstand von 1 und 3)
anfangs etwa 1 cm betragen, während, falls unser Apparat
die Geschwindigkeit des Teilchens auf das Zehnfache erhöht,
der Abstand der letzten Gitter (z.B. 97 und 99) 10 cm
betragen müsste. Arbeiten wir mit Spannungen von 30 000 Volt
(maximal Sp.) so würde mit 100 Gittern die kinetische Energie des Jons auf so hohe Werte kommen, wie beim Durchlaufen
eines einfachen statischen Feldes von etwa 2 Millionen Volt.

Man kann auch zwei Gittersysteme verwen-
den,von denen jede allein so betrieben wird, wie oben beschrieben und die gegeneinander mit einer Phasenverschie-

-- 6 --

bung von 90 Grad schwingen.Die beiden Gittersysteme liegen dann in Bezug aufeinander, so dass sich die Phase
von Nachbar zu Nachbar um je 90 Grad verschiebt, sodass
auf diese Weise ein von links nach recht (beschleunigt)
bewegtes elektrisches Feld praktisch allein vorhanden ist.

       In Figur 1 ist 10 eine Pumpe, 11 eine
Gleich - oderWechselstromquelle zum Betriebe der Kanal-
strahlröhre 8; 13 ist die Anode dieser Röhre, 12 die
durchbohrte Kathode.

       Patentansprüche:

-- 7 --

Patentansprüche:

1.) Apparat oder Verfahren zur Beschleunigung elektrisch beschleunigbarer Korpuskeln, wie z. B.
Elektronen oder Jonen, dadurch gekennzeichnet, dass der Korpuskel durch mehrere Raumabschnitte fliegt, in welchen Raumabschnitten der Richtung nach wechselnde elektrische Felder
aufrechterhalten werden und zwar so, dass wenigstens für
einen Teil der Korpuskel in mehreren Raumabschnitten zur Zeit
des Verweilens dieser Korpuskeln in dem betreffenden Raumabschnitte die Richtung des dort herrschenden elektrischen
Feldes gerade so ist, dass die kinetische Energie des flie-
genden Korpuskels in dem betreffenden Raumabschnitt vermehrt
wird.

2.) Apparat oder Verfahren nach Anspruch 1,
dadurch gekennzeichnet, dass in einzelnen Raumabschnitten
die Feldrichtung periodisch umgekehrt wird.

3.) Apparat oder Verfahren nach Anspruch
1 oder 2, dadurch gekennzeichnet , dass an der Grenze der
Raumabschnitte Leiter z.B. Gitter angeordnet sind, an die
ein hochfrequentes Potential gelegt ist.

4.) Apparat oder Verfahren nach Anspruch
1, 2 oder 3 , dadurch gekennzeichnet, dass die Länge des
Weges der Korpuskeln in den hintereinander liegenden Abschnit-

-- 8 --

ten verschieden ist und zwar so, dass die Länge dieses
Weges in Abschnitten, die vom Korpuskel später erreicht
werden, grösser ist.

        5.) Apparat oder Verfahren nach An-
spruch 3 oder 4 , dadurch gekennzeichnet, dass die benach-
barten Gitter der Phase nach entgegengesetzt , alle grad-
zahligen Gitter untereinander und alle ungradzahligen
Gitter untereinander aber in Phase schwingen.

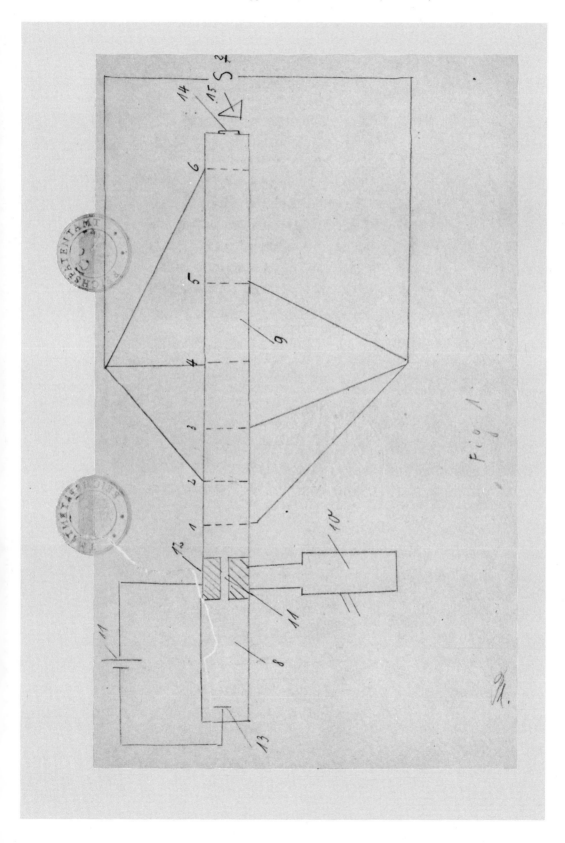

B e s c h e i n i g u n g .

Herr Dr. Leo Szilard in Berlin, Prinzregentenstr. 95,
hat im Deutschen Reiche ein Gesuch um ein Erfindungspatent
für

" Korpuskularstrahlröhre "

am 5.Januar 1929 beim Reichspatentamt hinterlegt.

Die angehefteten Stücke sind eine richtige und ge-
naue Wiedergabe der ursprünglichen Unterlagen dieser
Patentanmeldung.

Berlin, den 28.Juli 1932.

R e i c h s p a t e n t a m t .

S.89 288 VIII a/21 g.

S 89 288 VIII/21 g

Dr. Leo S z i l a r d , Berlin -Wilmersdorf,

"Korpuskularstrahlröhre."

            Die Erfindung betrifft einen Apparat
oder Verfahren zur Beschleunigung von Korpuskularstrahlen, (wie z.B. Kanalstrahlen oder Kathodenstrahlen) welches gestattet,die kinetische Energie der im Strahl flie-
genden Korpuskel auf einen hohen Betrag zu bringen,ohne
entsprechend hohe elektrische Potentialunterschiede auftreten zu lassen.

            Es wird erfindungsgemäss der Korpuskular-
strahl durch ein Kraftfeld,am besten durch ein Magnetfeld,
gezwungen, annähernd in einem Kreise, genauer in einer
Spirale zu laufen und wird durch elektrische Kräfte annähernd tangential zu seiner Bahn beschleunigt (genauer die
mittlere Beschleunigung liegt annähernd tangential zur
Bahn der Korpuskel).Am besten wird ein homogenes und wäh-

-- 2 --

rend der in Frage kommenden Zeit möglichst zeitlich konstantes Magnetlfeld verwendet,um die Korpuskularstrahlen aufzuwickeln.Würden dann die Korpuskularstrahlen nicht elektrisch beschleunigt werden, so könnten sie sich in diesem Magnetfeld in zueinander konzentrischen Kreisen bewegen,wobei (····) die Geschwindigkeit in den äusseren Kreisen grösser,in den inneren Kreisen kleiner sein muss, und zwar muss (solange die Relativitätskorrektion noch nicht in Frage kommt, also beispielsweise bei nicht extrem raschen Prothonenstrahlen) diese Geschwindigkeit dem Raius des Kreises proportional sein.

Teilt man nun die als flache Kreisscheibe ausgebildete Korpuskularstrahlröhre, in welcher wie eben angedeutet sich die Korpuskularstrahlen zunächst je nach ihrer Geschwindigkeit in verschiedenen zu einander konzentrischen Kreisen bewegen könnten, durch passende Leiter, z.B. durch radial angeordnete Gitter in Sektoren,so kann man in einfacher Weise die Korpuskularstrahlen in der Röhre elektrisch beschleunigen.Dies geschieht so, dass man mit Hilfe der Gitter in den einzelnen Sektoren der Richtung nach rasch wechselnde elektrische Felder aufrecht erhält und zwar so, dass das fliegende Korpuskel,welches in einem Sektor durch das dort gerade herrschende elektrische Feld eine Beschleunigung erfährt,diesen Sektor verlässt,bevor sich dort das elektrische Feld umkehrt (oder kurz nachdem diese Umkehr erfolgt ist) und im nächsten Sektor wieder ein elektrisches Feld vorfindet,das ihn beschleunigt.Man erreicht dies,indem

-- 3 --

man die Hälfte der Gitter an die eine Klemme und die Hälfte
der Gitter an die andere Klemme einer hochfrequenten elektrischen Spannungsquelle legt, so dass benachbarte Gitter
stets an den entgegengesetzten Klemmen liegen.Dabei muss die
Frequenz so gewählt sein,dass die Zeit, die ein Korpuskel
braucht, um von einem Gitter zum nächsten zu gelangen,
gleich einer halben Periode ist; diese Zeit berechnet sich
als das Produkt aus dem Winkel, den die beiden Gitter miteinander bilden, und aus dem Radius des Kreises,auf welchem
näherungsweise, wie oben angedeutet, die Korpuskel sich bewegen können, dividiert durch die Geschwindigkeit der Korpus-
kel.Es ist entscheidend,dass diese Zeit auf diese Weise vom
Radius unabhängig ausfällt, da Geschwindigkeit und Radius
des Kreises zueinander proportional sind.Dies hat zur Folge,
dass für dieselbe Frequenz die oben angegebene Bedingung
sowohl für die im Innern langsam kreisende wie für die weiter aussen rasch kreisende Korpuskel erfüllt ist.

      Schiessen wir alo einen Kanalstrahl
in unsere Vorrichtung,so dass sie sich auf einem Kriese von
kleinem Radius um die Achse der Vorrichtung bewegt, und legen wir die Wechselspannung an die Gitter an, so werden die
Ionen beschleunigt durch die elektrischen Felder und der Radius
des Kreises vergrössert sich mit wachsender Geschwindigkeit
(Die Ionen bewegen sich ungefähr in einer Spirale),so passie-
sen dass die Ionen stets in denselben Zeitabständen die be -
nachbarten Gitter passieren und auf diese Weise dauernd wei-

-- 4 --

ter beschleunigt werden,bis der Radius der Bahn so gross geworden ist, dass die Korpuskeln durch eine passend angeordnetes
Fenster aus der Röhre herausfliegen.

Die Geschwindigkeit, mit der die Korpuskularstrahlen zu Anfang versehen werden müssen, ergibt sich aus
dem Winkel des Sektors und aus der Frequenz der elektrischen
Schwingung, die die Gitter steuert.Es schadet aber nichts,wenn
die Korpuskularstrahlen zum Teil etwas grössere Geschwindigkeit haben, die betreffenden Korpuskeln werden dann auf einem
zur Achse nicht symmetrischen Kreise zu laufen beginnen und
dadurch trotzdem in Tritt kommen.

Figur 1 zeigt ein Ausführungsbeispiel
im Schema gezeichnet. 1 ist die flache Röhre, die als Kreisscheibe ausgebildet ist und sich zwischen den Polen 2 und 3
eines Elektromagneten befindet. 4,5,6,7,8 und 9 sind die Gitter, welche die Scheibe in 6 Sektoren teilen.Die Gitter 4,6 und 8

liegen an der einen, die Gitter 5,7 und 9 an der anderen Klem-
me einer Senderöhre oder eines Löschfunkensystems und werden mit einer Schwingung gesteuert,deren Wellenlänge sich zwischen 10 cm und 10 m befindet. 10 ist eine Kanalstrahl- oder
Kathodenstrahlröhre, in welcher die Korpuskularstrahlen erzeugt, beschleunigt und bei 11 in die Röhre 1 hineingeschossen
werden.Die Bahn der Korpuskularstrahlen in der Röhre 1 ist
durch eine Spirale schematisch angedeutet.12 ist ein Lenardfenster, durch welches die Korpuskularstrahlen die Röhre 1
verlassen.In der Röhre 1 herrscht selbstverständlich hohes
Vakuum, doch ist die zur Aufrechterhaltung des Vakuums die-
nende Pumpe in der Figur nicht eingezeichnet.

-- 5 --

Wenn man das beschleunigende elektrische
Feld nicht wie zuletzt beschrieben mit radial stehenden
Gittern und raschen elektrischen Schwingungen erzeugt und
im besonderen, wenn man Elektronen beschleunigen will,kann
es vorteilhaft sein, zur Krümmung der Bahn nicht ein Magnetfeld sondern ein radial stehendes elektrisches Feld zu
verwenden.

Es ist vorteilhaft,die Quelle der Korpuskularstrahlen (den Glühdraht, der die Elektronen emittiert
oder die Mündung eines Kanalstrahlendrohres) so anzuordnen,
dass die Korpuskularstrahlen nicht vollkommen senkrecht zu
den magnetischen Kraftlinien die Röhre 1 betreten,sondern
eine geringe Neigung nach oben besitzen.Auf diese Weise verlaufen sie dann nicht in eine Ebene ,sondern erheben sich
spiralig, allerdings in einer Spirale, die mit jeder Windung
flacher und flacher verläuft in Richtung der Achse.

Figur 2 zeigt ein von Figur 1 wesentlich
verschiedenes Ausführungsbeispiel im Schema gezeichnet.Hier
wird das elektrische Feld, welches den Korpuskeln,hier am
besten Elektronen, die hohe kinetische Energie zu erteilen
hat, durch den geschlossenen Eisenkern 22, welcher durch die
aus einer einzigen Windung bestehenden Wicklung 20 magneti-
siert wird, induziert.In diese Wicklung wird ein hochfre-
quenter Wechselstrom, z.B.   bis   Perioden pro Sekunde
geschickt,und es entstehen in der Korpuskularstrahlenröhre 21
sich kreisförmig schliessende elektrische Kraftlinien,in denen die Richtung der Kraft sekundlich entsprechend oft umge-
kehrt.Durch die Wicklungen 23 und 24 fliesst Gleichstrom

-- 6 --

und es wird auf diese Weise ein Magnetfeld in der Korpuskularstrahlenröhre 21 aufrecht erhalten, welches den
Elektronen in der Röhre ermöglicht, auf Kreisen zu laufen,
welche den Eisenkern 22 umschliessen.Solche kreisende
Elektronen werden durch das induzierte elektrische Feld
beschleunigt, und zwar stets annähernd in Richtung der
Bahntangente, und es vergrössert sich in dem Masse wie
die Geschwindigkeit wächst der Radius des Kreises,auf dem
die Elektronen laufen oder genauer, die Elektronen laufen auf einer Spirale von innen nach aussen.Sie kommen
dadurch bei einer bestimmten Geschwindigkeit dazu,auf
die Antikathode 25 aufzutreffen,wo sie Röntgenstrahlen
auslösen.Die Antikathode ist leitend in der Glühkathode
26 verbunden, welche die Elektronenquelle bildet.

Um die Wirkungsweise richtig zu ver-
stehen, sei folgendes numerisches Beispiel angegeben.
Die Periodenzahl der induzierenden Wechselspannung sei
, der mittlere Umfang der Elektronenbahnen sei 10cm,
die Windungsspannung sei 3000 V_0lt effektiv.Dann ergibt
sich die Spannung, die der erreichten Geschwindigkeit
der Elektronen entsprechen würde,wenn diese nicht vor-
zeitig an die Antikathode auftreffen würden,aus dem Produkt von Windungsspannung und d er Zeitdauer der halben
Periode der induzierten Wechselspannung,indem man diese
durch den Umfang der Elektronenbahnen dividiert und mit
der mittleren Elektronengeschwindigkeit multipliziert.
Es ergibt sich so eine Spannung von 500 000 Volt.Da die

-- 7 --

Elektronen schon etwa mit 100 000 Volt entsprechender
Geschwindigkeit auf die Antikathode auftreffen sollen,
(wir wollen unserem Beispiel eine solche Anordnung zugrunde legen) so erreichen etwa 80% der während der
Halbperiode emittierten Elektronen mit der gewünschten
Geschwindigkeit die Antikathode.Dieser letztere Umstand
ist für den Wirkungsgrad der Röntgenröhre von entschei-
dender Bedeutung und dadurch bedingt,dass ein praktisch
konstantes Feld und nicht ein vom hochfrequenten Strom
hervorgerufenes Feld die Krümmung der Elektronenbahnen
regelt.

Man kann die Elektronen dadurch in
der Höhe der Antikathode halten (also in einer Ebene
senkrecht zum Magnetfeld),dass man ein inhomogenes Magnetfeld verwendet, und zwar inhomogen in dem Sinne,dass
das Magnetfeld mit wachsender Entfernung von der Achse
etwas schwächer wird.Dies lässt sich z.B. durch geeig-
nete Polschuhe erzielen und in <u>Figur 3</u> versinnbildlicht.

<u>P a t e n t a n s p r ü c h e :</u>

1.) Verfahren oder Apparat zur Be-
schleunigung von Korpuskularstrahlen, dadurch gekennzeichnet,dass die Korpuskeln durch ein zeitlich praktisch konstantes bezw. nur langsam veränderliches Magnetfeld aufgewickelt und durch elektrische Kräfte annähernd tangential
zu ihrer Bahn beschleunigt werden.

-- 8 --

2.) Verfahren oder Apparat zur Be-
schleunigung von Korpuskularstrahlen , dadurch gekennzeichnet, dass in mehreren Raumabschnitten der Richtung nach rasch
wechselnde elektrische Felder aufrecht erhalten werden,am besten mit Hilfe von Gittern, und dass ein zeitlich verhältnismässig langsam veränderliches Kraftfeld, z.B. ein konstantes Magnetfeld, die Bahn der Korpuskularstrahlen krümmt,
wobei die Beschleunigung durch das elektrische Feld annähernd
in Richtung der Bahntangente erfolgt.

3.) Verfahren oder Apparat nach An-
spruch 2, dadurch gekennzeichnet, dass ein Korpuskel wiederholt durch denselben Raumabschnitt fliegt.

4.) Verfahren oder Apparat zur Be-
schleunigung von Korpuskularstrahlen, im besonderen Kathodenstrahlen, bei welchem die Korpuskeln durch ein zeitlich praktisch konstantes bezw. nur langsam veränderliches elektrisches
oder magnetisches Kraftfeld aufgewickelt und durch elektrische
Kräfte im Sinne der Erhöhung ihrer kinetischen Energie beschleunigt werden,dadurch gekennzeichnet,dass das letztgenannte
elektrische Kraftfeld von einem zeitlich rasch veränderlichen
Magnetfeld z.B. einem durch einen hochfrequenten elektrischen
wechselstrom magnetisierten Eisenkern induziert ist,welches Magnetfeld bezw. welcher Eisenkern von den Korpuskeln umkreist
wird.

5.)Korpuskularstrahlenröhre im beson-
deren als Röntgenröhre mit Antikathode ausgebildet nach Anspruch
4,dadurch gekennzeichnet,dass der induzierende Eisenkern sich
ausserhalb der Röhre befindet.

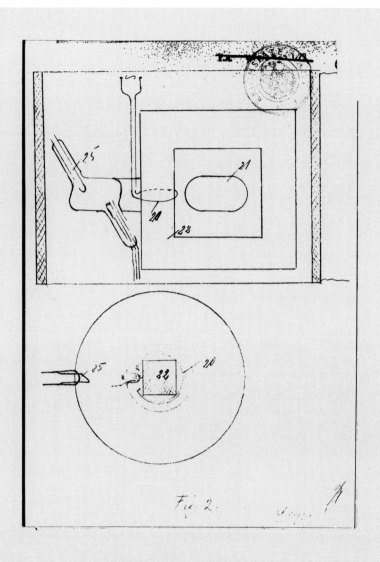

Fig. 2.

<u>Form P. Ack. 4.</u>

THE PATENT OFFICE,

No.**5730**

25, Southampton Buildings,
Chancery Lane, London, W.C. 2.

Date

Received documents purporting to be the Application and

Provisional Specification of   *L. Szilard*

which have been numbered and dated as above.

M. F. LINDLEY,

Comptroller-General.

N.B.—Unless a Complete Specification is left on an Application for a Patent within TWELVE MONTHS
from the date of application (or with extension fee, 13 months), the Application is deemed
to be abandoned.   The investigation as to novelty prescribed by the Patents Acts, 1907
and 1932, is made only when a Complete Specification has been left.
The number and date of this Application must be quoted on the Complete Specification
and Drawings (if any), as well as in any correspondence relative thereto.

Wt 2952   144 Bks/8/33   Wt & Sons Ltd   431f/1385—4

Patents Form—No. 1.

PATENTS & DESIGNS ACTS, 1907 to 1932.

(To be accompanied by two copies of Patents Form No. 2 or
of Patents Form No. 3.)

APPLICATION FOR PATENT.

(a) Here insert (*in full*) name, address, and nationality of applicant or applicants (including the actual inventor).

(a) I (or We) *Leo Szilard,*
citizen of Germany and subject
of Hungary, 6 Halliwick Rd
Muswell Hill, London N.10.
~~2nd Strand Palace Hotel, Strand,~~
~~London,~~ do hereby

5730

21 FEB 1934

declare that I am (or we are) in possession of an invention the title of

(b) Here insert title of invention.

which is (b) *Asynchronous and Synchronous*
Transformers for Particles

(c) State here who is or are the inventor or inventors.

that (c) *I (Leo Szilard)*

claime..to be the true and first inventor......thereof, and that the same

is not in use by any other person or persons to the best of my (or our)

knowledge and belief; and I (~~or we~~) humbly pray that a Patent may be

granted to me (~~or us~~) for the said invention.

Dated the *20* day of *February*, 193 *34*

(d) To be signed by applicant or applicants and, in the case of a Firm, by each partner.

(d) *Leo Szilard*

NOTE.—One of the two forms on the back hereof, or a separate authorisation
of agent, should be signed by the applicant or applicants.

To the Comptroller,
    The Patent Office, 25, Southampton Buildings,
        Chancery Lane, London, W.C.2

Y9430) Wt 13359/4327 30,000(2) 7/33 H & Sp Gp 102

Patents Form No. 2.

T⁹ be issued with Patents Forms Nos. 1, 1ₐ, 1c, 1c, 1c***** or 1ᴅ.**

PATENTS & DESIGNS ACTS, 1907 to 1932.

PROVISIONAL SPECIFICATION.
(To be furnished in Duplicate.)

15. Informa[...]

(a) Here insert title verbally agreeing with that in the application form

5730

21 FEB 1934

(b) Here insert (in full) name, address and nationality of applicant or applicants as in application form.

(a) *Asynchronous and Synchronous Transformer or Particle*

(b) I (or ~~We~~) *Leo Szilard a citizen of Germany and a subject of Hungary*

6. Halliwick Rd ~~and Strand Place~~
Muswell Hill ~~dated, Strand~~
London N. 10. ~~London~~

do hereby declare the nature of this invention to be as follows :—

(c) Here begin description of the nature of the invention. The continuation of the specification should be upon wide-ruled paper of the same size as this form, on one side only, with a margin of one inch and a half on the left-hand part of the paper. The specification and the duplicate thereof must be signed at the end and dated (thus): "Dated the day of 19 ."

(c)

(1) *Where application is made through a Solicitor, Patent Agent, or other authorised representative.*

I (or We) hereby appoint...

of ....

to act for me (or us) in respect of the within application for a Patent, and request that all notices, requisitions, and communications relating thereto may be sent to him (or them) at the above address.

Dated the.............day of.........................., 193....

* To be signed by applicant or applicants.

* ....

....

....

....

-----

(2) *Where application is made without an Agent (Rule* **7**).

I (or We) hereby request that all notices, requisitions, and communications in respect of the within application may be sent to

Miss Simpson 6. Hallowick Rd Muswell Hill at* *London N·10·*

* The address must be in the United Kingdom.

Dated the *20* day of *February*, 193*4*

† *Leo Szilard*

† To be signed by applicant or applicants.

Asynchronous and Synchronous Transformers for Particles.

     The invention concerns methods and apparatus
for the production of fast charged particles, e.g. electrons
or protons.   All these methods, described below, are based
on multiple acceleration, i.e. the velocity of the particle
is exceeding the maximum voltage which arises between any
two parts of the apparatus.   We shall have to deal with
two different methods - the method of the asynchronous
transformer, and the method of the synchronous transformer.
In the first case we shall deal with single action and
multiple action transformers, and we shall start by dealing
with the asynchronous transformer method which is based
on the acceleration of a charged particle in the electric
field induced around the *a* changing magnetic flux.
    SIC.

     Fig. 1 shows the principle of a method in which
a magnetic flux is produced in an iron core 1, and the
flux is rapidly changing its magnitude.   This flux is
produced through the coil 2, this coil being built so as
to consist of one or two windings only, and a rapidly
changing electric current being sent through the coil kma 2.
    SIC.
If an electron encircles in its path several times the iron
core while the flux in the core is changing its value the
energy of the electron will increase at each revolution.
The value of this increase in energy is determined by the

-2-

voltage induced by the changing flux in a single winding
round the core.   In order to compel the electron to encircle
the core 1 a magnetic field 3 can be sustained round this
core which will bend the path of the electron round it.
If one uses a stationary field for this purpose one has the
advantage of being able to build up the magnet 4 out of
thick sheets or a block and is not compelled to use thin
sheets.   Another advantage of a stationary field is the
lowax voltage which can be used in generating the field.
The field xxxxx 3 is generated by means of the coils 5 and 6.
          However, if one uses a stationary magnetic field
as described in figure 1 the radius of curvature of the
path of the electron necessarily increases as the momentum
of the electron increases,   For instance if the momentum
of the electron increases in the ratio 1 to 10 this radius
also increases in the ratio 1 to 10.   It is not possible
to remedy this disadvantage by using a field the strength
of which increases with increasing r.   Such a field would
render the plane of the revolving electron unstable.
Fig. 2 will show what remedy can be employed to prevent a
large increase in radius.
          The maxiumu energy which can be transmitted
to the revolving electron with this arrangement is only
determined by the maximum value of the induction in the
core, the cross section of the core and (for large energies

-3-

where the velocity of the electron approaches the velocity
of light) the mean circumference of the path of the revolving
electron.   It is therefore important to keep this circumference as small as possible and an increase of the radius
of the path of the revolving electron ought therefore to
be avoided.

Fig. 2 shows a magnet system which will bring
about a slight variation of the radius only, when the
momentum of the electron increases.   In this magnetic
field the revolving electron emerges intermittently
from a strong field into a gap in which the field is weaker
or nil and enters again into the next section of the strong
field.   10 is the centre of the arrangement. 11 and
12 are poles of magnets which generate a magnetic field
perpendicular to the surface of the pole (and perpendicular
to the plane of the diagram).   The angles α 11, α 12
and α 10 are equal, and the adjacent sides of the magnets
11 and 12 are parallel.   There are further magnets
not shown in the diagram, which form together a symmetrical
figure round the centre 10; this is indicated in Fig. 2b.
An electron of a given momentum will encircle the centre
10 at a definite distance r from the centre in such a way
that the path enters at right angles the left side of the
magnet 11 , is bent round in the field of the magnet 11

-4-

by the angle α, leaves the other side of the magnet 11
at right angles, goes through the gap 30, enters at right
angles the left side of the magnet 12, is bent round by
the angle α in the magnetic field of the magnet 12, and
leaves the right side of the magnet 12 at right angles again,
and so on.   13 is a coil energised by an electric current,
that generates a **magnetic** field across the air gaps of the
individual magnets 11, 12 and so on.   One coil 13 can
serve several magnets, but the coil must not surround the
iron core 1$. In most cases it will have advantages
to use at least two coils (1 coil for each half of the
magnet system.)

  If a practically uniform magnetic field were
to be used in the gaps of the individual magnets 11, 12
and so on, the radius r of the electron path would only
vary slightly with increasing momentum of the electron i.e.
according to the equation $\dfrac{r_2 - R}{r_1 - R} = \dfrac{p_2}{p_1}$ circle
In this equation R is the radius of the ~~circle/circular path~~ ~~SIC.~~
having its centre at point 10, at the periphery of which
circle the corners of the magnets 11, 12 etc. are arranged.
p, and p$_2$ represent the momentum of the electron at the
beginning and at the end ($p = \dfrac{m\,v}{\sqrt{1 - v^2/c^2}}$).   In practice
instead of a uniform magnetic field one will use a field,
the strength of which will slightly decrease with increasing r,

-5-

and therefore, a slightly modified equation will be used.

Fig. 3 shows the magnet 11.   15 is the air gap
across which a stationary magnetic field is maintained by
means of the coil 13.   This coil 13 is formed by an upper
coil 16 and a lower coil 17, leaving the neighbourhood of
the plane of symmetry AB in which the electron revolves
unobstructed, thereby leaving a free space for the path of
the electron emerging from the magnet system at the end of
the accelerating process.

The magnetic field in the area 15 is not uniform,
but decreases with increasing r.   Such a field stabilises
the path of the electron in the neighbourhood of the plane
of symmetry AB.   In order to indicate such a decreasing
field the width of the area 15 is shown to be increasing
with increasing r in Fig. 3.

45 indicates a glowing filament that may be used
as a source of electrons, the electrons ejected by the
filaments to be accelerated subsequently in the magnet
system.

The outer part of the air gap indicated by the
number 18 shows a rapidly increasing width of the gap with
increasing r in such a way that the magnetic field across
the gap should decrease in this area stronger than $\frac{1}{r-R}$.
The purpose of this is to draw the circular path of the
electron into a spiral, thereby making it easier to lead the

-6-

electron beam at a definite spot out of the magnet system,
An electric condensor 19 or other suitable means may serve
for the purpose.   This will be more fully explained with
the help of Fig. 5.

          The path of the electron in the space 20 between
the metal plates 21 and 22 is protected from xx the electric
field induced by the changing flux in the core.   Therefore
the electron is only accelerated when emerging from between
the plates 21 and 22, and about to enter the space between
two similar plates 23 and 24.   This is shown in Fig. 4.

          In Fig. 4 one half of the circular path 27 of the
electron lies between the metal plates 21 and 22, and the
other half between the metal plates 23 and 24.   All these
plates constitute equipotential surfaces and the electron
is accelerated by a strong electrifß field in the gaps 28
and 29 between the two pairs of plates 21/22 and 23/24.
This electric field is induced by a changing flux in the
transformer core 26.   This core is laminated as shown in
ßlC𝚡𝚎𝚡𝚡 Fig. 4a, and the two pairs of plates are both connected
to the middle sheet of the core.   The electric lines of
force generated between the two pairs of plates are shown in
Fig. 4b.   Such a field has the effect that any electron
which has deviated out of the plane of symmetry AB is bent
back towards that plane so that the path of the electron is
stabilised in the neighbourhood of that plane.

-7-

In Fig. 6. the effect of a strongly decreasing
magnetic field at the outer edge of the system is described.
Between the two circles 30 and 31 the magnetic field

x   decreasing stronger than $\dfrac{1}{\rho}$ whereas within the

circle 30 the magnetic field, though it may decrease with

x   increasing ρ it does not decrease as strongly as that.
The path of the electron within the circle 30 deviates
from a closed circular path round the centre 10 only
insofar as the electron is accelerated by the electric
field which leads to a steady increase of the radius of
curvature of the path.   However, as the electron passes
beyond the circle 30 and enters into the region between
the circles 30 and 31 no closed circular path is any longer
possible, and the path of the electron will be drawn out
into a spiral 32, as shown in Fig. 6.   Accordingly it is
possible to have an electric condensor built of a thin
plate 33 and another plate 34 and compel thereby the electron
to leave the magnet system at a definite spot after passing
through this condensor, without losing most of the electrons
through collisions with the condensor plate 33.   If the path 32
of the electron 32 were not drawn out into a spiral by means
of the strongly decreasing magnetic field in the region
between the circles 30 and 31 many electrons would get lost
by striking the plate 33.   Even if the path is drawn out

-8-

into a spiral the plate 33 must be thin or else a large

number of electrons get lost by striking the plate.

Of course the electric acceleration of the electron has

also the effect of drawing out the path into a spiral and

this effect can in certain cases be sufficient in itself

i.e. if the voltage induced by the changing flux of the

core 26 is extremely large.   Fig. 6b shows a cross

section of the magnet 36 indicating an increasing width

of the gap with increasing f; increasing strongly between

the circles 30 and 31.

Fig. 7 shows another feature of the same type

of magnetic field.   If a multiple action transformer is

to be used, as will be described later on, it is essential

to bring the electron from the periphery into the magnet

system, and keep it within the magnet system for many

revolutions during which it is accelerated by electric fields.

The magnet system in Fig. 6 and 7 on the one hand differs

from the magnet system in Fig. 3 on the other hand insofar

as the electric coils 40 and 41 and the magnet cores

42 and 43 are inside the magnet gap area in Fig. 6 and 7

and not outside of that area as in Fig. 3.

Fig. 7 also shows the metal plates 21 and 23,

and the gaps 28 and 29 between the parts of plates 21/22

and 23/24 as described in Fig. 4.   If an electron is shot

-9-

into the air gap of the magnet system from the periphery
it is drawn into the magnet field as indicated by the
curved path 44.   However, the electron would leave the
magnet system again if it were not for the accelerating
action of the electric field in the gaps 28 and 29.
This accelerating action will have the effect of bringing
the path of the electron to approximate a circle having
its centre near point 10.   This circle will, according
to the initial energy of the electron, be situated
within the circle 30.   As the energy of the electron
gradually increases owing to the electric acceleration,
the radius of the path gradually increases and the electron
emerges from within the circle 30. as described previously
in Fig. 6a.

In Fig. 8 another aspect of the arrangement
is described.   It is necessary that the path of the
electron within the magnet system should lie in vacuum
and a simple method of achieving this is shown in Fig. 8.
Fig. 8a shows a flat metal box (a flat tube) composed of
two pieces 50 and 51 which are joined, having a piece of
insulating material 52 and 53 at each of the two junctures.
An unobstructed annular space 54 is formed within the box,
and the box is evacuated. The top wall of the fxi flat
box or tube 50 corresponds to the metal plate 21 and the
bottom of it corresponds to the metal plate 22 in Fig. 4.

-10-

The top wall of the flat box or tube 51 corresponds
to 23, and the bottom wall of it corresponds to 24 in
Fig. 4.   Fig. 8b shows that the magnet system also
consists of two insulated halves 55 and 56, and this is
indicated by electric lines of force being drawn in the
gap 57 between the halves 55 and 56.

        If an alternating magnetic flux within an
iron core is used as described in Fig. 1 for the
acceleration of electrons, and one wishes to increase
the total energy transmitted to the electron one has to
increase the cross section of the iron core.   However,
if the cross section is multiplied by four the circumference is doubled and therefore the maximum energy of
the electron is only doubled.   The weight of the
apparatus would accordingly increase with the square
of the transmitted energy.   (This rule holds for
fast electrons the velocity of which is practically
not changing any longer with increasing energy).

        One can obtain a less heavy apparatus by
utilising several half cycles of the changing flux.
In order to do that the electron has to encircle different
iron cores during subsequent half cycles of the changing
flux or if it encircles several times the (during several
subsequent cycles) the same core it must do so in the

-11-

proper sense of revolution.   An example for this is
shown in Fig. 9 which indicates an apparatus subsequently
referred to as a single phase multiple action transformer.
60 and 61 are two iron cores forming a transformer frame.
This frame is magnetised by means of alternating current
which is passed through a coil 62 consisting of only one
winding.   A circular magnet system indicated by 63
(the magnet systems are not drawn in Fig. 9c) surrounds
the core 60.   An electron is shot into the magnet 63
in a manner described in Fig. 7.   The electron enters
the magnet about the time when the voltage induced by
the changing flux of the core 60 passes through zero
(i.e. about the time when the flux passes its maximum)
and this electron encircles the core 60 many times,
gradually gaining energy as the flux decreases, while the
induced voltage has a positive value.   About the time
when the flux in the core 60 reaches its minimum value
and the induced voltage again approaches zero the electron
leaves the magnet 63 and passes into the magnet 64 which
encircles the core 61. The strength of the magnetic
field in the gap of the magnet 63 determines at what
energy value the electron will leave 63. It is therefore
possible to adjust the strength of the magnet so that this
should happen at the proper time.   The electron is in

-12-

a similar manner accelerated within the magnet 64 and
so on within other magnets which alternately surround
the cores 60 and 61.   The number of magnets to be used
is only limited by the length of the transformer frame
which can be made so as to suit the purpose.

Fig. 10a and b illsutrate a certain appect
of such a single phase multiple action transformer.
The two curves shown in Fig. 10 give the voltage round
the cores 60 and 61 plotted against time. At the point 1
the electron enters the first magnet surrounding core 1
and at point 2 the electron has accumulated sufficient
energy to leave the first magnet and to enter the second
magnet in which it is further accelerated.   At point 3
the electron leaves in a similar manner the second magnet
in order to enter into the third magnet and so on.

Whether Fig. 10a or b is the more appropriate
illustration of the multiple action depends on the strength
of the magnetic fields generated in the magnets, since
the strength of xhxk fix those fields determin at what
energy level the electron leaves the respective magnet.

Fig. 11 shows a two-phase multiple action
transformer.   70 and 71 are two transformer frames
magnetised at 90° phase difference.   72 is a magnet
system surrounding the core 73 of the transformer 70.

-13-

The electron after encircling the core 73 in the magnet
72 leaves the magnet 72 and enters the magnet 74 which
encircles the core 75 of the transformer 71.   After
this the electron enters the magnet 76 which encircles
the core 77 of the transformer 70. The next magnet
78 encircles the core 79, then follows 80 encircling
the core 73 once more, then follows the magnet 81
encircling the core 75.   The position of the planes in
which the electrons revolve is indicated in Fig. 11a and b
by marking the slope by arrows.

    Fig. 12 illustrates the action of such a
two-phase transformer.   The four curves represent the
voltage induced around the four cores (having 90° phase
difference in relation to eachother) as a function of time.
An electron enters at point 1 the magnet 1) and leaves it
at point 2.   It enters the magnet 2) at point 2 and
leaves it at point 3 .   It enters the magnet 3) at point
3 and leaves it at point 4.   It enters magnet 4) at point
4 and leaves it at point 5 and so on.   Magnet 1), and
        and so on
magnet 4)/surround the first iron core.   Magnet 2),
magnet 5) and so on surround the second iron core etc. etc.
The multiple action illustrated in Fig. 12 has a remarkable
stabilising effect on the phase at which the passing of the
electron from one magnet into the next magnet occurs.

-14-

Should an electron for instance enter the magnet 1) at the
point 1´ the time required to reach the energy level at
w..ich it can leave the magnet) 1) will be shorter than a
quarter part of a cycle and it will therefore leave the
magnet 1) at a point 2´ which is nearer to point 2 than 1´
was to point 1.   Therefore the magnet 2) will already be
entered with a smaller phase difference than it has been the
case with magnet 1).   In this way the phase ~~difference~~
                                                        SIC.
will gradually ~~readh~~ its normal value.

　　　　　The two-phase transformer has a certain
advantage over the single-phase transformer insofar as when
the electron enters the magnet of the two-phase transformer
it is at once strongly accelerated by the electric field
in the gaps 28 and 29.   This is of importance as described
in Fig. 7 in preventing the electron from leaving the magnet
immediately after its entrance .

　　　　　Fig. 13 shows an iron core to be used for
the acceleration of electrons.   This core is wound from
a thin iron sheet.   If one uses iron sheets below 1/10th mm.
e.g. .03 mm., it is quite possible to obtain 1,000 volts
per revolution using an induction of B = 10,000 Gauss and a
cross section of about 10 cm X 10 cm.   Using very thin
sheets an iron core built up of tese sheets by winding the
sheets is preferable to an iron core built up in the usual
way.

-15-

The methods described hereafter are based
on synchronous acceleration.   Instead of using an iron core
in Fig. 4 it is also possible to use an extremely high frequent
electric oscillator, and connect 21/22 to one pole and 23/24
to the other pole of the oscillator.   If the frequency
of the oscillator corresponds to the time of revolution of
the electron the electron will be accelerated in the gaps
28 and 29.   It is essential to keep the electric capacity
of the plates 21/22 and 23/24 small, otherwise very heavy
currents would flow into these plates and it would be
difficult to maintain a sufficiently high potential in the
plates.   It is therefore essential that the magnet system
as shown in Fig. 8b should be built of two insulated parts,
55 and 56, and that the part 55 should be kept at the same
alternating potential as the plates 21/22 (tube 50), and
similarly that the part 56 should be kept at the same
alternating potential as the plates 23/24 (tube 51). On the
other hand an alternating potential will appear between the
two halves of the magnet system 55 and 56, and this is indicated
by the lines of electric force drawn in the gap 57.   If an
arrangement as shown in Fig. 2 is used and if R has a value of
two metres, an oscillation of about six metres wave-length has
to be used for the acceleration of the electrons (this holds
when the initial velocity of the electron is already close to

-16-

the velocity of light.   During the accelerating process
the time of revolution will increase as the momentum of the
electron increases (again we assume that the initial energy
was sufficiently high, otherwise the time of revolution would
decrease in the beginning).   Therefore the wave-length of the
applied high frequency oscillation should not be constant but
should increase during a period of time which is required
for the acceleration of the electron within the magnet.   If
for example the magnetic field is so adjusted that for the
initial revolutions of the electron r - R is equal to one cm.
and if the initial momentum of the electron is increased 10
times during the acceleration of the electron, then for the
last revolutions of the electron within the magnet r - R
will be equal to 10 cm.   Accordingly the time of revolution
(R being equal 100 cm.) will gradually increase by about 10%
and therefore the wave-length should gradually increase by 10%.

            The change of the wave-length is very small
for a single revolution.   It is important to note that the
time required for the acceleration of the electron so as to
bring the momentum up to a certain value depends on the phase
relation which characterises the passage of the electron through
the gaps 28 and 29 in Fig. 4a.   This fact leads to a
considerable freedom in the rate at which the high frequency
oscillation may change its frequency within wide limits within
which it is effective in accelerating the electron.   The

oscillation having a changing frequency will stabilise the phase

-17-

of the passage of the electron across the gaps 28 and 29.

Fig. 14 shows the potential difference between
the two pairs of plates 21/22 and 23/24 as a function of
time.   The increase in energy which the electron gets
at each revolution when passing through the gaps 28 and 29
is not determined by the maximum value of the oscillating
voltage alone, but also determined by the phase of the
passage across the gap.   If the increase of energy per
revolution is too large then the time of revolution increases
too rapidly and the passage of the electron across the gaps
28 and 29 is shifted nearer and nearer to the maxima of
the oscillation. (This holds when the initial energy
of the electron is sufficiently large so that its velocity
practically does not change any longer.)   Fig. 14b illustrates
how the passage of the electron across the gaps 28 and 29
moves nearer to the maxima of the oscillation if the energy
of the electron lags behind so that the time of revolution
is too short.   Fig. 14a illustrates how the passage of
the electron across the gaps lags behind the maxima if the
electron has an access of energy and accordingly a too long
period of revolution.

If one has to accelerate protons or electrons
the initial velocity of which is small as compared to the
velocity of light an electric oscillation with an increasing
frequency has to be used.   Whereas in Fig. 14 the electron

-18-

passage across the gaps followed the maxima of the oscillation it will now preceed those maxima.

If the initial velocity of the particle is small the velocity reached towards the end of the accelerating process however is approaching the velocity of light an oscillation has to be used, the frequency of which first increases and later deminishes. The voltage must be so adjusted that about the time when the frequency reaches its maximum the passage of the particle should coincide with the maxima (i.e. the passage should first follow and later on preceed the maxima).

One way of producing an oscillation, the frequency of which changes slowly is to insert into the oscillating ~~xixxixx~~ circuit a condenser the capacity of
SIC.
which is changing periodically. Such a changing condensor is shown in Fig. 15. It is built in the following way. Two discs facing eachother rotate very rapidly in the opposite sense round acommon axis. The two rotating surfaces which face eachother are equipped with teeth and therefore the capacity of the two discs towards eachother varies rapidly. A quartz plate 71 is inserted between the two discs 70 and 72.

The same effect on the frequency which can be brought about by a condensor periodically changing its

-19-

capacity can also be brought about by a condensor in series
with a changing resistance.   If one uses two valves
one can build up with these a changing resistance of a
suitable nature.   The valves can be controlled by applying
a periodically changing voltage to their grids.

20

The following are essential features of the invention, relating to the asynchron transformer method:-

1.　The use of ~~the~~ a stationary magnetic field for bending round the path of the charged particle encircling the changing flux.

2.　The insertion of gaps in the field of the magnet system which encircles the changing flux.

3.　The use of equipotential surfaces around the changing flux which limit the accelerating electric field to a few gaps, thereby stabilising the path of the revolving charged particle in the neighbourhood of the plane of symetry.

4.　The use of a magnet system, having the magnetic field decreasing with increasing radius.

5.　Having part of the magnetic field decreasing stronger than $\dfrac{1}{r-R} = \dfrac{1}{s}$ so as to make circular paths of the charged particle in stable and to draw them out into a spiral.

6.　Electric or magnetic fields to lead the particle at a definite spot out of this unstable region.

7.　Multiple acceleration having the particle revolving around a different flux in subsequent half ~~sizes.~~ *cycles.*

8.　The use of *polyphase* ~~polyface~~ multiple accelerators.

-2- ~

9.   Bringing the charged particle from the periphery into the magnet system at a point which is about 90° removed from the accelerating spot gaps, and removing the charged particle at the periphery after having increased its energy within the magnet system.

10.   Building up the iron core by winding up a thin iron sheet.

11.   Coils producing the stationary field not surrounding the changing flux.

12.   These coils being built of two parts leaving free the neighbourhood the plane of revolution.

13.   The magnet system built of two halves insulated from eachother, each of these halves being at the same potential as the corresponding paths of plates 21-22 and 22-23.   The two halves of the magnet system having an electric field between them.

The following are essential features of the invention, relating to the synchronous transformer method:

1.   A circular magnetic system with gaps in the field.

2.   A magnetic field decreasing with increasing r so as to stabilise the path of the charged particle in the neighbourhood of the plane of symetry.

3.   The application of a high-frequency oscilation the frequency of which is slowly changing.

-32-

4. The voltage of this oscillation adjusted in such a way
that there should be a time lag between the passage of the
particle through the accelerating gap and the maximum of
the oscilation.

5. The circular magnetic system built of two halves
oscilating against each other and not oscilating against
the respective pairs of plates 21-22 or 23-24.

Dated the day af 20th of February
1934

Leo Szilard.

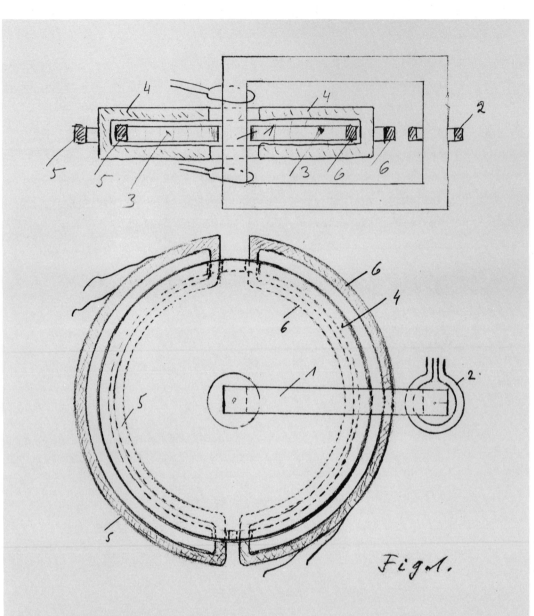

Fig. 1.

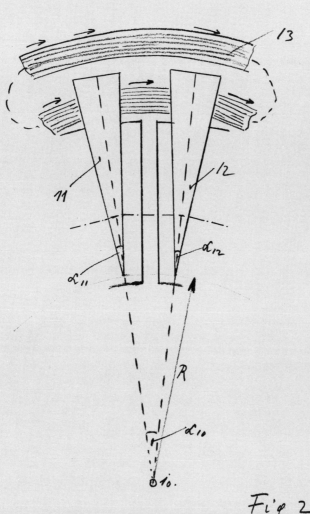

Fig 2

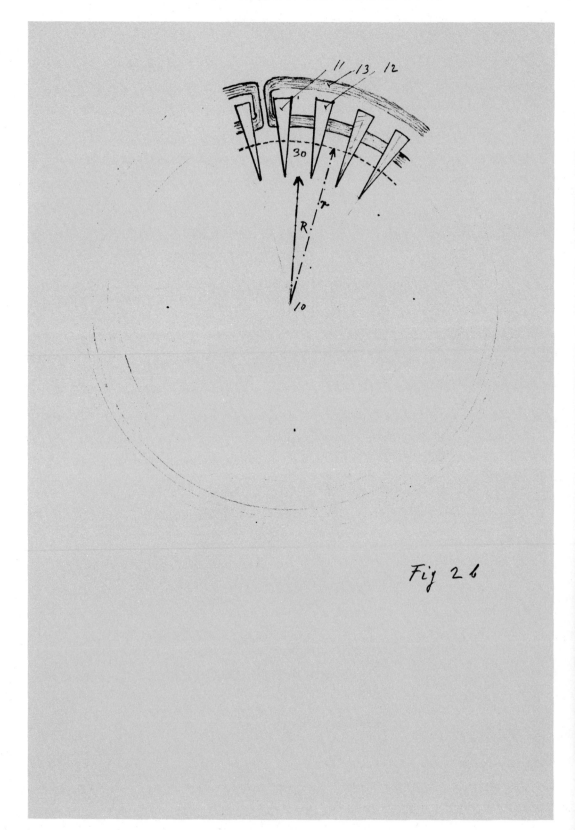

Fig 26

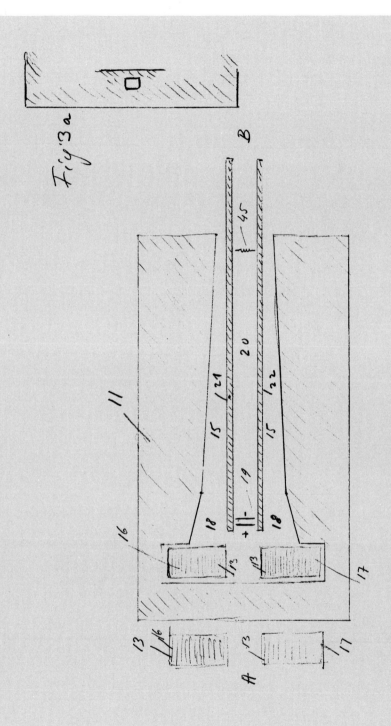

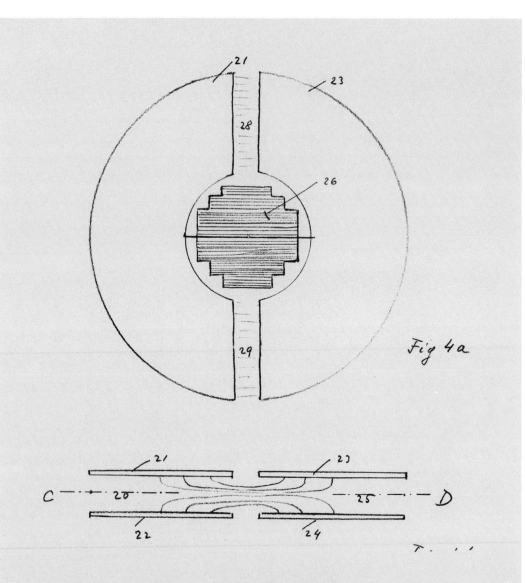

Fig 4a

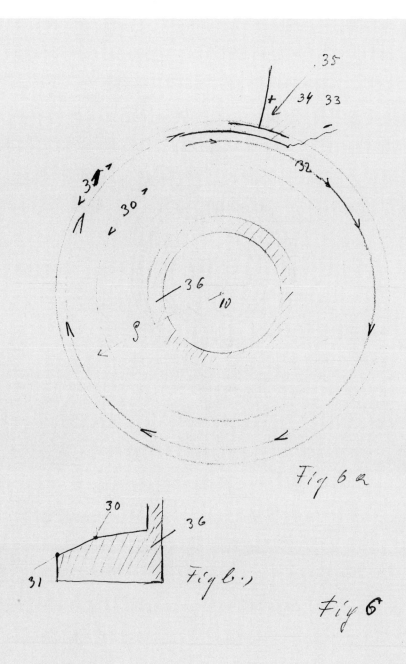

Fig 6a

Fig 6b)

Fig 6

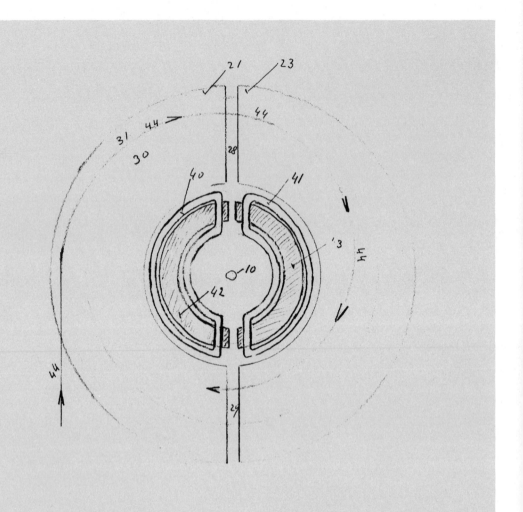

Fig 7.

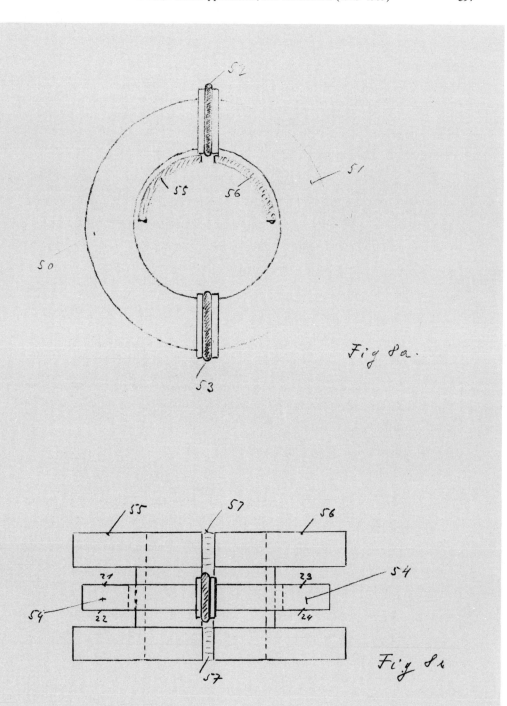

Fig 8a.

Fig 8b

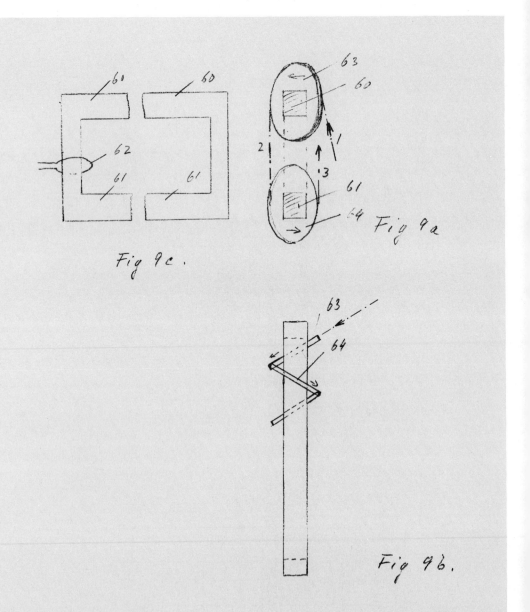

Fig 9c.

Fig 9a

Fig 9b.

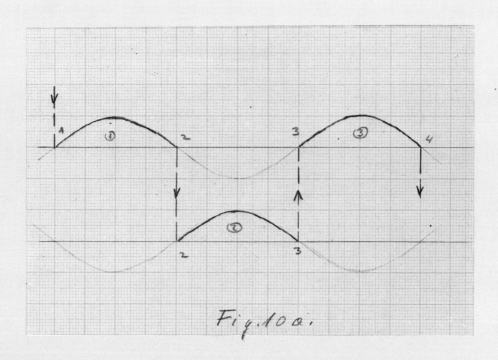

Fig. 10 a.

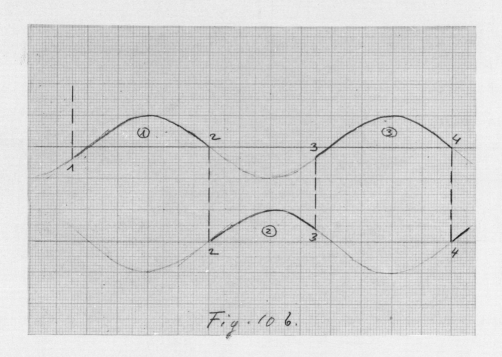

Fig. 10 b.

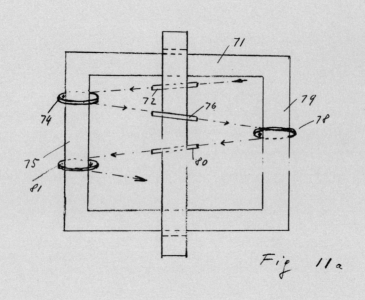

Fig 11 a

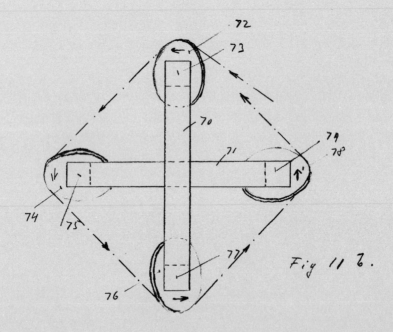

Fig 11 b.

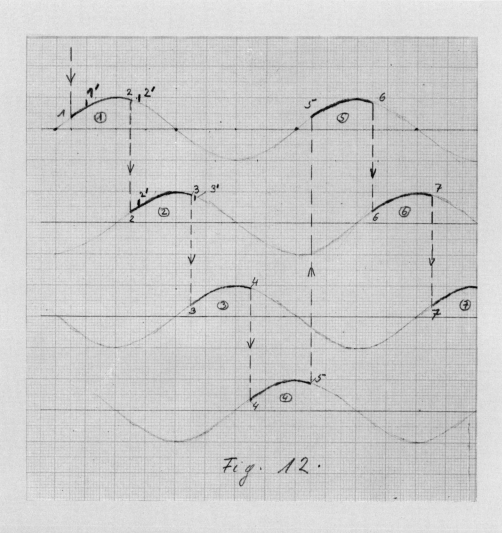

Fig. 12.

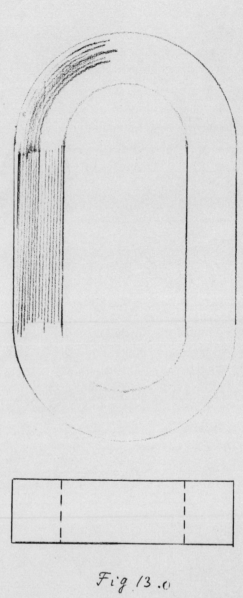

Fig 13.0

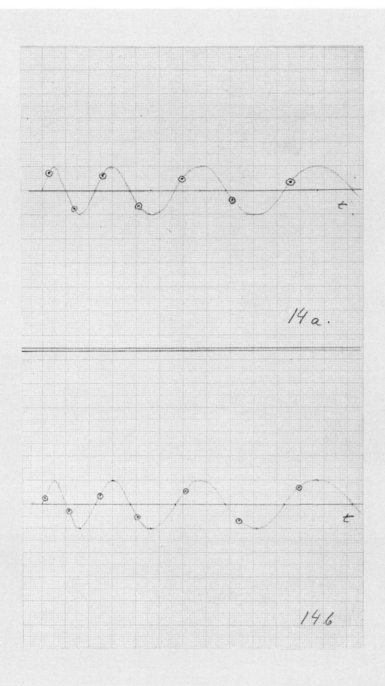

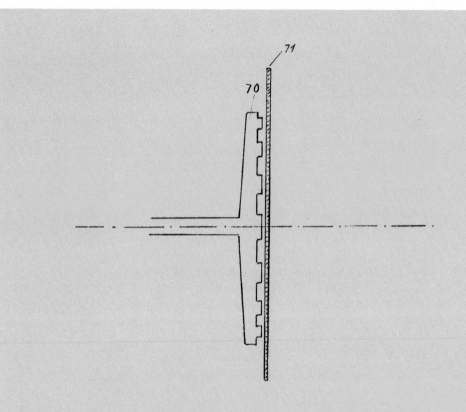

Fig. 15.

Form P. Comp. 1.

Telegraphic Address: PATENT OFFICE, LONDON.
Telephone No.: HOLBORN 8721.

THE PATENT OFFICE,

25, SOUTHAMPTON BUILDINGS,

LONDON, W.C. 2.

Zilard...'s Application for Patent.

Numbered _7640_

Dated _12 March 1934_

1 0 APR 1935 _____193__

The Comptroller has to inform you that the desired extension of time for one month for leaving a Complete Specification in respect of the application for a Patent, numbered and dated as above, has been allowed.

~~Under this extension the Complete Specification must be left~~ within ten months ~~*from the date of the application for a Patent.*~~

~~No further extension is obtainable.~~

To _G. A. L. Marriott Esq._

(R9953) Wt 212744/2189 5,000(3) 2/32 H & Sp Gp112

Note

Owing to the fragility and uneven quality of the original, which did not permit satisfactory photographic reproduction, we reproduce in the following an accurate copy of the original patent application. Of course, unlike the original, the hand-written corrections are from our hand rather than Leo Szilard's, but we have other-wise reproduced the original without change, including errors (e.g., the formula on page 614 should have read $D\, d^2(rs)/dr^2 + A(rs) = 0$). The original is in the files of Leo Szilard.

The Editors

-i-

Transmutation of chemical elements

The invention concerns methods and apparatus for the production of
nuclear transmutation leading to the generation of radio-active bodies,
to the storage of energy by means of the generation of radio-active bodies
and the utilisation of the energy which has thus been stored for the production of heat and power, further to the liberation of nuclear energy
and the utilisation of the liberated energy.

1) Generation of radio/-active bodies . It is not new to produce
elements capable of spontaneous transmutation by bombarding certain
elements with fast charged nuclei,for instance by bombarding carbon with
protons or aluminium,bor and magnesium with helium ions (particles).
However, the radio-active elements produced by the bombardment of these
light elements with protons or alpha particles, have a short existance
(they disintegrate spontaneously in a time shorter than a few hours to
half their amount), and it is not possible to use these charged nuclei
for the transmutation of the heavier elements with good efficiency as
the ionisation loss gets too large.   It is, however, possible to produce
with good efficiency(both from light and heavy elements) radio active
substances a quantity of which decomposes to one half of the original
amount in a period of time exceeding 24 hours, if a thick layer of substance is exposed to a penetrating radiation which is emited when collisions between heavy hydrogen (diplogen) atoms (or nuclei) and light
elements,including heavy hydrogen (diplogen or so-called deuterium) itself,
are produced.

Fig 1 shows an example of a suitable arrangement. 11 is an electrical discharge tube ejecting a beam 12 of fast diplogen ions. The ions
fall on a substance 13 consisting of for instance gasous diplogen or a
diplogen compound or lithium, causing transmutation, i.e. a nuclear
reaction of the diplogen ion with an atom of the target.   The substance 13 is surrounded by a layer 14 containing the element which we
with to transmute into a radio-active element.   In order to have a good
efficiency,the thickness of the layer 14 has to be sufficiently large,

-2-

compared with the mean free path of the neutron, for this transmutation.

Fig 2. shows the electrical discharge tube referred to in fig. 1.
It is a high voltage positive ray tube. There is an auxiliary positive
ray tube on top of the high voltage tube. 11 is the anode, 15 the cathode
of this auxiliary tube. Diplogen is admitted through the tube 13 and pumped away through 14.

The transmutation of elements into radio active bodies under the
influence of neutrons can be demonstrated even before one knows which
elements will transmute into radio active bodies, if one prepares a
mixture of all suitable elements ~~leaving~~ leaving out the radio active elements, but
including Uranium and Thorium (from which the beta active products have
been removed) and exposes this mixture to a neutron radiation. The mixture shows after exposure radioactivity decaying with a large number of
half-life periods the relative intensity of which depends on the composition of the mixture and on the time of irradiation.

These are the essential features of the method for the production
of radio active substances which are disintegrating slowly:
Light elements are bombarded by each other, especially diplogen is brought
into collision with other light elements or with diplogen itself. Uncharged particles of a mass of the order of magnitude of the mass of a proton
are emitted as a consequence of the collisions between nuclei of light
elements. Such uncharged nuclei penetrate even substances containing
the heavier elements without ionization losses and cause the formation of
radio active substances in a layer which is exposed to them with good
efficiency if the thivkness of the layer is of the order of magnitude of
[the mean free path of the neutron.]
~~a metre.~~ We have therefore two steps of good efficiency in series: the
production of uncharged nuclei by collision between light elements (the
ionization losses are small because the elements have a small atomic
number and therefore a small nuclear charge and a small number of electrons per atom) and the production of radio active substances by means of
the uncharged nuclei (the ionization losses are practically absent even
in case of passage through heavy elements.)

-3- *a b?*

The method described hitherto was characterized by shooting a particle through matter which is at rest. As described, a diplon (a diplogen nucleus) shot into diplogen at rest will in a large proportion of cases lose its energy by ionizing the diplogen and cause no transmutation in those cases.

If we were to maintain a very large concentration of energy in a space filled with atoms of such elements which will suffer transmutations, if the atom (nuclei) strike each other at that temperature (which corresponds to the energy concentration maintained) then the following would hold good: the energy transmitted to the electrons by the moving nuclei would be continuously retransmitted to the nuclei. It is sufficient to maintain a suitable energy concentration for a fraction of a second, One can do so by shooting charged particles which have been accelerated in an electrical discharge tube through a space in which diplogen alone, or lithium hydrid (or other compounds of hydrogen and lithium) or other combinations of hydroden or diplogen with a third light element are present. If we use an electric condensor [and discharge it] in a fraction of a second across the discharge tube we can introduce (especially if we use several discharge tubes which are operated simultaneously) a very large energy in a very short period of time into the "transmutation space" filled, e.g. with diplogen. As "heating rays" we can use protons or heavier ions, or we can use cathod rays. We can easily estimate how much energy must be stored in the electric condensor in order to have sufficient supply of energy to heat up 1 cubic cm of diplogen.

If nuclear reaction of diplogen with itself is enforced through heating up diplogen with an electric discharge, a neutron radiation is emitted which can be used for the generation of radio active bodies as described above. This method is described in the following:

4.

In Figure 6, 27 is the window of the high voltage tube through which the fast electrons are ejected. The electrons hit the rotating anticathode 30 which is covered with lead or tungston (W) 31. This anticathode is water-cooled, the water entering the rotating body through the axis 35. 32 is a beryllium block in which a space has been left for the rotating anticathode and for tha path of the cathode rays 33 between the window 27 and the anticathode. This beryllium block may for instance have a size of 25 cm. x 25 cm. The voltage used to operate the electron tube may be three million volt. The beryllium block is surrounded by a block 34 of the element which we wish to transmute into a radio-active elenent. For instance, iodine or arsenic or any other element that is suitable. While it is advisable to use metalic beryllium in the block 32 the element in the block 34 may be present in the form of an organic compound in order to make an isotopic separation possible after irradiation. The dimensions of block 34 may, for instance, be 50 cm. x 50 cm.

Fast electrons have a similar action on beryllium as hard x-rays, a fraction of this action may be due to the direct action of the fast electrons on the beryllium. In view of the fact that hard x-rays generate fast electrons in the beryllium, part of their action can be due to fast electrons. In any case, we do not wish to differentiate here between the action of fast electrons and hard x-rays, and while we think it likely that the direction action of hard x-rays on the beryllium plays the major part in the liberation of neutrons, we wish to envisage the following modification of our method: The electrons of the dish arge tube fall instead of lead on beryllium which can be put into the place of the lead costing 31 of the rotating anticathode 30 in Figure 6.

5.

Many elements transmute when bombarded by slow neutrons into
their own radio-active isotope and it requires a special method
chemically to separate the radio-active element from its irradiated
isotope. We can achieve such a separation by irradiating a suitable chemical compound of the said element. Those atoms of our
element which transmute into a radioactive atom are thrown out of
the compound and ɪkȷkɪkuɪ will subsequently be called "free" If we
choose a compound which in the circumstances does not interchange
the atoms of our element bound within the compound with the "free"
atoms which are their isotopes we can chemically separte the "free"
atoms from the compound and thereby separate the radioactive isotope
from the irradiated element. Compounds in which the element in which
we are interested are bound direct to carbon are very often suitable.
For instance, in the case of iodine compounds like iodoform or
ethyl iodide can be irradiated and after irradiation the radioactive isotope can be concentrated by separating the "free" iodine
from the iodoform or the ethyl iodide. In order to protect radioactive iodine a small amount of ordinary iodine can be dissolved in
the organic iodine compound before irradiation or after irradiation
but before separation.

In the following we shall deal with methods and apparatus for the production of energy and generation of radio-active bodies by means of chain re-action. In order to maintain such a chain an initial radiation of neutrons is generated by one of the methods described further above. If the neutrons enter a space which has the proper shape and size and which is filled with a proper combination of elements their energy or the *in* number, or both, can be increased through their interaction with the substance which fills the chain reaction space. The interaction of a neutron with matter can lead to the liberation of further neutrons - these newly liberated neutrons liberate again in their turn further neutrons so that we can have a chain reaction in which a large number of neutrons are liberated, the total number of which is determined by the geometry of the arrangement.

Figure 7 and 8 show such a chain reaction apparatus. A neutron radiation, the initial radiation, is generatd by the high voltage canal ray tube 1 (shown in greater detail in figure 2.) This tube generates fast deutons which strike the target 28 which contains deuterium. The neutron radiation emerging from 28 acts on the matter 3 which fills the spherical transmutation space. The composition of this matter 3 will be discussed further below and is such that *a* kik chain reaction is released by the neutrons. The pumps 120, 121 and 122 pump a liquid for instance water or mercury through the pipe systems 107, 110, 111 thereby cooling the transmutation area 3 and driving the heated liquid through the boiler 126. The boiler supplies steam to a power plant. The neutrons emerging from the sphere 3 act on a layer 9 which is composed of an element gk that will transmute into a radio-active body (which is suitable for the storage of energy.)

An essentially different way of introducing the initial radiation into the chain reaction chamber is the

7.

arrangement shown in Figure 9. 401 is the cathode ray
tube described in Figure 1. 402 is a sheet of a heavy
element for instance Pb, or U in which penetrating radiation (hard X-rays) is generated with an extremely good
efficiency if the electrons have a voltage about or over
one million volt. This efficiency increases very rapidly
with the voltage, and is much higher than it could be ex-
pedteddfrom the experience based on ordinary x-ray work.
The thickness of the sheet 402 is such as to enable the
generated penetrating radiation to penetrate through this
sheet and act on the transmutation chamber 106 (in Figure
8). Nevertheless the sheet can be sufficiently thick
to utilize more than half of the energy of the cathode
rays. The X-rays emerging from sheet 402 penetrate the
layer 3 and can liberate ~~efficient particles~~ *[neutrons]* either from
the layer 3 or from a substance 407 placed in the inter-
ior of the layer 3.) For instance, 'f we have ~~diplogen~~ *[beryllium]*
present in 403 or in 3, neutrons will be ~~jkikjjk~~ liberated
by X-rays.e These neutrons can then maintain a chain reaction as discussed further above and further below.
The advantage of using X-rays as an initial radiation is
the following: The x-rays penetrate through a perfectly
closed layer 3 into the interior of the layer and therefore
a leak of neutrons from the interior can be avoided. This
is specially important if we have to deal with a neutron
chain in which no multiplicator action is involved. In
such cases x-rays may be used with advantage as initial
radiation especially in view of the unexpectedly large
efficiency of the x-ray production by means of fast electrons acting on heavy elements.
        In the simplest case, when neutrons alone form/ the links
of the chain, we shall demonstrate in the following
the importance of the shape and the size of the transmutation space. If we have a closed spherical layer of
material in which the chain reaction takes place the
inner radius (r) of which is large compared with the

8

mean free path$\int$ of the ~~sffknkmmkkksxxikkskskskfkd~~ neutrons
which maintain the chain, the density (s) of the neutrons
will with good approximation be given as a function of the
radius (r) by the following equation:

D d (rs) / dr $+$ A (rs) = 0

D and A are determined by: the mean free path of the neutrons
a; the mean velocity of the neutrons w; the factor of the
multiplicating action f which says how many collisions of a$\mathbf{k}$
~~electron~~ *neutron* are needed in the average in order to produce one
new ~~electron~~ *neutron*.   A = w / af $;$ D = aw / 3 $;$ $\sqrt{\dfrac{D}{A}} = \dfrac{a\sqrt{f}}{\sqrt{3}}$

We are interested in the critical thickness of the
spherical layer for which the gredient of the density S vanishes   If the thickness $(r_2 = r_1)$ approaches l, we can maintain with a very weak source of initial radiation in the
interior of the inner surface of the spherical layer a very
strong chain reaction and we can easily get one thousand or
more times more neutrons emerging from the chain reaction
layer than the number of the neutrons forming the initial
radiation. If the outer surface $(r = r_2)$ of the spherical
layer were to stand free in space the density s would be
zero for that surface and the critical value l would be
given by $l_o = \pi/2 \sqrt{D / A.}$ If the outer surface is covered
by some material, for instance if the transmutation layer
is immersed into water or covered by lead the critical value
l is reduced. This is due to the back scattering by water
or lead and also to the fact that the neutrons are slowed
down in the water and their mean free path is thereby reduced.

It is important to prevent neutrons from escaping out
of the interior of the inner surface of the spherical layer
and also from being absorbed in the interior. If the initial radiation is generated by apparatus placed into the
interior of the sphere the material used should be so
selected as to lead to a minimum of absorption.

If the thickness is larger than the critical value l
we can produce an explosion.

9.

We shall now discuss the composition of the matter in which the chain reaction is to be maintained. We wish to distinguish three main types of chains.

(a) Pure neutron chains, in which the links of the chain are formed by neutrons of the mass number 1 alone. Such chains are only possible in the presence of a metastable element. A metastable element is an element the mass of which (packing fraction) is sufficiently high to allow its disintergration into its parts under liberation of energy. Elements like uranium and thorium are examples of such (metastable) elements; these two elements reveal their metastable nature by emitting alpha particles. Other elements may be metastable without revealing their nature in this way. Whether an element is metastable or not can be determined by means of the mass spectrograph. If, for instance, the value obtained by Bainbridge for beryllium by means of the mass spectrograph which appears to be generally accepted at present is really valid, we have to conclude that beryllium is a metastable element and can disintergrate into parts with the liberation of energy, one of the parts set free in its disintergration being a neutron. ~~Centraiiskgjskfjdksidkjgksiskkn dksifkdjfkkisdfjskenenx xnfjdksiajsidkfjdksiajsxkxjk~~

If we have an element which is metastable but the disintergration of which is inhibited and if this inhibition can be lifted in a collision with a neutron we shall call such an element an inhibited metastable element. ~~Ifxnsx~~ If an inhibited metastable element "A" is exposed to neutrons, we may have the following reaction.

"A" + n = "B" + n + Energy

The element "A" transmutes into an element "B" which has the same atomic number and mass number and energy is transmitted to the neutron. The element "B" may break up into an element "C" and a neutron, the element "C" having the same atomic number as "B". The element "C" may or may not break up into further ~~again~~ parts. If the interaction of a neutron with the element "A" leads in this way to an increase in the number of neutrons (the newly created neutrons would together with the original neutrons continue to interact with the elements "A" thereby forming the links of a chain ~~sking~~ reaction.) We shall call element "A" a metastable multiplicator.

10.

All particles which have a mass ~~number~~ approximately equal to the
mass of the proton ~~can~~ or a multiple thereof can play a similar role
as the neutrons provided they carry no charge or a negative charge ~~but~~
~~idiskxighxifjxixjjxxx~~ and we shall call allthose particles heavy non-
positive particles. Protons,deutons and other positive particles can
not be used as links of chain reactions. All pure chains in which one
and the same heavy non-positive particles forms the links of the chain
must necessarily make use of inhibited metastable elements. The simplest
non-positive heavy particles apart from the neutron are the neutron with
the
x/mass number 2 and the negative proton.

(b) Chains in which two different types of heavy non-positive particles ~~xkxigxaxx~~ alternate. Such chains need not necessarily make use
of metastable elements. For instance, an element "D" may be so chosen
that when interacting with a neutron (of mass number 1) a double-neutron
(a particle with the atomic number 0 and the mass number 2) should arise
and when interacting with a double-neutron a neutron should arise, and
that these reactions should have a positive energy balance. If such
a chain could be maintained in deuterium each link would liberate between 4 to 5 m.e.v.

$$\left. \begin{array}{l} H\ (2) + n\quad (1)\ =\ n\ (2)\ +\ H\ (1) \\ H\ (2) + n\quad (2)\ =\ H\ (3)\ +\ n\ (1) \end{array} \right\}\ 4.5.\ m.e.v.$$

Such a chain can also be maintained in mixtures of two different elements "E" and "F" which show the reaction:

$$"F\ (m)" + n\ (1)\ =\ n\ (2)\ +\ "G\ (m-1)"$$

$$"E\ (k)" + n\ (2)\ =\ "K\ (k\ 1)"\ +\ n\ (1)$$

We shall call an element "F" which reacts with an heavy non-positive
particle and ~~xxffixx~~ transmutes into an element the mass number of
which is one less,a converter element. An element "E" which reacts
                                                      an
with a heavy non-positive particle and transmutes into/element the mass
numbe r of which is increased by one, reducer element. Most elements
which yield protons when bombarded by deutons can be used as reducer
elements. Beryllium can act as a converter element. In order to have
a chain action in which the number of neutrons increases we must have
apart from the converter and the reducer a multiplicator element which
either splits up double-neutrons into two neutrons or from which neutrons
are liberated in a process in which the interacting non-positive heavy
particle is not captured.

11.

"**C**"- Chain reactions in which a heavy non-positive particle, for instance, a neutron, and a sygma quantum alternate.

Many
/elements which capture a neutron emit a radiation which carries away the energy liberated in the capture process. While the nature of this radiation is not yet established beyond doubt (a large fraction of it may consist in a gamma radiation) it can be shown from the ~~fixd~~ laws of thermo-dynamic equilibrium that this radiation,which we shall call sygma radiation, can liberate neutrons from elements and the cross section of this process (which is the inverse process of the capture) can be calculated. Some elements emit two sygma quanta if they capture a neutron and can act therefore as multiplicators in a chain reaction. If we have a mixture of elements (even pure elements have to be considered as mixtures of their isotpes) we can choose the components of the mixture so that one ~~xiswed~~ of the components "K" captures neutrons and emits two sygma quanta of the energies E_1 and E_2 another component "L" absorbs the quanta of energy E_1 and emits neutrons which are again captured by "K" and lead again to the iremission of sygma quanta; a third component "M" absorbs the quanta of energy E_2 and also emits neutrons which too will be captured by "K".

In order to have a large absorption coefficient for the sygma quanta we ought to choose such elements "L" and "M" which have a ~~new~~ ~~kffx~~ resonance for quanta of the energies E_1 and E_2 respectively. Similarly "K" may have resonance for the capture of the neutrons liberated by the sygma radiations from the other components or else ~~tknin~~ ~~kinn~~ the neutrons may be slowed down ~~by~~ for instance by elastic collisions in hydrogen and "K" may have resonance for the neutrons which have been slowed down to its resonance level (which may be at zero energy).

One possible combination of the resonance levels is that "K","L" and "M" have all resonances at zero energy of the neutron ("K" for capture, "L" and "M" for liberation of neutrons) and that E_1 and E_2 are slightly above the binding energies of the neutron in "L" and "M" respectively. Examples for elements which have a resonance for the capture of
[and the like.]
neutrons at zero energy are cadmium, mercury ~~et cetera~~. Other such elements, like for instance - rhodium, transmute into a radioactive element if they capture a neutron. If a radioactive element is formed one of the two sygma (gamma) quanta may be emitted with a considerable time lag,corresponding to the half life period of the radioactive element.

12.

By maintaining a chain/action in combination with means for leading

away and utilizing the heat set free in the transmutation process energy

can be produced and utilized for power production. In the same way,

by heating up deuterium by means of an electrical discharge as described

in that part of this specification which relates to Figures 3 and 4

in combination with means for leading away and utilizing the heat set

free in the transmutation process energy can be produced and utilized

for power production.

-13- [Having now particular described + ascertained the nature of my
Claims. said invention, and in what manner the same is to be
performed, I declare that what I claim is:]

1) Method and apparatus for the generation of radioactive elements,

characterized by a neutron radiation, emitted from a space in which

a nuclear transmutation process leading to the liberation of neutrons,

is maintained, and by the exposure of an element to the said neutron radi

ation which element transmutes into a radioactive element under the in-

fluence of the said neutron radiation.

2) Method and apparatus according to claim 1 characterized by the said

nuclear transmutation process leading to the liberation of neutrons

being a nuclear reaction unknown of diplogen (deuterium) with diplogen

or other light elements, or other light elements with eachother.

3) Method and apparatus according to claim 2 characterized by the said

nuclear reaction between light elements,being maintained through the

action of fast light ions,generated by an electrical device, for instance

high-voltage canal ray tube, on a target containing light elements; for

instance through the action of fixi diplogen in canal rays on a target

containing diplogen.

4) Method and apparatus according to claim 2 characterized by the said

nuclear reaction between light elements being enforced by means of

heating up suddenly a space which contains diplogen or other light

elements through an electrical discharge in which energy,which has been

stored,is suddenly released.

5) Method and apparatus according to claim 1 characterized by the a

nuclear transmutation process leading to the liberation of neutrons

being maintained through the action of x-rays, generated for instance

by means of a high voltage electron tube, on suitable elements,for in-

stance Beryllium.

6) Method and apparatus according to claim 1 characterized by a nuclear

transmutation process leading to the liberation of neutrons being main-

tained through the action of cathode rays, generated for instance by

means of a high voltage electron tube, on suitable elements, for

instance Beryllium.

-14-

7) Method and apparatus for the generation of radioactive elements according to claim 1, characterized by the exposure of an element to the said neutron radiation, which element transmutes into its own radioactive isotope, in the form of a chemical compound, which is adapted for the chemical separation of the radioactive element from its non-radioactive isotope.

8) Methods and apparatus for the production of energy, characterized by a nuclear reaction between light elements being enforced by means of heating up suddenly a space which contains diplogen or other light ele - ments through an electrical discharge in which energy, which has been stored, is suddenly released.

9) Method and apparatus for the production of radioactive elements or energy, characterized by the generation of an initial radiation, for instance a neutron radiation and exposed to this radiation a body, so composed that a chain xkxkgkxixkg reaction of neutrons is maintained by the initial radiation.

10) Method and apparatus for the production of radioactive elements according to claim 9, characterized by the exposure of an element, which element transmutes into a radioactive element under the influence of neutrons, to the neutron radiation generated in the said body in which a xhx chain reaction of neutrons is maintained.

11) Method and apparatus for the production of radioactive elements or energy according to claim 9, characterized by the said body being so composed that a chain of heavy non-positive particles is maintained.

12) Method and apparatus for the production of radioactive elements or energy according to claim 9, characterized by the said body containing a converter element, a reducer element, and a multiplicator.

13) Method and apparatus for the production of radioactive elements or energy according to claim 9, characterized by the said body containing Beryllium.

-15-

14) Method and apparatus for the production of radioactive elements or
energy according to claim 9, characterized by the said body being so
composed that a chain reaction in which neutrons and sygma particles
alternate, is maintained.

15) Method and apparatus for the production of radioactive elements or
energy according to claim 9, characterized by the said body containing
an element "K" which emits more than one sygma quanta for each cap -
tured neutron and one or more elements "L", "M" ..... which absorb
strongly the sygma quanta emitted by "K" and eject neutrons in doing so.

16) Method and apparatus according to claims 9 and 10, characterized by
the use of a hydrogen containing substance, for instance water, for
scattering the neutrons, for instance by surrounding the whole space
in which transmutation takes place, by water.

[17). A method of and apparatus for the transmutation of chemical elements substantially as hereinbefore described and illustrated in the accompanying drawings   Dated the   day of   1935]

PATENT SPECIFICATION

Application Date: March 12, 1934.   No. 7840/34.

„   „   July 4, 1934.   No. 33540/35.

(No. 33540/35 being divided out of Application No. 19721/34.)

Application Date: Sept. 20, 1934.   No. 27050/34.

One Complete Specification Left: April 9, 1935.

(Under Section 16 of the Patents and Designs Acts, 1907 to 1932.)

Specification Accepted: Dec. 12, 1935.

440,023

PROVISIONAL SPECIFICATION

No. 7840 A.D. 1934.

Improvements in or relating to the Transmutation of Chemical Elements

I, LEO SZILARD, a citizen of Germany and Hungary, c/o Claremont Haynes & Co., Vernon House, Bloomsbury, Square, London, W.C.1, do hereby declare the nature of this invention to be as follows:—

It has been demonstrated that if atoms or nuclei, e.g. hydrogen atoms (or protons), heavy hydrogen atoms, referred to from now onwards as diplogen, (or diplogen ions, referred to from now onwards as diplons) etc. are shot at chemical elements, a definite fraction of these shooting particles will cause transmutation in many elements. (How large this fraction is will depend on the nature of the element, the nature of the shooting particle, and its velocity.) If one uses the above mentioned particles and shoots them on light or heavy hydrogen lithium (6) or lithium (7) or other elements a certain proportion of the particles lose their energy through ionizing the substance through which they are shot, and only a fraction of the shooting particles will meet a nucleus of the substance before losing so much energy that the shooting particle is unable to cause transmutation in nuclei which it meets. Of these particles which meet a nucleus in their path (while still being in possession of a sufficiently large fraction of their initial energy) again only a further fraction will be able to penetrate the nucleus, (will be able to cause a transmutation); if the shooting particles are positively charged they are repulsed by the positively charged nucleus, and the probability of their penetrating the nucleus is a function of their relative velocity.

This probability rises rapidly with increasing velocity of the shooting particle and eventually reaches unity at a velocity which depends both on the nature of the shooting particle and the nature of the bombarded element.

However, even if this probability is equal to unity one still has to face the fact that a shooting particle has to travel for instance in air a large distance in order to encounter a nuclear collision (which may cause transmutation), but due to the energy loss which it suffers through ionizing the air its range is comparatively small if its initial velocity corresponds to several million volts energy. Only a fraction of the above mentioned shooting particles can therefore produce transmutation if shot into air or other substances or similar characteristics concerning ionization losses and nuclear collisions.

In accordance with the present invention radio-active bodies are generated by bombarding suitable elements with neutrons, which can be produced in various ways.

In accordance with one feature of the present invention nuclear transmutation leading to the liberation of neutrons and of energy may be brought about by heating up a small area filled with suitable elements very suddenly to high temperature by means of an electric discharge.

RADIO ACTIVE SUBSTANCES.

It is possible to produce elements capable of spontaneous transmutation by bombarding certain elements with fast charged nuclei, for instance by bombarding carbon with protons or aluminium, boron and magnesium with helium ions (particles). However, most of the radio active elements produced by the bombardment of these light elements with protons or alpha particles have a short existence (they disintegrate spontaneously in a time shorter than a few hours to half their amount), and it is not possible to use these charged nuclei for the transmutation of the heavier elements with good efficiency as the ionization loss gets too large. It is, however, possible to produce with good efficiency (both from lighter and heavier elements) radio active sub-

[Price 1/-]

2   440,023

stances a quantity of which decomposes to one half of the original amount in a period of time exceeding 24 hours, if a thick layer of substance is exposed to a

5 penetrating radiation which is emitted when collisions between heavy hydrogen (diplogen) atoms (or nuclei) and light elements including heavy hydrogen, (itself diplogen) are produced.

10   In the accompanying drawings Figure 1 shows an example of a suitable arrangement. 11 is an electrical discharge tube ejecting a beam 12 of fast diplogen ions. The ions fall on a substance 13 consisting

15 for instance of gasous diplogen or a diplogen compound or lithium, causing transmutation. The substance 13 is surrounded by a layer 14 of about 1 metre thickness and an average density above 1

20 gram per cubic cm.
      If a substance X which is mixed out of compounds of a large number of elements in such a way that it should contain in about equal molecular quantity (the

25 number of grams proportional to the atomic weight), all the elements (either free or in compounds) of the periodical system, with the exception of the rare gases, Rhenium, Masurium, and the radio

30 active elements (but including uranium, thorium and all the alkali elements) is exposed to the penetrating radiation of substance 13 (the acting agents of which radiation apparently are uncharged

35 nuclei the mass of which are about equal to the mass of a proton or multiples of the mass of a proton), the substance X contains (after the exposure to the radiation) spontaneously transmuting elements, the

40 time of disintegration of which is large as compared to the time of disintegration of most of the before mentioned radio active elements which can be produced by the direct bombardment of light elements

45 with fast charged particles.
      The efficiency can further be increased by measuring which of the 70 elements in substance X do not contribute substantially to the radio active element which

50 I want to produce and by eliminating those elements from the substance out of which I compose the layer. Such a measurement can be easily carried out by means of methods that are well-known in

55 every physical laboratory.
      These are the some features of this method for the production of radio active substances which are disintegrating slowly:

60   Light elements are bombarded by each other, especially diplogen is brought into collision with other light elements or with diplogen itself. Uncharged particles of a mass of the order of magnitude of the

65 mass of a proton are emitted as a con-

sequence of the collisions between nuclei of light elements. Such uncharged nuclei penetrate even substances containing the heavier elements without ionization losses and cause the formation of radio active 70 substances in a layer which is exposed to them with good efficiency if the thickness of the layer is of the order of magnitude of the mean free path of the neutrons for this process, or larger. I have therefore 75 two steps of good efficiency in series: the production of uncharged nuclei by collision between light elements (the ionization losses are small because the elements have a small atomic number and 80 therefore a small nuclear charge and a small number of electrons per atom), and the production of radio active substances by means of the uncharged nuclei (the ionization losses are practically absent 85 even in case of passage through heavy elements).

The production of such radio active substances with good efficiency might lead to a system of energy storage and such 90 substances can be used as accumulators for driving cars, vessels, aeroplanes and so on. They would be indeed accumulators of a very small weight.

METHODS AND APPARATUS FOR CAUSING 95
TRANSMUTATION.

The methods described hitherto were characterised by shooting a particle through matter which is at rest. As described, a diplon (a diplogen nucleus) 100 shot into diplogen at rest will in a large proportion of cases lose its energy by ionizing the diplogen and cause no transmutation in those cases. A further limitation of that method lies in the 105 difficulty of producing as large ion currents as would be needed for industrial purposes. The following method is free from these limitations:

If I were to maintain a very large con- 110 centration of energy in a space filled with atoms of such elements which will suffer transmutation, if the atoms (nuclei) strike each other at that temperature (which corresponds to the energy concen- 115 tration maintained) then the following would hold good: the energy transmitted to the electrons by the moving nuclei would be continuously retransmitted to the nuclei. It is sufficient to maintain a 120 suitable energy concentration for a fraction of a second. One can do so by shooting charged particles which have been accelerated in an electrical discharge tube through a space in which 125 diplogen alone, or lithium hydride (or other compounds of hydrogen and lithium) or other combinations of hydrogen or diplogen with a third light element are present. If I use an electric condenser to 130 store electrical energy and discharge this condenser in a fraction of a second across the discharge tube I can introduce (especially if I use several discharge tubes which are operated simultaneously) a very large energy in a very short period of time into the " transmutation space " filled e.g. with diplogen. As " heating rays " I can use protons or heavier ions or I can use cathode rays. I can easily estimate how much energy must be stored in the electric condenser in order to have a sufficient supply of energy to heat up sufficiently 1 cubic cm. of diplogen (an amount of diplogen that would fill 1 cubic cm. under ordinary pressure and temperature).

Fig. 2 shows an illustration of the method. 41 is a discharge tube, 42 an electrode in this discharge tube, 43 a thin metal window hermetically sealing the vacuum in the discharge tube and allowing the passage of charged particles from the discharge tube into the interior of the vessel 44. A number of other discharge tubes similar to tube 41 can be placed around the vessel 44 in a position similar to that of 41, and all these tubes can be operated simultaneously. Each tube may have a separate set of electrical condensers and all these condensers may be discharged across the corresponding discharge tube simultaneously by using an electric impulse to bridge all the spark gaps simultaneously. Fig. 3 shows how the tube 41 is connected through a spark gap 51 to a condenser which is fed by direct current through the chokes 53 and 54.

The corpuscular rays of all the discharge tubes are focused on a small area 45 the " transmutation area " the volume of which is only a small fraction, e.g. of the order of one millionth to one thousandth of a millionth of the volume of the vessel 44. The ratio of the two volumes to be chosen will depend on the density of matter in the transmutation area; one can e.g. replace 1 cubic cm. of diplogen gas (having a density corresponding to ordinary pressure and temperature) by a solid compound of diplogen which will hold the same amount of diplogen in a few cubic millimetres. In the interior of the vessel 44 a low pressure is maintained but sufficient gas is left (if cathode rays are used) to compensate the negative space charge during the discharge; to the same effect electric fields can be maintained between the conductors 46 and 47 which will remove part of the secondary electrons produced by the cathode rays during the discharge.

The duration of the discharge can be chosen so as to suit the purpose and may be anything between 10^{-5} sec. and 10^{-7} sec. If I had for instance 1 cubic cm. of diplogen gas enclosed in a small container having very thin walls in the transmutation area 45 and if I had a pressure of one atmos. before the discharge I shall have an enormous pressure after the discharge which would burst not only the thin walls of the container but would burst much thicker walls. However, the displacements during the explosion are proportionate to the square of the time that has passed since the starting of the explosion, and it is therefore possible to maintain for a short period of time a sufficiently high density of the matter (e.g. of the diplogen gas) in the transmutation area, if the " heating up " of this area is brought about by the corpuscular rays in a sufficiently short time. It is important to maintain a sufficiently high density in order to have a sufficient number of collisions (to have a sufficient chance for a collision causing transmutation) between the nuclei while the energy concentration is high in the transmutation area. From this point of view it is an advantage to start with a very high density such as one gets if one uses liquid or solid compounds of diplogen (or liquid or solid forms of the other light elements that one may use). If the " heating up " occurs in a short time the heat losses through radiation and conduction are tolerable. As the vessel 44 has a very large volume the rise of pressure, in it after the explosion, is tolerable.

One possible way of heating up electrically a definite area is by means of fast particles which have been accelerated electrically (while being electrically charged) in vacuum. The fast particles represent energies corresponding to 100,000 to ten million volts, and could theoretically heat up the transmutation area to anything between 100,000 to one billion degrees centigrade. An essential feature of a suitable method is to produce the heat in a very short time by applying to a discharge tube, adapted to eject the said fast particles. a high voltage for a very short time. This can be effected by storing the energy for the discharge mechanically, magnetically or by means of an electric condenser.

If an electric condenser is used it is possible to apply a very high voltage and thereby to store a very large energy by loading the condenser in a very short time, e.g. 10^{-4} to 10^{-5} sec. and discharging the condenser through the discharge tube in a time shorter than 10^{-5} sec. For a time of 10^{-4} to 10^{-5} sec. a condenser built for the purpose is able to stand a much higher voltage than it

4   440,023

would be able to stand for a longer time.
The following are some features of the described processes.
1. The heating up of an area by means of rays produced by the acceleration of charged particles in vacuum.
2. High voltage and large currents to be applied for the production of the said rays in order to heat up the said area in an extremely short time.
3. Diplogen or the combinations of: diplogen and lithium, or hydrogen and lithium, or hydrogen and other light elements, or diplogen and other light elements in the said area, or other combinations of two light elements.
4. The selection of the size of the radiated area and the density of matter in it, so that the charged particles pro-

duced in the transmutation process should get absorbed within the area to that extent that their contribution to the heating up of the area should be substantial.
A thick sheet of substance, for instance a combination of heavy elements exposed to the penetrating radiation (neutrons) of the first or second transmutation area having such elements in the sheet as will transmute into radio-active elements.
In connection with point 2 the use of means for storing energy for the purpose of discharging a large amount of energy all of a sudden in a discharge tube; for instance such means as electric condensers.

Dated the 10th day of March, 1934.
LEO SZILARD.

PROVISIONAL SPECIFICATION
No. 33540 A.D. 1935.

Improvements in or relating to the Transmutation of Chemical Elements

I, LEO SZILARD, Citizen of Germany and Subject of Hungary, c/o Claremont Haynes & Co., of Vernon House, Bloomsbury Square, London, W.C.1, do hereby declare the nature of this invention to be as follows:—
The invention relates to a process and to apparatus adapted for the generation of radio-active bodies by means of neutrons.
In accordance with the present invention neutrons are generated through the action of X-rays on matter or through the direct action of fast cathode rays on matter.
Neutrons are liberated from some elements, for instance Beryllium, if they are exposed in an electric discharge to the

action of electrons. For instance if we expose them to the action of cathode rays of a couple of million volts neutrons are liberated from Beryllium.
Instead of exposing the substance which I wish to transmute to the direct action of the electron I shall in some cases expose it with advantage to the action of the X-rays which are generated if electrons travel through matter especially through heavy elements like Bi, Pb, Hg, Th, U etc.

Dated the 4th day of July, 1934.
CLAREMONT HAYNES & CO.,
Vernon House, Bloomsbury Square,
W.C.,
Applicant's Solicitors.

PROVISIONAL SPECIFICATION
No. 27050 A.D. 1934.

Improvements in or relating to the Transmutation of Chemical Elements

I, LEO SZILARD, a citizen of Germany, and a subject of Hungary, c/o Claremont Haynes & Co., of Vernon House, Bloomsbury Square, London, W.C.1, do hereby declare the nature of this invention to be as follows:—
This invention has for its object the production of radio active bodies and the storage of energy through the production of such bodies.

In accordance with the present invention neutrons are liberated from matter by X-rays or fast cathode rays, for instance from Beryllium.
In accordance with one feature of this invention radio active bodies are generated by means of these neutrons.
In accordance with another feature of this invention the radio active element which is produced by means of neutrons

440,023   5

which are generated in this or in some other way is separated from its bombarded isotope by being produced through the irradiation of a chemical compound of the element which chemical compound does not exchange the atoms of this element within the compound against the atoms of the same element outside the compound.

Radio-active bodies can be generated by bombarding certain elements with neutrons from a Radon-α-Ray Beryllium source. The process involved is of different type in different elements. Some elements transmute when bombarded by such neutrons into a radio-active element which has an atomic number lower than the original element. I conclude that such a transmutation is not possible, if heavy elements are bombarded by slow neutrons. I conclude this for the following reason:—if in the primary process which leads to the formation of a radio-active atom, the neutron disappears and a proton or an alpha particle or another positive nucleus is ejected, one gets a radio-active atom, the chemical properties of which are different from the original atom as the atomic number of the new element is lower. I predict that such a process will not take place if the energy of the neutron and the atomic number of the bombarded element are such that the positive particles which were to be ejected could not penetrate against the Coulomb field in the inverse process. Therefore I predict that if I induce radio-activity by slow neutrons, (as for instance the neutrons generated by the bombardment of Diplogen by Diplogen canal rays of less than a million volt energy) in heavy elements, the new element will have no lower atomic number than the original element. Such induced activities in some heavy elements with slow neutrons have been observed to be processes of good efficiency. As a change in atomic number does not occur I conclude that there is a change in the mass number of the bombarded element, and I predict that in the known cases the change consists not in a decrease but in an increase of the mass number. I can further predict that a radiation subsequently called sygma radiation is emitted by such elements (i.e. by elements which show no chemical change, but show an induced activity having the same chemical properties as the original elements when bombarded by neutrons as for instance, iodine, arsenic, gold, bromine, rhodium, iridium etc.) when bombarded by neutrons. This sygma radiation carries away the excess energy liberated in the process of the capture of the bombarding neutron. A fraction of this sygma radiation consists of " hard X-rays," but I wish to leave it

open what portion of the sygma radiation consists of " hard X-rays " and this fraction might vary from element to element. I further predict that the sygma radiation can liberate neutrons from certain elements and also " hard X-rays " can do so. Further a similar action can be exercised by fast electrons in processes in which the fast electron is not captured. They can further produce radio-active bodies indirect (and this is most important) by ejecting from an element neutrons which generate radio-active bodies in the same element or in another element mixed with the first element, or exposed to the radiation of the first element.

Radio-active bodies can be generated according to this invention by the action of fast electrons or the action of hard X-rays which are emitted by matter if fast electrons fall on matter (especially if they fall on heavy elements like Bi, Pb, Hg, Th, U).

The said " hard X-rays " generate neutrons which re-act with the element itself and transmute it into a radio-active body. (The neutron may be ejected by the " hard X-rays " from one atom and be absorbed by another atom of the element). Whether this type of action is present can be seen by investigating the action of the said " hard X-rays " on thin sheets of the element; thin sheets will show no effect of this type.

Certain " mixtures " of two or more elements will show an action of the X-rays or cathode rays of the following type:—neutrons will be emitted by one of the components of the mixture under the influence of the " hard X-rays " and another component of the mixture will transmute into a radio-active body under the influence of the said neutral radiation.

Examples are combinations of an element like Beryllium which yields neutrons under the influence of " hard X-rays " with other elements, especially with such elements which show an induced activity when bombarded by neutrons without showing a chemical change in the process. Such elements can be selected either by radiating the elements mentioned above one by one with a Radon-α Ray Beryllium neutron source and investigating the chemical properties of the induced activity or alternatively by radiating the elements one by one with slow neutrons and selecting those elements which show an induced activity under the influence of slow neutrons. Slow neutrons can for the purpose of this investigation be generated by bombarding Fluorine with a Radon-α Ray source or by bombarding Diplogen with Diplogen canal rays). If I produce radioactive bodies by irradiating one of those

6   440,023

elements which, like iodine, transmute into their own radioactive isotope, it requires a special method chemically to separate the radioactive element from its irradiated isotope. According to the invention I can achieve such a separation by irradiating a chemical compound of the said element. Those atoms of our element which transmute into a radioactive atom are thrown out of the compound and will subsequently be called "free." If I choose a compound which in the circumstances does not interchange the atoms of our element bound within the compound with the "free" atoms which are their isotope, we can chemically separate the "free" atoms from the compound and thereby separate the radioactive isotope from the irradiated element. Compounds in which the element in which we are interested are bound direct to carbon are very often suitable. For instance in the case of iodine compounds like iodoform or ethyl iodide can be irradiated and after irradiation the radio-active isotope can be concentrated by separating the "free" iodine from the iodoform or the ethyl iodide. In order to protect radioactive iodine a small amount of ordinary iodine can be dissolved in the organic iodine compound before irradiation or after irradiation but before separation.

This principle of isotopic separation can also be applied if the element transmutes into a radioactive element which is chemically different from the bombarded element. Though in this case there is no inherent difficulty of chemically separating the radioactive element it may be convenient in certain cases to proceed as if the radioactive and bombarded element were isotopes.

An impulse generator is adapted to produce intermittent voltage up to 10 million volts, which can be transmitted to cathode of a discharge tube through the spark gap, the anode of which is connected to the earth. The fast electrons can emerge through the metal window (which is connected to the anode) and are hitting a body used as an anticathode which yields "hard X-rays" with very good efficiency if it is built of Bi, Pb or some other heavy element. If the anticathode is surrounded by a thick sheet containing some of the elements radioactive bodies are generated in this sheet.

The generation of radio-active bodies is due to different types of action, mentioned above.

A particularly efficient way of producing radioactive bodies by using high voltage electron-tubes is based on the method described above. A possible combination is to use beryllium as one component and another element, for instance iodine or arsenic or some other suitable element as the second component. It is of course not necessary actually to mix the two elements but only to expose the second element to the radiation which is emitted by the beryllium under the influence of the excitation of the beryllium by the X-rays or fast electrons produced by the high voltage electron tubes.

In Figure 1 of the accompanying drawing, 27 is the window of the high voltage tube through which the fast electrons are ejected. The electrons hit the rotating anticathode 30 which is covered with lead or tungsten (W.) 31. This anticathode is water-cooled, the water entering the rotating body through the axis 35. 52 is a beryllium block in which a space has been left for the rotating anticathode and for the path of the cathode rays 33 between the window 27 and the anticathode. This beryllium block may for instance have a size of 25 cm. × 25 cm. × 25 cm. The voltage used to operate the electron tube may be three million volt. The beryllium block is surrounded by a block 34 of the element which I wish to transmute into a radioactive element, for instance, iodine, or arsenic or any other element that is suitable. While it is advisable to use metallic beryllium in the block 52 the element in the block 34 may be present in the form of an organic compound in order to make an isotopic separation possible after irradiation. The dimensions of block 34 may for instance be 50 cm. × 50 cm. × 50 cm.

If heavy hydrogen or compounds of heavy hydrogen like lithium hydride or heavy water are used instead of beryllium it is advisable to use higher tensions for the generation of the X-rays in order to get a higher efficiency.

Since the output is within certain limits proportionate to the thickness of material which the X-rays have to travel in beryllium, and also to the thickness of material which the rays emitted by the beryllium have to travel in the material in which I wish to induce radioactivity it is essential to choose the dimensions both of the beryllium and of the material which we wish to activate. If I have only a limited amount of the material which I wish to activate (limited from the point of view of keeping down the capital investment and the waste of material which accompanies the chemical separation of the radioactive material) it may be advisable to reverse the arrangement shown in Figure 2 in the following way: 34 the outer block should be made of beryllium and 52 the inner block should be made of the material which I wish to activate.

440,023   7

Fast electrons have a similar action on beryllium as hard X-rays, a fraction of this action may be due to the generation of hard X-rays in the beryllium but another fraction may be due to the direct action of the fast electrons on the beryllium. In view of the fact that hard X-rays generate fast electrons in the beryllium, part of their action might be due to fast electrons. In any case I do not wish to differentiate here between the action of fast electrons and hard X-rays, and while I think that the direct action of hard X-rays on the beryllium plays the major part in the liberation of neutrons I wish to envisage the following modification of our method: the electrons of the discharge tube fall instead of lead on beryllium which can be put into the place of the lead coating 31 of the rotating anticathode 30 in Figure 2.

The material which one wishes to activate is not used up appreciably. For instance if I irradiate ethyl iodide with neutrons and separate the active iodine by removing it from the ethyl iodide through shaking the ethyl iodide with water in the presence of a reducing agent I can prevent the ethyl iodide from being mixed with the water and easily separate the water containing the active iodine from the ethyl iodide.

After separating the radioactive element one can build up chemical compounds especially organic compounds from it.

Some features of the invention are the generation of radioactive bodies by means of high voltage electron tubes and the chemical separation of the radioactive element based on the irradiation of the chemical compound of the element which does not exchange the atoms of this element within the compound against the atoms of the same element outside the compound. A further feature is the combination of an X-ray tube with beryllium and a second element which will transmute into a radioactive element under the influence of the rays emitted by beryllium which is exposed to the X-rays.

Dated this 20th day of September, 1934.
LEO SZILARD.

COMPLETE SPECIFICATION

Improvements in or relating to the Transmutation of Chemical Elements

I, LEO SZILARD, a citizen of Germany, and a subject of Hungary, c/o Claremont Haynes & Co., of Vernon House, Bloomsbury Square, London, W.C.1, do hereby declare the nature of this invention and in what manner the same is to be performed, to be particularly described and ascertained in and by the following statements: —

The invention concerns methods and apparatus for the production of nuclear transmutations leading to the generation of radio-active bodies.

It is not new to produce elements capable of spontaneous transmutation by bombarding certain elements with fast charged nuclei, for instance by bombarding carbon with protons or aluminium, boron and magnesium with helium ions (α particles). However, the radio-active elements produced by the bombardment of these light elements with protons or alpha particles, have mostly a short existence (they disintegrate spontaneously in a time shorter than a few hours to half their amount) and it is not possible to use these charged nuclei for the transmutation of the heavier elements with good efficiency as the ionisation loss gets too large.

According to this invention radioactive elements are generated by neutron radiation emitted from a space in which a nuclear transmutation process leading to the liberation of neutrons, is maintained, and by the exposure of an element to the said neutron radiation which element transmutes into a radioactive element under the influence of the said neutron radiation.

It is possible to produce with good efficiency (both from light and heavy elements) radio active substances a quantity of which decomposes to one half of the original amount in a period of time exceeding 24 hours, if a thick layer of substance is exposed to a penetrating radiation which is emitted when collisions between heavy hydrogen (diplogen) atoms (or nuclei) and light elements, including heavy hydrogen (diplogen, also called deuterium) itself, are produced.

In the accompanying drawings Fig. 1 shows an example of a suitable arrangement. 11 is an electrical discharge tube ejecting a beam 12 of fast diplogen ions. The ions fall on a substance 13 consisting of for instance gaseous diplogen or a diplogen compound or lithium, causing transmutation, i.e. a nuclear reaction of the

8   440,023

diplogen ion with an atom of the target. The substance 13 is surrounded by a layer 14 containing the element which we wish to transmute into a radio-active. element.
5 In order to have a good efficiency, the thickness of the layer 14 has to be sufficiently large, compared with the mean free path of the neutron.

Fig. 2 shows the electrical discharge tube
10 referred to in Fig. 1. It is a high voltage positive ray tube. There is an auxiliary positive ray tube on top of the high voltage tube. 11 is the anode, 15 the cathode of this auxiliary tube. Diplogen is ad-
15 mitted through the tube 13 and pumped away through 14.

The transmutation of elements into radio active bodies under the influence of neutrons can be demonstrated even before
20 one knows which elements will transmute into radio active bodies, if one prepares a mixture of all suitable elements leaving out the natural radio active elements, but including uranium and thorium (from
25 which the beta active products have been removed) and exposes the mixture to a neutron radiation. The mixture shows after exposure radio activity decaying with a large number of half-life periods
30 the relative intensity of which depends on the composition of the mixture and on the time of irradiation.

The following are the essential features of one method for the production of radio
35 active substances according to the present invention. Light elements are bombarded by each other, especially diplogen is brought into collision with other light elements or with diplogen itself. Un-
40 charged particles of a mass of the order of magnitude of the mass of a proton are emitted as a consequence of the collisions between nuclei of light elements. Such uncharged particles penetrate even sub-
45 stances containing the heavier elements without ionisation losses and cause the formation of radio active substances in a layer which is exposed to them with good efficiency if the thickness of the layer is
50 of the order of magnitude of the mean free path of the neutron or larger. I have therefore two steps of good efficiency in series: the production of uncharged nuclei by collision between light elements
55 (the ionization losses are small because the elements have a small atomic number and therefore a small nuclear charge and a small number of electrons per atom) and the production of radio active substances
60 by means of the uncharged nuclei (the ionization losses are practically absent even in case of passage through heavy elements) liberated in a nuclear reaction between light elements. The nuclear re-
65 actions between the light elements can be

brought about either by the bombardment of a target containing light elements with a beam of fast light atoms or by heating up a small space containing a light element.   70

The technique for carrying out this method described in connection with Fig. 2 was characterised by shooting a particle through matter which is at rest. As described a diplon (a diplogen nucleus) shot 75
into diplogen at rest will in a large proportion of cases lose its energy by ionizing the diplogen and cause no transmutation in those cases. Another technique for carrying out this method is described in 80
the following.

If one maintains a very large concentration of energy in a space filled with atoms of such elements which will suffer transmutation, if the atoms (nuclei) strike 85
each other at that temperature which corresponds to the energy concentration maintained then the following would hold good: the energy transmitted to the electrons by the moving nuclei would be con- 90
tinuously retransmitted to the nuclei. It is sufficient to maintain a suitable energy concentration for a fraction of a second. One can do so by shooting charged particles which have been accelerated in 95
an electrical discharge tube through a space in which diplogen alone, or lithium hydride (or other compounds of hydrogen and lithium) or other combinations of hydrogen or diplogen with a third light 100
element are present. If we use an electric condenser and discharge it in a fraction of a second across the discharge tube we can introduce (especially if we use several discharge tubes which are operated simul- 105
taneously) a very large energy in a very short period of time into the " transmutation space " filled, e.g. with diplogen. As " heating rays " we can use protons or heavier ions, or we can use cathode 110
rays. We can easily estimate how much energy must be stored in the electric condenser in order to have sufficient supply of energy to heat up 1 cubic cm. of diplogen.   115

The neutron radiation generated by bringing diplogen into nuclear reaction with itself through heating it up with an electric discharge can be utilised for the generation of radio active bodies as de- 120
scribed above.

Fig. 3 illustrates the technique of generating neutrons by heating up suddenly diplogen with an electric discharge. In Fig. 3 52 is an electrical con- 125
denser which is charged up through the chokes 53 and 54 to a high electrical potential. When the potential reaches a critical value the electrical energy stored in this condenser is suddenly discharged 130

through the spark gap 51 (which is connected to the cathode 42) and the discharge tube 41 which contains diplogen.

One possible form of the discharge tube 41 is shown in Fig. 4. In Fig. 4 the discharge tube 41 takes the form of a cathode ray tube. 52 is an electrical condenser which is charged up through the chokes 53 and 54 to a high electrical potential. When the potential reaches a critical value the electrical energy stored in this condenser is suddenly discharged through the spark gap 51 (which is connected to the cathode 42) and the cathode ray tube 41. A thin metal window hermetically seals off the vacuum in the cathode ray tube 41 but allows the passage of the cathode rays from the discharge tube 41 into the interior of a spherical container in the centre of which is placed a small container 44. The corpuscular rays generated by one or more such cathode ray tubes are focused on this small container 44 which may contain diplogen.

Another new method for the generation of radio active bodies is based on the fact that X-rays and also fast electrons can liberate neutrons from certain elements, for instance from beryllium. This is described in connection with Fig. 5 and Fig. 6.

Fig. 5 shows an arrangement suitable for the production of hard X-rays, in which 1 is the primary of a transformer, the secondary 2 of which is connected to the points 3 and 4. 3 is connected to the cathode 8 of the rectifier tube 5 and to the anode 7 of the rectifier tube 6. Point 4 is connected to the cathode 9 of the rectifier tube 10 and to the anode 11 of the rectifier tube 12. The cathode 13 and 14 are connected to each other and to earth. The anode 15 and 16 are connected to point 17, and this point is connected to the pole 18 of the impulse generator 20, the pole 19 of which is connected to earth. The impulse generator 20 is built of condensers 21, resistances 22 and spark gaps 23.

This impulse generator is adapted to produce intermittent voltage up to 10 million volts, transmitted to the discharge tube 24 through the spark gap 25. 26 is the cathode of the discharge tube, the anode 27 of which is connected to earth. The fast electrons emerge through the metal window 27 (which is the anode as well) and hit a body 28. This body is used as an anticathode and yields hard X-rays with very good efficiency if it is built of Bi, Pb or some other heavy element.

The generation of radio active bodies by means of neutrons which have been liberated from some suitable element by X-rays is shown in Fig. 6. In Fig. 6 27 is the window of the high voltage tube through which the fast electrons are ejected. The electrons hit the rotating anticathode 30 which is covered with lead or tungsten (W) 31. This anticathode is water-cooled, the water entering the rotating body through the axle 35. 52 is a beryllium block in which a space has been left for the rotating anticathode and for the path of the cathode rays 33 between the window 27 and the anticathode. This beryllium block may for instance have a size of 25cm. × 25cm. × 25cm. The voltage used to operate the electron tube may be three million volts. The beryllium block is surrounded by a block 34 of the element which we wish to transmute into a radio-active element. For instance, iodine or arsenic or any other element that is suitable. While it is advisable to use metallic beryllium in the block 52 the element in the block 34 may be present in the form of a suitable chemical compound in order to make an isotopic separation possible after irradiation. The dimensions of block 34 may, for instance, be 50cm. × 50cm. × 50cm.

Fast electrons have a similar action on beryllium to hard X-rays, a fraction of this action may be due to the direct action of the fast electrons on the beryllium. In view of the fact that hard X-rays generate fast electrons in the beryllium, part of their action can be due to fast electrons. In any case, I do not wish to differentiate here between the action of fast electrons and hard X-rays, and while I think it likely that the direct action of hard X-rays on the beryllium plays the major part in the liberation of neutrons, I wish to envisage the following modification of my method: The electrons of the discharge tube fall instead of on lead on beryllium which can be put into the place of the lead coating 31 of the rotating anticathode 30 in Fig. 6.

Many elements transmute when bombarded by slow neutrons into their own radio-active isotope and it requires a special method chemically to separate the radio-active element from its irradiated isotope. One can achieve such a separation by irradiating a suitable chemical compound of the said element. Those atoms of this element which transmute into a radio active atom are thrown out of the compound and will subsequently be called " free." If one chooses a compound which in the circumstances does not interchange the atoms of this element bound within the compound with the " free " atoms which are their isotopes one can chemically separate the " free " atoms from the compound and thereby separate the

10   440,023

radio active isotope from the irradiated element. Compounds in which the element, the radio active isotope of which one wants to separate, is bound direct to carbon are often very suitable. For instance, in the case of iodine compounds like iodoform or ethyl iodide can be irradiated and after irradiation the radioactive isotope can be concentrated by separating the " free " iodine from the iodoform or the ethyl iodide. In order to protect radio-active iodine a small amount of ordinary iodine can be dissolved in the organic iodine compound before irradiation, or after irradiation but before separation.

Having now particularly described and ascertained the nature of my said invention, and in what manner the same is to be performed, I declare that what I claim is:—

1. A method for the generation of radioactive elements, characterized by a neutron radiation, emitted from a space in which a nuclear transmutation process leading to the liberation of neutrons, is maintained, and by the exposure of an element to the said neutron radiation which element transmutes into a radioactive element under the influence of the said neutron radiation.

2. A method according to Claim 1 characterised by the said nuclear transmutation process leading to the liberation of neutrons being a nuclear reaction of diplogen (deuterium) with diplogen or other light elements, or other light elements with each other.

3. A method according to Claim 2 characterized by the said nuclear reaction between light elements, being maintained through the action of fast light ions, generated by an electrical device, for instance a high-voltage canal ray tube, on a target containing light elements; for instance through the action of diplogen canal rays on a target containing diplogen.

4. A method according to Claim 2 characterized by the said nuclear reaction between light elements being enforced by means of heating up suddenly a space which contains diplogen or other light elements through an electrical discharge in which energy, which has been stored, is suddenly released.

5. A method according to Claim 1 characterized by a nuclear transmutation process leading to the liberation of neutrons being maintained through the action of X-rays, generated for instance by means of a high voltage electron tube, on suitable elements, for instance beryllium.

6. A method according to Claim 1 characterized by a nuclear transmutation process leading to the liberation of neutrons being maintained through the action of cathode rays, generated for instance by means of a high voltage electron tube, on suitable elements, for instance beryllium.

7. A method for the generation of radioactive elements according to Claim 1, characterized by the exposure of an element in the form of a chemical compound to the said neutron radiation, which element transmutes into its own radioactive isotope, the compound being chosen for the separation of the radioactive element from its non-radioactive isotope.

8. A method of generating radioactive elements substantially as hereinbefore described with reference to the accompanying drawings.

9. Apparatus when used for carrying out the methods claimed in any of the Claims 1 to 8, substantially as described with reference to the accompanying drawings.

Dated the 9th day of April, 1935.
CLAREMONT HAYES & CO.,
Vernon House, Sicilian Avenue,
Bloomsbury Square, W.C.1,
Applicant's Solicitors.

Leamington Spa: Printed for His Majesty's Stationery Office, by the Courier Press.—1936.

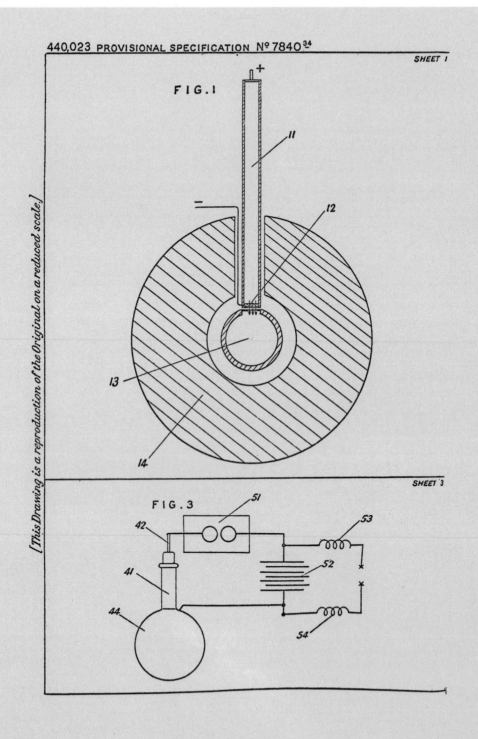

440,023 PROVISIONAL SPECIFICATION N° 7840^{34}

SHEET 1

FIG.1

+

11

−

12

13

14

SHEET 3

FIG.3

51

42

41

44

53

52

54

[This Drawing is a reproduction of the Original on a reduced scale.]

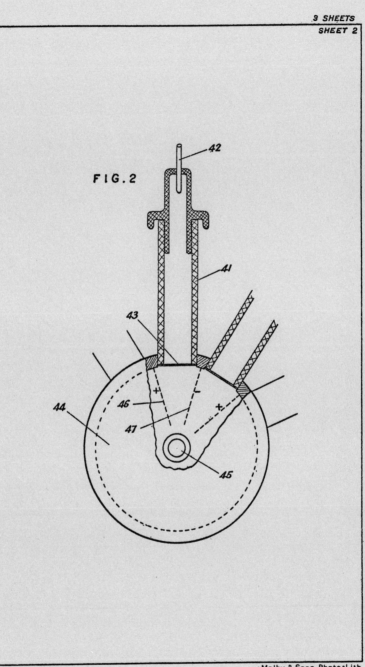

FIG.2

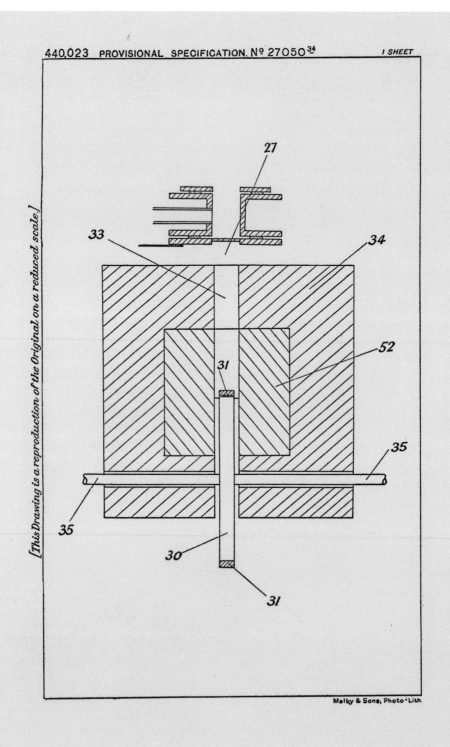

Malby & Sons, Photo-Lith.

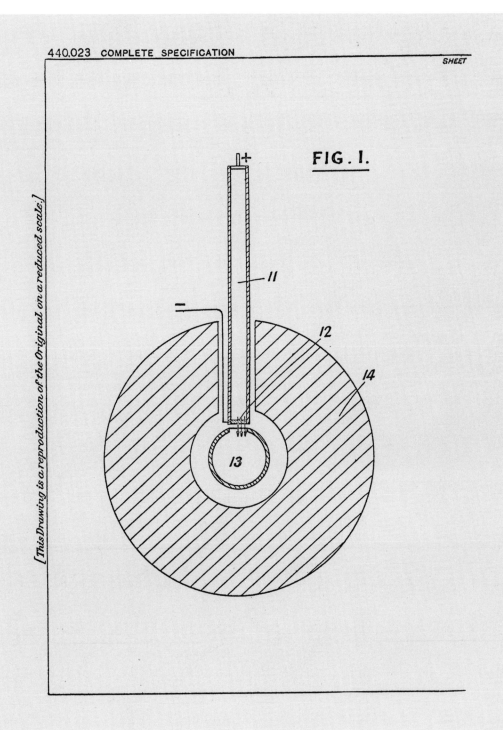

FIG. I.

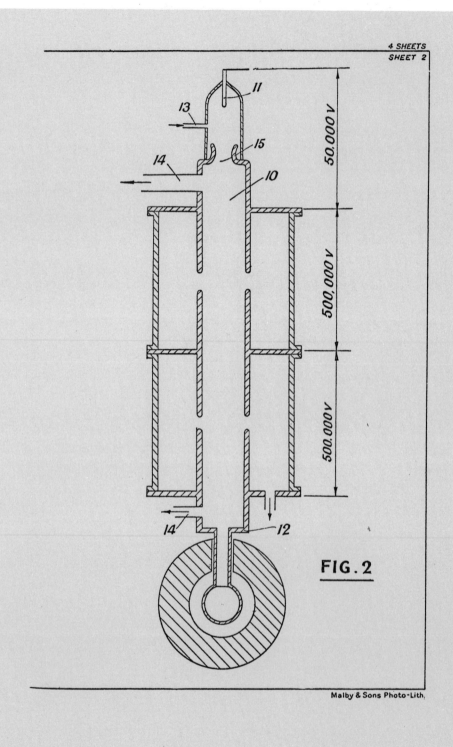

FIG. 2

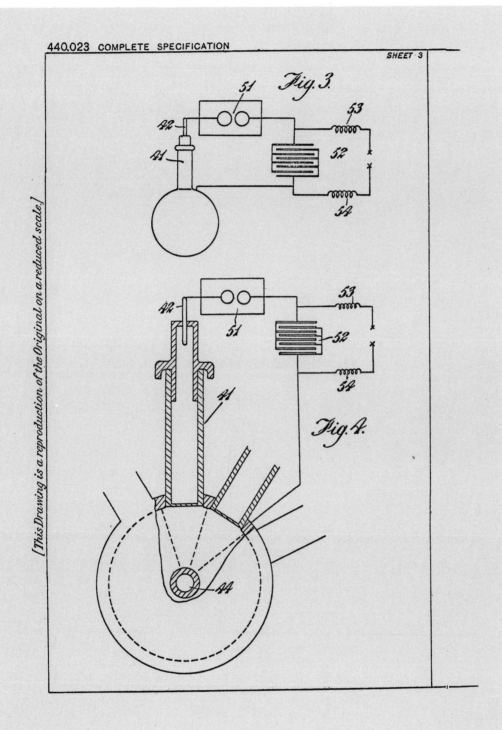

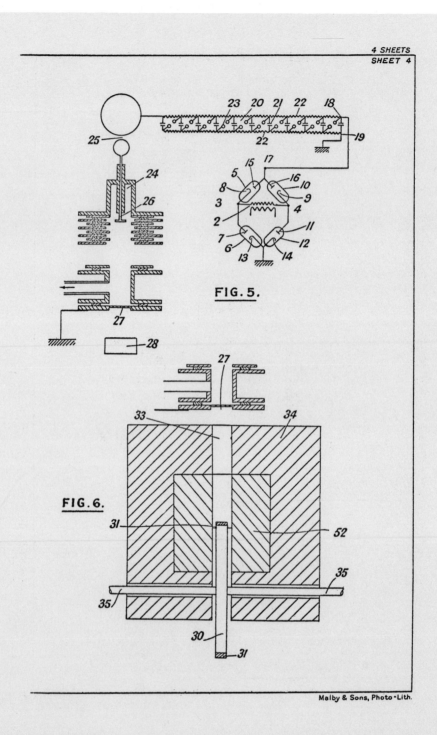

FIG. 5.

FIG. 6.

Malby & Sons, Photo-Lith.

PATENT SPECIFICATION

630,726

Application Date: June 28, 1934.   No. 19157/34.

„   „   July 4, 1934.   No. 19721/34.

One Complete Specification left (under Section 16 of the Patents and Designs Acts, 1907 to 1946): April 9, 1935.

Specification Accepted: March 30, 1936 (but withheld from publication under Section 30 of the Patent and Designs Acts 1907 to 1932)

Date of Publication: Sept. 28, 1949.

Index at acceptance: —**Class 39(iv),** P(1:2:3x).

PROVISIONAL SPECIFICATION
No. 19157 A.D. 1934.

Improvements in or relating to the Transmutation of Chemical Elements

I, LEO SZILARD, a citizen of Germany and subject of Hungary, c/o Claremont Haynes & Co., of Vernon House, Bloomsbury Square, London, W.C.1, do hereby
5 declare the nature of this invention to be as follows:—

This invention has for its object the production of radio active bodies the storage of energy through the production
10 of such bodies and the liberation of nuclear energy for power production and other purposes through nuclear transmutation.

In accordance with the present inven-
15 tion nuclear transmutation leading to the liberation of neutrons and of energy may be brought about by maintaining a chain reaction in which particles which carry no positive charge and the mass of which
20 is approximately equal to the proton mass or a multiple thereof form the links of the chain.

I shall call such particles in this specification " efficient particles."
25 A way of bringing about efficiently transmutation processes is to build up transmutation areas choosing the composition and the bulk of the material so as to make chain reactions efficient and
30 possible, the links of the chain being " efficient particles."

One example is the following. The chain transmutation contains an element C, and this element is so chosen that an
35 efficient particle x when reacting with C may produce an efficient particle y, and the efficient particle y when reacting with C may produce either an efficient particle x or another efficient particle which in its
40 turn is directly or indirectly when reacting with C capable of producing x. The bulk of the transmutation area, on the other hand, must be such that the linear dimensions of the area should sufficiently

[*Price 2/-*]

exceed the mean free path between two 45 successive transmutations within the chain. For long chains composed of, say, 100 links the linear dimensions must be about ten times the mean free path.

I shall call a chain reaction in which 50 two efficient particles of different mass number alternate a " doublet chain." An example for a doublet chain which is a neutron chain would be the following reaction, which might be set up in a mix- 55 ture of a " neutron reducer element " (like lithium (6) or boron (10) or preferably some heavy " reducer " element), and a " neutron converter element " which yields $n(2)$ when bombarded by 60 $n(1)$. An example for such a chain in which carbon acts as reducer and beryllium acts as converter would be the following:

$$C(12) + n(2) = C(13) + n(1)$$
$$Be(9) + n(1) = " Be(8) " + n(2)$$ 65

(" Be(8) " need not mean an existing element, it may break up spontaneously).

One can very much increase the efficiency of the hitherto mentioned 70 neutron chain reactions by having a " neutron multiplicator " O mixed with the elements which take part in the chain reaction. A neutron multiplicator is an element which either splits up $n(2)$ into 75 $n(1) + n(1)$ or an element which yields additional neutrons for instance $n(1)$ when bombarded by $n(1)$. A multiplicator need not be a meta-stable element. Beryllium may be a suitable multipli- 80 cator

$$Be(9) + n(1) = " Be(8) " + n(1) + n(1)$$

An efficient particle disappears (and a

2   630,726

chain is therefore interrupted if this happens in a chain reaction) if a neutron reacts with a nucleus in such a way that the neutron disappears and a positive 5 particle for instance a proton or an alpha particle is emitted. I can suppress the production of a positive particle when bombarding the element by neutrons by choosing the element and the neutron 10 energy so that the positive particle, the creation of which has a potential possibility, should not have sufficient energy at its disposal to penetrate in the inverse process the nucleus of that element. In 15 order to avoid such an occurrence in my chain reactions I shall use as reducers, converters and multiplicators the heaviest elements which are otherwise satisfactory.

In the accompanying drawings Figure 20 1 and 2 show one example for utilising neutron chains for power production and the generation of radio-active bodies. 101 is a high voltage positive ray tube generating fast light ions like diplons or 25 helium ions which cause by striking diplogen or beryllium in 102 the emission of a penetrating radiation (neutrons).The radiation emerging from 102 acts on the material 103 which forms a sphere

around 102. This material is such that a 30 chain reaction, preferably accompanied by the action of a multiplicator is released. For instance one can have a sphere 103 the dimensions of which are so chosen that the energy liberated in it should be a 35 multiple of the energy input. The pumps 120, 121 and 122 pump a liquid for instance water or mercury through the pipe systems 107, 110, 111 thereby cooling the transmutation area 103 and driving the 40 heated liquid through the boiler 126. The boiler supplies steam to a power plant. The neutrons emerging from the sphere 103 act on a layer 104 which is composed of an element T that will transmute into 45 a radio-active body which is suitable for the storage of energy. The element T need not be present as a free element but can preferably be present in the form of a compound soluble in water; that makes 50 it easier to separate the radio active bodies formed in the process. A third layer 105 contains an element V that will absorb the neutrons $|n(1)|$ under liberation of energy (Li). 106 is a heat insulating 55 layer.

Dated this 28th day of June, 1934.
                    LEO SZILARD.

PROVISIONAL SPECIFICATION
No. 19721 A.D. 1934.

Improvements in or relating to the Transmutation of Chemical Elements

I, LEO SZILARD, citizen of Germany and Subject of Hungary, c/o Claremont, Haynes & Co., of Vernon House, Blooms-60 bury Square, London, W.C.1, do hereby declare the nature of this invention to be as follows:—

The invention relates to a process and to apparatus adapted for power produc-65 tion, the storage of power through the generation of radio-active bodies, and the generation of radio-active bodies in general by means of the generation of neutrons (particles which carry no charge 70 and the mass of which is roughly equal to the mass of a proton or a multiple thereof). If I use the name " efficient particles " I mean neutrons or other particles (the mass of which is roughly equal to 75 the mass of the proton or a multiple thereof) which carry no positive charge and are efficient either because they can travel a long way through matter without being stopped like neutrons or they have 80 a shorter range but are able to react with a positive nucleus after having been stopped. I shall discus both the generation of efficient particles and also their use in chain reactions.

In accordance with the present inven-85 tion a chain reaction leading to the liberation of neutrons and of energy is maintained in a body, the geometrical proportions of which are so chosen that a good efficiency of the process be obtained, 90 through the introduction of an initial radiation, for instance a neutron radiation.

According to one feature of the invention such a neutron radiation is generated 95 through the action of X-rays on matter.

According to another feature of the invention such a neutron radiation is generated through the action of fast cathode rays on matter.   100

Neutrons are liberated from same elements, for instance beryllium, if they are exposed in an electric discharge to the action of electrons. For instance if we expose them to the action of cathode rays 105 of a couple of million volts neutrons are liberated from beryllium.

Instead of exposing the substance which I wish to transmute to the direct action of the electron I shall in some 110 cases expose it to the action of the penetrating radiation which is generated if electrons travel through matter especially through heavy elements like Bi, Pb, Hg, Th, U etc.

In the accompanying drawings.

5 Figure 1 shows an arrangement suitable for the production of fast electrons. 1 is the primary of a transformer, the secondary 2 of which is connected to the points 3 and 4. 3 is connected to the cathode 8 10 of the rectifier tube 5 and to the anode 7 of the rectifier tube 6. Point 4 is connected to the cathode 9 of the rectifier tube 10 and to the anode 11 of the rectifier tube 12. The cathodes 13 and 14 are con- 15 nected to each other and to the earth. The anodes 15 and 16 are connected to point 17, and this point is connected to the pole 18 of the impulse generator 20, the pole 19 of which is connected to earth. The 20 impulse generator 20 is built of condensers 21, resistances 22 and spark gaps 23.

This impulse generator is adapted to produce intermittant voltage up to 10 25 million volts, transmitted to the discharge tube 24 through the spark gap 25. 26 is the cathode of the discharge tube, the anode 27 of which is connected to the earth. The fast electrons emerge through 30 the metal window 27 (which is the anode as well) and are hitting a body 28.

Figure 2 shows how the radiation emitted by a body 28 (in Figure 1) which is exposed to the action of fast electrons 35 can act as the initial radiation for a chain reaction. In Figure 2, 1 is an electrical discharge tube which generates fast electrons. These electrons enter through the narrow tube 2 into the interior of, a 40 spherical layer 3 which is formed by a substance in which a chain reaction can be maintained, the links of the chain being efficient particles, in the presence of an initial radiation emitted by 28. The 45 tube 2 is evacuated and the electrons emerge from it through the window (a thin aluminium sheet) 4. The space 5 in the interior of the spherical layer 3 can be evacuated. If the voltage of the 50 cathode rays hitting 28 is sufficient to liberate neutrons from 28, (for instance if one uses voltages of about or over one million volt and uses diplogen or compounds of diplogen for instance a diplo- 55 gen-lithium compound to form the body 28) one gets a neutron radiation as initial radiation which can maintain a chain reaction in the layer 3. It is essential to prevent that neutrons should easily escape 60 from the space 5 through the discharge tube 1 and it may therefore be necessary to surround the whole discharge tube with a wall, the thickness of which may be, if necessary, several meters. If this wall 65 is built from a material containing heavy

elements which have a large cross-section for neutron collisions the thickness of the wall may be less than for a wall built of light elements.

If I use instead of a cathode ray tube a 70 tube which ejects fast diplons or helium ions I can also generate an initial radiation of neutrons if I expose to those ions a body 28 (in Figure 2) which is composed of diplogen or beryllium, and can 75 in certain cases prefer this is as an alternative solution.

An essentially different way of introducing the initial radiation into the chain reaction chamber is the arrangement 80 shown in Figure 3. 401 is the cathode ray tube described in Figure 1. 402 is a sheet of a heavy element for instance Pb, or U in which a penetrating radiation (hard X-rays) is generated with good efficiency if 85 the electrons have a voltage about or over one million volt. This efficiency increases very rapidly with the voltage, and is much higher than it could be expected from the experience based on ordinary X- 90 ray work. The thickness of the sheet 402 is such as to enable the generated penetrating radiation to penetrate through this sheet and act on the transmutation chamber 106 (in Figure 4). Nevertheless 95 the sheet can be sufficiently thick to utilise more than half of the energy of the cathode rays. The X-rays emerging from sheet 402 penetrate the layer 3 and can liberate efficient particles either from 100 the layer 3 or from a substance 407 placed in the interior of the layer 3. The heat liberated in 3 and 407 of Fig. 4 can be utilised as shown in Fig. 2, 3 and 407 forming the interior of the transmutation 105 chamber 106. These neutrons can then maintain a chain reaction as discussed further above and further below. The advantage of using X-rays as an initial radiation is the following: the X-rays 110 penetrate through a perfectly closed layer 3 into the interior of the layer and therefore a leak of neutrons from the interior can be avoided. This is specially important if one has to deal with a 115 neutron chain in which no multiplicator action is involved. In such cases X-rays may be used with advantage as initial radiation especially in view of the unexpectedly large efficiency of the X-ray 120 production by means of fast electrons acting on heavy elements.

Figure 2 shows features some of which are needed in neutron chains. The layer 3 contains some suitable elements to 125 maintain a chain. I refer to Figure 2 of my application 19157/34. in which 107, 108, and 109 form a tube system through which water or mercury is circulated by means of the pump 120. The liquid leav- 130

4   630,726

ing 109 is lead through a boiler 126 in the
tube system 123 and transmits its heat to
the boiler, the steam produced being used
for power production. Another tube
5 system 110 is operated by the pump 121
and is heated by the layer 9 composed of
a material which will transmute into a
radio-active body under the influence of
the radiation emerging from layer 3.
10 Pump 122 pumps liquid through 11
along the outer surface of the transmutation area and through the boiler 126
through the pipe system 125.
 If I have a chain reaction with a multi-
15 plicating action i.e. if the number of
efficient particles increases along the
chain I can reach very high efficiency for
the production of heat or radio-active
bodies. If I have a closed spherical layer
20 of material in which the chain reaction
takes place the inner radius (r_1) of which
is large as compared with the mean free
path (a) of the efficient particles which
maintain the chain, the density (s) of the
25 efficient particle will with good approximation be given as a function of the
radius (r) by the following equation:

$$D\ d(rs)/dr + A(rs) = 0$$

D and A are determined by: the mean
30 free path of the efficient particles a; the
mean velocity of the efficient particles w;
the factor of the multiplicating action f
which says how many collisions of an
efficient particle are needed in the average
35 in order to produce one new efficient
particle.

$$A = w/af; \quad D = aw\ 3; \sqrt{\frac{D}{A}} = \frac{a\sqrt{b}}{\sqrt{3}}$$

I am interested in the critical thickness
l of the spherical layer for which the
40 gradient of the density s vanishes. If the
thickness ($r_2 - r_1$) approaches l I can
maintain with a very weak source of
initial radiation in the interior of the
inner surface of the spherical layer a very
45 strong chain reaction and I can easily get
one thousand or more times more efficient
particles emerging from the chain reaction layer than the number of the efficient
particles forming the initial radiation. If
50 the outer surface ($r = r_2$) of the spherical
layer were to stand free in space the density s would be zero for that surface and
the critical value l_0 would be given by

$$l_0 = \pi/2 . \sqrt{D/A}.$$

55 If the outer surface is covered by some
material, for instance if the transmutation layer is immersed into the earth or

into water or covered by some cheap heavy
material for instance lead the critical
value l_0 is smaller. Accordingly one can 60
economise if an expensive material is
used to maintain the chain reaction in
the layer by covering that layer and reducing its thickness.
 It is important to prevent efficient par- 65
ticles from escaping out of the interior of
the inner surface of the spherical layer
and also from being absorbed in the interior. If the initial radiation is generated by apparatus placed into the interior 70
of the sphere the material used should be
so selected as to lead to a minimum of
absorption.
 If the thickness is larger than the
critical value l_0 I can produce an 75
explosion.
 The differential equation which I have
given for s ceases to be a good approximation if f is small for instance one or two
but gives a fairly good approximation if 80
f is large for instance one or two
hundreds.
 Some features of described processes
are:
 1. Production of heat or power or pro- 85
duction of radio-active bodies by causing
transmutation through exposing elements
or mixtures of elements to an electric discharge especially fast cathode rays. The
exposure of an element that will yield 90
when bombarded by electrons efficient
particles especially neutrons; beryllium
being an example.
 2. Transmutation as under 1 caused by
the penetrating radiation generated by 95
the action of fast electrons on heavy elements like Pb or U (X-rays).
 3. The maintainance of a chain reaction in a closed for instance spherical
layer, the initial radiation being gener- 100
ated according to 1 or 2 in such a way
in the interior of the spherical layer or
within the spherical layer itself that
efficient particles should not be able to
escape through an opening from within 105
the interior space surrounded by the
chain reaction layer.
 4. The chain reaction layer being surrounded by a large bulk of material which
is cheaper than the chain reaction 110
material. The surrounding material being a heavy element like lead or a light
element which does not absorb neutrons
and which does not convert them into
positive particles. 115
 5. The maintainance of chain reactions in a layer forming a closed body
for instance a sphere, the thickness of
the layer being slightly less than the
critical thickness. 120

Dated this 4th day of July, 1934.
LEO SZILARD.

630,726   5

COMPLETE SPECIFICATION

Improvements in or relating to the Transmutation of Chemical Elements

I, LEO SZILARD, a citizen of Germany, a subject of Hungary, C/o Claremont Haynes & Co., of Vernon House, Bloomsbury Square, London, W.C.1, do hereby
5 declare the nature of this invention and in what manner the same is to be performed, to be particularly described and ascertained in and by the following statement:—
10 The invention concerns methods and apparatus for the production of nuclear transmutations leading to the generation of radio-active bodies, to the liberation of nuclear energy and the utilisation of the
15 liberated energy.
According to this invention radio active elements or energy or both are generated by means of neutron isotopes produced by means of a chain reaction in
20 a body in which chain reaction neutron isotopes of differing mass number take part. (I have reason to believe that apart from neutrons which carry no charge and have a mass approximately equal to the
25 proton mass heavier isotopes of the neutron exist which particles carry no charge and has a mass number approximately equal to a multiple of the proton mass.)
30 The generation of radio active bodies by the neutron isotopes may be indirect i.e. these neutron isotopes may generate radiations which do not consist of neutron isotopes and which radiations produce
35 radio active elements.
There are several radiations arising out of chain reaction which may generate radio-active bodies, for instance, radiation consisting of neutrons of mass num-
40 ber 1; radiation consisting of neutrons of mass number higher than 1 (multiple neutrons) and gamma radiation. I wish to make it clear that methods and apparatus for the generation of radio-active bodies
45 by means of neutrons of mass number 1, without chain reactions, in itself is not claimed, and does not form part of the subject matter of this specification. It forms part of and is claimed in my speci-
50 fication number 440,023.
In the chain reactions to be described below, energy is liberated in the form of heat and can be utilized for power production by making use of the heat
55 liberated in the chain reaction. Through the generation of radio-active bodies energy is being stored and gradually liberated in the form radiations which can easily be transformed into heat which

heat can be utilised for power production. 60 Furthermore, the energy stored in the form of radio-active bodies can also be more directly utilised for the generation of electricity since radio-active bodies emit electrically charged particles and 65 thereby may directly generate electrical energy.
In the following I shall deal with methods and apparatus for the production of energy and the generation of radio- 70 active bodies by means of chain reactions. In order to maintain such a chain an initial radiation of neutrons may be generated, for instance by one of the methods described in my Specification Number 75 440,023. If the neutrons enter a space which has the proper shape and size and is filled with the proper combination of elements the energy or the number of the neutrons, or both, can be greatly in- 80 creased through their interaction with the substance which fills the chain reaction space. The interaction of a neutron with matter can lead to the liberation of a multiple neutron—this multiple neutron 85 liberates in its turn one or more neutrons of mass number 1, which in their turn liberate again multiple neutrons. In this way we can maintain a chain reaction in which a large number of neutrons and 90 multiple neutrons are liberated, the total number being determined by the geometry of the arrangement.
Figs. 1 and 2 show such a chain reaction apparatus. A neutron radiation, the 95 initial radiation, is generated by the high voltage canal ray tube 1, Fig. 1. This tube generates fast deuterons which strike the target 28 which contains deuterium. The neutron radiation emerging from 28 100 acts on the matter 3 which fills the spherical transmutation space. The composition of this matter 3 will be discussed further below and is such that a chain reaction is released by the neutrons. The 105 pumps 120, 121, and 122, Fig. 2 pump a liquid for instance water or mercury through the pipe systems 107, 110, 111, Figs. 1 and 2 thereby cooling the transmutation area 3. Fig. 1, and driving the 110 heated liquid through the boiler 126, Fig. 2. The boiler may supply steam to a power plant. The neutrons emerging from the sphere 3 act on a layer 9, Fig. 1 which is composed of an element that will 115 transmute into a radio-active body.
An essentially different way of introducing the initial radiation into the chain

reaction chamber is the arrangement shown in Fig. 3. 1 is the cathode ray tube 402 is a sheet of heavy element for instance Pb, or U in which a penetrating
5 radiation (hard X-rays) is generated with good efficiency if the electrons have a voltage of over one million volts. This efficiency increases very rapidly with the voltage, and is much higher than could be
10 expected from the experience based on ordinary X-ray work. The thickness of the sheet 402 is such as to enable the generated penetrating radiation to penetrate through this sheet and act on the
15 transmutation chamber consisting of the layers 407 and 3 (for the cooling of this chamber and the utilisation of the heat generated in it I refer to Fig. 2, 106 in Fig. 3 is to be identified with 106 in Fig.
20 2). Nevertheless the sheet can be sufficiently thick to utilize more than half the energy of the cathode rays. The X-rays emerging from sheet 402 penetrate the layer 3 and can liberate neutrons either
25 from the layer 3 or from a substance 407 placed in the interior of the layer 3. For instance, if beryllium is present in 407 or in 3, neutrons will be liberated by X-rays. These neutrons can then maintain a chain
30 reaction as discussed further above and further below. The advantage of using X-rays as an initial radiation is the following: The X-rays penetrate through a perfectly closed layer 3 into the interior
35 of the layer and therefore a leak of neutrons from the interior can be avoided.

I shall demonstrate in the following the importance of the shape and the size of
40 the transmutation space. I assume that the chain reaction takes place in a closed spherical layer of material the inner radius (r_1) of which is large compared with the mean free path (a) of the
45 neutrons (or other particles which are involved in maintaining the chain). In the simplest case the density (s) of the neutrons will with good approximation be given as a function of the radius (r) by
50 the following equation:

$$D.\ d(rs)/dr + A.\ (rs) = 0$$

D and A are determined by: the mean free path (a) of the neutrons; the mean velocity of the neutrons w; the factor of
55 the multplicating action f which says how many collisions of a neutron are needed in the average in order to produce one new neutron.

$$A = w/af;\quad D = aw/3;\ \sqrt{\frac{D}{A}} = \frac{a\sqrt{f}}{\sqrt{3}}$$

60 The critical thickness of the spherical layer for which the gradient of the density s vanishes for the internal radius (r_1) has a bearing on the present invention as will now be shown. If the thickness of the spherical layer $(r_2 - r_1)$ r_2 and r_1 are 65 the external and internal radii respectively) approaches a certain critical thickness L one can maintain with a very weak source of initial radiation in the interior of the inner surface of the spherical 70 layer a very strong chain reaction and one can easily get one thousand or more times more neutrons emerging from the chain reaction layer than the number of the neutrons forming the initial radia- 75 tion. If the outer surface $(r = r_2)$ of the spherical layer were to stand free in space the density s would be zero for that surface and the critical value L would be given by 80

$$L = \pi/2\sqrt{D/A}.$$

If the outer surface is covered by some material, for instance if the transmutation layer is immersed in water or other hydrogen containing substance or covered 85 by lead the critical value L is reduced. This is due to the back scattering by water or lead and also to the fact that the neutrons are slowed down in the water and their mean free path is thereby re- 90 duced.

It is important to prevent neutrons from escaping out of the interior of the inner surface of the spherical layer and also from being absorbed in the interior. 95 If the initial radiation is generated by apparatus placed in the interior of the sphere the material used should be so selected as to lead to a minimum of absorption. 100

If the thickness of the layer is larger than the critical thickness L the number of neutrons would go on increasing indefinitely and such an increase is only stopped when the heat which is liberated 105 in the process causes the spherical layer to explode.

The differential equation which I have given above and from which we have derived the value for the critical thick- 110 ness L does not give the correct description of the density of the neutrons in a chain reaction nor does it give the correct value for L. In order to get the correct equation we have obviously to distinguish 115 between the mean free path a of the neutron for a collision and its factor f which says how many collisions of a neutron are needed in the average in order to produce a multiple neutron 120 on the one hand, and on the other hand between the mean free path a_2 of the multiple neutron and its factors f_2 and f_3 of its multiplying action which says how many 125

collisions on the average of the multiple neutron are needed in order to produce one or two new neutrons respectively etc. The only purpose of putting down the
5 above simplified equations was to demonstrate the general type of behaviour of chain reactions with multiplying action and to show the existence of a critical thickness L. The simplified equation is
10 an approximation of the correct equation if many collisions of the neutron are needed to generate a multiple neutron but few collisions of the multiple neutron are needed to generate two neutrons.
15   I shall now discuss the composition of the matter in which the chain reaction is to be maintained. It is essential that two isotopes of the neutron should take part in the reaction in order to obtain a
20 chain. (Neutron isotopes are particles which have no charge and the mass of which is roughly equal to the mass of the proton or a multiple thereof. I have reason to believe that heavy neutron iso-
25 topes, the mass of which is approximately a multiple of the proton mass exist).
   A mixture of two elements " E " and " F " can be so chosen that element " F " (the converter element) when it re-
30 acts with a simple neutron should transmute into an element the mass number of which is lower and generate a multiple neutron; on the other hand element " E " (the reducer element) should when it re-
35 acts with a multiple neutron transmute into an element the mass number of which is increased and generate a simple neutron. In order to have a chain reaction in which the number of neutrons
40 increases it is necessary that apart from the converter and the reducer element there should be present a multiplicator element that is to say one from which neutrons are liberated by neutrons in a
45 process in which the interacting neutron is not captured or alternatively a multiplicator element which generates two neutrons from a multiple neutron.
   I wish to give the following indication
50 of which elements may be used as converter " F ": The fact that an element ejects a multiple neutron, for instance a tetra neutron (a neutron of mass number 4), when bombarded by simple neutrons,
55 can be revealed in certain circumstances by the fact that it becomes radio-active through neutron bombardment, and that the generated radio-active element is an isotope of the bombarded element itself.
60 For instance, if indium is bombarded by fast neutrons (of less than 8 M.E.V. energy, but more than 100,000 E.V. energy) a radio-active isotope of indium is generated, which decays with a $4\frac{1}{2}$ h.
65 period. This indicates that one stable in-

dium isotope captures a neutron, and a multiple neutron is ejected, leading to a radio-active indium isotope of mass number 112. A radio-active indium iso-
70 tope of mass number 112 arises if the stable indium isotope 115 captures the neutron and ejects a tetra neutron. Only very few elements will eject a tetra neutron when bombarded by very slow
75 neutrons. The number of elements which can eject a tetra neutron increases with the kinetic energy of the bombarding simple neutron. Not all the elements reveal this fact by an appreciable radio-
80 activity, therefore a more general method can be employed to investigate each element separately. This more general method is based on the detection of the ejected tetra neutron. The ejected tetra
85 neutron can be detected through the transmutation which it causes in various elements which are exposed to it. Such transmutations reveal their presence in two different ways; either through radio-
90 activity induced in the element which is exposed to the tetra neutron, or through the ejection of charged particles (proton or alpha-particle etc.), from the element which is exposed to the tetra neutron.
95 The ejection of such charged particles can be observed by means of an ionization chamber, a Wilson cloud chamber or a photographic plate which contains the element, which transmutes when exposed
100 to the tetra neutron.
   I further wish to give some indication as to which elements may be used as reducer element " E," from which a multiple neutron liberates a simple
105 neutron, and a multiplicator element, from which a multiple neutron liberates two simple neutrons.
   A lower limit for the mass of the tetra neutrons can be deduced from consider-
110 ing two radio-active elements, of which the lighter one arises from the heavier one, through two beta transformations and one alpha transformation. If the mass of the tetra neutron were smaller
115 than the mass differences of these two radio-active elements, the heavier elements would spontaneously have to eject the tetra neutron, and would thus spontaneously transmute into the lighter
120 element.
   By applying this consideration to the known radio-active elements, we obtain as a lower limit for the mass of the tetra neutron about 4.014. While the slow
125 neutron will eject a tetra neutron from only few elements, a tetra neutron having such a high mass will eject a neutron from most of the elements and will eject two neutrons from a number of elements.
130 In order to determine from which

8 630,726

elements it ejects two neutrons (multiplicator elements) we have to take each element in its turn, bombard it with tetra neutrons and either observe the number of simple neutrons which emerge, or observe the radio-activity induced in the bombarded element, and thereby identify the nature of this transmutation. Examples of multiplicator elements are beryllium, and certain heavy elements. Heavy multiplicator elements are as a rule preferable since they will emit no, or few, positively charged particles, and we can thereby avoid interruptions of the chain.

Other examples for elements from which neutrons can liberate multiple neutrons are uranium and bromine.

The value of the critical thickness " L " previously referred to, can be estimated for a spherically symmetrical body as follows: The mean free path for an elastic collision of the neutron is in many elements of the order of 5 cms. Every hundredth elastic collision may lead to the ejection of a tetra neutron, and every collision of the tetra neutron (mean free path of the order of 5 cms) may lead to the ejection of two simple neutrons. In these circumstances " L " will be of the order of magnitude of 50 cm.

By maintaining chain reaction in combination with means for leading away and utilizing the heat set free in the transmutation process energy can be produced and utilized for power production.

Having now particularly described and ascertained the nature of my said invention and in what manner the same is to be performed, I declare that what I claim is:—

1. A method of generating radio-active elements or energy or both by means of neutron isotopes produced by means of a chain reaction in a body in which chain reaction neutron isotopes of differing mass number take part.

2. A method according to Claim 1 characterised by a chain reaction in which a neutron of mass number 1 and a heavier neutron isotope take part.

3. A method according to Claim 1 or 2 characterised by the generation of an initial radiation which can consist of neutrons of mass number 1, and the exposure to this initial radiation of a body so composed that a chain reaction is caused by the initial radiation.

4. A method according to Claims 1, 2 or 3 characterised by the said body containing a converter element and a reducer element.

5. A method according to Claims 1, 2 or 3 characterised by the said body containing a converter element and a multiplicator element.

6. A method according to Claims 1, 2, or 3 characterised by the said body containing a converter, a reducer and a multiplicator element.

7. A method according to Claims 1, 2 or 3 characterised by the said body containing beryllium.

8. A method according to Claims 1, 2 or 3 characterised by the exposure of an element to the radiations generated in the said body which element transmutes into a radio-active element under the influence of the radiations generated by the chain reaction.

9. A method according to Claims 1, 2 or 3 characterised by the use of a hydrogen-containing substance, for instance water, for scattering the neutrons, for example by surrounding by water the whole body in which the transmutation takes place.

10. Improvements in or relation to the transmutation of chemical elements by means of a chain reaction as hereinbefore described and illustrated in the accompanying drawings.

11. An apparatus for carrying out the methods claimed in any of the Claims 1 to 9 substantially as hereinbefore described in the specification and shown in the accompanying drawings.

Dated the 9th day of April, 1935.
CLAREMONT HAYNES & CO.,
Vernon House, Sicilian Avenue,
Bloomsbury Square, W.C.,
Applicants Solicitors.

Leamington Spa: Printed for His Majesty's Stationery Office, by the Courier Press.—1949.
Published at The Patent Office, 25, Southampton Buildings, London, W.C.2, from which copies, price 2s. 0d. each (inland) 2s. 1d. (abroad) may be obtained.

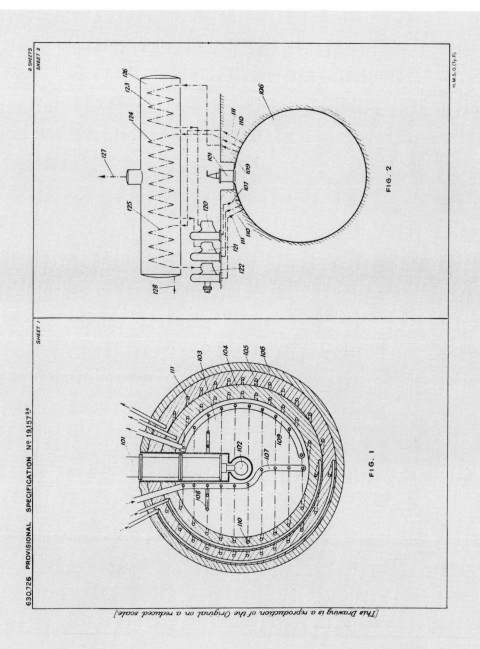

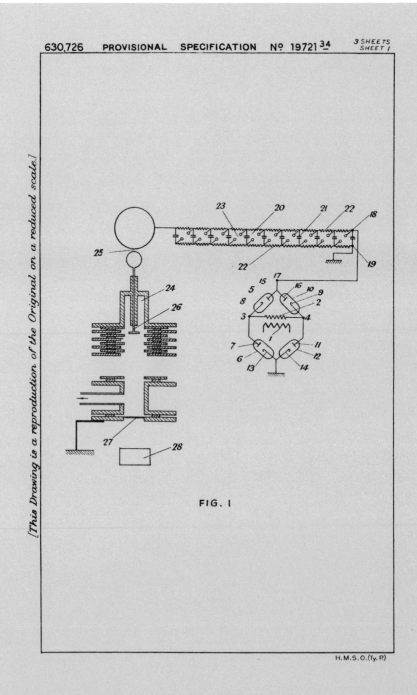

FIG. I

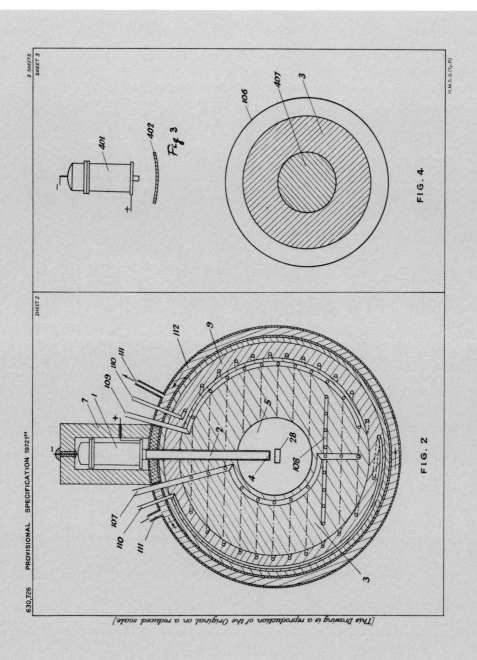

630,726   PROVISIONAL   SPECIFICATION 1972¹⁴

SHEET 2

3 SHEETS
SHEET 3

FIG. 2

FIG. 4

Fig 3

H.M.S.O.(Ty.P)

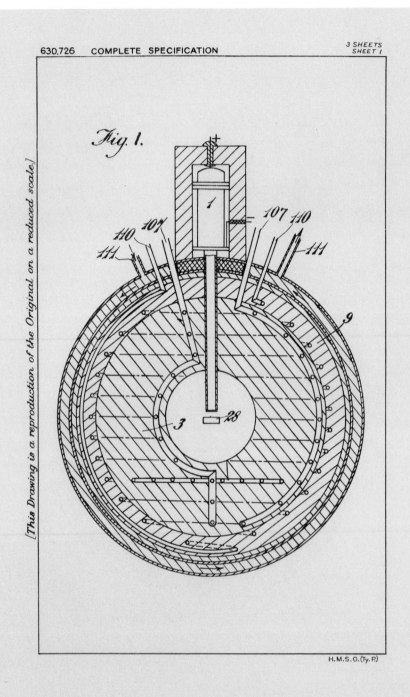

Fig. 1.

[This Drawing is a reproduction of the Original on a reduced scale.]

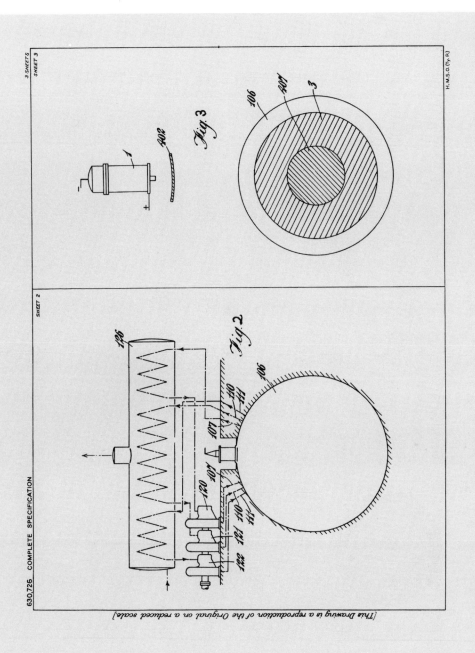

NUMBER (Series of 1935)   PATENT No. D-

265017   1939   DATED

DIV. 56   (EX'R'S BOOK)

LEO SZILARD

NEW YORK
NEW YORK

APPARATUS FOR NUCLEAR TRANSMUTATION
          ORIGINAL   RENEWED

APPLICATION FILED COMPLETE MAR 20 1939

This is to certify that annexed
hereto is a true copy from the records
of the United States Patent Office of
File Wrapper and Contents of the file
identified above.

By authority of the
COMMISSIONER OF PATENTS

Certifying Officer

Date June 19, 1968

NUMBER (Series of 1935)   PATENT No. D-

263017   **1939**   DATED

DIV. 56   (EX'R'S BOOK)

Name   LEO SZILARD

of   NEW YORK

State of   NEW YORK

Invention

APPARATUS FOR NUCLEAR TRANSMUTATION

	ORIGINAL	RENEWED
APPLICATION FILED COMPLETE	MAR 20 , 1939	
Petition, Specification, Oath, First Fee $30, 7 sheets Drawings, 1 EXTRA CLAIM $1	MAR 20 , 1939 MAR 20 1939	

Examined and passed for Issue _____, 193	Reexam'd and passed for Issue _____, 193
Exr. Div.	Exr. Div.
Notice of Allowance _____, 193 By Commissioner.	Notice of Allowance _____, 193 By Commissioner.
Final Fee _____, 193	Final Fee _____, 193

Attorney   PENNIE, DAVIS, MARVIN & EDMONDS   NEW YORK N. Y.

Associate Attorney

No. of Claims Allowed _____ *Print Claims* _____ in O.G. Class _____

Title as allowed

ENGLAND . MAR 12 1934

MAR-20-39   1 8 7 4 6 9   E   Check —   30.00

MAR-20-39   1 8 7 4 6 9   E   Check —   1.00

PENNIE, DAVIS, MARVIN AND EDMONDS

WILLIAM H. DAVIS
ARBA B. MARVIN - RETIRED
DEAN S. EDMONDS
FRANK E. BARROWS
W. BROWN MORTON
MERTON W. SAGE
WILLIS H. TAYLOR, JR.
ERNEST H. MERCHANT
GEORGE E. MIDDLETON
MORRIS D. JACKSON
R. MORTON ADAMS
RAYMOND F. ADAMS
LESLIE B. YOUNG
DANIEL V. MAHONEY

COUNSELLORS AT LAW

165 BROADWAY

NEW YORK

WASHINGTON OFFICES
NATIONAL PRESS BUILDING

ASSOCIATES

RAYMOND B. CANFIELD
BALDWIN GUILD
AMBROSE A. ARNOLD
S. HOWELL BROWN, JR.
LOUIS D. FORWARD
H. STANLEY MANSFIELD
GEORGE E. FAITHFULL
DONAL F. McCARTHY
HAL E. SEAGRAVES
JOHN HOXIE
JOHN T. FARLEY
CURT VON BOETTICHER, JR.
JAMES B. CHRISTIE
JAMES W. LAIST
CYRUS S. HAPGOOD
ROGER T. McLEAN
HAROLD A. TRAVER
KENNETH H. MURRAY
HARRY E. DOWNER
CARL E. RING

Washington, D.C.
March 20, 1939

Honorable Commissioner of Patents

Washington, D.C.

Sir:

    We are forwarding herewith, for filing, an

application of LEO SZILARD, for improvements in

APPARATUS FOR NUCLEAR TRANSMUTATION, executed

March 9, 1939.

    This application comprises the following:

Petition

Specification

Oath

7 sheets of drawings

    We are enclosing our check for $ 31.00,

covering the filing fee.

    Respectfully,

Pennie Davis Marvin Edmonds

Attorneys for Applicant.

enc.

263017

FORM 32 1M 5-38

PETITION

TO THE COMMISSIONER OF PATENTS:

263017

MAR 30 1939

Your petitioner   LEO SZILARD, a subject of Hungary,

PATENT OFFICE

residing at   420 West 116th Street,

in the County of   New York,   State of   New York,

whose Post Office address is   420 West 116th Street, New York,

New York,

prays that Letters Patent may be granted to him for improvements in

APPARATUS FOR NUCLEAR TRANSMUTATION

as set forth in the annexed specification.

And he hereby appoints William H. Davis, Dean S. Edmonds,

Frank E. Barrows, W. Brown Morton, Merton W. Sage, Willis H. Taylor,

Jr., Ernest H. Merchant, George E. Middleton, Morris D. Jackson,

R. Morton Adams, Raymond F. Adams, Leslie B. Young, Daniel V. Mahoney

(constituting the firm of Pennie, Davis, Marvin & Edmonds, 165 Broadway,

New York City, Registration No. 10839) and each of them, his Attorneys,

with full power of substitution and revocation, to prosecute this

application, to make alterations and amendments therein, to sign the

drawings, to receive the Patent, and to transact all business in the

Patent Office connected therewith.

Leo Szilard
_____
(Inventor's Full Name)

SPECIFICATION

TO ALL WHOM IT MAY CONCERN:

BE IT KNOWN, that I,   LEO SZILARD, a subject of Hungary,

residing at   420 West 116th Street,

in the County of   New York,   State of   New York,

have invented certain new and useful improvements in

APPARATUS FOR NUCLEAR TRANSMUTATION

and I do hereby declare the following to be a full, clear and exact

description of the invention, such as will enable others skilled

in the art to which it appertains, to make and use the same:

This application is a continuation in part of my copend-
ing application Serial No. 10,500, filed March 11, 1935.

263017

1   The invention concerns a process and apparatus for

2 the transmutation of elements and makes it possible to

3 bring about such a transmutation on a large scale.   The

4 production of power by means of the heat generated in the

5 process and the production of radio active elements on a

6 large scale are thus made possible by the invention.   The

7 invention teaches that it is possible to produce a nuclear

8 chain reaction and to maintain the stationary conditions in

9 such chain reactions.   It also teaches that it is possible

10 to have explosive bodies in which an explosion is brought

11 about at will by a sudden change in the distribution of

12 matter.

13   According to the invention such a chain reaction

14 may be maintained for instance by maintaining an initial

15 radiation of neutrons by means of one of the known methods

16 for the production of neutron radiations (for instance by

17 bombarding a lithium target with deuterons or by exposing

18 beryllium to X-rays or gamma-rays) and letting these neu-

19 trons fall on a body/suitably chosen substance, shape and

20 dimensions and thereby obtaining an increase in the number

21 and energy of the neutrons which increase is brought about

22 by their interaction with the said body.   Thus if the

23 proper conditions are observed the interaction of neutrons

24 with matter can lead to the liberation of further neutrons.

25 These newly liberated neutrons liberate again in their turn

26 further neutrons so that we have a chain reaction in which

27 a large number of neutrons are liberated, the total number

28 being limited only by the dimensions of the geometrical

29 arrangement.

30

-2-   263017

1   Such change is not only possible in the presence of a meta
2   stable element. A meta stable element is an element the
3   mass of which (packing fraction) is sufficiently high to
4   allow its disintegration into its parts and a liberation of
5   energy. Elements like uranium and thorium are examples of
6   such elements. It is possible to measure the packing
7   fraction of the elements by means of the mass spectograph
8   and thereby determine whether an element is meta stable or
9   not in the sense in which the word is used in this appli-
10   cation. If, for instance, the values which were generally
11   accepted in 1934 for helium and beryllium had been correct
12   (and we know today that they were false, at least in their
13   relation to each other), we have been correct in concluding
14   from these values which were accepted in 1934 that beryllium
15   is a meta stable element and can be disintegrated into parts
16   with a liberation of energy, one of the parts set free in
17   its disintegration being a neutron. If such an element which
18   is meta stable but the disintegration of which is inhibited
19   is exposed to neutrons, we may have a reaction in which a
20   neutron disappears and more than one neutron is emitted.
21   The additional neutrons would together with the number of
22   the original neutrons continue to interact with the meta
23   stable element thereby forming the links of a chain reaction.
24   In order to be able to utilize a nuclear reaction
25   in which an excess number of neutrons is liberated by neutrons
26   for the maintenance of a chain reaction, it is not sufficient
27   to measure the cross-section and other constants of the
28   reaction, but it is also necessary to be aware of the laws
29   which govern the neutron output of such reactions in function
30   of the geometrical conditions. Once the general laws, the

-3-

263017

1 type of behavior is known, the exact dimensions can be
2 easily determined in each particular case by actually
3 measuring the neutron output. Such experimental adjustments
4 can, however, only be made if the general type of behavior
5 is known.
6   In the following we shall by way of example demon-
7 strate certain general features of such a neutron chain
8 reaction in a special case. In this special case the
9 carrier of the chain reaction forms a spherical layer as
10 illustrated in Fig. 1.   In Fig. 1, 1 is a spherical layer
11 which contains a carrier of the chain reaction, 2 is a
12 neutron source which has fixed constant output of neutrons
13 r_1 is the inner radius of the spherical layer, and r_2 is
14 the outer radius of the spherical layer. In order to be
15 able to treat the problem as a diffusion problem, we choose
16 r_1 very much larger than the mean free path a of the neu-
17 trons produced in the chain reaction in the substance of

-3A-   263013

1 the chain reaction layer.   The neutrons liberated in the
2 chain reaction layer will have to make a number of collisions
3 with the nuclei which compose the chain reaction layer, and
4 we designate the number of collisions necessary for causing
5 transmutations in the carrier of the chain reaction layer
6 which lead on the average to the liberation of one addi-
7 tional neutron with f. We further assume that f is a large
8 number so that we may have conditions in which the well
9 known equations of diffusion can be applied.   With these
10 assumptions the density s of the neutron will with good
11 approximation be given within the chain reaction layer as
12 a function of the radius r by the following equation:
13 $(1) \quad D \frac{d^2(rs)}{dr^2} + A(rs) = 0$
14 D is a diffusion constant which is given by
15 $D = \frac{a\,v}{3}$
16 v is the mean velocity of the neutrons; a the mean free path
17 for scattering collisions within the chain reaction layer;
18 and f a number which says how many scattering collisions
19 a neutron has to make in the chain reaction layer in order
20 to produce on the average an additional neutron.
21   A stands for the number of neutrons produced per
22 c.c. of chain reaction layer in a second and its value is
23 given by $A = \frac{v}{af}$ so that $\sqrt{\frac{D}{A}} = \frac{a\sqrt{f}}{\sqrt{3}}$
24
25 For a given value of r_1 there is a certain value of r_2
26 and accordingly a certain value of $r_2 - r_1$ for which the
27 number of neutrons per second diffusing out of the spherical
28 layer into space becomes infinite for a finite neutron pro-
29 duction of the source. This value of $(r_2 - r_1)$ we may call
30 the critical thickness of the chain reaction layer. If the

-4-

1   thickness of the chain reaction layer is smaller than but

2   very close to the critical thickness, the neutron output

3   is very much larger than the neutron input and we may have

4   1000 or more times as many neutrons emerging from the chain

5   reaction layer as the neutron input of the neutron source 2.

6   The value of the critical thickness is a function of r_1

7   and of the boundary conditions for $r = r_1$ and $r = r_2$. If

8   the outer surface ($r = r_2$) of the spherical layer were to

9   stand free in space, the density s would be 0 for the outer

10 surface, and if there is no absorption of neutrons in the

11 hollow sphere containing the neutron source, i.e. if the

12 number of neutrons produced by the neutron source is equal

13 to the number of neutrons diffusing outwards from the

14 sphere $r = r_1$, we obtain for very large values of r_1 ($r_1 \gg \sqrt{\frac{D}{A}}$)

15 for the value of the critical thickness

16 (2) $\quad \int s\,\kappa = \frac{\pi}{2}\sqrt{\frac{D}{A}} = \frac{\pi}{2}\, a\, \frac{\sqrt{f}}{\sqrt{3}}$

17 The critical thickness for these conditions we shall call

18 further below the standard critical thickness of the sub-

19 stance which composes the chain reaction layer.

20   If the outer surface is covered by some material,

21 for instance if the transmutation layer is covered by lead

22 or immersed into water, the critical thickness is less than

23 the standard critical thickness. This us due to the back

24 scattering by lead or water, and, in the case of water,

25 the fact that the neutrons are slowed down by water and their

26 mean free path is thereby reduced plays an important role.

27   If the neutrons are allowed to escape out of the

28 hollow sphere containing the neutron source in the interior

29 of the spherical layer or if they are absorbed within this

30 hollow sphere, the critical thickness is increased.

-5-

1   If all the neutrons were captured in the hollow

2   sphere within $r = r_1$, the critical thickness would become

3   independent of the value of r_1 and have a value exactly

4   double the ~~normal~~ ^{standard} critical thickness provided that the

5   neutron density s remains 0 at the outer surface ($r = r_2$).

6   Obviously the above given diffusion equation holds

7   only for stationary solutions, that is for solutions where

8   the neutron density s is a function of r only and does

9   not vary with time. Not for all boundary conditions will

10 such stationary solutions exist. If we gradually increase

11 the thickness of the ~~critical~~ ^{chain reaction} layer we reach for any given

12 boundary conditions a thickness at which the neutron out-

13 flow becomes infinite for finite neutron production of the

14 neutron source.   This thickness is ^{the} critical thickness

15 of the arrangement. It can be calculated for every case in

16 the following way from given boundary conditions.

17   The solution of the above given diffusion equation

18 has the form of

19 (3) $$s r = C_1 \sin\left(\sqrt{\tfrac{v}{A}}\, r\right) + C_2 \cos\left(\sqrt{\tfrac{v}{A}}\, r\right)$$

20 For the boundary condition $s = 0$ for $r = r_2$ the solution

21 takes the form of

22 (4) $$s r = C \sin\left[(r_2 - r)\sqrt{\tfrac{v}{A}}\right]$$

23 or

24 (5) $$s = C\, \frac{\sin\left[(r_2 - r)\sqrt{\tfrac{v}{A}}\right]}{r}$$

25   If there is no absorption of neutrons in the

26 hollow sphere around the neutron source and if the neutron

27 source produces a fixed number N of neutrons per second,

28 the number of neutrons diffusing from the sphere $r = r_1$

29 into the chain reaction layer must be equal to N. On the

30 other hand, the number of neutrons thus diffusing through

263017

any sphere is given by

$$4 \pi r^2 D \frac{ds}{dr}$$

and it is therefore also

$$N = 4 \pi r_1^2 D \left[\frac{ds}{dr} \right]_{r=r_1}$$

For a value of $\underline{r} = \underline{r}_2$ for which $\frac{ds}{dr}$ becomes 0, the ratio of the number of neutrons diffusing through the outer surface $\underline{r} = \underline{r}_2$ to N becomes infinite. The corresponding value of $(r_2 - r_1)$ is therefore the critical thickness which can thus be calculated for values of $\underline{r}_1$. For very small values of $\underline{r}_1$ ($r_1 \ll \sqrt{\frac{D}{A}}$) it obviously becomes twice the standard critical thickness, i.e. $\pi \sqrt{\frac{D}{A}}$.

To take another example, if the boundary conditions are $\underline{s} = 0$ both for $\underline{r}_1$ and $\underline{r}_2$, then a glance at equation (5) shows that $\underline{r}_2 - \underline{r}_1 = \pi \sqrt{\frac{D}{A}}$, i.e. twice the critical thickness, quite independent of the value of $\underline{r}_1$.

Figs. 2 and 3 show such a chain reaction apparatus. A neutron radiation, the initial radiation, is generated by the high voltage canal ray tube 1 (shown in greater detail in Fig. 6). This tube generates fast deuterons which strike the target 28 which contains deuterium. The neutron radiation emerging from 28 acts on the matter 3 which fills the spherical transmutation space. The composition of this matter 3 will be discussed further below and is such that a chain reaction is released by the neutrons. The pumps 120, 121 and 122 pump a liquid, for instance water or mercury, through the pipe systems 107, 110, 111 thereby cooling the transmutation area 3 and driving the heated liquid through the boiler 126. The boiler supplies steam to a power plant. The neutrons emerging from the sphere 3 act on a layer 9 which is composed of an element xx that will trans-

-7-

1 mute into a radio-active body (which is suitable for the
2 storage of energy).
3   An essentially different way of introducing the
4 initial radiation into the chain reaction chamber is the
5 arrangement shown in Fig. 4. 401 is the cathode ray
6 tube described in Fig. 1. 402 is a sheet of a heavy
7 element for instance Pb, or U in which penetrating radia-
8 tion (hard X-rays) is generated with an extremely good
9 efficiency if the electrons have a voltage over one million
10 volts. This efficiency increases very rapidly with the
11 voltage, and is much higher than it could be expected
12 from the experience based on ordinary X-ray work. The
13 thickness of the sheet 402 is such as to enable the gener-
14 ated penetrating radiation to penetrate through this sheet
15 and act on the transmutation chamber 106 (in Fig. 3).
16 Nevertheless the sheet can be sufficiently thick to utilize
17 more than half of the energy of the cathode rays. The
18 X-rays emerging from sheet 402 penetrate the layer 3 and
19 can liberate neutrons either from the layer 3 or from a
20 substance 407 placed in the interior of the layer 3). For
21 instance, if we have beryllium present in 403 or in 3
22 neutrons will be liberated by X-rays. These neutrons can
23 then maintain a chain reaction as discussed further above
24 and further below. The advantage of using X-rays as an
25 initial radiation is the following: The X-rays penetrate
26 through a perfectly closed layer 3 into the interior of
27 the layer and therefore a leak of neutrons from the inter-
28 ior can be avoided. This is specially important if we have
29 to deal with a neutron chain in which no multiplicator
30 action is involved. In such cases X-rays may be used with

-8-

1 ~~advantage as initial radiation especially in view of the~~

2 unexpectedly large ~~efficiency of~~ the X-ray production by

3 ~~means of fast electrons acting on heavy elements.~~

4   We wish to discuss now certain features which

5 arise when the above teachings of this application are

6 applied to uranium. If uranium is exposed to neutrons of

7 a few million volts energy or to neutrons slowed down by

8 water to a few volts of energy or less, uranium emits

9 neutrons in two different ways. This can be demonstrated

10 with particular ease if uranium is exposed to slow neutrons

11 and the slow neutron stream hitting the uranium is suddenly

12 stopped. One then finds that uranium emits neutrons after

13 the stoppage of the slow neutron stream for another few,

14 perhaps ten to fifteen, seconds. This delayed neutron

15 emission is weak, i.e. there are less than perhaps thirty

16 neutrons emitted in this way per incident neutron which

17 disappears in reacting with the uranium. No chain reactions

18 could be based on this delayed neutron emission alone. On

19 the other hand, it is easily demonstrated that uranium emits

20 a large number of neutrons while it is exposed to the slow

21 neutron stream and the number of neutrons which are thus

22 instantaneously (within a fraction of a second) emitted

23 from uranium per number of incident neutrons which are

24 captured in the reaction with uranium is larger than one.

25 This fact makes it possible to use uranium as the carrier

26 of a chain reaction. The neutrons which are emitted from

27 uranium during the irradiation with slow neutrons can be

28 distinguished from the incident slow neutrons by virtue

29 of the fact that their velocity is much higher and ranges

30 somewhere between a few ten thousand volts to perhaps about

1 a million volts or so.

2   In spite of its small value the delayed neutron

3 emission is not without significance. In view of its exist-

4 ence we have - in order to be more precise - to distinguish

5 between the critical thickness for the total neutron emis-

6 sion (instantaneous and delayed emission together).

7   It is possible to use uranium in the chain reaction

8 layer not only as uranium metal or as uranium oxide but also

9 it is possible to use either of these mixed with a hydrogen

10 containing substance such as for instance water and it is

11 possible to use such a large concentration of the hydrogen

12 containing substance that most of the neutrons are slowed

13 down to a few volts before they react with the uranium in

14 releasing an additional number of neutrons (releasing more

15 than one neutron on the average for every neutron captured by

16 uranium). That the use of hydrogen containing substances in

17 such concentrations should be possible is very surprising

18 and unexpected and is in contradiction with the generally

19 accepted value of the capture cross-section of uranium

20 for neutrons slowed down by water to a few volts velocity

21 or less. The total cross-section of uranium has been

22 assumed on the basis of published measurements to be above

23 40×10^{-24}, i.e. the capture cross-section well above 30.

24 On the other hand, the cross-section for that transmutation

25 of uranium which yields on the average neutrons in excess

26 of one but certainly not in excess of 8 per transmutation is

27 known to be less than 3. In these circumstances most of the

28 neutrons which are slowed down to a few volts or less will be

29 captured by uranium without causing a transmutation of the type

30 which yields neutrons. Accordingly one would expect that

-10-

1 less than one neutron will be emitted on the average for

2 every neutron captured by standard uranium. (By "standard

3 uranium" we mean uranium in which the relative abundance

4 of the two isotopes 135 and 138 has the ratio which one

5 finds in uranium that occurs in nature, i.e. about 1:139.)

6 The only possible explanation is that for some reason or

7 other an error has been made by those who previously meas-

8 ured the total cross-section or uranium to be about 43×10^{-24}

9 though it is not possible to state through what mistake such

10 error occurred.

11   The mixing of uranium with a hydrogen containing

12 substance to form the chain reaction layer has the advantage

13 greatly to reduce the critical thickness and also the amount

14 of uranium required. The uranium or uranium oxide can either

15 be mixed with the hydrogen containing substance or the chain

16 reaction layer may be built up from alternating layers of

17 uranium and the hydrogen containing substance. As a

18 hydrogen containing substance water, paraffin wax or calcium

19 hydride appear to be suitable. If the hydrogen containing

20 substance is not mixed with uranium, but alternating layers

21 are used, the layers should be as thin as possible. In

22 particular the thickness of the hydrogen containing layer

23 should be as small as possible, and if paraffin, water, or

24 a hydrogen containing substance of about the same hydrogen

25 concentration is used, the thickness of the layer should not

26 exceed about 7 mm.

27   It is important that the ratio of the number of

28 grams of hydrogen to the number of grams of uranium in the

29 chain reaction layer shall not exceed a certain critical

30 value as otherwise no chain reaction can be maintained for

31 any thickness of the layer which contains the mixture.

--11--

1　This is due to the fact that hydrogen captures slow neutrons

2　and therefore competes with uranium. The capture cross-

3　section of the hydrogen atom is about one-third 10^{-24}

4　and the capture cross-section of uranium for the trans-

5　mutations which lead on the average to the liberation of

6　about one additional neutron is about 2 x 10^{-24} sq. c.

7　If these approximate values were exactly correct, no chain

8　reaction would be possible in a layer which contains

9　hydrogen and uranium in a ratio of 6 grams of hydrogen to

10　238 grams of uranium. It is certain that no chain reaction

11　is possible in a layer which contains 20 grams of hydrogen

12　for every 240 grams of uranium. On the other hand, it is

13　desirable in order to keep the critical thickness low

14　not to use too little hydrogen. A reasonable value for

15　the ratio of hydrogen to uranium which gives good results

16　and none too large critical thickness can be found if the

17　expression

18　$(7) \qquad X \sqrt{\dfrac{1}{2 - \frac{2}{6} X}}$

19　is plotted and a value of Xm is selected for which the

20　expression becomes maximum. If a mixture is prepared in

21　which the ratio of the number of hydrogen atoms to the

22　number of uranium atoms N_n/N_u has a value between

23　$(8) \qquad \frac{1}{2} X m \;\; \text{and} \;\; \dfrac{X m + 6}{2}$

24　and if such a mixture is used to build up the chain

25　reaction layer, satisfactory results will be obtained. As

26　stated before, alternating layers of the hydrogen contain-

27　ing substance and uranium or uranium oxide can be used

28　instead of a mixture.

29

30

-12-

263017

1   If instead of standard uranium, uranium is used in
2 which the isotope 235 has been concentrated so that its
3 relative abundance is larger than one part in 139, a larger
4 concentration of hydrogen can be used in the chain reaction
5 layer. The ratio of the hydrogen to the uranium concen-
6 tration can be increased by the same factor by which the
7 relative abundance of the uranium isotope 235 has been
8 increased.   Such an increase of the relative abundance of
9 the uranium isotope 235 can be achieved by subjecting
10 uranium hexa fluoride to one of the known diffusing pro-
11 cesses which lead to separation of molecules of different
12 molecular weight.
13   In such a chain reaction layer which contains such
14 large concentrations of hydrogen as indicated above, the
15 neutron emitted from the uranium in the chain reaction
16 is slowed down after traveling in the chain reaction layer
17 an average distance b from its origin. At the distance
18 b from its origin the neutron will therefore be so slow that
19 its mean free path a for scattering is much smaller than the
20 value b.   In paraffin wax this mean free path a would be
21 about 2-1/2 mm, and in the·chain reaction layer it will be
22 comewhat larger, i.e. by a factor k which gives the ratio
23 of the concentration of hydrogen in paraffin to the concen-
24 tration of hydrogen in the chain reaction layer.   This
25 reduction of the mean free path for scattering is due to
26 the large scattering cross-section of hydrogen for room
27 temperature neutrons.   In addition, the neutrons which
28 have been slowed down at an average distance b from their
29 origin have now, being slow, a large cross-section for
30 those transmutations of uranium which lead to the liber-

-13-   263017

1 ation of neutrons. The combined effect of the large scatter-

2 ing cross-section of the slow neutrons in the chain reaction

3 layer and the large transmutation cross-section of the slow

4 neutrons for uranium is that a neutron which is emitted by

5 uranium and which becomes slow at the average distance $\underline{b}$

6 from its point of origin will transmute a uranium nucleus

7 at a point which is at an average distance $\underline{a}$ from the point

8 at which the neutron became slow, and the distance $\underline{a}$ is

9 small compared to $\underline{b}$, so that the neutron will transmute

10 a uranium nucleus at a distance $\underline{b_1}$ not very different from

11 $\underline{b}$ from the point of its origin.

12   In these circumstances the previously given equation

13 no longer holds, and the critical thickness is approximately

14 given in the following way: For the stationary state the

15 neutron density within the spherical chain reaction layer

16 obeys now the equation:

17 (9) $$(\bar{b})\ \frac{d^2(r\delta)}{dr^2} + \frac{6(w_2 - w_0)}{w_1 + 2w}\ (r\delta) = 0$$

18   In this equation $\underline{w_0}$ is the probability for a slowed

19 down neutron to cause a transmutation of uranium in which

20 the slow neutron disappears and no fast neutron is emitted;

21 $\underline{w_1}$ is the probability for a slowed down neutron to cause a

22 transmutation of uranium in which the slow neutron disappears

23 and one fast neutron is emitted; $\underline{w_2}$ is the probability for a

24 slowed down neutron to cause a transmutation of uranium in

25 which the slow neutron disappears and two fast neutrons are

26 emitted.

27   For the special case: $\underline{w_2}$ = 1; $\underline{w_1}$ = o; $\underline{w_0}$ = o, the

28 above equation gives

29

30 10 $$(\bar{b})^2\ \frac{d^2(rs)}{dr^2} + 3(r\delta) = 0$$

The critical thickness is given by

standard

$$(11) \qquad \ell_{s}t = \frac{\pi}{3}\,\bar{\ell}\,\left(\frac{w_2 - w_0}{3\,(w_1 - 2\,w_2)}\right)^{1/2}$$

in the general case and by

$$(12) \qquad \ell_{s}t = \frac{\pi}{2}\,\bar{\ell}\,\sqrt{\frac{1}{3}}$$

L.h.

in the special case.

Obviously, the above diffusion equation pre-
supposes for its validity a small value of $\underline{w}_2$, but even
for large values of $\underline{w}_2$ it gives at least the order of
magnitude for the critical thickness.

The critical thickness will in practice always
be determined empirically for instance in the following
way: a neutron source is surrounded by the chain reaction
layer of an approximately correct thickness which is safely
below the critical thickness. The radiations emitted from
the chain reaction layer while exposed to this neutron
source are observed by means of an ionization chamber.
Then the thickness of the chain reaction layer is brought
closer to the critical thickness by gradually increasing
either the quantity of uranium or the quantity of hydrogen
containing substances mixed with the uranium. The amount
of ionizing radiation which is emitted is again observed
and the thickness of the chain reaction layer is again
brought closer to the critical thickness in the same way
as before. In this way, by observing the increase of the
emitted radiation as a function of the increasing effective thickness of the chain reaction layer the critical
thickness can be extrapolated from the observed curve by
plotting the intensity of the emitted neutron radiation
against the effective thickness of the chain reaction
layer. Instead of an ionization chamber which registers

-15-

1 the neutron intensity by means of recoil ions in the gas

2 of the chamber, induced activity caused by the neutrons

3 can be used as a measure of the radiation intensity.

4 <u>Variation of Critical Thickness</u>

5   If slow neutrons are used the critical thickness can

6 be increased by having a slow neutron absorber within the

7 hollow sphere in the center of the spherical arrangement.

8 If the inner radius of the spherical shell of the chain

9 reaction layer is much larger than the critical thickness

10 (to be accurate we have said the standard critical thick-

11 ness given by the above formulas), and if all slow neutrons

12 are absorbed, for instance by a cadmium layer covering the

13 inner surface of the spherical chain reaction layer, the

14 critical thickness of the arrangement is increased. By

15 removing such absorbing matter from the inside of the chain

16 reaction layer, the critical thickness may be reduced below

17 the actual thickness, and thus an explosion may be brought

18 about. The explosion will be all the more violent the

19 more quickly the absorbing substance is removed. A similar

20 increase in the critical thickness of a spherically sym-

21 metrical chain reaction layer can be brought about by

22 removing a section of the layer and thereby producing an

23 opening through which the neutrons can escape. For instance

24 a conical section corresponding to a few percent of the

25 spherical chain reaction layer can be so arranged as to be

26 easily moved out of its place and replaced, and thereby the

27 critical thickness may be reduced or increased.

28   Explosions may be brought about by a variation of

29 the critical thickness for the purpose of creating a large

30 amount of radio active elements. In such a case it is

-16-

1 desirable to have mild explosions. Mild explosions can

2 be brought about by slowly changing the space distri-

3 bution of matter and also by arranging layers so that

4 when the explosion sets in there shall be no rapid

5 removal of substances which capture neutrons and thereby

6 reduce the critical thickness. For instance, if the

7 chain reaction layer is based on fast neutrons and we have

8 a substance in the hollow sphere in the center of the chain

9 reaction layer which will slow down neutrons and absorb

10 the slow neutrons, for instance a solution of a boron

11 salt in water, the arrangement may be such that there shall

12 be no easy outflow for the water from the inside of the

13 spherical chain reaction layer towards the outside. On

14 the contrary, if a strong destructive explosion is wanted,

15 it can be brought about by providing for such outflow.

16 It is advisable to have explosions which for the purpose

17 of reducing radio active elements are so arranged that

18 the explosions shall take place in the middle of a large

19 water tank or below water. After the explosion the

20 scattered radio active material can then be collected

21 from the water.

22   Regulation

23   As we have seen, the ratio of neutron input to

24 neutron output becomes infinite for the critical thickness

25 of the arrangement. The neutron input is in practice

26 limited by the accuracy of the arrangement, since the

27 thickness of the chain reaction layer must be extremely

28 close to the critical thickness, and yet must remain below

29 it in order to avoid an explosion. Fortunately, it is

30 possible to overcome this difficulty by reason of the

-17-   263017

1 following fact:

2   In reality we have to deal not with one critical

3 thickness only, but with two different critical thicknesses

4 which we shall call the instantaneous critical thickness

5 and the delayed critical thickness. They arise by virtue

6 of the fact that, while the bulk of neutrons is emitted

7 instantaneously when uranium is transmuted by neutrons,

8 there is also a delayed emission of neutrons, the delay

9 being of the order of magnitude of a few seconds. If the

10 thickness of the chain reaction layer is larger than the

11 delayed critical thickness, but smaller than the instan-

12 taneous critical thickness, the neutron output increases

13 to infinity, but does not increase too rapidly. This

14 makes it possible that by moving an object which forms part·

15 of the arrangement and which has an influence on the critical

16 thicknesses (for instance, by having a slow neutron absorber

17 in the interior of the hollow sphere of the spherical chain

18 reaction layer, and by partially withdrawing it from there,

19 we can reduce the critical thickness, and in a similar way

20 we can also increase it by the opposite movement), we can

21 vary the critical thickness in time. We shall call objects

22 which are used in this way regulator objects, and accord-

23 ing to our invention the neutron output can be kept very

24 high without risking an explosion by moving the regulator

25 object in such a way that part of the time the critical

26 thickness for delayed emission should be below, and part

27 of the time it should be above, the real thickness of the

28 spherical layer. It is only necessary for safe functioning

29 to have an instrument which is sensitive to the emitted radi-

30 ation or the temperature of some part of the chain reaction

-18-

layer, and this instrument can control the position of
the regulator object.   Obviously, in order to have stable
functioning, the regulator object will have to be moved in
direction of an increase of the critical thickness with
increasing neutron radiation, and it has to be moved in the
opposite way with decreasing neutron radiation.   While the
thickness of the chain reaction layer will still have to be
accurately chosen, since it has to be within narrow limits,
i.e. between the critical thickness for instantaneous
neutron emission and the critical thickness for delayed
neutron emission, the latter being only slightly larger
than the former.   Yet the above described regulation makes
it possible to get a very much higher neutron output without
reaching an explosion.

        The critical thickness for delayed emission could
also be conveniently called the total critical thickness
because it corresponds to the total neutron emission, both
instantaneous and delayed.   It can be easily determined
empirically by varying the thickness of the chain reaction
layer and observing for each thickness the emitted neutron
radiation as a function of time.   Below the critical thickness for delayed emission the neutron radiation is a function
of time which resembles a growth curve in the field of radio
activity, i.e. it approaches an upper limit practically
reaching saturation after some time. Above the critical
thickness for delayed emission, but below the critical thickness for instantaneous emission, the observed neutron emission increases more and more rapidly with time, and the
arrangement has quickly to be changed in order to avoid
overheating.   The value of the critical thickness for

1 delayed emission is reached when one type of curve goes
2 over into the other, and at the critical thickness itself
3 the neutron intensity as a function of time is a straight
4 line.

1   We wish now to insert a description of Figs. 5, 6

2   and 7 to which we have referred in the preceding part of

3   the specification.

4   Fig. 5 shows an example of a suitable arrangement.

5   11 is an electrical discharge tube ejecting a beam 12 of

6   fast diplogen ions.*   The ions fall on a substance 13

7   consisting of for instance gaseous diplogen or a diplogen

8   compound or lithium, causing transmutation, i.e. a nuclear

9   reaction of the diplogen ion with an atom of the target.

10   The substance 13 is surrounded by a layer 14 containing

11   the element which we wish to transmute into a radio active

12   element.   In order to have a good efficiency, the thick-

13   ness of the layer 14 has to be sufficiently large, compared

14   with the mean free path of the neutron for this trans-

15   mutation.

16   Fig. 6 shows the electrical discharge tube referred

17   to in Fig. 5.   It is a high voltage positive ray tube.

18   There is an auxiliary positive ray tube on top of the high

19   voltage tube. 11 is the anode, 15 the cathode of this

20   auxiliary tube. Diplogen is admitted through the tube 13

21   and pumped away through

22   Fig. 7 shows an arrangement suitable for the

23   production of hard X-rays. 1 is the primary of a trans-

24   former, the secondary 2 of which is connected to the points

25   3 and 4.   3 is connected to the cathode 8 of the rectifier

26   tube 5 and to the anode 7 of the rectifier tube 6. Point 4

27   is connected to the cathode 9 of the rectifier tube 10 and

28   to the anode 11 of the rectifier tube 12. The cathodes 13

29   and 14 are connected to each other and to the earth.   The

30   anodes 15 and 16 are connected to point 17, and this point

*   also called diplons or deutons

-21-

1 is connected to the pole 18 of the impulse generator 20, the

2 pole 19 of which is connected to earth. The impulse gener-

3 ator 20 is built of condensers 21, resistances 22 and spark

4 gaps 23.

5   This impulse generator is adapted to produce inter-

6 mittent voltage up to 10 million volts, transmitted to the

7 discharge tube 24 through the spark gap 25. 26 is the

8 cathode of the discharge tube, the anode 27 of which is

9 connected to the earth. The fast electrons emerge through

10 the metal window 27 (which is the anode as well) and hit

11 a body 28. This body is used as an anticathode and yields

12 hard X-rays with very good efficiency if it is built of

13 Bi, Pb or some other heavy element.

I CLAIM:

1. Production of radio active elements comprising the step of maintaining a neutron radiation in a layer containing uranium, the thickness of the said layer being slightly below the critical thickness.

2. Production of radio active elements comprising the step of maintaining a neutron radiation in a layer containing uranium, the thickness of the said layer being slightly below the critical thickness, and the step of separating at intervals from the layer the radio active elements produced in the layer.

3. Production of radio active elements comprising the step of maintaining a neutron radiation in a layer containing uranium, the thickness of the said layer being slightly below the critical thickness, and exposing another element which can be made radio active by means of slow neutrons to the radiation of the chain reaction layer.

4. Production of radio active elements according to claims 1 to 3 comprising the step of maintaining a neutron radiation in a layer containing uranium or uranium oxide, the thickness of the layer being chosen so as to conform with equation (1) and given by equation (5) according to the prescription (6) respectively using values for the mean free path corresponding to a cross-section of uranium of about 6×10^{-24} and oxygen of about 3×10^{-24} and a value of $\underline{f}$ of about 60.

5. Production of radio active elements according to claim 4, the layer being composed of metallic uranium and having a thickness of about 30 to 70 centimeters for an inner radius $\underline{r}_1$ about 10 centimeters.

6. Production of radio active elements according
to claim 5, the spherical chain reaction layer of metallic
uranium being surrounded with a spherical layer of metallic
bismuth the thickness of which bismuth layer exceeds 50
centimeters and the thickness of the uranium layer being
below the values given in claim 5, the thickness of uranium
being reduced to about half if very large amounts of bismuth
are used.

7. Production of radio active elements according
to claims 1 to 3 in which a hydrogen containing substance,
such as for instance paraffin wax or water, is mixed with
uranium, the number of gram atoms of hydrogen having the
ratio to the number of gram atoms of uranium given by
equation (8)

8. Production of radio active elements according
to claim 7, the chain reaction layer being surrounded with
a spherical layer of bismuth metal exceeding in thickness
50 centimeters.

9. Production of radio active elements according
to claim 7, the spherical chain reaction layer being immersed
in water or another similar hydrogen containing substance,
for instance paraffin wax.

10. Production of radio active elements according
to claim 1 or claim 7 in which the uranium used has a different relative abundance of the isotopes 235 and 238 than
standard uranium, the relative abundance of uranium 235 being
increased by a factor of 3 or more.

11. Production of radio active elements comprising
the step of increasing the critical thickness of an arrangement comprising a layer of uranium, the said increase being

effected by changing the distribution of matter within or without the chain reaction layer and the increase carried to the point where the critical thickness of the arrangement is exceeded so that a chain reaction leading to an explosion takes place.

12. Production of radio active elements according to claim 11 in which the process described in claim 11 is carried out in a place surrounded by water or below water and the radio active elements produced through the explosion and scattered in the water are collected.

13. Production of power comprising the step of maintaining a neutron radiation in a layer containing uranium, the thickness of the said layer being slightly below the critical thickness.

14. Production of power according to claim 13 comprising the step of maintaining a neutron radiation in a layer containing uranium or uranium oxide, the thickness of the layer being chosen so as to conform with equation (1) and given by equation (5) according to the prescription (6) respectively using values for the mean free path corresponding to a cross-section of uranium of about 6×10^{-24} and oxygen of about 3×10^{-24} and a value of $\underline{f}$ of about 60.

15. Production of power according to claim 14, the layer being composed of metallic uranium and having a thickness of about 30 to 70 centimeters for an inner radius $\underline{r}_1$ about 10 centimeters.

16. Production of power according to claim 15, the spherical chain reaction layer of metallic uranium being surrounded with a spherical layer of metallic bismuth the thickness of which bismuth layer exceeds 50 centimeters and

the thickness of the uranium layer being below the
values given in claim 5, the thickness of uranium being
reduced to about half if very large amounts of bismuth are
used.

17. Production of power according to claim 13
in which a hydrogen containing substance, such as for
instance paraffin wax or water, is mixed with uranium, the
number of gram atoms of hydrogen having the ratio to the
number of gram atoms of uranium given by equation (8).

18. Production of power according to claim 17,
the chain reaction layer being surrounded with a spherical
layer of bismuth metal exceeding in thickness 50 centimeters.

19. Production of power according to claim 17,
the spherical chain reaction layer being immersed in water
or another similar hydrogen containing substance, for
instance paraffin wax.

20. Production of power according to claim 12
or claim 17 in which the uranium used has a different relative abundance of the isotopes 235 and 238 than standard
uranium, the relative abundance of uranium 235 being increased
by a factor of 3 or more.

21. Production of power comprising the step of
increasing the critical thickness of an arrangement comprising a layer of uranium, the said increase being effected
by changing the distribution of matter within or without the
chain reaction layer and the increase carried to the point
where the critical thickness of the arrangement is exceeded
so that a chain reaction leading to an explosion takes place.

IN TESTIMONY WHEREOF I affix my signature.

(signature)

O A T H

STATE OF NEW YORK   )
) ss:
COUNTY OF NEW YORK   )

    I, LEO SZILARD, the above-named petitioner, being duly sworn, depose and say that I am a subject of Hungary and reside at 420 West 116th Street, New York, New York and that I verily believe myself to be the original, first and sole inventor of the improvements in APPARATUS FOR NUCLEAR TRANSMUTATION as described and claimed in the annexed specification; that I do not know and do not believe that the same was ever known or used before my invention or discovery thereof, or patented or described in any printed publication in any country before my invention or discovery thereof;

    That as to those features of the said invention which are not disclosed in my prior application Serial No. 10,500, filed March 11, 1935, of which the present application is a continuation in part, I do not know and do not believe that they were patented or described in any printed publication in any country more than two years prior to this application or patented in any foreign country on an application filed by me or my legal representatives or assigns more than twelve months prior to this application or in public use or on sale in the United States for more than two years prior to this application; and that no application for patent on said features of this invention has been filed by me or my legal representatives or assigns in any foreign country;

    That as to those features of the said invention disclosed in this and in said prior application Serial No. 10,500, filed March 11, 1935, I do not know and do not believe that they were patented or described in any printed publication in any country more than two years before the filing of

said prior application or patented in any foreign country
on an application filed by me or my legal representatives
or assigns more than twelve months before the filing of
said prior application, or in public use or on sale in
the United States for more than two years before this
present application; and that no application for patent
on said features of the invention was filed by me or my
legal representatives or assigns in any foreign country
before the filing of said prior United States application
except as follows: England, March 12, May 9, June 14,
June 28, July 4, September 20 and September 25, 1934.

Subscribed and sworn to before me
this 9th day of March, 1939.

Notary Public

263017

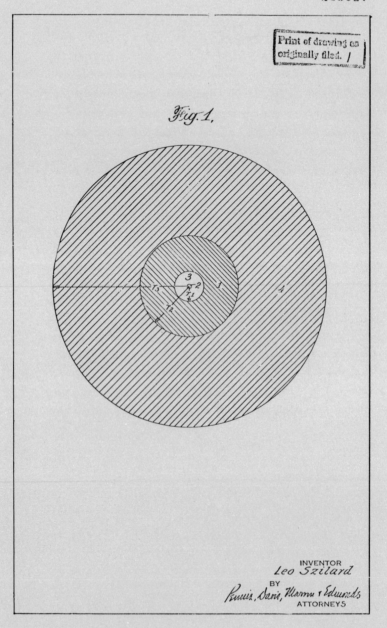

263017

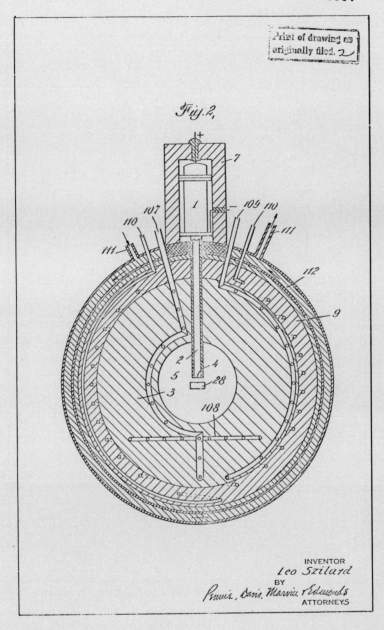

Fig. 2.

INVENTOR
Leo Szilard
BY
Brunis, Davis, Marvin Edmonds
ATTORNEYS

30   263017

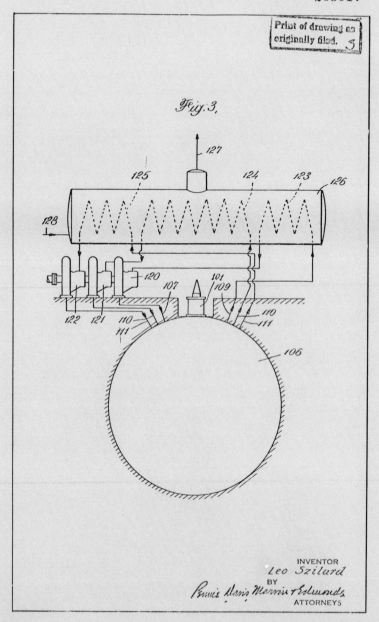

Fig. 3.

INVENTOR
Leo Szilard
BY
Pennie Davis Marvin + Edmonds
ATTORNEYS

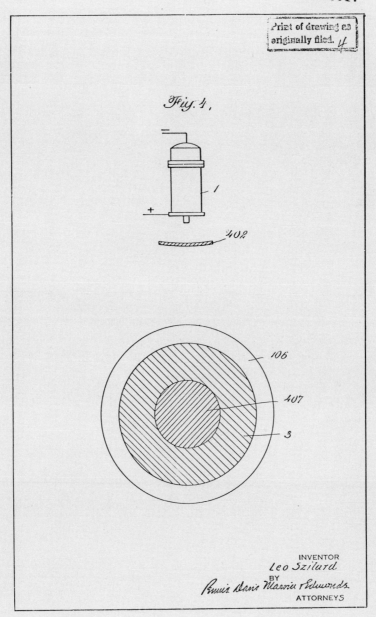

263017

Print of drawing as originally filed.

Fig. 4.

INVENTOR
Leo Szilard
BY
ATTORNEYS

263017

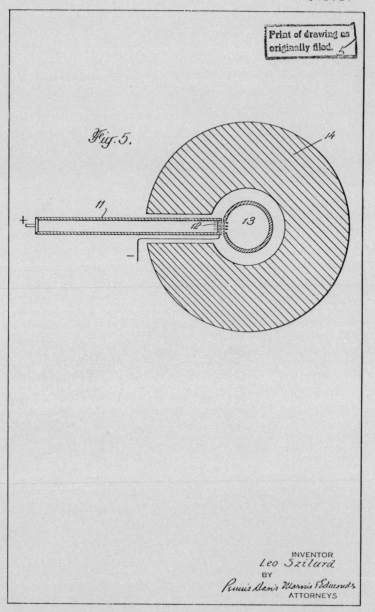

Print of drawing as
originally filed.

Fig. 5.

11

12

13

14

+

−

INVENTOR
Leo Szilard
BY
Pennie Davis Marvin Edmonds
ATTORNEYS

33   263017   R

263017

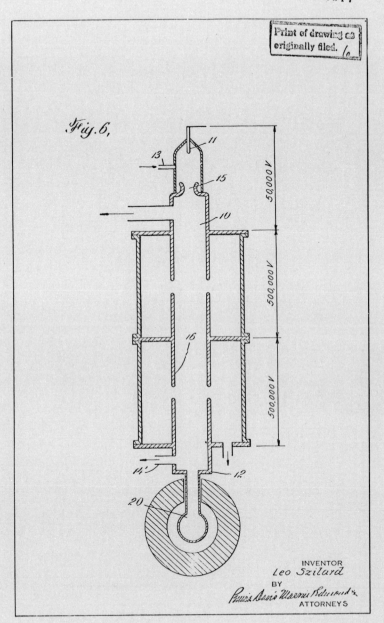

Fig. 6.

50,000 V

500,000 V

500,000 V

INVENTOR
Leo Szilard
BY
ATTORNEYS

34

263017

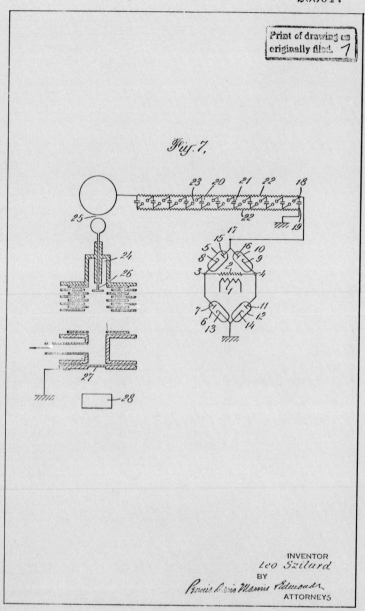

Fig. 7.

INVENTOR
Leo Szilard
BY
ATTORNEYS

35   263017

United States Patent Office

2,708,656

Patented May 17, 1955

1

2,708,656

NEUTRONIC REACTOR

Enrico Fermi, Santa Fe, N. Mex., and Leo Szilard, Chicago, Ill., assignors to the United States of America as represented by the United States Atomic Energy Commission

Application December 19, 1944, Serial No. 568,904

8 Claims. (Cl. 204—193)

The present invention relates to the general subject of nuclear fission and particularly to the establishment of self-sustaining neutron chain fission reactions in systems embodying uranium having a natural isotopic content.

Experiments by Hahn and Strassman, the results of which were published in January 1939, Naturwissenschaften, vol. 27, page 11, led to the conclusion that nuclear bombardment of natural uranium by slow neutrons causes explosion or fission of the nucleus, which splits into particles of smaller charge and mass with energy being released in the process. Later it was found that neutrons were emitted during the process and that the fission was principally confined to the uranium isotope U^{235} present as $\frac{1}{139}$ part of the natural uranium.

When it became known that the isotope U^{235} in natural uranium could be split or fissioned by bombardment with thermal neutrons, i. e., neutrons at or near thermal equilibrium with the surrounding medium, many predictions were made as to the possibility of obtaining a self-sustaining chain reacting system operating at high neutron densities. In such a system, the fission neutrons produced give rise to new fission neutrons in sufficiently large numbers to overcome the neutron losses in the system. Since the result of the fission of the uranium nucleus is the production of two lighter elements with great kinetic energy, plus approximately 2 fast neutrons on the average for each fission along with beta and gamma radiation, a large amount of power could be made available if a self-sustaining system could be built.

In order to attain such a self-sustaining chain reaction in a system of practical size, the ratio of the number of neutrons produced in one generation by the fissions, to the original number of neutrons initiating the fissions, must be known to be greater than unity after all neutron losses are deducted, and this ratio is, of course, dependent upon the values of the pertinent constants.

In the co-pending application of Enrico Fermi, Serial No. 534,129, filed May 4, 1944, and entitled "Nuclear Chain Reacting Systems," there is described and claimed a means and method of determining the neutron reproduction ratio for any type of uranium-containing structure, directly as a result of a simple measurement which can be performed with precision. Accurate values for all of the pertinent nuclear constants need not be known.

We have discovered certain essential principles required for the successful construction and operation of self-sustaining neutron chain reacting systems (known as neutronic reactors) with the production of power in the form of heat. These principles have been confirmed with the aid of measurements made in accordance with the means and method set forth in the above-identified application, and neutronic reactors have been constructed and operated at various power outputs, in accordance with these principles, as will be more fully brought out hereinafter.

In a self-sustaining chain reaction of natural uranium with slow neutrons, as presently understood, reactions occur involving the isotopes U^{238} and U^{235}. Thus, 92^{238}

2

is converted by neutron capture to the isotope 92^{239}. The latter is converted by beta decay to 93^{239} and this 93^{239} in turn is converted by beta decay to 94^{239}. Other isotopes of 93 and 94 may be formed in small quantities. By slow or thermal neutron capture, 92^{235}, on the other hand, can undergo nuclear fission to release energy appearing as heat and gamma and beta radiation, together with the formation of fission fragments appearing as radioactive isotopes of elements of lower mass numbers, and with the release of secondary neutrons.

The secondary neutrons thus produced by the fissioning of the 92^{235} nuclei have a high average energy, and must be slowed down to thermal energies in order to be in condition to cause slow neutron fission in other 92^{235} nuclei. This slowing down, or moderation of the neutron energy, is accomplished by passing the neutrons through a material where the neutrons are slowed by collision. Such a material is known as a moderator. While some of the secondary neutrons are absorbed by the uranium isotope 92^{238} leading to the production of element 94, and by other materials such as the moderator, enough neutrons can remain to sustain the chain reaction, when proper conditions are maintained.

Under these proper conditions, the chain reaction will supply not only the neutrons necessary for maintaining the neutronic reaction, but also will supply the neutrons for capture by the isotope 92^{238} leading to the production of 94, and excess neutrons for use as desired.

As 94 is a transuranic element, it can be separated from the unconverted uranium by chemical methods, and as it is fissionable by slow neutrons in a manner similar to the isotope 92^{235}, it is valuable, for example, for enriching natural uranium for use in other chain reacting systems of smaller overall size. The fission fragments are also valuable as sources of radioactivity.

The ratio of the fast neutrons produced in one generation by the fissions to the original number of fast neutrons in a theoretical system of infinite size where there can be no external loss of neutrons is called the reproduction of multiplication factor or constant of the system, and is denoted by the symbol K. For any finite system, some neutrons will escape from the periphery of the system. Consequently a system of finite size may be said to have a K constant, even though the value thereof would only exist if the system as built were extended to infinity without change of geometry or materials. Thus when K is referred to herein as a constant of a system of practical size, it always refers to what would exist in the same type of system of infinite size. If K can be made sufficiently greater than unity to indicate a net gain in neutrons in the theoretical system of infinite size, and then an actual system is built to be sufficiently large so that this gain is not entirely lost by leakage from the exterior surface of the system, then a self-sustaining chain reacting system of finite and practical size can be built to produce power and related by-products by nuclear fission of natural uranium. The neutron reproduction ratio in a system of finite size therefore differs from K by the external leakage factor, and by a factor due to the neutron absorption by localized neutron absorber, and the reproduction ratio must still be sufficiently greater than unity to permit the neutron density to rise exponentially with time in the system as built.

Progressive empirical enlargement of any proposed system for which the factor K is not accurately known, in an attempt to attain the overall size of a structure of finite size above which the rate of loss of neutrons by diffusion through the periphery of the structure is less than the rate of production of neutrons in the system, leads only to an expensive gamble with no assurance of success. The fact that K is greater than unity and the fact that the critical size is within practical limits must

2,708,656

3

be known rather accurately in advance, as otherwise a proposed structure having a K factor less than unity, or even a K factor greater than but sufficiently close to unity, would not sustain a chain reaction even if all of the urani-um in the world were included.

The earliest attempts to predict a structure capable of sustaining a chain reaction, using natural uranium, involved the use of fine uranium particles such as uranium oxide powder, dispersed in hydrogen in combined form as the slowing agent. However, these attempts were not successful, and analysis of experiments made has indicated that the neutron losses in such a system when natural uranium is used, can prevent a chain reaction from being sustained, irrespective of the size of the system.

However, in considering such experiments, it was found that neutron losses caused by absorption of neutrons by U²³⁸ which is present in natural or even enriched uranium, could be very substantially reduced by aggregating the uranium into bodies of substantial dimensions as compared to the uranium powder-hydrogen mixture previously suggested. It was also found that such aggregation will reduce resonance losses when a moderator such as graphite is used. This gain in neutrons, saved for use in the chain, has proved to be one of the major factors in obtaining a sufficiently low over-all neutron loss as to make possible the attainment of a self-sustaining chain reaction in various moderators, when other losses are also controlled.

During the interchange of neutrons in a system of finite size, comprising bodies of any size disposed in a neutron moderator, neutrons may be lost to the chain reaction in four ways:

1. By absorption or capture in the uranium content of the bodies without producing fission,

2. By absorption or capture in the moderator material itself,

3. By absorption or capture by the impurities present in both the uranium bodies and the moderator,

4. By leakage out of the system through the periphery thereof.

THE CHAIN FISSION REACTION

To illustrate the importance of the various factors entering into a chain reaction, we next describe the chain reaction process as it is presently understood to occur in any system of finite size utilizing natural uranium bodies dispersed in a graphite (a selected example) moderator at some position in the reactor where the neutron density is substantially constant. For better explanation, reference is here made to the diagram constituting Fig. 1 of the accompanying drawings, description of the remaining figures being more conveniently set forth in a subsequent part of this specification.

In Fig. 1, the letter

A represents a uranium body of any size from which fast neutrons are set free as a result of the fission process.

B represents a fast neutron loss due to leakage out of the system.

C represents a uranium body of any size in which both volume and surface resonance absorption of neutrons by U²³⁸ takes place, at resonance energies above thermal energy, leading to the formation of element 94.

D represents the number of neutrons reaching thermal energy.

E represents a thermal neutron loss by diffusion of thermal neutrons out of the system.

F represents a neutron loss caused by capture of neutrons by impurities in uranium, graphite, and controls.

G represents a neutron loss due to capture of thermal neutrons by the graphite as the thermal neutrons diffuse therethrough before entering uranium.

H represents the number of thermal neutrons entering uranium body.

I represents a uranium body of any size in which part of the thermal neutrons entering the body are absorbed by U²³⁸ leading to the formation of 94²³⁹, the remaining

4

thermal neutrons causing new fissions in U²³⁵ thereby producing fast neutrons, a few of which produce additional fast neutrons by fission of U²³⁸ atoms in the same body.

We will first consider the condition obtaining where thermal neutrons enter uranium body A. Some of these thermal neutrons will cause fission in the U²³⁵ content of the uranium body A to produce fast neutrons, the yield being at an average rate of about 2 neutrons per fission. As a result of this fission, fission fragments are released together with beta and gamma rays, thereby producing energy which, in the system, is manifested mostly by the heating of the uranium bodies with only a slight release of heat in the graphite. The actual average yield of fast neutrons by fission of U²³⁵ is slightly higher, e. g., by a few per cent, than the average of 2 mentioned above. Some of the fast neutrons released in the fission of U²³⁵ by the thermal neutrons of this example almost immediately produce fast fission of U²³⁸ in the same uranium body, with the production of additional fast neutrons.

The fast neutrons leaving the uranium body, for example $100n$ neutrons, enter the mass of moderator, travel therethrough, and through the uranium bodies over paths long in comparison with the spacing of the uranium bodies, to undergo successive collisions that slow them down. A substantial proportion of the fast neutrons are thus destined to be reduced, by about 100 elastic collisions apiece in the case of graphite and mostly in the moderator, to thermal energy. During this travel, before the neutrons arrive at thermal energies, a small percentage of the higher energy neutrons on the average may leak out of the system because of the finite size of the reactor, and be lost to the chain reaction. Furthermore, during the extremely irregular path of the neutrons while they are being slowed down by elastic collisions in the graphite, some of the neutrons will reach a uranium resonance absorption energy as they are about to enter a uranium body, such as C, and are absorbed immediately on or close to the surface of the uranium body. In addition some neutrons are reduced to resonance energy after entering the uranium body by an elastic collision with the uranium, and are therefore immediately absorbed within the uranium body. Irrespective of whether the neutron resonance absorption in U²³⁸ is on the surface, or in the volume of the uranium body, element 94 is produced by the resonance absorption according to the following process:

$$_{92}U^{238} + n \longrightarrow {}_{92}U^{239} + [6 \text{ m. e. v. of } \gamma \text{ rays, not necessarily all of one frequency}]$$

$$_{92}U^{239} \xrightarrow{23 \text{ min.}} {}_{93}Np^{239} + B\text{-}[1 \text{ m. e. v. } B\text{-}, \text{ no } \gamma \text{ rays}]$$

$$_{93}Np^{239} \xrightarrow{2.3 \text{ day}} {}_{94}Pu^{239} + B\text{-}[600 \text{ kv. upper } B\text{- energy limit. Also } 2\gamma \text{ rays, 400 kv. and 270 kv., about } \frac{1}{2} \text{ converted to electrons}]$$

A small amount of 94²⁴⁰ may also be found, formed by addition of a neutron to 94²³⁹. Capture of thermal neutrons by U²³⁸, as indicated in bodies A and C of Fig. 1, also results in production of element 94 by the same process.

The predominant isotope produced, 94²³⁹, is a long lived radioactive product with a half life of about 20,000 years.

A large percentage of the original fast neutrons escape resonance capture and fast neutron leakage, and are reduced to thermal energy within the system. Of these thermal neutrons, a small number on the average may leak by diffusion out of the system and be lost from the chain reaction, leaving the remainder of the thermal neutrons diffusing through the moderator in condition to produce fission if they promptly enter U²³⁵ or element 94 without being captured by any other material.

The fission reaction is as follows:

$$_{92}U^{235} + \text{neutron} = A + B + n \text{ neutrons (average)}$$

A="light" fission fragment, e. g., Br, Kr, Rb, Sr, Y, Zr,

2,708,656

53

for example, under manual control of the motors oy an operator watching the neutron density values.

However, with any control system there is always the possibility of accident in moving the control rod, as, for example, if motors driving the control rod should respond improperly and drive the control rod completely out of the reactor and for some reason (such as power line failure) refuse to return the rod. To take care of such an emergency safety devices are often provided such as, for example, neutron absorbing safety rods, which are normally held completely out of the reactor while the reactor is operating, but which can be manually or automatically released at a predetermined neutron density to enter the reactor rapidly, to absorb enough neutrons to stop the reaction. Such rods are shown in Figs. 1, 25, 31, and 38, numerals **40**, **112**, **241***b* and **370***b* respectively. To prevent power supply failure from affecting operation of these safety rods, they are usually arranged to drop, or be pulled, by gravity into the reactor when released at a predetermined neutron density.

The use of such safety rods again brings out the importance of not permitting the reproduction ratio to greatly exceed 1.01 in an operating reactor at any time. With reactors operating at high power such as, for example, the water cooled reactor described herein, even a single doubling of the operating neutron density might be disastrous, and the safety rods must be tripped to enter into the reactor before such doubling takes place. At $r=1.01$ the neutron density will double in about ⅓ of a second. If the time of doubling became too short, the safety rods could not arrive in place before the neutron density reached undesired values. However, with minimum doubling times of several seconds, for example, the safety rods can readily operate to prevent abnormal neutron densities from being reached after the rods are tripped.

It has been pointed out above that K can change slightly because of temperature changes and because of changes in atmospheric pressure when reactors are open to the atmosphere. Such changes in K are usually small and therefore change critical and operating sizes very slightly. The resultant changes in reproduction ratio are correspondingly small and may be compensated for by relatively short movements of the control rod.

However, K may change for other reasons during operation, and in high power reactors may change by such a large amount that if not properly compensated, the reproduction ratio may fall below unity during operation. These changes in K are at least partly due to the formation of reaction by-products in the reactor, either radioactive or stable. As the production of both radioactive and stable reaction products is a function of rate of irradiation of the uranium, the effect of these by-products on neutron reproduction becomes most important in reactors operating at high neutron densities.

Several relatively long term changes in K may be expected. Isotope 94²³⁹ is known to produce, on the average, a fraction more neutrons per fission than U²³⁵. Consequently, as U²³⁵ is used up by fission, and U²³⁸ is converted into 94²³⁹, the K constant may be expected to rise. To offset this rise in K there is an accumulation of long lived or stable fission products in the uranium that are the ends of fission fragment decay chains.

In reactors operating at high neutron densities, however, radioactive elements of exceedingly high capture cross section may be formed relatively quickly in the uranium as an intermediate element in the decay chains of the fission fragments and this formation can change K during operation. One of the most important of these decay chains is believed to be the 135 fission chain starting with

Te(short)→I(6.6 hr.)→Xe(9.4 hr.)
　　　　　　　　→Cc(20–30 yr.)→barium

the parenthetical times indicating half lives. The neutron absorption of tellurium, iodine, cesium and barium

54

is relatively unimportant, but the neutron capture cross section of radioactive xenon¹³⁵ has been measured to be about $2,500,000 \times 10^{-24}$ cm.², many times larger than that of stable gadolinium for example, the cross section of which is about $30,000 \times 10^{-24}$ cm.². Upon absorption of a neutron, xenon¹³⁵ shifts to xenon¹³⁶ an element of relatively small capture cross section. The change in K corresponds in period, to the xenon¹³⁵ appearance, and decay.

The rate of production of the Te is a function of the neutron density in which the uranium is immersed, and therefore dependent upon the power at which reactors of given type are operated. The radioactive xenon¹³⁵ is produced with a noticeable effect on the reaction a few hours after the reaction is started and the effect is, of course, greater as the neutron density is increased and maintained. The xenon¹³⁵ effect in high power reactors can be summarized as follows.

The reaction is started by withdrawing the control rod. The neutron density rises at a rate determined by the reproduction ratio and the effect of the delayed neutrons, until some predetermined neutron density is attained. The control rod is then placed in the unity reproduction ratio position and the reaction is stabilized at the power desired. During this time radioactive iodine is formed, decaying to xenon¹³⁵. As more and more iodine decays, more and more xenon¹³⁵ is formed, this xenon¹³⁵ absorbing sufficient neutrons to reduce the reproduction ratio below unity. This absorption also converts the xenon¹³⁵ to xenon¹³⁶ which has no excessive capture cross section. The neutron density drops. If no compensation were made for this drop by the rod the density might drop until background conditions prevailed, and then the reaction might automatically start up as the xenon¹³⁵ decayed. Normally the neutron density drop is compensated for by removal of the control or equivalent rod to a new position where the reproduction ratio is again above unity. A neutron density rise occurs, bringing the density back to its former level. Again, more xenon¹³⁵ is formed and the process is repeated until an equilibrium condition is reached where the xenon¹³⁵ formed is transmuted by neutron absorption and by decay into isotopes of lower capture cross section, as fast as it is being formed. In the meantime, the control rod (or equivalent) has to be withdrawn by an amount thereby removing from the reactor neutron absorbers at least equal in effect to the absorption caused by the xenon¹³⁵.

It should also be pointed out that this xenon¹³⁵ effect will be present when shifting from a power where the effect is stabilized, to a higher power. The shift can be made and the reaction stabilized at the new power for a time, because the iodine formed from the new fissions has not had time to produce a significant amount of additional xenon¹³⁵. As the new xenon¹³⁵ is formed from the decay of the newly formed iodine, the reproduction ratio will again drop and must again be increased by withdrawal of absorbers from the reactor.

However, the reduction in K due to the xenon¹³⁵ equilibrium amount present when the neutron density is theoretically infinity in the reactor, is believed to be about .03, which means that to obtain a rise in neutron density to any desired density up to infinity, the rod would have to be eventually removed by an amount corresponding to an increase in the reproduction ratio of about .03, and somewhat less than .03 when finite densities are to be obtained, in accordance with the density desired the size of the reactor would have to be big enough to provide the increase. For example, in a continuously operated water cooled U-graphite reactor such as herein described, the reproduction ratio decreases due to xenon¹³⁵ at equilibrium is about .0012 at 10,000 kilowatts; .009 at 100,000 kilowatts and would be about .013 and .020 at 200,000 kilowatts and 500,000 kilowatts respectively. As before stated reactors ordinarily are not built sufficiently large in size to provide maximum repro-

2,708,656

55

duction ratios of over 1.01 with all rods removed. However, if power outputs over 100,000 kilowatts are desired, the reactor must have its critical and operating sizes calculated as set forth herein using a final K constant decreased by the xenon[135] factor for the power desired even though the amount of reduction is over .01. In other words a significant impurity will be added during operation at high powers.

This may lead to the requirement for a reactor of such size that, if it did not acquire xenon[135] during operation, could attain a maximum reproduction ratio of over 1.01 with all rods removed, and which, before a substantial amount of the iodine produced from fission decayed into xenon[135], could in consequence attain a dangerous neutron density if all rods were removed.

Such a reactor can be adequately safeguarded by the use of "shim" or limiting rods inserted in the reactor preferably to depths that will not permit a reproduction ratio of about 1.01 to be attained at any time during the operation of the reactor, even when the control rod is completely removed. Then, if a reproduction ratio of more than unity cannot be attained by outward movement of the control rod alone, due to the build-up of the xenon[135] effect, the shim rod can be withdrawn to compensate for the xenon[135] effect, but still be left in a position where the reproduction ratio cannot exceed 1.01, when the control rod is completely removed. Such shim rods are shown in Figs. 7, 25, 31 and 37, numerals 30, 150, 241a and 370a respectively.

It can thus be seen that compensation for the xenon[135] effect is obtained, first by considering the xenon[135] impurity factor for the power desired, as a reduction in K to determine a proper operating size, for a desired power, and second, by initially providing in the reactors, impurities that can be removed by amounts compensating for the xenon[135] equilibrium amount acquired at a given power output.

Care must be taken, when shutting down a high power reactor operating with a xenon[135] equilibrium, that sufficient neutron absorbers are inserted to prevent automatic start-up of the reaction after the xenon[135] has decayed to the point where it does not materially affect the operation of the system. As a practical matter, shut down should include the full insertion of all control, shim and safety rods into the reactor.

Reactors operating at a few hundred watts, and operated intermittently, as when for example, the reactors are shut down at night, are not significantly affected by xenon[135] poisoning. Such a reactor is exemplified by the reactor first described herein. Even in the air cooled reactor described herein, operating at from 500 to a few thousand kilowatts output, the xenon[135] effect is on the order of the temperature and pressure effects. However, in the D$_2$O reactor and in the water cooled reactor described herein, the effect is more pronounced and is compensated for as described.

Because of the fact that the xenon[135] effect does not become important for several hours, reactors having an operating size too small to provide full compensation for the xenon effect at elevated neutron densities when continuously operated, nevertheless can be operated intermittently to attain such elevated densities or even higher densities for short periods until the xenon[135] effect prevents further operation at those densities. As the xenon[135] effect enters the reaction and stops the reaction, the density will drop. However, by waiting until the xenon[135] decays to the point where the neutron reproduction ratio can again be made greater than unity, the reactor can again attain the desired neutron density level. Thus reactors with a small maximum reproduction ratio can be operated intermittently to attain for short periods, neutron densities far greater than could be continuously maintained.

However, the control rod can be calibrated in several ways for steady state conditions and the calibration is

56

adequate. As the effect per inch movement of control rod is greater for the portion of the control rod nearer the center of the reactor than it is for the portion near the edge (because of the larger neutron density at the center), a unit may be chosen so that for movement of the control rod, one of said units will always have the same effect on the reproduction ratio of the reactor without regard to the actual depth of said rod in the reactor. Such a unit is based on the conventional inch, and is sometimes called a "cinch." Any movement of the control rod the distance of one cinch has the same effect on the reproduction ratio of the reactor as a movement of the control rod one inch from the critical position.

The control rod can also be calibrated in terms of a unit known as the "inhour." One inhour is the distance that the control rod must be moved from the critical or balanced position to give the reactor a period of one hour. The period of a neutronic reactor is, by definition, the time necessary for the neutron intensity to increase by a factor of "e" ($e = 2.718$).

In measuring the period of the reactor, a correction is made for any change in atmospheric pressure if the reactor is open to the atmosphere. An increase in atmospheric pressure will cause an increase in the weight of air inside the reactor. Oxygen has a small danger coefficient and, therefore, does not absorb neutrons in great quantities, but nitrogen on the other hand, has larger danger coefficient and so the great quantity of nitrogen present in the reactor has a distinct effect on the reproduction ratio r. It has been found that a change in atmospheric pressure on the reactor first described herein, is equal to 0.323 inhour for a change of one millimeter of mercury from the standard atmospheric pressure of 760 millimeters of mercury. At higher powers, a factor corresponding to the changes in operating density must be taken into account in calibration if measurements are made after the xenon effect appears.

USES OF NEUTRONIC REACTORS

In the descriptions of the various neutronic reactors given herein, only a little has been said regarding the uses of the reactors. As such uses are many, only a few of the most important will be mentioned here.

All of the reactors described herein are primarily extremely powerful neutron and gamma ray sources. When used as neutron sources, materials to be made radioactive can be placed in or close to the periphery of the reactors, and radioactive isotopes produced, for example, in large quantities, as the materials are there exposed to the entire energy spectrum of the neutrons in or escaping from the reactor. As the leakage from the D$_2$O-uranium reactor is even larger than the neutron leakage from the graphite-uranium reactors, a relatively larger neutron flux can be intercepted on the exterior of this type of reactor.

One example of isotope production by exposure to reactor generated neutrons, followed by transmutation, is the manufacture of U[233] from the thorium 232, for which process thorium 233 can first be produced from thorium 232, the extent of the reaction being dependent upon the product of the slow neutron density and the time of exposure. Thorium 233 then decays to form protoactinium 233 and thence to uranium 92[233] which is valuable as a fissionable material similar in its action to U[235] and 94[239]. The reaction is as follows:

$$_{90}Th^{232} + n \longrightarrow {}_{90}Th^{233} + \text{gamma rays}$$

$$_{90}Th^{233} \xrightarrow[(-\beta^-)]{23.5 \text{ min.}} {}_{91}Pa^{233} \xrightarrow[(-\beta^-)]{27.4 \text{ days}} {}_{92}U^{233}$$

As a further example of transmutation, radioactive carbon may be produced by allowing the neutrons leaving the reactor to react with nitrogen in compound form. This reaction gives rise to carbon of mass 14 which is radioactive, and can be separated chemical-

2,708,656

57

ly from the nitrogenous compounds. Such radioactive carbon is suitable for medical and physiological uses as it may be incorporated in organic compounds and used as tracers in living organisms. Well 21 can also be filled with graphite blocks and the resultant pier projected upwardly through the top. The top of this graphite pier provides a strong thermal neutron source and such a pier is known as a thermal neutron column.

It can thus be seen that the neutrons normally escaping from the reactors need not be lost, but can be put to work, and the neutrons escaping from the reactors can be utilized to produce transmutations or isotopes from elements placed in, or surrounding all parts of the active portion of the reactor. For example, the number of neutrons radiated from the external surface of a uranium-graphite reactor when operating at ten thousand kilowatts power output in the form of heat, is approximately 1.8×10^{16} neutrons per second.

In utilizing the output of reactors, internal shafts such as shaft 26 and tube 109b play an important role. They extend to the vicinity of the center of the reactors where the highest neutron densities exist, and intense neutron bombardment of materials inserted into the bottom of these shafts will take place, particularly at high reactor powers. Furthermore, the shafts act to collimate the fast neutrons released inside of the reactors and a high density collimated beam of neutrons emerges through the external aperture, projected outwardly. Such a collimated beam, having a far greater fast neutron density than any neutron beam heretofore produced, can be utilized outside of the pile for nuclear research in all of its aspects. The number of neutrons escaping from these shafts is several times the number escaping over an area of the external surface of the reactors equal to the cross section of the shafts.

In addition, extremely high energy gamma rays are emitted during nuclear fission. These rays also escape through the shafts to the exterior of the reactors and can there be used for taking radiographs through large castings, for example, with relatively short exposures, during high power operation of the reactors. The neutrons coming from the reactors can be screened out of the gamma ray beam by the use of relatively thin sheets of materials having high neutron absorption cross sections without substantially reducing the gamma ray intensity. In addition, a bismuth filter has been found to effectively reduce the gamma rays, without substantial interference with the neutron beam. Thus shaft 26 and tube 109b can be used either to produce a high intensity collimated neutron beam, or to produce a high energy beam of gamma rays, as desired, both for use outside the reactors.

In addition, all of the reactors described are also extremely useful in testing materials for neutron absorption and neutron production. Using one or more of the removable stringers 36a for example, in the uncooled graphite-uranium reactor the reactor can be balanced at a given neutron density with uranium bodies of known constants in the stringer. The stringer is then withdrawn and new uranium bodies substituted for those withdrawn. The stringer is reinserted into the pile, and the neutron intensity brought to the original value. The change in position of the control rod for the balance condition, when corrected for atmospheric pressure and temperature, will at once tell whether or not the newly inserted bodies are better or worse when used as elements in the system, than those removed. From the results obtained, calculations can be made as to systems incorporating the new bodies. The effect of changes in size, impurities, coatings and temperature on the chain reaction, can similarly be determined. In the latter case, uranium lumps can be heated and inserted to determine the effect of temperature on the reaction as measured by change in the position of the control rod. However, the stringer

58

method of determining the effect of changes in pile construction is no part of the present invention. Similarly, materials can be tested by insertion in tube 109b in the D₂O reactor or in the coolant channels in the other reactors.

The reactors described herein, in addition to being high power neutron generators, are capable of producing the products of the neutronic reaction in quantities related to the power at which they are operated, in that at least a portion of the uranium bodies are removable from the reactors after exposure to the reaction for varying periods of time. By proper chemical treatment the 94^{239} and fission products can be recovered from the removed irradiated uranium bodies and thereafter utilized as desired.

With modifications, the reactors herein described can also be used as sources of power in useful form. The D₂O moderated reactors can be operated under pressure at an elevated temperature, with continuous removal of the heated D₂O for flashing into D₂O steam for operation of low pressure turbines. Enriched uranium with a light water moderator can be operated in the same manner. The gas cooled reactors, when cooled with helium under pressure, for example, can be used to produce steam by passing the heated helium through heat exchangers. Diphenyl when used as a coolant in the liquid cooled reactor can be heated above the boiling point of water and then used in heat exchangers to produce steam. With proper design, neutronic reactors can also be operated to produce steam directly in tubes passing through the reactor, utilizing the heat of vaporization for cooling the reactor, and the resultant steam for power.

While the theory of the nuclear chain fission mechanism in uranium set forth herein is based on the best presently known experimental evidence, we do not wish to be bound thereby, as additional experimental data later discovered may modify the theory disclosed.

What is claimed is:

1. A neutronic reactor which comprises a moderator of graphite and natural uranium rods disposed in a geometric pattern therein, the size of the rods and the volume ratio of moderator to uranium being within the area encompassed by the $k=1.00$ curve of Figure 3, the purity of the graphite and the uranium and the total mass thereof being sufficient to sustain a chain reaction.

2. A neutronic reactor which comprises a moderator selected from the group consisting of heavy water and graphite and bodies of a thermal neutron fissionable material selected from the group consisting of natural uranium and natural uranium oxide disposed in a geometric pattern therein, each body being surrounded by moderator and the moderator being in a substantially continuous phase, the shape of the bodies and the radius of the bodies and the volume ratio of moderator to thermal neutron fissionable material being within the area encompassed by the $k=1.00$ curve of Figures 2 through 6, the purity of the moderator and the thermal neutron fissionable material and the total mass thereof being sufficient to sustain a chain reaction.

3. A neutronic reactor which comprises a moderator of graphite and bodies of natural uranium in the form of spheres disposed in a geometric pattern therein, each body being surrounded by moderator and the moderator being in a substantially continuous phase, the radius of the bodies and the volume ratio of moderator to uranium being within the area encompassed by the $k=1.00$ curve of Figure 2, the purity of the moderator and the uranium and the total mass thereof being sufficient to sustain a chain reaction.

4. A neutronic reactor which comprises a moderator of graphite and bodies of natural uranium oxide in the form of spheres disposed in a geometric pattern therein, each body being surrounded by moderator and the moderator being in a substantially continuous phase, the radius of the bodies and the volume ratio of moderator to uranium oxide being within the area encompassed by

2,708,656

59

the $k=1.00$ curve of Figure 4, the purity of the moderator and the uranium oxide and the total mass thereof being sufficient to sustain a chain reaction.

5. A neutronic reactor which comprises a moderator of graphite and bodies of natural uranium oxide in the form of rods disposed in a geometric pattern therein, each body being surrounded by moderator and the moderator being in a substantially continuous phase, the radius of the bodies and the volume ratio of moderator to uranium oxide being within the area encompassed by the $k=1.00$ curve of Figure 5, the purity of the moderator and the uranium oxide and the total mass thereof being sufficient to sustain a chain reaction.

6. A neutronic reactor which comprises a moderator of heavy water and bodies of natural uranium in the form of rods disposed in a geometric pattern therein, each body being surrounded by moderator and the moderator being in a substantially continuous phase, the radius of the bodies and the volume ratio of moderator to uranium being within the area encompassed by the $k=1.00$ curve of Figure 6, the purity of the moderator and the uranium and the total mass thereof being sufficient to sustain a chain reaction.

7. In a neutronic reactor having an active portion comprising a moderator of graphite having dispersed therein uranium containing U^{235} and U^{238}, the improved construction wherein the uranium is aggregated in the form of bodies substantially free of moderator and of neutron absorbers other than U^{238}, said bodies being in the moderator, geometrically spaced therein, and surrounded by the moderator, the moderator being in a substantially continuous phase, said bodies having all dimensions thereof at least 0.5 centimeter, the purity of the moderator and the uranium, the size and spacing of the bodies of uranium in the moderator, and the total mass of uranium and moderator being sufficient to sustain a chain reaction.

60

8. In a neutronic reactor having an active portion comprising a mass of moderator selected from the group consisting of graphite and heavy water, having dispersed therein a thermal neutron fissionable material containing a thermal neutron fissionable isotope and an isotope having a resonance absorption for neutrons, the improved construction wherein the thermal neutron fissionable material is aggregated in the form of bodies substantially free of moderator and of neutron absorbers other than said latter istope, said bodies being in the moderator, geometrically spaced therein, and surrounded by the moderator, the moderator being in a substantially continuous phase, said bodies having all dimensions thereof at least 0.5 centimeter, the purity of the moderator and the thermal neutron fissionable material, the size and spacing of the bodies of fissionable material in the moderator, and the total mass of fissionable material and moderator being sufficient to sustain a chain reaction.

References Cited in the file of this patent

UNITED STATES PATENTS

2,206,634	Fermi et al.	_____	July 2, 1940

FOREIGN PATENTS

114,150	Australia	_____	May 2, 1940
114,151	Australia	_____	May 3, 1940
233,011	Switzerland	_____	Oct. 2, 1944
861,390	France	_____	Oct. 28, 1940
648,293	Great Britain	_____	Jan. 3, 1951

OTHER REFERENCES

Power, July 1940, page 58. Copy in 204–154.2.

Kelly et al.: Physical Review 73, 1135–9 (1948). Copy in Patent Office Library (204/154.2).

Flugge: Naturwissenschaften, vol. 27, pages 402–410 (1939). Copy in Patent Office Library (204/154.2).

List of Patents and Disclosures

List of Published Patents

Germany

DEUTSCHES REICH

AUSGEGEBEN
AM 26. JULI 1924

REICHSPATENTAMT
PATENTSCHRIFT
— № 399056 —
KLASSE 21g GRUPPE 16
(S 64002 VIII/21g⁹)

399056

Dr. Leo Szilard in Berlin-Dahlem und Dipl.-Jng. Imre Patai in Budapest.
Für Röntgenstrahlen empfindliche Zelle.
Patentiert im Deutschen Reiche vom 9. Oktober 1923 ab.

DEUTSCHES REICH

AUSGEGEBEN AM
3. JUNI 1929

REICHSPATENTAMT
PATENTSCHRIFT
№ 476812
KLASSE 31c GRUPPE 26
S 72993 VI/31c
Tag der Bekanntmachung über die Erteilung des Patents: 8. Mai 1929

Dr. Leo Szilard in Berlin-Dahlem
Verfahren zum Gießen von Metallen in Formen unter Anwendung elektrischer Ströme
Patentiert im Deutschen Reiche vom 20. Januar 1926 ab

DEUTSCHES REICH

AUSGEGEBEN AM
27. JULI 1929

REICHSPATENTAMT

PATENTSCHRIFT

№ 480 037

KLASSE **17** c GRUPPE 3

S 82114 I/17 c

Tag der Bekanntmachung über die Erteilung des Patents: 4. Juli 1929

Dr. Leo Szilard in Berlin-Wilmersdorf

Kältespeicher

Patentiert im Deutschen Reiche vom 9. Oktober 1927 ab

DEUTSCHES REICH

AUSGEGEBEN AM
28. MÄRZ 1930

REICHSPATENTAMT

PATENTSCHRIFT

№ 494 810

KLASSE **17** a GRUPPE 12

S 86025 I/17 a

Tag der Bekanntmachung über die Erteilung des Patents: 13. März 1930

Dr. Leo Szilard in Berlin-Wilmersdorf

Intermittierend wirkende Kältemaschine mit getrenntem Kocher und Absorber

Patentiert im Deutschen Reiche vom 13. Juni 1928 ab

DEUTSCHES REICH

AUSGEGEBEN AM
6. AUGUST 1930

REICHSPATENTAMT

PATENTSCHRIFT

№ 504 545

KLASSE **21** g GRUPPE 12

S 67727 VIII a/21 g

Tag der Bekanntmachung über die Erteilung des Patents: 24. Juli 1930

Dr. Leo Szilard in Berlin-Dahlem

Entladungsröhre mit Steuerung des Anodenstromes

Patentiert im Deutschen Reiche vom 13. November 1924 ab

DEUTSCHES REICH

AUSGEGEBEN AM
27. SEPTEMBER 1930

REICHSPATENTAMT

PATENTSCHRIFT
№ 508486
KLASSE **17a** GRUPPE 13/*03*
S 82485 I/17a
Tag der Bekanntmachung über die Erteilung des Patents: 11. September 1930

Dr. Leo Szilard in Berlin-Wilmersdorf

Kältemaschine

Patentiert im Deutschen Reiche vom 1. November 1927 ab

DEUTSCHES REICH

AUSGEGEBEN AM
23. DEZEMBER 1930

REICHSPATENTAMT

PATENTSCHRIFT
№ 515054
KLASSE **21**g GRUPPE 12
S 66997 VIIIa/21 g
Tag der Bekanntmachung über die Erteilung des Patents: 11. Dezember 1930

Dr. Leo Szilard in Berlin-Dahlem

Entladungsröhre, bei welcher eine Gasentladung als Elektronenquelle dient

Patentiert im Deutschen Reiche vom 4. September 1924 ab

DEUTSCHES REICH

AUSGEGEBEN AM
1. DEZEMBER 1933

REICHSPATENTAMT

PATENTSCHRIFT
№ 531581
KLASSE **17a** GRUPPE 3₀₄
S 92040 I/17a
Tag der Bekanntmachung über die Erteilung des Patents: 30. Juli 1931

531581

Dr. Leo Szilard in Berlin-Wilmersdorf

Pumpe, insbesondere zur Verdichtung von Gasen und Dämpfen in Kältemaschinen

Patentiert im Deutschen Reiche vom 4. Juni 1929 ab

DEUTSCHES REICH

AUSGEGEBEN AM
21. SEPTEMBER 1931

REICHSPATENTAMT
PATENTSCHRIFT
№ 533945
KLASSE 17a GRUPPE 3
S 91304 I/17a
Tag der Bekanntmachung über die Erteilung des Patents: 3. September 1931

Dr. L. Szilard in Berlin-Wilmersdorf

Pumpe

Patentiert im Deutschen Reiche vom 25. April 1929 ab

___

DEUTSCHES REICH

AUSGEGEBEN AM
3. FEBRUAR 1932

REICHSPATENTAMT
PATENTSCHRIFT
№ 543214
KLASSE 17a GRUPPE 3
S 92043 I/17a
Tag der Bekanntmachung über die Erteilung des Patents: 14. Januar 1932

543214

Dr. Leo Szilard in Berlin-Wilmersdorf
Vorrichtung zur Bewegung von flüssigen Metallen
Patentiert im Deutschen Reiche vom 4. Juni 1929 ab

___

DEUTSCHES REICH

AUSGEGEBEN AM
1. NOVEMBER 1933

REICHSPATENTAMT
PATENTSCHRIFT
№ 548136
KLASSE 17a GRUPPE 3 04
S 95968 I/17a
Tag der Bekanntmachung über die Erteilung des Patents: 24. März 1932

Dr. Leo Szilard in Berlin-Wilmersdorf
Kältemaschine
Patentiert im Deutschen Reiche vom 7. Januar 1931 ab

___

DEUTSCHES REICH

AUSGEGEBEN AM
27. JULI 1933

REICHSPATENTAMT

PATENTSCHRIFT

№ 554959

KLASSE **17a** GRUPPE 3 04

S 83372 I/17a

Tag der Bekanntmachung über die Erteilung des Patents: 30. Juni 1932

Dr. Leo Szilard in Berlin-Wilmersdorf und Dr. Albert Einstein in Berlin

Vorrichtung zur Bewegung von flüssigem Metall, insbesondere zur Verdichtung
von Gasen und Dämpfen in Kältemaschinen

Patentiert im Deutschen Reiche vom 28. Dezember 1927 ab

———

DEUTSCHES REICH

AUSGEGEBEN AM
9. DEZEMBER 1933

REICHSPATENTAMT

PATENTSCHRIFT

№ 555141

KLASSE **17a** GRUPPE 3 04

S 92042 I/17a

Tag der Bekanntmachung über die Erteilung des Patents: 30. Juni 1932

Dr. Leo Szilard in Berlin-Wilmersdorf

Vorrichtung zur Bewegung von flüssigen Metallen

Patentiert im Deutschen Reiche vom 4. Juni 1929 ab

———

DEUTSCHES REICH

AUSGEGEBEN AM
28. JULI 1933

REICHSPATENTAMT

PATENTSCHRIFT

№ 555413

KLASSE **17a** GRUPPE 3 04

S 88745 I/17a

Tag der Bekanntmachung über die Erteilung des Patents: 7. Juli 1932

Dr. Albert Einstein in Berlin und Dr. Leo Szilard in Berlin-Wilmersdorf

Pumpe, vorzugsweise für Kältemaschinen

Patentiert im Deutschen Reiche vom 4. Dezember 1928 ab

———

DEUTSCHES REICH

AUSGEGEBEN AM
2. SEPTEMBER 1933

REICHSPATENTAMT
PATENTSCHRIFT
№ 556536
KLASSE **17**a GRUPPE **3**04
S 92025 I/17a
Tag der Bekanntmachung über die Erteilung des Patents: 21. Juli 1932

Dr. Leo Szilard in Berlin-Wilmersdorf

Kältemaschine

Patentiert im Deutschen Reiche vom 4. Juni 1929 ab

DEUTSCHES REICH

AUSGEGEBEN AM
16. SEPTEMBER 1933

REICHSPATENTAMT
PATENTSCHRIFT
№ 556535
KLASSE **17**a GRUPPE 3/04
E 40538 I/17a
Tag der Bekanntmachung über die Erteilung des Patents: 21. Juli 1932

Dr. Albert Einstein in Berlin und Dr. Leo Szilard in Berlin-Wilmersdorf

Pumpe, vorzugsweise für Kältemaschinen

Zusatz zum Patent 555 413

Patentiert im Deutschen Reiche vom 15. April 1930 ab
Das Hauptpatent hat angefangen am 4. Dezember 1928.

DEUTSCHES REICH

AUSGEGEBEN AM
13. APRIL 1933

REICHSPATENTAMT
PATENTSCHRIFT
№ 561904
KLASSE **17**a GRUPPE 3/04
17a E 64. 30
Tag der Bekanntmachung über die Erteilung des Patents: 29. September 1932

Dr. Albert Einstein in Berlin und Dr. Leo Szilard in Berlin-Wilmersdorf

Kältemaschine

Patentiert im Deutschen Reiche vom 15. April 1930 ab

DEUTSCHES REICH

AUSGEGEBEN AM
20. SEPTEMBER 1933

REICHSPATENTAMT

PATENTSCHRIFT

№ 562 040

KLASSE **21** d² GRUPPE **18**₀₁

S 85876 VIII b/21 d²   6. Oktober 1932.

Tag der Bekanntmachung über die Erteilung des Patents: 29. September 1932

Dr. Leo Szilard in Berlin-Wilmersdorf und Dr. Albert Einstein in Berlin

Elektromagnetische Vorrichtung zur Erzeugung einer oszillierenden Bewegung

Patentiert im Deutschen Reiche vom 1. Juni 1928 ab

DEUTSCHES REICH

AUSGEGEBEN AM
8. APRIL 1933

REICHSPATENTAMT

PATENTSCHRIFT

№ 562 300

KLASSE **17**a GRUPPE **1**/₀₁

E 40537 I/17 a

Tag der Bekanntmachung über die Erteilung des Patents: 6. Oktober 1932

Dr. Albert Einstein in Berlin und Dr. Leo Szilard in Berlin-Wilmersdorf

Kältemaschine

Patentiert im Deutschen Reiche vom 15. April 1930 ab

DEUTSCHES REICH

AUSGEGEBEN AM
26. OKTOBER 1932

REICHSPATENTAMT

PATENTSCHRIFT

№ 562 523

KLASSE **17**a GRUPPE **14**

S 99398 I/17 a

Tag der Bekanntmachung über die Erteilung des Patents: 6. Oktober 1932

Dr. Leo Szilard in Berlin-Wilmersdorf

Absperrorgan

Patentiert im Deutschen Reiche vom 26. Juni 1931 ab

DEUTSCHES REICH

AUSGEGEBEN AM
1. NOVEMBER 1933

REICHSPATENTAMT

PATENTSCHRIFT

№ 562 898

KLASSE **17**a   GRUPPE **13**₀₁

17a S 194. 30

Tag der Bekanntmachung über die Erteilung des Patents: 13. Oktober 1932

Dr. Leo Szilard in Berlin-Wilmersdorf

Wärmeübertrager

Patentiert im Deutschen Reiche vom 10. September 1930 ab

DEUTSCHES REICH

AUSGEGEBEN AM
30. MAI 1933

REICHSPATENTAMT

PATENTSCHRIFT

№ 563 403

KLASSE **17**a   GRUPPE **3**₀₁

S 82663 I/17a

Tag der Bekanntmachung über die Erteilung des Patents: 29. Oktober 1932

Dr. Leo Szilard in Berlin-Dahlem und Dr. Albert Einstein in Berlin

Kältemaschine

Patentiert im Deutschen Reiche vom 13. November 1927 ab

DEUTSCHES REICH

AUSGEGEBEN AM
16. JULI 1934

REICHSPATENTAMT

PATENTSCHRIFT

№ 564 680

KLASSE **17**a   GRUPPE **3**/₁₀₄

S 95969 I/17a

Tag der Bekanntmachung über die Erteilung des Patents: 3. November 1932

Dr. Leo Szilard in Berlin-Wilmersdorf

Kältemaschine

Patentiert im Deutschen Reiche vom 7. Januar 1931 ab

DEUTSCHES REICH

AUSGEGEBEN AM
4. JULI 1933

REICHSPATENTAMT

PATENTSCHRIFT

№ 565 614

KLASSE **17**a GRUPPE **3**₀₄

E 39852 I/17a

Tag der Bekanntmachung über die Erteilung des Patents: 17. November 1932

Dr. Albert Einstein in Berlin und Dr. Leo Szilard in Berlin-Wilmersdorf

Kompressor

Patentiert im Deutschen Reiche vom 11. September 1929 ab

———

DEUTSCHES REICH

AUSGEGEBEN AM
15. SEPTEMBER 1933

REICHSPATENTAMT

PATENTSCHRIFT

№ 568 680

KLASSE **17**a GRUPPE **3**₀₄

S 95967 I/17a

Tag der Bekanntmachung über die Erteilung des Patents: 5. Januar 1933

Dr. Leo Szilard in Berlin-Wilmersdorf

Stator für Kältemaschinen

Patentiert im Deutschen Reiche vom 7. Januar 1931 ab

———

DEUTSCHES REICH

AUSGEGEBEN AM
17. MÄRZ 1934

REICHSPATENTAMT

PATENTSCHRIFT

№ 570 959

KLASSE **17**a GRUPPE **3**₀₄

S 92041 I/17a

Tag der Bekanntmachung über die Erteilung des Patents: 2. Februar 1933

Dr. Leo Szilard in Berlin-Wilmersdorf

Vorrichtung zur Bewegung von flüssigem Metall

Patentiert im Deutschen Reiche vom 4. Juni 1929 ab

———

DEUTSCHES REICH

AUSGEGEBEN AM
2. AUGUST 1933

REICHSPATENTAMT

PATENTSCHRIFT

№ 581 780

KLASSE 17a GRUPPE 3₀₄

S 95926 I/17a

Tag der Bekanntmachung über die Erteilung des Patents: 13. Juli 1933

Dr. Leo Szilard in Berlin-Wilmersdorf

Kompressor, im besonderen für Kältemaschinen

Patentiert im Deutschen Reiche vom 3. Januar 1931 ab

DEUTSCHES REICH

AUSGEGEBEN AM
29. JUNI 1933

REICHSPATENTAMT

PATENTSCHRIFT

№ 579 679

KLASSE 21g GRUPPE 12₀₃

21g S 717. 30

Tag der Bekanntmachung über die Erteilung des Patents: 15. Juni 1933

Dr. Leo Szilard in Berlin-Wilmersdorf

Entladungsröhre, bei welcher eine Gasentladung als Elektronenquelle dient

Patentiert im Deutschen Reiche vom 4. September 1924 ab

DEUTSCHES REICH

AUSGEGEBEN AM
16. JULI 1934

REICHSPATENTAMT

PATENTSCHRIFT

№ 600 162

KLASSE 42g GRUPPE 11₀₁

S 98694 IX/42g

Tag der Bekanntmachung über die Erteilung des Patents: 28. Juni 1934

Dr. Leo Szillard in Berlin-Wilmersdorf

Verfahren zum elektrochemischen Aufzeichnen von Sprechströmen

Patentiert im Deutschen Reiche vom 17. Mai 1931 ab

Erteilt auf Grund des Ersten Überleitungsgesetzes vom 8. Juli 1949
(WiGBl. S. 175)

BUNDESREPUBLIK DEUTSCHLAND

AUSGEGEBEN AM
13. JUNI 1957

DEUTSCHES PATENTAMT

PATENTSCHRIFT

№ 965 522

KLASSE 21g   GRUPPE 37 01

INTERNAT. KLASSE H 05 j ———

S 18863 VIII c / 21 g

Dr. Leo Szilard, New York, N.Y. (V. St. A.)

Mikroskop

Patentiert im Gebiet der Bundesrepublik Deutschland vom 4. Juli 1931 an
Patentanmeldung bekanntgemacht am 20. Dezember 1956
Patenterteilung bekanntgemacht am 29. Mai 1957
Die Schutzdauer des Patents ist nach Gesetz Nr. 8 der Alliierten Hohen Kommission verlängert

---

United Kingdom

AMENDED SPECIFICATION.

Reprinted as amended under Section 8 of the Patents and Designs Acts, 1907 and 1919.

PATENT SPECIFICATION

Convention Date (Germany): Sept. 3, 1924.   **239,518**

Application Date (in United Kingdom): Sept. 2, 1925.   No. 21,972/25.

Complete Accepted: Aug. 26, 1926.

COMPLETE SPECIFICATION (AMENDED).

Improvements in or relating to Electron Discharge Tubes.

[We, Siemens-Schuckertwerke Gesellschaft mit beschränkter Haftung, of Berlin-Siemensstadt, Germany, a German company (part Assignees of Dr. Leo Szilard, of 16, Faradayweg, Berlin-Dahlem, Germany, of Hungarian nationality), and Dr. Leo Szilard, of 16, Faradayweg, Berlin-Dahlem, Germany, of Hungarian nationality, do hereby declare the nature of this invention and in what manner the same is to be performed, to be particularly described and ascertained in and by the following statement:-]

PATENT SPECIFICATION

Convention Date (Germany): Dec. 16, 1926.   **282,428**

Application Date (in United Kingdom): Dec. 16, 1927.   No. 34,096/27.

Complete Accepted: Nov. 15, 1928.

COMPLETE SPECIFICATION.

Improvements relating to Refrigerating Apparatus.

We, ALBERT EINSTEIN, of 5, Haberland-strasse, Berlin, W. 30, Germany, a citizen of Switzerland, and LEO SZILARD, of 95, Prinzregentenstrasse, Berlin-Wilmersdorf, formerly of Faradayweg 16, Berlin-Dahlem, Germany, a Hungarian citizen, do hereby declare the nature of this invention and in what manner the same is to be performed, to be particularly described and ascertained in and by the following statement:—

[This invention relates to refrigerating apparatus having a refrigerant evaporated in the evaporator by the introduction of a pressure equalising auxiliary medium thereinto and separated from said medium by the absorption of the latter and condensation of the refrigerant as described in British Patent Specification No. 250,983.]

NOTE.—*The application for a Patent has become void.*
This print shows the Specification as it became open to public inspection.

PATENT SPECIFICATION

Convention Date (Germany): Dec. 29, 1926. **282,808**

Application Date (In United Kingdom): Dec. 29, 1927. No. 35,201 / 27.

Complete not Accepted.

COMPLETE SPECIFICATION.

Refrigerating Machines in which the Pumping of Liquid is Effected by Intermittently Increasing the Vapour Pressure.

We, Dr. LEO SZILARD, of 95, Prinzregentenstrasse, Berlin-Wilmersdorf, Germany, and Professor Dr. ALBERT
EINSTEIN, of Haberlandstrasse 5, Berlin,
5 Germany, both German citizens, do
hereby declare the nature of this invention and in what manner the same is to
be performed, to be particularly described
and ascertained in and by the following
10 statement:—

NOTE.—*The application for a Patent has become void.*
This print shows the Specification as it became open to public inspection
under Section 91 (3) (a) of the Acts.

PATENT SPECIFICATION

Convention Date (Germany): Jan. 24, 1927. **284,222**

Application Date (in United Kingdom): Jan. 24, 1928. No. 2268 / 28.

Complete not Accepted.

COMPLETE SPECIFICATION.

Refrigerating Machine with Organic Solvent.

We, Professor Dr. ALBERT EINSTEIN, of
5, Haberlandstrasse, Berlin, Germany,
and Dr. LEO SZILARD, of 95, Prinzregentenstrasse, Berlin - Wilmersdorf,
5 Germany, both German citizens, do
hereby declare the nature of this
invention and in what manner the same
is to be performed, to be particularly
described and ascertained in and by the
10 following statement:—

RESERVE COPY.

PATENT SPECIFICATION

Convention Date (Germany): July 14, 1927.   **293,865.**

Application Date (In United Kingdom): July 10, 1928.   No. 20,043 / 28.

Complete Accepted: May 30, 1929.

COMPLETE SPECIFICATION.

Improvements in Refrigerating Processes and Apparatus.

I, LEO SZILARD, formerly of Faradayweg 16, Berlin-Dahlem, but now of
Prinz Regentenstrasse, 95, Berlin-Wilmersdorf, Germany, of Hungarian
5 nationality, on my own behalf and as
Assignee of ALBERT EINSTEIN, of 5,
Haberlandstrasse, Berlin, Germany, of
German nationality, do hereby declare
the nature of this invention and in what
10 manner the same is to be performed, to be
particularly described and ascertained
in and by the following statement:—

NOTE.—*The application for a Patent has become void.* RESERVE COPY.
*This print shows the Specification as it became open to public inspection under
Section 91 (3) (a) of the Acts.*

PATENT SPECIFICATION

Convention Date (Germany): Oct. 17, 1927.   **298,950**

Application Date (In United Kingdom): Oct. 16, 1928.   No. 29,915 / 28.

Complete not Accepted.

COMPLETE SPECIFICATION.

Improvements in or relating to Electric Condensers.

Dr. LEO SZILARD, of 95, Prinzregentenstrasse, Berlin-Wilmersdorf, Germany, do
hereby declare the nature of this invention and in what manner the same is to
5 be performed, to be particularly described
and ascertained in and by the following
statement:—

PATENT SPECIFICATION

Convention Date (Germany): Oct. 31, 1927.　　**299,783**

Application Date (In United Kingdom): Oct. 29, 1928.　No. 31,364 / 28.

Complete Accepted: Jan. 23, 1930.

COMPLETE SPECIFICATION.

Improvements in and relating to Refrigerating Machines.

I, Dr. Leo Szilard, of 95, Prinz Regentenstrasse, Berlin-Wilmersdorf, Germany, of Hungarian Nationality, do hereby declare the nature of this inven-
5 tion and in what manner the same is to be performed, to be particularly described and ascertained in and by the following statement:—

PATENT SPECIFICATION

Convention Date (Germany): Dec. 27, 1927.　　**303,065**

Application Date (in United Kingdom): Dec. 24, 1928.　No. 38,091 / 28.

Complete Accepted: May 26, 1930.

COMPLETE SPECIFICATION.

Electrodynamic Movement of Fluid Metals particularly for Refrigerating Machines.

We, Prof. Albert Einstein, of Swiss Nationality, of 5, Haberlanderstrasse, Berlin, Germany, and Dr. Leo Szilard, of Hungarian Nationality, of 95, Prinz-
5 regentenstrasse, Berlin-Wilmersdorf, Germany, do hereby declare the nature of this invention and in what manner the same is to be performed, to be particularly described and ascertained in and by
10 the following statement:—

RESERVE COPY

PATENT SPECIFICATION

Convention Date (Germany): Dec. 3, 1928.

Application Date (in United Kingdom): Dec. 3, 1929. No. 37,004/29.

Complete Accepted: March 3, 1931.

344,881

COMPLETE SPECIFICATION.

Pump, especially for Refrigerating Machines.

We, Prof. Dr. ALBERT EINSTEIN, of 5, Haberlandstrasse, Berlin, Germany, of Swiss Nationality, and Dr. LEO SZILARD, of 95, Prinz Regentenstrasse, Berlin-Wilmersdorf, Germany, of Hungarian Nationality, do hereby declare the nature of this invention and in what manner the same is to be performed, to be particularly described and ascertained in and by the following statement :—

PATENT SPECIFICATION

Application Date: March 12, 1934. No. 7840/34.

„   „   July 4, 1934.   No. 33540/35.

(No. 33540/35 being divided out of Application No. 19721/34.)

Application Date: Sept. 20, 1934. No. 27050/34.

One Complete Specification Left: April 9, 1935.

(Under Section 16 of the Patents and Designs Acts, 1907 to 1932.)

Specification Accepted: Dec. 12, 1935.

440,023

PROVISIONAL SPECIFICATION
No. 7840 A.D. 1934.

Improvements in or relating to the Transmutation of Chemical Elements

I, LEO SZILARD, a citizen of Germany and Hungary, c/o Claremont Haynes & Co., Vernon House, Bloomsbury, Square, London, W.C.1, do hereby declare the nature of this invention to be as follows :

PATENT SPECIFICATION

630,726

Application Date: June 28, 1934.   No. 19157/34.
   „   „   July 4, 1934.   No. 19721/34.

One Complete Specification left (under Section 16 of the Patents and
Designs Acts, 1907 to 1946): April 9, 1935.

Specification Accepted: March 30, 1936 (but withheld from publication
under Section 30 of the Patent and Designs Acts 1907 to 1932)

Date of Publication: Sept. 28, 1949.

Index at acceptance: —**Class 39(iv)**, P(1:2:3x).

PROVISIONAL SPECIFICATION
No. 19157 A.D. 1934.

Improvements in or relating to the Transmutation of Chemical Elements

I, LEO SZILARD, a citizen of Germany
and subject of Hungary, c/o Claremont
Haynes & Co., of Vernon House, Bloomsbury Square, London, W.C.1, do hereby
5 declare the nature of this invention to be
as follows:—
   This invention has for its object the
production of radio active bodies the
storage of energy through the production
10 of such bodies and the liberation of
nuclear energy for power production and
other purposes through nuclear transmutation.

PATENT   SPECIFICATION

817,571

Date of Application and filing Complete Specification: May 10, 1945.
No. 11801/45.

Application made in United States of America on May 8, 1944.

Complete Specification Published: Aug. 6, 1959.

(Under Section 6(1) (a) of the Patents &c. (Emergency) Act, 1939 the proviso to Section 91 (4) of the Patents and Designs Acts, 1907 to 1942 became operative on
April 4, 1957.)

Index at acceptance:—**Class 39(4)**, C4B2, P2.

International Classification:—G21.

COMPLETE SPECIFICATION
Means for Testing Materials

We, UNITED KINGDOM ATOMIC ENERGY
AUTHORITY, of London, a British Authority,
do hereby declare the nature of this invention,
and in what manner the same is to be per-
5 formed, to be particularly described in and
by the following statement:—
   The present invention relates generally to
the subject of nuclear physics and more particularly to means for determining neutron
10 absorption in graphite, or other neutron slowing material.

PATENT SPECIFICATION

817,751

Date of Application and filing Complete Specification: Feb. 1, 1945.

No. 2617/45.

Application made in United States of America on Dec. 19, 1944.

Complete Specification Published: Aug. 6, 1959.

(Under Section 6 (1) (a) of the Patents &c. (Emergency) Act, 1939 the proviso to Section 91 (4) of the Patents and Designs Acts, 1907 to 1946 became operative on April 4, 1957.

Index at acceptance:—Classes 1(3), A1(D28 : G36D28) ; 39(4), C1(A : B), C2(A2 : B1 : B2 : C1 : C2 : F : H : J), C3(C : E : F), C4(A1 : A2 : B2), C5C, Q ; 82(1), I4A(2 : 3X : 4C) ; 82(2), E3 ; and 83(2), A187.

International Classification:—B23p. C01g. C22b. C23g. G21.

COMPLETE SPECIFICATION

Nuclear Chain Reactions

We, UNITED KINGDOM ATOMIC ENERGY AUTHORITY, of London, a British Authority, do hereby declare the nature of this invention and in what manner the same is to be
5   performed, to be particularly described and ascertained in and by the following statement: —

PATENT SPECIFICATION

DRAWINGS ATTACHED

825,521

Date of Application and filing Complete Specification May 22, 1946.

No. 15575/46.

Application made in United States of America on May 29, 1945.

Complete Specification Published Dec. 16, 1959.

(Under Section 12 of the Atomic Energy Act, 1946 the proviso to Section 91 (4) of the Patents and Designs Acts, 1907 to 1946 became operative on April 14, 1958).

Index at acceptance: —Class 39(4), C(1A : 2A2 : 2B1 : 2B3A : 2C1 : 2D : 3A : 3D : 3E : 3F : 4A1 : 4B2 : 4B3), Q.

International Classification: —G21.

COMPLETE SPECIFICATION

Gas-Cooled Nuclear Reactor

We, UNITED KINGDOM ATOMIC ENERGY AUTHORITY, of London, a British Authority, do\hereby declare the nature of this invention and in what manner the same is to be per-
5   formed, to be particularly described and ascertained in and by the following statement: —

Netherlands

Uitgegeven 15 December 1928.

Dagteekening octrooi 16 October 1928.

Auteursrecht voorbehouden.

OCTROOIRAAD

NEDERLAND

OCTROOI N°. 19092.

KLASSE 21 *g*.   GROEP 21.

SIEMENS-SCHUCKERTWERKE GESELLSCHAFT MIT BESCHRÄNKTER HAFTUNG, te Berlijn–Siemensstadt en Dr. LEO SZILARD, te Berlijn–Dahlem.

Ontladingsbuis voor regelbare ontlading.

Aanvrage 31155 Ned., ingediend 2 September 1925, 2 u. 38 m. n.m.; openbaar gemaakt 15 Juni 1928; voorrang van 3 September 1924 af (Duitschland).

Uitgegeven 15 November 1933.

Dagteekening 17 October 1933.

C. D. 621.57

Auteursrecht voorbehouden.

OCTROOIRAAD

NEDERLAND

OCTROOI N°. 31163.

KLASSE 17 *a*. 20.

Prof. Dr. ALBERT EINSTEIN, en Dr. LEO SZILARD, beiden te Berlijn.

Werkwijze voor het comprimeeren van den damp van het koudmakend middel in een koelmachine en koelmachine, geschikt voor de toepassing van deze werkwijze.

Aanvrage 44262 Ned., ingediend 27 December 1928, 14 u. 52 m.; openbaar gemaakt 15 Juni 1933, voorrang van 27 December 1927 af, voor de conclusies 1, 3 en 4, en van 3 December 1928 af, voor conclusie 2, (Duitschland).

United States

Patented Jan. 1, 1929.   **1,697,210**

UNITED STATES PATENT OFFICE.

LEO SZILARD, OF BERLIN, GERMANY, ASSIGNOR TO SIEMENS-SCHUCKERTWERKE AKTIENGESELLSCHAFT, OF BERLIN-SIEMENSSTADT, GERMANY, A CORPORATION OF GERMANY.

DISCHARGE TUBE.

Application filed April 20, 1925, Serial No. 24,575, and in Germany September 3, 1924.

Patented June 4, 1929.   **1,715,874**

UNITED STATES PATENT OFFICE.

LEO SZILARD, OF BERLIN-DAHLEM, GERMANY, ASSIGNOR TO SIEMENS-SCHUCKERTWERKE GESELLSCHAFT MIT BESCHRÄNKTER HAFTUNG, OF SIEMENSSTADT, NEAR BERLIN, GERMANY, A CORPORATION OF GERMANY.

DISCHARGE TUBE.

Application filed October 28, 1925, Serial No. 65,394, and in Germany November 5, 1924.

Patented Nov. 11, 1930   **1,781,541**

UNITED STATES PATENT OFFICE

ALBERT EINSTEIN, OF BERLIN, AND LEO SZILARD, OF BERLIN-WILMERSDORF, GER-
MANY, ASSIGNORS TO ELECTROLUX SERVEL CORPORATION, OF NEW YORK, N. Y., A
CORPORATION OF DELAWARE

REFRIGERATION

Application filed December 16, 1927, Serial No. 240,566, and in Germany December 16, 1926.

Patented June 13, 1939   **2,161,985**

UNITED STATES PATENT OFFICE

2,161,985

**PROCESS OF PRODUCING RADIO-ACTIVE
ELEMENTS**

Leo Szilard, New York, N. Y.

Application March 11, 1935, Serial No. 10,500
In Great Britain March 12, 1934

9 Claims. (Cl. 204—31)

United States Patent Office

2,708,656
Patented May 17, 1955

1

2,708,656

NEUTRONIC REACTOR

Enrico Fermi, Santa Fe, N. Mex., and Leo Szilard, Chicago, Ill., assignors to the United States of America as represented by the United States Atomic Energy Commission

Application December 19, 1944, Serial No. 568,904

8 Claims. (Cl. 204—193)

United States Patent Office

2,778,792
Patented Jan. 22, 1957

1

2,778,792

METHOD FOR UNLOADING REACTORS

Leo Szilard, Chicago, Ill., assignor to the United States of America as represented by the United States Atomic Energy Commission

Application April 19, 1946, Serial No. 663,452

1 Claim. (Cl. 204—154)

United States Patent Office

2,796,396
Patented June 18, 1957

1

2,796,396

METHOD OF INTERMITTENTLY OPERATING A NEUTRONIC REACTOR

Leo Szilard, Chicago, Ill., assignor to the United States of America as represented by the United States Atomic Energy Commission

Application April 16, 1946, Serial No. 662,512

2 Claims. (Cl. 204—154)

United States Patent Office

2,798,847
Patented July 9, 1957

1

2,798,847

METHOD OF OPERATING A NEUTRONIC REACTOR

Enrico Férmi and Leo Szilard, Chicago, Ill., assignors to the United States of America as represented by the United States Atomic Energy Commission

Application December 1, 1952, Serial No. 323,452

1 Claim. (Cl. 204—154)

United States Patent Office

2,807,581
Patented Sept. 24, 1957

1

2,807,581

NEUTRONIC REACTOR

Enrico Fermi, Santa Fe, N. Mex., and Leo Szilard, Chicago, Ill., assignors to the United States of America as represented by the United States Atomic Energy Commission

Application October 11, 1945, Serial No. 621,838

2 Claims. (Cl. 204—193.2)

United States Patent Office

2,825,689
Patented Mar. 4, 1958

1

2,825,689

NEUTRONIC REACTOR AND FUEL ELEMENT THEREFOR

Leo Szilard, Chicago, Ill., and Gale J. Young, Hawthorne, N. Y., assignors to the United States of America as represented by the United States Atomic Energy Commission

Application April 25, 1946, Serial No. 664,732

2 Claims. (Cl. 204—193.2)

United States Patent Office

2,832,733
Patented Apr. 29, 1958

1

2,832,733
**HEAVY WATER MODERATED NEUTRONIC
REACTOR**
Leo Szilard, Chicago, Ill., assignor to the United States of
America as represented by the United States Atomic
Energy Commission
Application April 23, 1946, Serial No. 664,145
1 Claim. (Cl. 204—193.2)

United States Patent Office

2,836,554
Patented May 27, 1958

1

2,836,554
AIR COOLED NEUTRONIC REACTOR
Enrico Fermi, Santa Fe, N. Mex., and Leo Szilard, Chi-
cago, Ill., assignors to the United States of America
as represented by the United States Atomic Energy
Commission
Application May 29, 1945, Serial No. 596,465
3 Claims. (Cl. 204—193.2)

United States Patent Office

2,872,401
Patented Feb. 3, 1959

1   2

2,872,401
JACKETED FUEL ELEMENT
Eugene P. Wigner and Leo Szilard, Chicago, Ill.
Edward C. Creutz, Pittsburgh, Pa., assignors t
United States of America as represented by the U
States Atomic Energy Commission
Application May 8, 1946, Serial No. 668,110
2 Claims. (Cl. 204—193.2)

United States Patent Office

2,886,503
Patented May 12, 1959

1

2,886,503

**JACKETED FUEL ELEMENTS FOR GRAPHITE
MODERATED REACTORS**

Leo Szilard, Chicago, Ill., Eugene P. Wigner, Princeton,
N.J., and Edward C. Creutz, Santa Fe, N. Mex., assignors to the United States of America as represented by
the United States Atomic Energy Commission

Application February 20, 1946, Serial No. 649,080

4 Claims. (Cl. 204—193.2)

United States Patent Office

2,986,510
Patented May 30, 1961

1

2,986,510

MASSIVE LEAKAGE IRRADIATOR

Eugene P. Wigner, Princeton, N.J., Leo Szilard, Chicago,
Ill., Robert F. Christy, Santa Fe, N. Mex., and Francis
Lee Friedman, Chicago, Ill., assignors to the United
States of America as represented by the United States
Atomic Energy Commission

Filed May 14, 1946, Ser. No. 669,524

1 Claim. (Cl. 204—193.2)

United States Patent Office

3,103,475
Patented Sept. 10, 1963

1

3,103,475
REACTOR
Leo Szilard, Chicago, Ill., assignor to the United States
of America as represented by the United States
Atomic Energy Commission
Filed Sept. 20, 1946, Ser. No. 698,334
3 Claims. (Cl. 204—193.2)

List of Known Patent Applications That Did Not Issue into Patents

Germany

S. 89 028, filed Dec. 17, 1928
Beschleunigung von Korpuskeln.*

S. 89 172, filed Dec. 24, 1928
Verfahren zur Beschleunigung des Zerfalls schwach aktiver oder radioaktiver Substanzen
S. 89 288, filed Jan. 5, 1929
Korpuskularstrahlröhre*
S. 99 399, filed June 25, 1931
Einrichtung zur Dichtigkeitspruefung von luftdicht zu verschliessenden Apparaten, inbesondere von Kältemaschinen mit dünnwandigen Behältern
S. 105 894, filed August 19, 1932
Intermittierend wirkende Absorptions-Kältemaschine

United Kingdom

5730/34, filed Feb. 21, 1934
Asynchronous and Synchronous Transformers for particles*

7839/34, filed March 12, 1934
Reproduction of Books†
The invention concerns methods and apparatus for the reproduction of publications; it relates to photographical reproduction of books on strips of film or paper (or sheets of film or paper) and to arrangements of the pages forming the reproduction of a book on the film which will enable the reader to find quickly the individual page or the individual book for which he is looking, as well as other means which serve the latter purpose.
   The invention makes it possible to offer the reading public on a film roll, the length of which need not exceed 250 m, a library of 1,000 volumes. A special catalogue delivered with the film roll enables the reader to find for each book on his roll the index number which indicates the position of that particular book on the roll. By using this index number and making use of a device which is an object of this invention the reader can let the film run through a projector and stop it when the first page of the required book is in front of the lens of the projector, so that the first page of the book will appear on the screen. The reader can then easily bring any page of the book in which he is interested in front of the lens. [Excerpt]

10516/34, filed April 6, 1934
Reproduction of Sound†

26134/35, filed Sept. 1935

Reproduction of Sound† (modified version)
The invention concerns a method for registering sound, for instance speech or music by photographic methods, for instance on a strip of film, and reproducing it for

---

* Reproduced earlier in this part.
† Copy of application in Szilard files.

instance by sending light through the film and using a photocell to convert the variations of light into sound. [Excerpt]

6954/38, filed March 1938
Image Multiplier*
This is a photoelectric device: the image of an object is projected on a semi-transparent photo cathode. The electrons emitted from any one point of that cathode are accelerated in vacuum and have to pass through two auxiliary electrodes and a grid before reaching a fluorescent screen (with James L. Tuck). [Excerpt]

United States

263,017, filed March 20, 1939
Apparatus for Nuclear Transmutation†

264,263, filed Dec. 29, 1951
Process for Producing Microbial Metabolites

320,816, filed Nov. 15, 1952
Caffeine Containing Products and Method for their Preparation
The invention relates to the effect of caffeine containing products on mammalian tissue cells.

* Draft of application in Szilard files.
† Reproduced earlier in this part.

List of Known Disclosures in Szilard Files[*]

Jan. 27, 1940

Chain Reaction in a Lattice of Spheres of Uranium Embedded in Graphite

We shall consider the balance of neutron emission and absorption in a system composed of uranium and a light element, the latter serving the purpose of slowing down the neutrons.... In order to make a chain reaction possible carbon is a much better element to use for slowing down the neutrons than hydrogen.... We wish to conclude that it is possible to maintain a chain reaction in a lattice of spheres of uranium embedded in graphite, and that it does not take enormous masses of uranium to reach the point of divergence at which nuclear transmutation will go on with an intensity which is limited only by the necessity of avoiding overheating. [Dated photocopies (22 pages) in the Szilard files, postmarked January 27, 1940.]

June 28, 1940

Use of Beryllium

It would appear that the chances of a chain reaction with slow neutrons in a system essentially composed of uranium and carbon could be considerably improved by having a lattice of spheres of uranium metal embedded in graphite and each sphere surrounded with a spherical shell of beryllium metal. [Photocopies of three memoranda are in the Szilard files, dated June 28, July 4, Nov. 3, 1940; postmarked Nov. 14, 1940]

Dec. 9, 1940

Method of Cooling

1. No cooling liquid inside the graphite-uranium system.
2. Cooling by liquid bismuth, the bismuth surrounding the uranium spheres; the bismuth flowing in graphite channels and not in iron pipes.
3. Cooling by some cooling liquid, for instance a bismuth-lead compound containing 60% of bismuth, melting at 126°, flowing inside a uranium tube inside a uranium cylinder. This method can be used only if cylindrical bodies of uranium are embedded in graphite. In this arrangement liquid mercury could be used instead of liquid bismuth or bismuth alloys; also perhaps water. Note: melting point of bismuth 322°; melting point of lead 326°. A Pb–Sn alloy containing 70% Sn melts at 185°. There may be suitable Sn–Pb–Bi alloys. Boiling point of bismuth is at 1470°. Boiling point of lead is at 1613° .... [From proposed patent application in Szilard's files.]

March 25, 1943

Chain Reaction Using Fast Neutron Fission

This invention concerns a potentially chain reacting system which comprises a quantity of an element like uranium which undergoes fission under the action of fast neutrons but does not undergo fission under the action of slow neutrons, (such an element will be called an element of the first category,) and a quantity of an element like U^{235} or element 94^{239} or U^{233} which is capable of undergoing fission under the action of slow neutrons, (such an element will be called an element of the second category)... An element of the second category is produced from the element of the first category by the capture of neutrons which have been slowed down below the fission threshhold of the element of the first category....

According to this invention, a chain reaction can be maintained, stabilized, and controlled in a system containing such a reactive composition of elements, and it is

* These disclosure descriptions are excerpted in Szilard's own words.

not necessary to slow down the neutrons which form the links of the chain reaction by means of some lighter element. [From proposed patent application in Szilard's files.]

Aug. 3, 1946

Method and Apparatus for Liquid Extraction
A method and apparatus for liquid extraction in which the two solvents are contained in an annular gap in a rotating body and have a different tangential velocity while they are passed through the rotating body in a counter-current flow.
[Dated photocopies of disclosure in Szilard files, postmarked Aug. 3, Oct. 1, 1946, May 17, 19, 1949.]

September 28 and October 7, 1946

Miscellaneous Inventions
The following inventions were described in letters by Szilard to the University of Chicago:

1. A method and apparatus for producing an air stream by means of rotating bodies, the air stream being perpendicular to the axis of the rotating bodies. This principle may be used in building an electric fan and also may be used in the designing of airplanes. In the case of airplanes, the air stream created would be directed vertically downwards and might perform the same function as at present performed by the propeller of the helicopter. The axis of the rotating bodies would however be horizontal and might coincide with the axis of the propeller which drives the airplane forward.

2. A method for separating compounds which differ in molecular weight without differing in chemical composition. This method is based on a difference in diffusion velocity and operates without the use of barriers. While the method utilizes the rotating system, it is not based on centrifugation.

3. A method for accelerating electrically charged particles by the betatron principle which permits the charged particle to be accelerated not only through one-half phase of the alternating current which is used for excitation of the magnet but permits the utilization of a number of half-cycles.

4. A method for growing micro-organisms in bulk on surfaces rather than in suspension, which may be applicable in the production of antibiotic substances. It is characterized by having a large volume filled with porous ceramic bodies, leaving air spaces free between these bodies. The nutrient solution is absorbed in the porous bodies and micro-organisms are permitted to grow on the surfaces of the porous bodies.

July 7, 1948

Image Multiplier
Similar to British Patent Application No. 6954/38

April 8, 1953

Adding Machine
An adding machine built like a mechanical pencil, which may be carried in the pocket and can be used, for instance, for adding up during the day the caloric value of the various meals taken.
[Dated photocopies of disclosure in Szilard files, postmarked April 8, 1953.]

June 29, 1954

Cigarette Holder

If a cigarette is smoked through this holder, it will cut down the amount of combustion products which are absorbed by the smoker per unit time (but not necessarily per cigarette smoked). This is accomplished by inhaling a mixture of the air which passes through the burning cigarette with the air which enters the cigarette-holder through an opening that forms a by-pass.

[Dated photocopies of disclosure in Szilard files, postmarked June 29, 1954.]

Sept. 30, 1954

Process and Apparatus for De-Salting Sea Water

This invention relates to the de-salting of sea water by freezing.

[Dated photocopies of disclosure in Szilard files, postmarked Oct. 4–12, 1954.]

March 18, 1955

Process and Apparatus for Growing Algae

Algae may be grown in a layer of sweet water which floats above a mass of salt water.

[Disclosure in Szilard files, signed and witnessed March 18, 1955.]

Nov. 15, 1956

Therapeutic Dairy Products

It is possible to prepare dairy products such as "milk", "cream" and "cheese" which are almost indistinguishable in taste from the commercial dairy products such as milk, cream and cheese, but which differ from them inasmuch as a high percentage—at least 80%—of the milk fats has been removed and in its place a slightly smaller, equal or even larger amount of a vegetable oil with a high iodine value (iodine value higher than 60) has been substituted.

[Dated photocopies of disclosure in Szilard files, postmarked Nov. 15, 1956.]

Jan. 10, 1957

Agents for Producing Disease Resistance

A product adapted to be used as a DR agent for a specific disease which consists of a mixture of a quantity of the killed infectious agent of the disease (or an extract thereof which contains the relevant antigens) and a quantity of antiserum specific for this infectious agent (or a fraction of such antiserum which contains the relevant antibodies), the ratio of antibodies to antigens being large enough for the antibodies to saturate the antigens.

[Dated photocopies of disclosure in Szilard files, postmarked Jan. 14, 15, 18, 1957.]

Aug. 14, 1957

Gramophone and Gramophone Record

A gramophone containing one or more tuned circuits of audio-frequency (sonic or supersonic), means for rotating the turntable at at least two speeds, and means for switching the speed from the lowest to a higher speed when the said tuned audio-frequency circuits are energized.

Gramophone records on which is recorded one or more sonic signals (in an audible frequency or in a supersonic frequency).

[Signed disclosure in Szilard files, witnessed Aug. 14, 1957.]

Jan. 16, 1959

Thermo-Electric Generator

Various ideas for improving an apparatus for the generation of electric power from heat. This apparatus consists of an assembly of individual thermo-electric generators.

Each of the individual units . . . consists of an outer tube which is kept hot and to which we shall refer as the "hot cathode," and a concentric inner tube. . . . In the annular gap between the two tubes there is a vacuum in which there is maintained a low vapor pressure of an alkali metal, preferably caesium. [Dated photocopies of disclosures in Szilard files, postmarked from Jan. 16 to Feb. 13, 1959. A March 18, 1959 expansion computes the enhanced power output due to the alkali vapor.]

Correspondence Relating to Patents

Szilard Letter to O.S.* (Translation) (Undated)

Dear Herr Professor,

Enclosed I am sending you "reprints" which will perhaps amuse you. These are two patent applications† which contain the methods since developed by Lawrence. The first application was submitted before the publication of the corresponding paper by Wideroe and contains the method for the production of fast protons by means of rapidly changing electric fields. The second application was submitted shortly afterwards and contains the use of a stationary magnetic field for winding up a proton beam in an oscillating electric field. The two papers by Lawrence are of a later date, but Lawrence had, of course, no knowledge of these unpublished applications. Of course, the merit lies in the carrying out and not in the thinking out of the experiments, and I am sending you these "reprints" only to enable you to praise me, if necessary, with a better conscience.

With kindest regards,

Very sincerely yours,
[Signed] Leo Szilard

P.S. If you wish to look at the applications at all, you will find that the first three to four pages of the second application are the most illuminating ones.

* We believe O.S. is probably Otto Stern—Eds.
† The two applications referred to are German S89 028 and S89 288, reproduced earlier in this part—Eds.

Szilard Letter to Fermi (March 13, 1936)

c/o The Clarendon Laboratory,
Parks Road,
Oxford.

13th March 1936.

Dear Professor Fermi,

I have been intending to write to you for some time in order to thank you for the last manuscript which you sent me, and also in order to tell you about ideas which have been put forward about selective absorption in the course of the last two months. They came simultaneously from Bohr, who was here for a visit, and from Wigner and Breit, with whom I have been in correspondence. You certainly saw Bohr's paper in *Nature* and I am going to send you a manuscript from Breit and Wigner as soon as I get it back from Cambridge.

I have to write to you now about the question of practical applications of modern nuclear physics, though I am by no means certain that such practical applications of importance at present really exist. Since however I received a letter from Mr. Giannini (a copy of which I enclose), I wish to tell you why I have applied for certain patents in this field and what I propose to do with them.

I feel that I must not consider these patents as my private property and that if they are of any importance, they should be controlled with a view of public policy. I see no objection to a commercial exploitation of some such patents, but I believe that the income (if there is any substantial income) should not be used for private purposes, but rather for financing further research or, if the income is very large, for other constructive purposes.

I know of one precedent for such procedure which was fairly successful. Some years ago, Cottrell, whom you perhaps know, took out in the U.S.A. certain patents and formed a research corporation to which he handed over his patents for commercial exploitation of the products. The profits are being used entirely for the promotion of further research.

In 1928 I formed the mistaken view that artificial disintegration would be developed in the course of a few years and would soon lead to practical application of very great importance. At that time I filed three patents which described the methods for the production of fast protons which later on were developed and published by Lawrence, one of them being the cyclotron, and also described the production of radio-active elements by bombardment of fast protons and alpha-particles. All these patents have subsequently been abandoned.

In 1933 I again formed the view that practical applications of very great importance are impending. Whether this view is correct, I could not say. It seemed to me that the production of radio-active elements by neutrons might have some importance and I filed a patent on this subject, after Joliot's discovery, on March 12th, 1934. A number of other applications followed, but it remains to be seen whether any of them have real practical significance.

When I filed all these patents, my intention was to hand them over, as soon as they turn out to be important, to a research corporation which could be set up at any moment in one form or another and which could use its funds for promoting further research. I personally do not think very much of producing radio-active elements for medical purposes and I should not like to be responsible for inducing manufacturers to embark upon such an enterprise at present. On the other hand, it is conceivable that certain applications of very great importance might materialise in a not too distant future and I have lately been taking, rightly or wrongly, a rather optimistic view on this subject. I have been writing a memorandum on the subject, which I might send you in due course of time and am sufficiently optimistic to feel justified in proposing that a fund should be created of about £5,000 and used for further experiments. It seems to me that such a fund should be used up in the course of the next three years and used in three different ways. First of all, it should be used for salaries of young physicists who could carry on systematic investigations which would fit in well with our present work, but on which we cannot embark for lack of time. Secondly, it should be used for hiring radium to be used for some such experiments on the basis of a very favourable offer of radium which I have received in this connection. Thirdly, such funds should be used for enabling any of us to move from one laboratory to another whenever such a movement is justified from the point of view of apparatus which is present in one laboratory and lacking in another.

Though I do not know whether my efforts to raise such a fund will be successful, I should be glad if you could let me know whether you would care to share the responsibility for controlling such a fund.

I understand from Giannini's letter that you have applied for certain patents. You might perhaps have similar ideas about the commercial exploitation of your patents. Should you, however, no longer control these patents, or should you have other intentions with them, that would in no way affect the present issue. One should not give at present too much significance to any single patent in this field. If important applications should materialise in the future, some importance might, however, be attached to the co-operation of those who work in this field and also to their willingness to take responsibility in this matter.

Forgive me please for writing to you such a long and somewhat boring letter. I should appreciate any comment which you care to make on this subject; Giannini will leave England on March 19th, and if I hear from you before this date, I could discuss the matter with him before he leaves.

With best wishes,

Yours sincerely,
[Signed] Leo Szilard

Szilard Letter to Segrè (April 1, 1936)

c/o Clarendon Laboratory,
Parks Road,
Oxford.

1st April, 1936.

Dear Segrè,

Many thanks for your letter of March 21st. Please let me know when you reach a decision in Rome about the question of "patents." Could you please convey to the others some of the following points of view and suggestions which I am tentatively putting forward. Perhaps you could bear them in mind when you are discussing this thing in Rome at Easter.

Point *1*. Let us first envisage the possibility that practical applications in the field of nuclear physics will become so important that an attempt to exercise some measure of control over them through disinterested scientists will appear to be justified. In order to achieve this, some sort of association could then be brought into existence to which we could all hand over our patents, so that most of us should be able to remain more or less aloof.

Such an association need not follow the example of the Research Corporation (New York) by mixing manufacturing activities and the promotion of science. It could confine its activities to bringing about a co-operation between industry and scientific research along the following lines:

The association could grant non-exclusive rights for the use of its patents to manufacturers under the condition that they contribute to a fund which is used for promoting further research and on the further condition that the manufacturer does not block the way for others by patents of his own. The funds which are available could be used for carrying out systematic investigations in University laboratories which fit in well with the work already carried on in such laboratories. The results of such investigations ought to be automatically available to all industrialists who contribute their share to the research fund of the association. Perhaps it is possible to avert in this way competitive research into nuclear physics in industrial laboratories.

2. It may very well be that no important applications of nuclear physics may arise and in this case none of us will be willing to take much trouble about these patents. Unless, however, we are willing to take the trouble involved in their proper administration and in the proper administration of the funds which would be forthcoming, I for my part would rather withdraw the patents which I have taken out, than let them float about in an irresponsible way.

At present we do not know whether or not the practical applications will be sufficiently important for us to go out of our way and exercise some sort of control over the patents. The question therefore arises what should be done until we can decide about the proper course of action.

It seems to me that in the meantime we could ask some men like Chadwick, Cockroft or Fermi, or at least two of them, to accept the responsibility for whatever action is in the meantime required and jointly to decide each issue which arises.

If we think that it is justified to raise funds for further research we may attempt to do so and I have been in touch with some private persons who may or may not be willing to contribute towards such a fund. In my personal opinion we might feel justified in suggesting that a fund of £5,000 should be created and that this fund should be spent on research in the course of the next three years. It should be used to carry out investigations which fit in well with our present work, but which have a more direct bearing on possible practical applications. There are three main ways in which such a fund could be used for the present:

a) For hiring radium and providing certain laboratories which are badly off in this respect with steady sources of neutrons (radium *properly* mixed with beryllium).

b) For salaries of young physicists who could carry out certain systematic measurements in one of the laboratories in which such work is already in progress.

c) For enabling any of us to move from one laboratory to another if this is justified from the point of view of apparatus which is present in one laboratory, lacking in the other and needed for the particular experiments which now appear to be of interest.

Please ask Fermi when you see him in Rome to let me know if he would care to share the responsibility for the decisions which may now be required until we either withdraw the patents or find some definite form for their administration.

The question of the patents for which you are no longer free is hardly of primary importance. We ought, however, to bear in mind that it must be awkward for any scientist to have a personal income from such patents, while other scientists, who also could have taken out such patents, refrain from doing so. It is not customary to take out patents on scientific discoveries, and it is hardly desirable to act against such an unwritten law unless one has reason to think that a departure is justified by unique circumstances.

Naturally, customs are different in different countries and you have at any rate really discovered something, while I have mostly taken out patents on subsequent discoveries of other people. Our cases are different, and I do not think it right for me to have any financial advantages or any other privileges through patents which are connected with nuclear physics.

Please give my kind regards to all and thank your wife for her excellent German typing.

Yours sincerely,
[Signed] Leo Szilard

Szilard Letter to C. S. Wright, British Admiralty, with Introductory Note
(February 26, 1936)

February 26th, 1936.

Dr. C. S. Wright.
Department of Scientific Research & Experiment,
Admiralty, S.W.1.

Dear Wright,

I daresay you remember my ringing you up about a man working here who had a patent which he thought ought to be kept secret. I enclose a letter from him on the subject as you suggested. I am naturally somewhat less optimistic about the prospects than the inventor, but he is a very good physicist and even if the chances were a hundred to one against it seems to me it might be worth keeping the thing secret as it is not going to cost the Government anything.

With apologies for this hasty note, believe me,

Yours sincerely,*

———

c/o The Clarendon Laboratory,
Parks Road,
Oxford.

26 February, 1936

Dr. C. S. Wright,
Department of Scientific
Research & Experiment,
Admiralty,
London. S.W.1

Sir,

I wish to draw your attention to the fact that I have applied for a British Patent,† Application No. [19, 157] and you might find it advisable to prevent its publication.

* [Probably from Prof. F. A. Lindemann (later Lord Cherwell)—Eds.]
† Reproduced earlier in this part.

The object of this Patent has nothing to do with instruments of war, but it contains information which could be used in the construction of explosive bodies based on processes described in the Specification. Such explosive bodies would be very many thousand times more powerful than ordinary bombs, and in view of the disasters which could be caused by their use on the part of certain Powers which might attack this country, it appears very undesirable that such information should be published through the medium of this Patent.

I understand that you would be able to intervene and prevent the publication of this Patent only if I assign the Patent to the Admiralty. If a form can be found which will enable me to retain my freedom of action concerning the applications which have nothing to do with instruments of war, I shall be pleased to assign the Patent to the Admiralty. No financial obligation on the part of the Admiralty would arise out of such a transaction.

I am fully aware of the fact that if a successful private manufacture is set up on the basis of this Patent, information will leak out sooner or later. It is in the very nature of this invention that it cannot be kept secret for a very long time, and my only concern is that the processes should be developed in this country a few years ahead of certain other countries. This purpose would be served by keeping the Patent secret; and we cannot aim at anything more.

I have to add that we have no certainty for the time being whether the processes described can be put into operation. I have, however, observed and investigated certain anomalies which indicate that this may very well be the case. At present there are still two alternative explanations for these anomalies, and it may take a year before we can decide in favour of one or the other.

I have so far delayed the publication of this Patent as far as possible by post-dating the original application, and asking for extensions of time. The last date for the acceptance of the Patent is now March 28th, and this cannot be postponed further.

I shall be glad to be at your disposal in London should you wish to see me regarding the details of the Patented Processes. My Patent Solicitor—Mr. Champneys, of Claremont Haynes & Co., Vernon House, Sicilian Avenue, Bloomsbury Square, London, Telephone No. Holborn 8811—would also be pleased to give you any information on this matter.

Yours very truly,
[Signed] Leo Szilard

Name Index